Essential Topics in Combinatorics

Essential Topics in Combinatorics

Edited by **Lucas Lincoln**

NY RESEARCH PRESS

New York

Published by NY Research Press,
23 West, 55th Street, Suite 816,
New York, NY 10019, USA
www.nyresearchpress.com

Essential Topics in Combinatorics
Edited by Lucas Lincoln

International Standard Book Number: 978-1-63238-184-2 (Hardback)

Printed in China.

Contents

	Preface	**VII**
Chapter 1	**An Algebraic Representation of Graphs and Applications to Graph Enumeration** Ângela Mestre	**1**
Chapter 2	**Initial Ideals of Tangent Cones to the Richardson Varieties in the Orthogonal Grassmannian** Shyamashree Upadhyay	**15**
Chapter 3	**Bounds for the Largest Laplacian Eigenvalue of Weighted Graphs** Sezer Sorgun	**34**
Chapter 4	**On Bondage Numbers of Graphs: A Survey with Some Comments** Jun-Ming Xu	**42**
Chapter 5	**Graphs with no $K_{3,3}$ Minor Containing a Fixed Edge** Donald K. Wagner	**76**
Chapter 6	**Some New Classes of Open Distance-Pattern Uniform Graphs** Bibin K. Jose	**83**
Chapter 7	**The m-Path Cover Polynomial of a Graph and a Model for General Coefficient Linear Recurrences** John P. McSorley and Philip Feinsilver	**90**
Chapter 8	**Embedding Structures Associated with Riordan Arrays and Moment Matrices** Paul Barry	**103**
Chapter 9	**Bounds on the Size of the Minimum Dominating Sets of Some Cylindrical Grid Graphs** Mrinal Nandi, Subrata Parui and Avishek Adhikari	**110**
Chapter 10	**The Terminal Hosoya Polynomial of Some Families of Composite Graphs** Emeric Deutsch and Juan Alberto Rodríguez-Velázquez	**123**
Chapter 11	**Decomposition Formulas for Triple q-Hypergeometric Functions** Thomas Ernst	**127**

Chapter 12 **On the Cardinality of the T_0-Topologies on a Finite Set** 141
Messaoud Kolli

Chapter 13 **On Normal q-Ary Codes in Rosenbloom-Tsfasman Metric** 148
R. S. Selvaraj and Venkatrajam Marka

Chapter 14 **Crystal Bases as Tuples of Integer Sequences** 153
Deniz Kus

Chapter 15 **Generalized Pattern-Matching Conditions for $C_k \wr S_n$** 170
Sergey Kitaev, Andrew Niedermaier, Jeffrey Remmel and Manda Riehl

Permissions

List of Contributors

Preface

Basic combinatorial concepts and enumerative results appeared throughout the ancient world. In the Middle Ages, there was progress in combinatorics, but it was not at a rapid pace. And finally, in the second half of 20th century, tremendous growth was observed in the field of combinatorics. This growth was linked to new connections and applications to other fields, ranging from algebra to probability, from functional analysis to number theory, etc.

At its core, this is a field of pure mathematics. Algebra, topology, geometry and probability find a strong relevance here. Additionally, there are numerous sub-divisions in this field like enumerative combinatorics which is the most classical area of this field, followed by analytic combinatorics, partition theory, graph theory, design theory, finite geometry, order theory, matroid theory, extremal combinatorics, probabilistic combinatorics, algebraic combinatorics, geometric combinatorics, topological combinatorics, arithmetic combinatorics and more.

In the current scenario, there's a sizable overlap between combinatorics and physics. These and other such rapid developments made in this field over the past few decades makes it an extremely interesting field to study.

This book includes details and data of all such latest findings sent in by experts who have been involved in the domain for a sustained period of time. I especially wish to acknowledge the contributing authors, without whom a work of this magnitude would clearly not have been realizable. I thank all our authors for allocating much of their scarce time to this project. Not only do we appreciate their participation, but also their adherence as a group to the time parameters set for this publication.

Editor

An Algebraic Representation of Graphs and Applications to Graph Enumeration

Ângela Mestre

Centro de Estruturas Lineares e Combinatórias, Universidade de Lisboa, Av. Prof. Gama Pinto 2, 1649-003 Lisboa, Portugal

Correspondence should be addressed to Ângela Mestre; mestre@cii.fc.ul.pt

Academic Editor: Xueliang Li

We give a recursion formula to generate all the equivalence classes of connected graphs with coefficients given by the inverses of the orders of their groups of automorphisms. We use an algebraic graph representation to apply the result to the enumeration of connected graphs, all of whose biconnected components have the same number of vertices and edges. The proof uses Abel's binomial theorem and generalizes Dziobek's induction proof of Cayley's formula.

1. Introduction

As pointed out in [1], generating graphs may be useful for numerous reasons. These include giving more insight into enumerative problems or the study of some properties of graphs. Problems of graph generation may also suggest conjectures or point out counterexamples. The use of generating functions (or functionals) in the enumeration or generation of graphs is standard practice both in mathematics and physics [2–4]. However, this is by no means obligatory since any method of manipulating graphs may be used.

Furthermore, the problem of generating graphs taking into account their symmetries was considered as early as the 19th century [5] and more recently for instance, in [6]. In particular, in quantum field theory, generated graphs are weighted by scalars given by the inverses of the orders of their groups of automorphisms [4]. In [7, 8], this was handled for trees and connected multigraphs (with multiple edges and loops allowed), on the level of the symmetric algebra on the vector space of time-ordered field operators. The underlying structure is an algebraic graph representation subsequently developed in [9]. In this representation, graphs are associated with tensors whose indices correspond to the vertex numbers. In the former papers, this made it possible to derive recursion formulas to produce larger graphs from smaller ones by increasing by 1 the number of their vertices or the number of their edges. An interesting property of these formulas is that of satisfying alternative recurrences which relate either a tree or connected multigraph on n vertices with all pairs of their connected subgraphs with total number of vertices equal to n. In the case of trees, the algorithmic description of the corresponding formula is about the same as that used by Dziobek in his induction proof of Cayley's formula [10, 11]. Accordingly, the formula induces a recurrence for $n^{n-2}/n!$, that is, the sum of the inverses of the orders of the groups of automorphisms of all the equivalence classes of trees on n vertices [12, page 209].

For simplicity, here by *graphs* we mean simple graphs. However, our results generalize straightforwardly to graphs with multiple edges allowed. One instance of an algorithm for finding the biconnected components of a connected graph is given in [13]. Our goal here is rather to generate all the equivalence classes of connected graphs so that they are decomposed into their biconnected components and have the coefficients announced in the abstract. To this end, we give a suitable graph transformation to produce larger connected graphs from smaller ones by increasing the number of their biconnected components by one unit. This mapping is then used to extend the recurrence of [7] to connected graphs. This new recurrence decomposes the graphs into their biconnected components and, in addition, can be generalized to restricted classes of connected graphs with specified biconnected components. The proof proceeds as suggested in [7]. That is, given an arbitrary equivalence class whose representative is a graph on m edges, say, G, we show that every one of the m edges of the graph G adds $1/(m\cdot|\text{aut}\, G|)$ to the coefficient of G. To this end, we use the fact that labeled vertices are held fixed under any automorphism.

Moreover, in the algebraic representation framework, the result yields a recurrence to generate linear combinations of tensors over the rational numbers. Each tensor represents a connected graph. As required, these linear combinations have the property that the sum of the coefficients of all the tensors representing isomorphic graphs is the inverse of the order of their group of automorphisms. In this context, tensors representing generated graphs are factorized into tensors representing their biconnected components. As in [7–9], a key feature of this result is its close relation to the algorithmic description of the computations involved. Indeed, it is easy to read off from this scheme not only algorithms to perform the computations, but even data structures relevant for an implementation.

Furthermore, we prove that when we only consider the restricted class of connected graphs whose biconnected components all have, say, p vertices and r edges, the corresponding recurrence has an alternative expression which relates connected graphs on μ biconnected components with all the p-tuples of their connected subgraphs with total number of biconnected components equal to $\mu - 1$. This induces a recurrence for the sum of the inverses of the orders of the groups of automorphisms of all the equivalence classes of connected graphs on μ biconnected components with that property. The proof uses an identity related to Abel's binomial theorem [14, 15] and generalizes Dziobek's induction proof of Cayley's formula [10].

This paper is organized as follows. Section 2 reviews the basic concepts of graph theory underlying much of the paper. Section 3 contains the definitions of the elementary graph transformations to be used in the following. Section 4 gives a recursion formula for generating all the equivalence classes of connected graphs in terms of their biconnected components. Sections 5 and 6 review the algebraic representation and some of the linear mappings introduced in [7, 9]. Section 7 derives an algebraic expression for the recurrence of Section 4 and for the particular case in which graphs are such that their biconnected components are all graphs on the same vertex and edge numbers. An alternative formulation for the latter is also given. Finally, Section 8 proves a Cayley-type formula for graphs of that kind.

2. Basics

We briefly review the basic concepts of graph theory that are relevant for the following sections. More details may be found in any standard textbook on graph theory such as [16].

Let A and B denote sets. By $[A]^2$ we denote the set of all the 2-element subsets of A. Also, by 2^A we denote the power set of A, that is, the set of all the subsets of A. By card A we denote the cardinality of the set A. Furthermore, we recall that the symmetric difference of the sets A and B is given by $A\Delta B := (A \cup B) \setminus (A \cap B)$.

Here, a *graph* is a pair $G = (V, E)$, where $V \subset \mathbb{N}$ is a finite set and $E \subseteq [V]^2$. Thus, the elements of E are 2-element subsets of V. The elements of V and E are called *vertices* and *edges*, respectively. In the following, the vertex set of a graph G will often be referred to as $V(G)$, the edge set as $E(G)$. The cardinality of $V(G)$ is called the *order* of G, written as $|G|$. A vertex v is said to be *incident* with an edge e if $v \in e$. Then, e is an edge *at* v. The two vertices incident with an edge are its *endvertices*. Moreover, the *degree* of a vertex v is the number of edges at v. Two vertices v and u are said to be *adjacent* if $\{v, u\} \in E$. If all the vertices of G are pairwise adjacent, then G is said to be *complete*. A graph G^* is called a *subgraph* of a graph G if $V(G^*) \subseteq V(G)$ and $E(G^*) \subseteq E(G)$. A *path* is a graph P on $n \geq 2$ vertices such that $E(P) = \{\{v_1, v_2\}, \{v_2, v_3\}, \dots, \{v_{n-1}, v_n\}\}$, $v_j \in V(P)$ for all $j = 1, \dots, n$. The vertices v_1 and v_n have degree 1, while the vertices $v_2, \dots, v_{n-1}$ have degree 2. In this context, the vertices v_1 and v_n are *linked* by P and called the *endpoint* vertices. The vertices $v_2, \dots, v_{n-1}$ are called the *inner* vertices. A *cycle* is a graph C on $n > 2$ vertices such that $E(C) = \{\{v_1, v_2\}, \{v_2, v_3\}, \dots, \{v_{n-1}, v_n\}, \{v_n, v_1\}\}$, $v_j \in V(C)$ for all $j = 1, \dots, n$, every vertex having degree 2. A graph is said to be *connected* if every pair of vertices is linked by a path. Otherwise, it is *disconnected*. Given a graph G, a maximal connected subgraph of G is called a *component* of G. Furthermore, given a connected graph, a vertex whose removal (together with its incident edges) disconnects the graph is called a *cutvertex*. A graph that remains connected after erasing any vertex (together with incident edges) (resp. any edge) is said to be *2-connected* (resp. *2-edge connected*). A 2-connected graph (resp. 2-edge connected graph) is also called *biconnected* (resp. *edge-biconnected*). Given a connected graph H, a *biconnected component* of H is a maximal subset of edges such that the induced subgraph is biconnected (see [17, Section 6.4] for instance). Here, we consider that an isolated vertex is, by convention, a biconnected graph with no biconnected components.

Moreover, given a graph G, the set $2^{E(G)}$ is a vector space over the field $\mathbb{Z}_2$ such that vector addition is given by the symmetric difference. The *cycle space* $\mathscr{C}(G)$ of the graph G is defined as the subspace of $2^{E(G)}$ generated by all the cycles in G. The dimension of $\mathscr{C}(G)$ is called the *cyclomatic number* of the graph G. We recall that $\dim \mathscr{C}(G) = \text{card} E(G) - |G| + c$, where c denotes the number of connected components of the graph G [18].

We now introduce a definition of labeled graph. Let L be a finite set. Here, a *labeling* of a graph G is a mapping $l : V(G) \to 2^L$ such that $\cup_{v\in V(G)} l(v) = L$ and $l(v) \cap l(v') = \emptyset$ for all $v, v' \in V(G)$ with $v \neq v'$. In this context, L is called a *label* set, while the graph G is said to be *labeled with* L or simply a *labeled* graph. In the sequel, a labeling of a graph G will be referred to as l_G. Moreover, an *unlabeled* graph is one labeled with the empty set.

Furthermore, an *isomorphism* between two graphs G and G^* is a bijection $\varphi : V(G) \to V(G^*)$ which satisfies the following conditions:

(i) $\{v, v'\} \in E(G)$ if and only if $\{\varphi(v), \varphi(v')\} \in E(G^*)$,

(ii) $L \cap l_G(v) = L \cap l_{G^*}(\varphi(v))$.

Clearly, an isomorphism defines an equivalence relation on graphs. In particular, an isomorphism of a graph G onto itself is called an *automorphism* (or *symmetry*) of G.

3. Elementary Graph Transformations

We introduce the basic graph transformations to change the number of biconnected components of a connected graph by one unit.

Here, given an arbitrary set X, let $\mathbb{Q}X$ denote the free vector space on the set X over $\mathbb{Q}$, the set of rational numbers. Also, for all integers $n \geq 1$ and $k \geq 0$ and label sets L, let

$$V^{n,k,L} = \Big\{G : |G| = n, \dim \mathscr{C}(G) = k,\ G \text{ is labeled by } l_G : V(G) \longrightarrow 2^L\Big\}. \quad (1)$$

Furthermore, let

(i) $V_{\text{conn}}^{n,k,L} = \{G \in V^{n,k,L} : G \text{ is connected}\}$,

(ii) $V_{\text{biconn}}^{n,k,L} = \{G \in V^{n,k,L} : G \text{ is biconnected}\}$.

In what follows, when $L = \emptyset$ we will omit L from the upper indices in the previous definitions.

We proceed to the definition of the elementary linear mappings to be used in the following. Note that, for simplicity, our notation does not distinguish between two mappings defined both according to one of the following definitions, one on $V^{n,k,L}$ and the other on $V^{p,q,L'}$ with $n \neq p$ or $k \neq q$ or $L \neq L'$. This convention will often be used in the rest of the paper for all the mappings given in this section. Therefore, we will specify the domain of the mappings whenever confusion may arise.

(i) *Adding a biconnected component to a connected graph*: let L be a label set. Let G be a graph in $V_{\text{conn}}^{n,k,L}$. Let $V(G) = \{v_i\}_{i=1,\dots,n}$. For all $i = 1, \dots, n$, let $\mathscr{K}_i$ denote the set of biconnected components of G such that $v_i \in V(H)$ for all $H \in \mathscr{K}_i$. Let $\mathscr{L}$ denote the set of all the ordered partitions of the set $\mathscr{K}_i$ into p disjoint sets: $\mathscr{L} = \{z_i := (\mathscr{K}_i^{(1)}, \dots, \mathscr{K}_i^{(p)}) : \cup_{l=1}^{p} \mathscr{K}_i^{(l)} = \mathscr{K}_i \text{ and } \mathscr{K}_i^{(l)} \cap \mathscr{K}_i^{(l')} = \emptyset\ \forall l, l' = 1, \dots, p \text{ with } l \neq l'\}$. Furthermore, let $\mathscr{J}$ denote the set of all the ordered partitions of the set $l_G(v_i)$ into p disjoint sets: $\mathscr{J} = \{w_i := (l_G(v_i)^{(1)}, \dots, l_G(v_i)^{(p)}) : \cup_{l=1}^{p} l_G(v_i)^{(l)} = l_G(v_i) \text{ and } l_G(v_i)^{(l)} \cap l_G(v_i)^{(l')} = \emptyset\ \forall l, l' = 1, \dots, p \text{ with } l \neq l'\}$. Finally, let $\widehat{G}$ be a graph in $V_{\text{biconn}}^{p,q}$ such that $V(G) \cap V(\widehat{G}) = \emptyset$. (In case the graph $\widehat{G}$ does not satisfy that property, we consider a graph G' instead such that $G' \cong \widehat{G}$ and $V(G) \cap V(G') = \emptyset$. We will not point this out explicitly in the following.) Let $V(\widehat{G}) = \{u_l\}_{l=1,\dots,p}$. In this context, for all $i = 1, \dots, n$, define

$$r_i^{\widehat{G}} : \mathbb{Q}V_{\text{conn}}^{n,k,L} \longrightarrow \mathbb{Q}V_{\text{conn}}^{n+p-1,k+q,L}$$
$$G \longmapsto \sum_{z_i \in \mathscr{L}; w_i \in \mathscr{J}} G_{w_i}^{z_i}, \quad (2)$$

where the graphs $G_{w_i}^{z_i}$ satisfy the following:

(a) $V(G_{w_i}^{z_i}) = V(G) \setminus \{v_i\} \cup V(\widehat{G})$,

(b) $E(G_{w_i}^{z_i}) = E(\widehat{G}) \cup \{\{x, y\} \in E(G) : v_i \notin \{x, y\}\}$

$$\bigcup_{l=1}^{p} \Big\{\{x, u_l\} : \{x, v_i\} \in E(G),\ x \in \cup_{H^* \in \mathscr{K}_i^{(l)}} V(H^*)\Big\}, \quad (3)$$

(c) $l_{G_{w_i}^{z_i}}|_{V(G)\setminus\{v_i\}} = l_G|_{V(G)\setminus\{v_i\}}$ and $l_{G_{w_i}^{z_i}}(u_l) = l_G(v_i)^{(l)}$ for all $l = 1, \dots, p$.

The mappings $r_i^{\widehat{G}}$ are extended to all of $\mathbb{Q}V_{\text{conn}}^{n,k,L}$ by linearity. For instance, Figure 1 shows the result of applying the mapping $r_i^{C_4}$ to the cutvertex of a 2-edge connected graph with two biconnected components, where C_4 denotes a cycle on four vertices.

Furthermore, let $X \subseteq V_{\text{biconn}}^{p,q}$. Given a linear combination of graphs $\vartheta = \sum_{G \in X} \alpha_G G$, where $\alpha_G \in \mathbb{Q}$, we define

$$r_i^{\vartheta} := \sum_{G \in X} \alpha_G r_i^{G}. \quad (4)$$

We proceed to generalize the edge contraction operation given in [16] to the operation of contracting a biconnected component of a connected graph.

(ii) *Contracting a biconnected component of a connected graph*: let L be a label set. Let G be a graph in $V_{\text{conn}}^{n,k,L}$. Let $\widehat{G} \in V_{\text{biconn}}^{p,q,L'}$ be a biconnected component of G, where $L' = \cup_{v \in V(\widehat{G})} l_G(v)$. Define

$$c_{\widehat{G}} : V_{\text{conn}}^{n,k,L} \longrightarrow V_{\text{conn}}^{n-p+1,k-q,L}; \quad G \longmapsto G^*, \quad (5)$$

where the graph G^* satisfies the following:

(a) $V(G^*) = V(G) \setminus V(\widehat{G}) \cup \{v\}$, where $v := \min\{v' \in \mathbb{N} : v' \notin V(G) \setminus V(\widehat{G})\}$,

(b) $E(G^*) = \{\{x, y\} \in E(G) : \{x, y\} \cap E(\widehat{G}) = \emptyset\}$

$$\cup \Big\{\{x, v\} : \{x, y\} \in E(G) \setminus E(\widehat{G}) \text{ and } y \in V(\widehat{G})\Big\}, \quad (6)$$

(c) $l_{G^*}|_{V(G)\setminus V(\widehat{G})} = l_G|_{V(G)\setminus V(\widehat{G})}$ and $l_{G^*}(v) = \cup_{u \in V(\widehat{G})} l_G(u)$.

For instance, Figure 2 shows the result of applying the mapping c_{C_4} to a 2-edge connected graph with three biconnected components.

We now introduce the following auxiliary mapping.

Let L be a label set. Let G be a graph in $V^{n,k,L}$. Let $V(G) = \{v_i\}_{i=1,\dots,n}$. Let L' be a label set such that $L \cap L' = \emptyset$. Also, let $\mathscr{J}$ denote the set of all the ordered partitions of the set L' into n disjoint sets: $\mathscr{J} = \{y := (L'^{(1)}, \dots, L'^{(n)}) : \cup_{j=1}^{n} L'^{(j)} = L' \text{ and } L'^{(i)} \cap L'^{(j)} = \emptyset\ \forall i, j = 1, \dots, n \text{ with } i \neq j\}$. In this context, define

$$\xi_{L'} : \mathbb{Q}V^{n,k,L} \longrightarrow \mathbb{Q}V^{n,k,L\cup L'}; \quad G \longmapsto \sum_{y \in \mathscr{J}} G_y, \quad (7)$$

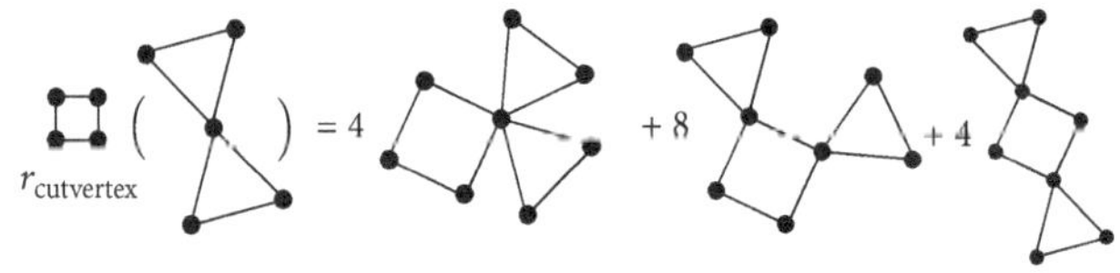

FIGURE 1: The linear combination of graphs obtained by applying the mapping $r_i^{C_4}$ to the cutvertex of the graph consisting of two triangles sharing one vertex.

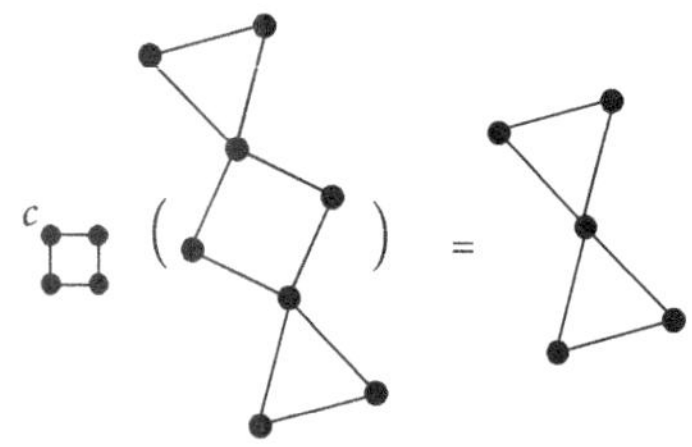

FIGURE 2: The graph obtained by applying the mapping c_{C_4} to a graph with three biconnected components.

where the graphs G_y satisfy the following:

(a) $V(G_y) = V(G)$,

(b) $E(G_y) = E(G)$,

(c) $l_{G_y}(v_i) = l_G(v_i) \cup L'^{(i)}$ for all $i = 1, \ldots, n$ and $v_i \in V(G)$.

The mapping $\xi_{L'}$ is extended to all of $\mathbb{Q}V^{n,k,L}$ by linearity.

4. Generating Connected Graphs

We give a recursion formula to generate all the equivalence classes of connected graphs. The formula depends on the vertex and cyclomatic numbers and produces larger graphs from smaller ones by increasing the number of their biconnected components by one unit. Here, graphs having the same parameters are algebraically represented by linear combinations over coefficients from the rational numbers. The key feature is that the sum of the coefficients of all the graphs in the same equivalence class is given by the inverse of the order of their group of automorphisms. Moreover, the generated graphs are automatically decomposed into their biconnected components.

In the rest of the paper, we often use the following notation: given a group H, by $|H|$ we denote the order of H. Given a graph G, by aut$G$ we denote the group of automorphisms of G. Accordingly, given an equivalence class $\mathscr{A}$, by aut$\mathscr{A}$ we denote the group of automorphisms of all the graphs in $\mathscr{A}$. Furthermore, given a set $W \subseteq V^{n,k,L}$, by $\mathscr{E}(W)$ we denote the set of equivalence classes of all the graphs in W.

We proceed to generalize the recursion formula for generating trees given in [7] to arbitrary connected graphs.

Theorem 1. *For all $p > 1$ and $q \geq 0$ suppose that $\beta_{biconn}^{p,q} := \sum_{G' \in V_{biconn}^{p,q}} \sigma_{G'} G'$ with $\sigma_{G'} \in \mathbb{Q}$, is such that for any equivalence class $\mathscr{A} \in \mathscr{E}(V_{biconn}^{p,q})$ the following holds: (i) there exists $G' \in \mathscr{A}$ such that $\sigma_{G'} > 0$, (ii) $\sum_{G' \in \mathscr{A}} \sigma_{G'} = 1/|\text{aut}\mathscr{A}|$. In this context, given a label set L, for all $n \geq 1$ and $k \geq 0$, define $\beta_{conn}^{n,k,L} \in \mathbb{Q}V_{conn}^{n,k,L}$ by the following recursion relation:*

$$\beta_{conn}^{1,0,L} := G, \quad \text{where } G = (\{1\}, \emptyset), \ l_G(1) = L,$$

$$\beta_{conn}^{1,k,L} := 0 \quad \text{if } k > 0,$$

$$\beta_{conn}^{n,k,L} := \frac{1}{k+n-1} \times \sum_{q=0}^{k} \sum_{p=2}^{n} \sum_{i=1}^{n-p+1} \left((q+p-1)\, r_i^{\beta_{biconn}^{p,q}} \left(\beta_{conn}^{n-p+1,k-q,L} \right) \right). \tag{8}$$

Then, $\beta_{conn}^{n,k,L} = \sum_{G \in V_{conn}^{n,k,L}} \alpha_G G$ with $\alpha_G \in \mathbb{Q}$. Moreover, for any equivalence class $\mathscr{C} \in \mathscr{E}(V_{conn}^{n,k,L})$, the following holds: (i) there exists $G \in \mathscr{C}$ such that $\alpha_G > 0$, (ii) $\sum_{G \in \mathscr{C}} \alpha_G = 1/|\text{aut}\,\mathscr{C}|$.

Proof. The proof is very analogous to the one given in [7] (see also [8, 19]). □

Lemma 2. *Let $n \geq 1$ and $k \geq 0$ be fixed integers. Let L be a label set. Let $\beta_{conn}^{n,k,L} = \sum_{G \in V_{conn}^{n,k,L}} \alpha_G G$ be defined by formula (8). Let $\mathscr{C} \in \mathscr{E}(V_{conn}^{n,k,L})$ denote any equivalence class. Then, there exists $G \in \mathscr{C}$ such that $\alpha_G > 0$.*

Proof. The proof proceeds by induction on the number of biconnected components μ. Clearly, the statement is true for $\mu = 0$. We assume the statement to hold for all the equivalence classes in $\mathscr{E}(V_{\text{conn}}^{n-p+1,k-q,L})$ with $p = 2, \ldots, n-1$ and $q = 0, \ldots, k$, whose elements have $\mu - 1$ biconnected components. Now, suppose that the elements of $\mathscr{C} \in \mathscr{E}(V_{\text{conn}}^{n,k,L})$ have μ biconnected components. Let $\beta_{\text{biconn}}^{p,q} = \sum_{G' \in V_{\text{biconn}}^{p,q}} \sigma_{G'} G'$ with $\sigma_{G'} \in \mathbb{Q}$. Recall that by (4) the mappings $r_i^{\beta_{\text{biconn}}^{p,q}}$ read as

$$r_i^{\beta_{\text{biconn}}^{p,q}} := \sum_{G' \in V_{\text{biconn}}^{p,q}} \sigma_{G'} r_i^{G'}. \tag{9}$$

Let G denote any graph in $\mathscr{C}$. We proceed to show that a graph isomorphic to G is generated by applying the mappings $r_i^{\beta_{\text{biconn}}^{p,q}}$ to a graph $G^* \in V_{\text{conn}}^{n-p+1,k-q,L}$ with $\mu - 1$ biconnected components and such that $\nu_{G^*} > 0$, where ν_{G^*} is the coefficient of G^* in $\beta_{\text{conn}}^{n-p+1,k-q,L}$. Let $\widehat{G} \in V_{\text{biconn}}^{p,q,L'}$ be any biconnected component of the graph G, where $L' = \cup_{v \in V(\widehat{G})} l_G(v)$.

Contracting the graph $\widehat{G}$ to the vertex $u := \min\{u' \in \mathbb{N} : u' \notin V(G) \setminus V(\widehat{G})\}$ yields a graph $c_{\widehat{G}}(G) \in V_{\text{conn}}^{n-p+1,k-q,L}$ with $\mu - 1$ biconnected components. Let $\mathcal{D} \in \mathcal{E}(V_{\text{conn}}^{n-p+1,k-q,L})$ denote the equivalence class such that $c_{\widehat{G}}(G) \in \mathcal{D}$. By the inductive assumption, there exists a graph $H^* \in \mathcal{D}$ such that $H^* \cong c_{\widehat{G}}(G)$ and $\nu_{H^*} > 0$. Let $v_j \in V(H^*)$ be the vertex mapped to u of $c_{\widehat{G}}(G)$ by an isomorphism. Relabeling the graph $\widehat{G}$ with the empty set yields a graph $G' \in V_{\text{biconn}}^{p,q}$. Applying the mapping $r_j^{G'}$ to the graph H^* yields a linear combination of graphs, one of which is isomorphic to G. That is, there exists $H \cong G$ such that $\alpha_H > 0$. □

Lemma 3. *Let $n \geq 1$ and $k \geq 0$ be fixed integers. Let L be a label set such that $\operatorname{card}L \geq n$. Let $\beta_{conn}^{n,k,L} = \sum_{G \in V_{conn}^{n,k,L}} \alpha_G G$ be defined by formula* (8). *Let $\mathcal{C} \in \mathcal{E}(V_{conn}^{n,k,L})$ be an equivalence class such that $l_G(v) \neq \emptyset$ for all $v \in V(G)$ and $G \in \mathcal{C}$. Then, $\sum_{G \in \mathcal{C}} \alpha_G = 1$.*

Proof. The proof proceeds by induction on the number of biconnected components μ. Clearly, the statement is true for $\mu = 0$. We assume the statement to hold for all the equivalence classes in $\mathcal{E}(V_{\text{conn}}^{n-p+1,k-q,L})$ with $p = 2, \ldots, n-1$ and $q = 0, \ldots, k$, whose elements have $\mu - 1$ biconnected components and the property that no vertex is labeled with the empty set. Now, suppose that the elements of $\mathcal{C} \in \mathcal{E}(V_{\text{conn}}^{n,k,L})$ have μ biconnected components. By Lemma 2, there exists a graph $G \in \mathcal{C}$ such that $\alpha_G > 0$, where α_G is the coefficient of G in $\beta_{\text{conn}}^{n,k,L}$. Let $m := \operatorname{card}E(G)$. Therefore, $m = k + n - 1$. By assumption, $l_G(v) \neq \emptyset$ for all $v \in V(G)$ so that $|\operatorname{aut}\mathcal{C}| = 1$. We proceed to show that $\sum_{G \in \mathcal{C}} \alpha_G = 1$. To this end, we check from which graphs with $\mu - 1$ biconnected components, the elements of $\mathcal{C}$ are generated by the recursion formula (8), and how many times they are generated.

Choose any one of the μ biconnected components of the graph $G \in \mathcal{C}$. Let this be the graph $\widehat{G} \in V_{\text{biconn}}^{p,q,L'}$, where $L' = \cup_{v \in V(\widehat{G})} l_G(v)$. Let $m' := \operatorname{card} E(\widehat{G})$ so that $m' = q + p - 1$. Contracting the graph $\widehat{G}$ to the vertex u with $u := \min\{u' \in \mathbb{N} : u' \notin V(G) \setminus V(\widehat{G})\}$ yields a graph $c_{\widehat{G}}(G) \in V_{\text{conn}}^{n-p+1,k-q,L}$ with $\mu - 1$ biconnected components. Let $\mathcal{D} \in \mathcal{E}(V_{\text{conn}}^{n-p+1,k-q,L})$ denote the equivalence class containing $c_{\widehat{G}}(G)$. Since $l_{c_{\widehat{G}}(G)}(v) \neq \emptyset$ for all $v \in V(c_{\widehat{G}}(G))$, we also have $|\operatorname{aut}\mathcal{D}| = 1$. Let $\beta_{\text{conn}}^{n-p+1,k-q,L} = \sum_{G^* \in V_{\text{conn}}^{n-p+1,k-q,L}} \nu_{G^*} G^*$ with $\nu_{G^*} \in \mathbb{Q}$. By Lemma 2, there exists a graph $H \in \mathcal{D}$ such that $H \cong c_{\widehat{G}}(G)$ and $\nu_H > 0$. By the inductive assumption, $\sum_{G^* \in \mathcal{D}} \nu_{G^*} = 1$. Now, let $v_j \in V(H)$ be the vertex which is mapped to u of $c_{\widehat{G}}(G)$ by an isomorphism. Let $G' \in V_{\text{biconn}}^{p,q}$ be the biconnected graph obtained by relabeling $\widehat{G}$ with the empty set. Also, let $\mathcal{A} \in \mathcal{E}(V_{\text{biconn}}^{p,q})$ denote the equivalence class such that $G' \in \mathcal{A}$. Apply the mapping $r_j^{G'}$ to the graph H. Note that every one of the graphs in the linear combination $r_j^{G'}(H)$ corresponds to a way of labeling the graph G' with $l_{c_{\widehat{G}}(G)}(u) = L'$. Therefore, there are $|\operatorname{aut}\mathcal{A}|$ graphs in $r_j^{G'}(H)$ which are isomorphic to the graph G. Since none of the vertices of the graph H is labeled with the empty set, the mapping $r_j^{G'}$ produces a graph isomorphic to G from the graph H with coefficient $\alpha_G^* = \nu_H > 0$. Now, formula (8) prescribes to apply the mappings $r_i^{G'}$ to the vertex which is mapped to u by an isomorphism of every graph in the equivalence class $\mathcal{D}$ occurring in $\beta_{\text{conn}}^{n-p+1,k-q,L}$ (with non-zero coefficient). Therefore,

$$\begin{aligned}\sum_{G \in \mathcal{C}} \alpha_G^* &= |\operatorname{aut}\mathcal{A}| \cdot \sum_{G^* \in \mathcal{D}} \nu_{G^*} \\ &= |\operatorname{aut}\mathcal{A}|,\end{aligned} \tag{10}$$

where the factor $|\operatorname{aut}\mathcal{A}|$ on the right hand side of the first equality is due to the fact that every graph in the equivalence class $\mathcal{D}$ generates $|\operatorname{aut}\mathcal{A}|$ graphs in $\mathcal{C}$. Hence, according to formulas (8) and (4), the contribution to $\sum_{G \in \mathcal{C}} \alpha_G$ is m'/m. Distributing this factor between the m' edges of the graph $\widehat{G}$ yields $1/m$ for each edge. Repeating the same argument for every biconnected component of the graph G proves that every one of the m edges of the graph G adds $1/m$ to $\sum_{G \in \mathcal{C}} \alpha_G$. Hence, the overall contribution is exactly 1. This completes the proof. □

We now show that $\beta_{\text{conn}}^{n,k,L}$ satisfies the following property.

Lemma 4. *Let $n \geq 1$ and $k \geq 0$ be fixed integers. Let L and L' be label sets such that $L \cap L' = \emptyset$. Then, $\beta_{conn}^{n,k,L \cup L'} = \xi_{L'}(\beta_{conn}^{n,k,L})$.*

Proof. Let $\xi_{L'} : V^{n,k,L} \to V^{n,k,L \cup L'}$ and $\widetilde{\xi}_{L'} : V^{n-p+1,k-q,L} \to V^{n-p+1,k-q,L \cup L'}$ be defined as in Section 3. The identity follows by noting that $\xi_{L'} \circ r_i^{\beta_{\text{biconn}}^{p,q}} = r_i^{\beta_{\text{biconn}}^{p,q}} \circ \widetilde{\xi}_{L'}$. □

Lemma 5. *Let $n \geq 1$ and $k \geq 0$ be fixed integers. Let L be a label set. Let $\beta_{conn}^{n,k,L} = \sum_{G \in V_{conn}^{n,k,L}} \alpha_G G$ be defined by formula* (8). *Let $\mathcal{C} \in \mathcal{E}(V_{conn}^{n,k,L})$ denote any equivalence class. Then, $\sum_{G \in \mathcal{C}} \alpha_G = 1/|aut(\mathcal{C})|$.*

Proof. Let G be a graph in $\mathcal{C}$. If $l_G(v) \neq \emptyset$ for all $v \in V(G)$, we simply recall Lemma 3. Thus, we may assume that there exists a set $V' \subseteq V(G)$ such that $l_G(V') = \{\emptyset\}$. Let L' be a label set such that $L \cap L' = \emptyset$ and $\operatorname{card} L' = \operatorname{card} V'$. Relabeling the graph G with $L \cup L'$ via the mapping $l : V(G) \to 2^{L \cup L'}$ such that $l|_{V(G) \setminus V'} = l_G|_{V(G) \setminus V'}$ and $l(v) \neq \emptyset$ for all $v \in V'$ yields a graph $G' \in V_{\text{conn}}^{n,k,L \cup L'}$ such that $|\operatorname{aut} G'| = 1$. Let $\mathcal{D} \in \mathcal{E}(V_{\text{conn}}^{n,k,L \cup L'})$ be the equivalence class such that $G' \in \mathcal{D}$. Let $\beta_{\text{conn}}^{n,k,L \cup L'} = \sum_{G' \in V_{\text{conn}}^{n,k,L \cup L'}} \nu_{G'} G'$ with $\nu_{G'} \in \mathbb{Q}$. By Lemma 3, $\sum_{G' \in \mathcal{D}} \nu_{G'} = 1$. Suppose now that there are exactly T distinct labelings $l_j : V(G) \to 2^{L \cup L'}$, $j = 1, \ldots, T$ such that $l_j|_{V(G) \setminus V'} = l_G|_{V(G) \setminus V'}$, $l_j(v) \neq \emptyset$ for all $v \in V'$, and $G_j \in \mathcal{D}$, where G_j is the graph obtained by relabeling G with l_j. Clearly, $\nu_{G_j} = \alpha_G > 0$, where ν_{G_j} is the coefficient of G_j in $\beta_{\text{conn}}^{n,k,L \cup L'}$. Define $f_j : G \mapsto G_j$ for all $j = 1, \ldots, T$. Now, repeating the

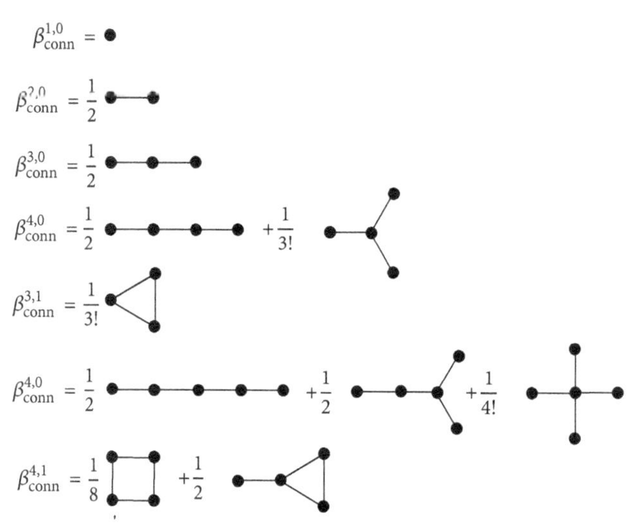

Figure 3: The result of computing all the pairwise non-isomorphic connected graphs as contributions to $\beta^{n,k}_{\text{conn}}$ via formula (8) up to order $n+k \leq 5$. The coefficients in front of graphs are the inverses of the orders of their groups of automorphisms.

same procedure for every graph in $\mathscr{C}$ and recalling Lemma 4, we obtain

$$\sum_{G' \in \mathscr{D}} \nu_{G'} = \sum_{j=1}^{T} \sum_{G \in \mathscr{C}} \nu_{f_j(G)} = T \sum_{G \in \mathscr{C}} \alpha_G = 1. \tag{11}$$

That is, $\sum_{G \in \mathscr{C}} \alpha_G = 1/T$. Since $|\text{aut}\,\mathscr{C}| = T$, we obtain $\sum_{G \in \mathscr{C}} \alpha_G = 1/|\text{aut}\,(\mathscr{C})|$.

This completes the proof of Theorem 1. □

Figure 3 shows $\beta^{n,k}_{\text{conn}}$ for $1 \leq n+k \leq 5$. Now, given a connected graph G, let $\mathscr{V}_G$ denote the set of biconnected components of G. Given a set $X \subset \cup_{p=2}^{n} \cup_{q=0}^{k} V^{p,q}_{\text{biconn}}$, let $V^{n,k}_X := \{G \in V^{n,k}_{\text{conn}} : \mathscr{E}(\mathscr{V}_G) \subseteq \mathscr{E}(X)\}$ with the convention $V^{1,0}_X := \{(\{1\}, \emptyset)\}$. With this notation, Theorem 1 specializes straightforwardly to graphs with specified biconnected components.

Corollary 6. *For all $p > 1$ and $q \geq 0$ suppose that $\gamma^{p,q}_{biconn} := \sum_{G' \in X^{p,q}} \sigma_{G'} G'$ with $\sigma_{G'} \in \mathbb{Q}$ and $X^{p,q} \subseteq V^{p,q}_{biconn}$, is such that for any equivalence class $\mathscr{A} \in \mathscr{E}(X^{p,q})$, the following holds: (i) there exists $G' \in \mathscr{A}$ such that $\sigma_{G'} > 0$, (ii) $\sum_{G' \in \mathscr{A}} \sigma_{G'} = 1/|aut\,\mathscr{A}|$. In this context, for all $n \geq 1$ and $k \geq 0$, define $\gamma^{n,k}_{conn} \in \mathbb{Q}V^{n,k}_{conn}$ by the following recursion relation:*

$$\gamma^{1,0}_{conn} := G, \quad \text{where } G = (\{1\}, \emptyset),$$

$$\gamma^{1,k}_{conn} := 0 \quad \text{if } k > 0,$$

$$\gamma^{n,k}_{conn} := \frac{1}{k+n-1} \times \sum_{q=0}^{k} \sum_{p=2}^{n} \sum_{i=1}^{n-p+1} \left((q+p-1)\, r_i^{\gamma^{p,q}_{biconn}} \left(\gamma^{n-p+1,k-q}_{conn} \right) \right). \tag{12}$$

Then, $\gamma^{n,k}_{conn} = \sum_{G \in V^{n,k}_X} \alpha_G G$, where $\alpha_G \in \mathbb{Q}$ and $X := \cup_{p=2}^{n} \cup_{q=0}^{k} X^{p,q}$. Moreover, for any equivalence class $\mathscr{C} \in \mathscr{E}(V^{n,k}_X)$ the following holds: (i) there exists $G \in \mathscr{C}$ such that $\alpha_G > 0$ (ii) $\sum_{G \in \mathscr{C}} \alpha_G = 1/|\,aut\,\mathscr{C}|$.

Proof. The result follows from the linearity of the mappings r_i^G and the fact that larger graphs whose biconnected components are all in X can only be produced from smaller ones with the same property. □

5. Algebraic Representation of Graphs

We represent graphs by tensors whose indices correspond to the vertex numbers. Our description is essentially that of [7, 9]. From the present section on, we will only consider unlabeled graphs.

Let V be a vector space over $\mathbb{Q}$. Let $\mathsf{S}(V)$ denote the symmetric algebra on V. Then, $\mathsf{S}(V) = \bigoplus_{k=0}^{\infty} \mathsf{S}^k(V)$, where $\mathsf{S}^0(V) := \mathbb{Q}1$, $\mathsf{S}^1(V) = V$, and $\mathsf{S}^k(V)$ is generated by the free commutative product of k elements of V. Also, let $\mathsf{S}(V)^{\otimes n}$ denote the n-fold tensor product of $\mathsf{S}(V)$ with itself. Recall that the multiplication in $\mathsf{S}(V)^{\otimes n}$ is given by the *componentwise product*:

$$\cdot : \mathsf{S}(V)^{\otimes n} \times \mathsf{S}(V)^{\otimes n} \longrightarrow \mathsf{S}(V)^{\otimes n}; \left(s_1 \otimes \cdots \otimes s_n, s_1' \otimes \cdots \otimes s_n' \right) \longmapsto s_1 s_1' \otimes \cdots \otimes s_n s_n', \tag{13}$$

where s_i, s_j' denote monomials on the elements of V for all $i, j = 1, \ldots, n$. We may now proceed to the correspondence between graphs on $\{1, \ldots, n\}$ and some elements of $\mathsf{S}(V)^{\otimes n}$.

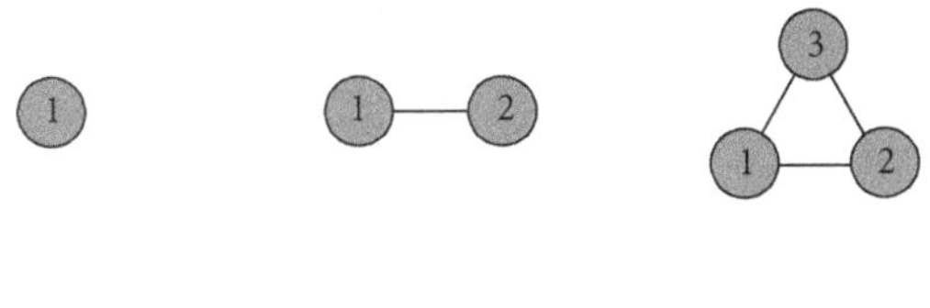

(a) (b) (c)

FIGURE 4: (a) An isolated vertex represented by $1 \in \mathsf{S}(V)$. (b) A 2-vertex tree represented by $R_{1,2} \in \mathsf{S}(V)^{\otimes 2}$. (c) A triangle represented by $R_{1,2} \cdot R_{1,3} \cdot R_{2,3} \in \mathsf{S}(V)^{\otimes 3}$.

First, for all $i, j = 1, \ldots, n$ with $i \neq j$, we define the following tensors in $\mathsf{S}(V)^{\otimes n}$.

$$R_{i,j} := 1^{\otimes i-1} \otimes v \otimes 1^{\otimes j-i-1} \otimes v \otimes 1^{\otimes n-j}, \tag{14}$$

where v is any vector different from zero. (As in Section 3, for simplicity, our notation does not distinguish between elements, say, $R_{i,j} \in \mathsf{S}(V)^{\otimes n}$ and $R_{i,j} \in \mathsf{S}(V)^{\otimes n'}$ with $n \neq n'$. This convention will often be used in the rest of the paper for all the elements of the algebraic representation. Therefore, we will specify the set containing consider the given elements whenever necessary.) Now, for all $i, j = 1, \ldots, n$ with $i \neq j$, let

(i) a tensor factor in the ith position correspond to the vertex i of a graph on $\{1, \ldots, n\}$,

(ii) a tensor $R_{i,j} \in \mathsf{S}(V)^{\otimes n}$ correspond to the edge $\{i, j\}$ of a graph on $\{1, \ldots, n\}$.

In this context, given a graph G with $V(G) = \{1, \ldots, n\}$ and $E(G) = \{\{i_k, j_k\}\}_{k=1,\ldots,m}$, we define the following algebraic representation of graphs.

(a) If $m = 0$, then G is represented by the tensor $1 \otimes \cdots \otimes 1 \in \mathsf{S}(V)^{\otimes n}$.

(b) If $m > 0$, then in $\mathsf{S}(V)^{\otimes n}$ the graph G yields a monomial on the tensors which represent the edges of G. More precisely, since for all $1 \leq k \leq m$ each tensor $R_{i_k,j_k} \in \mathsf{S}(V)^{\otimes n}$ represents an edge of the graph G, this is uniquely represented by the following tensor $S^G_{1,\ldots,n} \in \mathsf{S}(V)^{\otimes n}$ given by the componentwise product of the tensors R_{i_k,j_k}:

$$S^G_{1,\ldots,n} := \prod_{k=1}^{m} R_{i_k,j_k}. \tag{15}$$

Figure 4 shows some examples of this correspondence. Furthermore, let $X \subseteq \{1, \ldots, n'\}$ be a set of cardinality n with $n' \geq n$. Also, let $\sigma : \{1, \ldots, n\} \to X$ be a bijection. With this notation, define the following elements of $\mathsf{S}(V)^{\otimes n'}$:

$$S^G_{\sigma(1),\ldots,\sigma(n)} := \prod_{k=1}^{m} R_{\sigma(i_k),\sigma(j_k)}. \tag{16}$$

In terms of graphs, the tensor $S^G_{\sigma(1),\ldots,\sigma(n)}$ represents a disconnected graph, say, G', on the set $\{1, \ldots, n'\}$ consisting of a graph isomorphic to G whose vertex set is X and $n'-n$ isolated vertices in $\{1, \ldots, n'\} \setminus X$. Figure 5 shows an example.

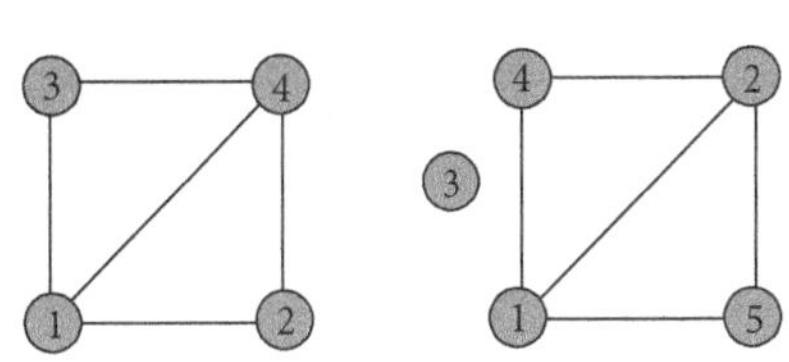

(a) (b)

FIGURE 5: (a) The graph represented by the tensor $S^G_{1,2,3,4} \in \mathsf{S}(V)^{\otimes 4}$. (b) The graph represented by the tensor $S^G_{1,5,4,2} \in \mathsf{S}(V)^{\otimes 5}$ associated with G and the bijection $\sigma : \{1,2,3,4\} \to \{1,2,4,5\}$; $1 \mapsto 1, 2 \mapsto 5, 3 \mapsto 4, 4 \mapsto 2$.

Furthermore, let $\mathsf{T}(\mathsf{S}(V))$ denote the tensor algebra on the graded vector space $\mathsf{S}(V)$: $\mathsf{T}(\mathsf{S}(V)) := \bigoplus_{k=0}^{\infty} \mathsf{S}(V)^{\otimes k}$. In $\mathsf{T}(\mathsf{S}(V))$ the multiplication

$$\bullet : \mathsf{T}(\mathsf{S}(V)) \times \mathsf{T}(\mathsf{S}(V)) \longrightarrow \mathsf{T}(\mathsf{S}(V)) \tag{17}$$

is given by concatenation of tensors (e.g., see [20–22]):

$$\begin{aligned}(s_1 \otimes \cdots \otimes s_n) \bullet (s'_1 \otimes \cdots \otimes s'_{n'}) \\ := s_1 \otimes \cdots \otimes s_n \otimes s'_1 \otimes \cdots \otimes s'_{n'},\end{aligned} \tag{18}$$

where s_i, s'_j denote monomials on the elements of V for all $i = 1, \ldots, n$ and $j = 1, \ldots, n'$. We proceed to generalize the definition of the multiplication $\bullet$ to any two positions of the tensor factors. Let $\tau : \mathsf{S}(V)^{\otimes 2} \to \mathsf{S}(V)^{\otimes 2}$; $s_1 \otimes s_2 \mapsto s_2 \otimes s_1$. Moreover, define $\tau_k := 1^{\otimes k-1} \otimes \tau \otimes 1^{\otimes n-k-1} : \mathsf{S}(V)^{\otimes n} \to \mathsf{S}(V)^{\otimes n}$ for all $1 \leq k \leq n-1$. In this context, for all $1 \leq i \leq n$, $1 \leq j \leq n'$, we define $\bullet_{i,j} : \mathsf{S}(V)^{\otimes n} \times \mathsf{S}(V)^{\otimes n'} \to \mathsf{S}(V)^{\otimes n+n'}$ by the following equation:

$$\begin{aligned}&(s_1 \otimes \cdots \otimes s_n) \bullet_{i,j} (s'_1 \otimes \cdots \otimes s'_{n'}) \\ &\quad := (\tau_{n-1} \circ \cdots \circ \tau_i)(s_1 \otimes \cdots \otimes s_n) \\ &\qquad \otimes (\tau_1 \circ \cdots \circ \tau_{j-1})(s'_1 \otimes \cdots \otimes s'_{n'}) \\ &\quad = s_1 \otimes \cdots \otimes \widehat{s_i} \otimes \cdots \otimes s_n \otimes s_i \\ &\qquad \otimes s'_j \otimes s'_1 \otimes \cdots \otimes \widehat{s'_j} \otimes \cdots \otimes s'_{n'},\end{aligned} \tag{19}$$

where $\widehat{s_i}$ (resp. $\widehat{s'_j}$) means that s_i (resp. s'_j) is excluded from the sequence. In terms of the tensors $S^G_{1,\ldots,n} \in \mathsf{S}(V)^{\otimes n}$ and $S^{G'}_{1,\ldots,n'} \in \mathsf{S}(V)^{\otimes n'}$ the equation previous yields

$$S^G_{1,\ldots,n} \bullet_{i,j} S^{G'}_{1,\ldots,n'} := S^G_{\sigma(1),\ldots,\sigma(n)} \cdot S^{G'}_{\sigma'(1),\ldots,\sigma'(n')}, \tag{20}$$

where $S^G_{\sigma(1),\ldots,\sigma(n)}, S^{G'}_{\sigma'(1),\ldots,\sigma'(n')} \in \mathsf{S}(V)^{\otimes n+n'}$, $\sigma(k) = k$ if $1 \leq k < i$, $\sigma(i) = n$, $\sigma(k) = k-1$ if $i < k \leq n$, and $\sigma'(k) = k+n+1$ if $1 \leq k < j$, $\sigma'(j) = n+1$, $\sigma'(k) = k+n$ if $j < k \leq n'$.

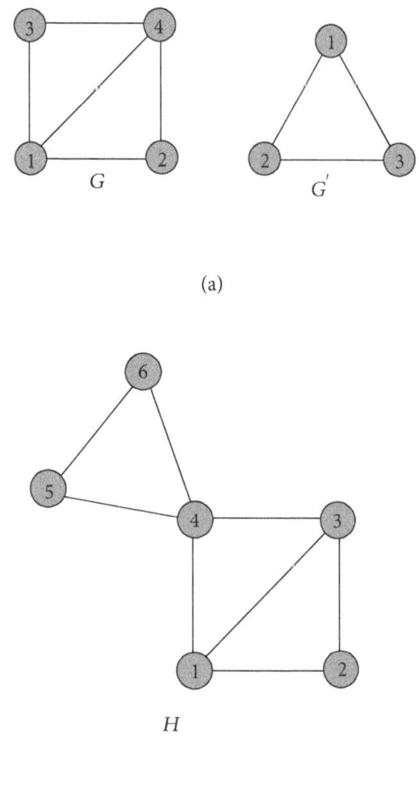

FIGURE 6: (a) The graphs represented by the tensors $S^G_{1,2,3,4} \in \mathsf{S}(V)^{\otimes 4}$ and $S^{G'}_{1,2,3} \in \mathsf{S}(V)^{\otimes 3}$. (b) The graph represented by the tensor $S^H_{1,\ldots,6} = S^G_{1,2,3,4} \diamond_{3,2} S^{G'}_{1,2,3} \in \mathsf{S}(V)^{\otimes 6}$.

Clearly, the tensor $S^G_{1,\ldots,n} \bullet_{i,j} S^{G'}_{1,\ldots,n'}$ represents a disconnected graph. Now, let $\cdot_i := 1^{\otimes i-1} \otimes \cdot \otimes 1^{\otimes n-i-1} : \mathsf{S}(V)^{\otimes n} \rightarrow \mathsf{S}(V)^{\otimes n-1}$. In $\mathsf{T}(\mathsf{S}(V))$, for all $1 \le i \le n$, $1 \le j \le n'$, define the following nonassociative and noncommutative multiplication:

$$\diamond_{i,j} := \cdot_n \circ \bullet_{i,j} : \mathsf{S}(V)^{\otimes n} \times \mathsf{S}(V)^{\otimes n'} \longrightarrow \mathsf{S}(V)^{\otimes n+n'-1}. \quad (21)$$

The tensor $S^G_{1,\ldots,n} \diamond_{i,j} S^{G'}_{1,\ldots,n'}$ represents the graph on $n + n' - 1$ vertices, say, H obtained by gluing the vertex i of the graph G to the vertex j of the graph G'. If both G and G' are connected, the vertex n is clearly a cutvertex of the graph H. Figure 6 shows an example.

6. Linear Mappings

We recall some of the linear mappings given in [9].

Let $\mathcal{B}_{n,k} \subset \mathsf{S}(V)^{\otimes n}$ denote the vector space of all the tensors representing biconnected graphs in $V^{n,k}_{\text{biconn}}$. Let $\mathcal{B} := \mathcal{B}_{1,0} \bigoplus \mathcal{B}_{2,0} \bigoplus_{n=3}^{\infty} \bigoplus_{k=1}^{k_{\max}(n)} \mathcal{B}_{n,k} \subset \mathsf{T}(\mathsf{S}(V))$, where $k_{\max}(n) = (n(n-3)/2) + 1$ is, by Kirchhoff's lemma [18], the maximum cyclomatic number of a biconnected graph on n vertices. Let the mapping $\Delta : \mathcal{B} \rightarrow \mathsf{T}(\mathsf{S}(V))$ be given by the following equations:

$$\Delta(1) := 1 \otimes 1,$$
$$\Delta\left(B^G_{1,\ldots,n}\right) := \frac{1}{n}\sum_{i=1}^{n} \Delta_i \left(B^G_{1,\ldots,n}\right) \quad \text{if } n > 1, \quad (22)$$

where G denotes a biconnected graph on n vertices represented by $B^G_{1,\ldots,n} \in \mathsf{S}(V)^{\otimes n}$. To define the mappings Δ_i, we introduce the following bijections:

$$\text{(i) } \sigma_i : j \longmapsto \begin{cases} j & \text{if } 1 \le j \le i, \\ j+1 & \text{if } i+1 \le j \le n, \end{cases}$$
$$\text{(ii) } \nu_i : j \longmapsto \begin{cases} j & \text{if } 1 \le j \le i-1, \\ j+1 & \text{if } i \le j \le n. \end{cases} \quad (23)$$

In this context, for all $i = 1, \ldots, n$ with $n > 1$, the mappings $\Delta_i : \mathsf{S}(V)^{\otimes n} \rightarrow \mathsf{S}(V)^{\otimes n+1}$ are defined by the following equation:

$$\begin{aligned} \Delta_i\left(B^G_{1,\ldots,n}\right) &:= B^G_{\sigma_i(1),\ldots,\sigma_i(n)} + B^G_{\nu_i(1),\ldots,\nu_i(n)} \\ &= B^G_{1,\ldots,\widehat{i+1},i+2,\ldots,n+1} + B^G_{1,\ldots,\hat{i},i+1\ldots,n+1}, \end{aligned} \quad (24)$$

where $\hat{i}$ (resp., $\widehat{i+1}$) means that the index i (resp., $i+1$) is excluded from the sequence. The tensor $B^G_{1,\ldots,\hat{i},\ldots,n+1}$ (resp., $B^G_{1,\ldots,\widehat{i+1},i+2,\ldots,n+1}$) is constructed from $B^G_{1,\ldots,n}$ by transferring the monomial on the elements of V which occupies the kth tensor factor to the $(k+1)$th position for all $i \le k \le n$ (resp., $i+1 \le k \le n$). Furthermore, suppose that $B^G_{\pi(1),\ldots,\pi(n)} \in \mathsf{S}(V)^{\otimes n'}$ and that the bijection π is such that $i \notin \pi(\{1,\ldots,n\}) \subset \{1,\ldots,n'\}$. In this context, define

$$\Delta_i\left(B^G_{\pi(1),\ldots,\pi(n)}\right) := B^G_{\nu_i(\pi(1)),\ldots,\nu_i(\pi(n))} \quad (25)$$

in agreement with $\Delta(1) := 1 \otimes 1$. It is straightforward to verify that the mappings Δ_i satisfy the following property:

$$\Delta_i \circ \Delta_i = \Delta_{i+1} \circ \Delta_i, \quad (26)$$

where we used the same notation for $\Delta_i : \mathsf{S}(V)^{\otimes n} \rightarrow \mathsf{S}(V)^{\otimes n+1}$ on the right of either side of the previous equation and $\Delta_i : \mathsf{S}(V)^{\otimes n+1} \rightarrow \mathsf{S}(V)^{\otimes n+2}$ as the leftmost operator on the left hand side of the equation. Accordingly, $\Delta_{i+1} : \mathsf{S}(V)^{\otimes n+1} \rightarrow \mathsf{S}(V)^{\otimes n+2}$ as the leftmost operator on the right hand side of the equation.

Now, for all $m > 0$, define the mth iterate of Δ_i, $\Delta_i^m : \mathsf{S}(V)^{\otimes n} \rightarrow \mathsf{S}(V)^{\otimes n+m}$, recursively as follows:

$$\Delta_i^1 := \Delta_i, \quad (27)$$

$$\Delta_i^m := \Delta_i \circ \Delta_i^{m-1}, \quad (28)$$

where $\Delta_i : \mathsf{S}(V)^{\otimes n+m-1} \rightarrow \mathsf{S}(V)^{\otimes n+m}$ in formula (28). This can be written in m different ways corresponding to the composition of Δ_i^{m-1} with each of the mappings Δ_j : $S(V)^{\otimes n+m-1} \rightarrow S(V)^{\otimes n+m}$ with $i \leq j \leq i+m-1$. These are all equivalent by formula (26).

Extension to Connected Graphs. We now extend the mappings Δ_i to the vector space of all the tensors representing connected graphs $\mathscr{B}^* := \bigoplus_{\mu=0}^{\infty} \mathscr{B}^{\diamond\mu}$, where $\mathscr{B}^{\diamond 0} = \mathbb{Q}1$. We proceed to define $\mathscr{B}^{\diamond\mu}$ for $\mu > 0$.

First, given two sets $A_n \subset S(V)^{\otimes n}$ and $B_p \subset S(V)^{\otimes p}$, by $A_n \diamond_{i,j} B_p \subset S(V)^{\otimes n+p-1}$ with $i = 1, \ldots, n$, $j = 1, \ldots, p$ we denote the set of elements obtained by applying the mapping $\diamond_{i,j}$ to every ordered pair $(a \in A_n, b \in B_p)$. Also, let $\mathscr{B}_2 := \mathscr{B}_{2,0}$ and $\mathscr{B}_n := \bigoplus_{k=1}^{k_{\max}(n)} \mathscr{B}_{n,k}$, $n \geq 3$. In this context, for all $\mu \geq 1$ and $n \geq \mu + 1$, define $\mathscr{B}_n^{*\mu}$ as follows:

$$\mathscr{B}_n^{*\mu} = \bigcup_{n_1+\cdots+n_\mu = n+\mu-1} \bigcup_{i_1=1}^{n_1} \cdots \bigcup_{i_{\mu-1}=1}^{n_{\mu-1}} \bigcup_{j_1=1}^{n_\mu+\cdots+n_2-\mu+2} \cdots \bigcup_{j_{\mu-1}=1}^{n_\mu} \mathscr{B}_{n_1} \diamond_{i_1,j_1} \left(\mathscr{B}_{n_2} \diamond_{i_2,j_2} \left(\cdots \left(\mathscr{B}_{n_{\mu-2}} \diamond_{i_{\mu-2},j_{\mu-2}} \left(\mathscr{B}_{n_{\mu-1}} \diamond_{i_{\mu-1},j_{\mu-1}} \mathscr{B}_{n_\mu}\right)\right)\cdots\right)\right). \tag{29}$$

Also, define

$$\mathscr{B}_n'^{*\mu} = \bigcup_{\pi \in S_n} \left\{ S^G_{\pi(1),\ldots,\pi(n)} \mid S^G_{1,\ldots,n} \in \mathscr{B}_n^{*\mu} \right\}, \tag{30}$$

where S_n denotes the symmetric group on the set $\{1, \ldots, n\}$ and $S^G_{\pi(1),\ldots,\pi(n)}$ is given by formula (16). Finally, for all $\mu \geq 1$, define

$$\mathscr{B}^{\diamond\mu} := \bigoplus_{n=\mu+1}^{\infty} \mathscr{B}_n'^{*\mu}. \tag{31}$$

The elements of $\mathscr{B}^{\diamond\mu}$ are clearly tensors representing connected graphs on μ biconnected components. By (16) and (21), these may be seen as monomials on tensors representing biconnected graphs with the componentwise product $\cdot : S(V)^{\otimes n} \times S(V)^{\otimes n} \rightarrow S(V)^{\otimes n}$ so that repeated indices correspond to cutvertices of the associated graphs. In this context, an arbitrary connected graph, say, G, on $n \geq 2$ vertices and $\mu \geq 1$ biconnected components yields

$$\prod_{a=1}^{\mu} B^{G_a}_{\sigma_a(1),\ldots,\sigma_a(n_a)}, \tag{32}$$

Where, for all $1 \leq a \leq \mu$, $B^{G_a}_{\sigma_a(1),\ldots,\sigma_a(n_a)} \in S(V)^{\otimes n}$, G_a is a biconnected graph on $2 \leq n_a \leq n$ vertices represented by $B^{G_a}_{1,\ldots,n_a} \in S(V)^{\otimes n_a}$, and $\sigma_a : \{1, \ldots, n_a\} \rightarrow X_a \subseteq \{1, \ldots, n\}$ is a bijection.

We now extend the mapping $\Delta := (1/n) \sum_{i=1}^{n} \Delta_i$ to $\mathscr{B}^*$ by requiring the mappings Δ_i to satisfy the following condition:

$$\Delta_i \left(\prod_{a=1}^{\mu} B^{G_a}_{\sigma_a(1),\ldots,\sigma_a(n_a)} \right) := \prod_{a=1}^{\mu} \Delta_i \left(B^{G_a}_{\sigma_a(1),\ldots,\sigma_a(n_a)} \right). \tag{33}$$

Given a connected graph G', the mapping Δ_i may be thought of as a way of (a) splitting the vertex i into two new vertices numbered i and $i+1$ and (b) distributing the biconnected components sharing the vertex i between the two new ones in all the possible ways. Analogously, the action of the mappings Δ_i^m consists of (a) splitting the vertex i into $m+1$ new vertices numbered $i, i+1, \ldots, i+m$ and (b) distributing the biconnected components sharing the vertex i between the $m+1$ new ones in all the possible ways.

We now combine the mappings Δ_i^{n-1} with tensors representing biconnected graphs. Let $n > 1$, $p \geq 1$ and $1 \leq i \leq p$ be fixed integers. Let $\pi_i : \{1, \ldots, n\} \rightarrow \{i, i+1, \ldots, i+n-1\}$; $j \mapsto j+i-1$ be a bijection. Let G be a biconnected graph on n vertices represented by the tensor $B^G_{1,\ldots,n} \in S(V)^{\otimes n}$. The tensors $B^G_{i,i+1,\ldots,i+n-1} := B^G_{\pi_i(1),\ldots,\pi_i(n)} \in S(V)^{\otimes n+p-1}$ (see formula (16)) may be viewed as operators acting on $S(V)^{\otimes n+p-1}$ by multiplication. In this context, consider the following mappings given by the composition of $B^G_{i,i+1,\ldots,i+n-1}$ with Δ_i^{n-1}:

$$B^G_{i,i+1,\ldots,i+n-1} \circ \Delta_i^{n-1} : S(V)^{\otimes p} \longrightarrow S(V)^{\otimes n+p-1}. \tag{34}$$

These are the analog of the mappings r_i^G given in Section 3. In plain English, the mappings $B^G_{i,i+1,\ldots,i+n-1} \circ \Delta_i^{n-1}$ produce a connected graph with $n+p-1$ vertices from one with p vertices in the following way:

(a) split the vertex i into n new vertices, namely, $i, i+1, \ldots, i+n-1$,

(b) distribute the biconnected components containing the split vertex between the n new ones in all the possible ways,

(c) merge the n new vertices into the graph G.

When the graph G is a 2-vertex tree, the mapping $R_{1,2} \circ \Delta$ coincides with the application $L = (\phi \otimes \phi)\Delta$ of [23] when ϕ acts on $S(V)$ by multiplication with a vector.

To illustrate the action of the mappings $B^G_{\pi_i(1),\ldots,\pi_i(n)} \circ \Delta_i$, consider the graph H consisting of two triangles sharing a vertex. Let this be represented by $B^{C_3}_{1,2,3} \cdot B^{C_3}_{3,4,5} \in S(V)^{\otimes 5}$, where C_3 denotes a triangle represented by $B^{C_3}_{1,2,3} = R_{1,2} \cdot R_{2,3} \cdot R_{1,3} \in S(V)^{\otimes 3}$. Let T_2 denote a 2-vertex tree represented by $B^{T_2}_{1,2} = R_{1,2} \in S(V)^{\otimes 2}$. Applying the mapping $R_{3,4} \circ \Delta_3$ to H yields

$$\begin{aligned} R_{3,4} \circ \Delta_3 \left(B^{C_3}_{1,2,3} \cdot B^{C_3}_{3,4,5} \right) &= R_{3,4} \cdot \Delta_3 \left(B^{C_3}_{1,2,3} \right) \cdot \Delta_3 \left(B^{C_3}_{3,4,5} \right) \\ &= R_{3,4} \cdot \left(B^{C_3}_{1,2,3} + B^{C_3}_{1,2,4} \right) \cdot \left(B^{C_3}_{3,5,6} + B^{C_3}_{4,5,6} \right) \\ &= R_{3,4} \cdot B^{C_3}_{1,2,3} \cdot B^{C_3}_{3,5,6} + R_{3,4} \cdot B^{C_3}_{1,2,3} \cdot B^{C_3}_{4,5,6} \\ &\quad + R_{3,4} \cdot B^{C_3}_{1,2,4} \cdot B^{C_3}_{3,5,6} + R_{3,4} \cdot B^{C_3}_{1,2,4} \cdot B^{C_3}_{4,5,6}. \end{aligned} \tag{35}$$

Figure 7 shows the linear combination of graphs given by (35) after taking into account that the first and fourth terms as well as the second and third correspond to isomorphic

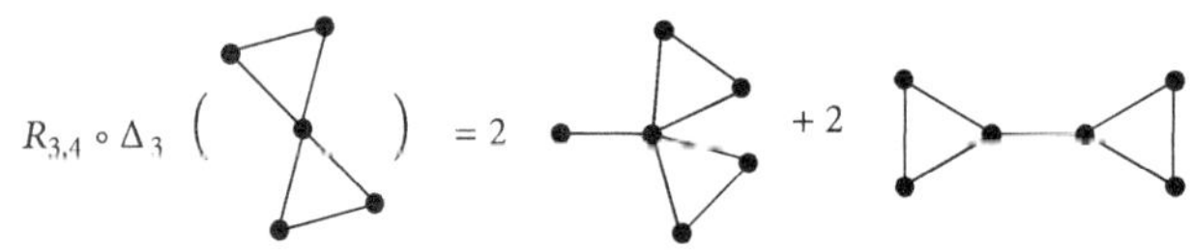

FIGURE 7: The linear combination of graphs obtained by applying the mapping $R_{3,4} \circ \Delta_3$ to the cutvertex of a graph consisting of two triangles sharing a vertex.

graphs. Note that 3 (resp. 4) is the only cutvertex of the graph represented by the first (resp. fourth) term, while 3 and 4 are both cutvertices of the graphs represented by the second or third terms.

7. Further Recursion Relations

Reference [7] gives two recursion formulas to generate all the equivalence classes of trees with coefficients given by the inverses of the orders of their groups of automorphisms. On the one hand, the main formula is such that larger trees are produced from smaller ones by increasing the number of their biconnected components by one unit. On the other hand, the alternative formula is such that for all $n \geq 2$, trees on n vertices are produced by connecting a vertex of a tree on i vertices to a vertex of a tree on $n-i$ vertices in all the possible ways. Theorem 1 generalizes the main formula to connected graphs. It is the aim of this section to derive an alternative formula for a simplified version of the latter.

Let G denote a connected graph. Recall the notation introduced in Section 4; $\mathscr{V}_G$ denotes the set of biconnected components of G and $\mathscr{E}(\mathscr{V}_G)$ denotes the set of equivalence classes of the graphs in $\mathscr{V}_G$. Given a set $X \subset \cup_{p=2}^{n} \cup_{q=0}^{k} V_{\text{biconn}}^{p,q}$, let $V_X^{n,k} := \{G \in V_{\text{conn}}^{n,k} : \mathscr{E}(\mathscr{V}_G) \subseteq \mathscr{E}(X)\}$ with the convention $V_X^{1,0} := \{(\{1\}, \emptyset)\}$. With this notation, in the algebraic setting, Corollary 6 reads as follows.

Theorem 7. *For all $p > 1$ and $q \geq 0$, suppose that $\Phi_{1,\ldots,p}^{p,q} := \sum_{G' \in X^{p,q}} \sigma_{G'} B_{1,\ldots,p}^{G'} \in \mathscr{B}_{p,q} \subset \mathsf{S}(V)^{\otimes p}$ with $\sigma_{G'} \in \mathbb{Q}$ and $X^{p,q} \subseteq V_{\text{biconn}}^{p,q}$, is such that for any equivalence class $\mathscr{A} \in \mathscr{E}(X^{p,q})$ the following holds: (i) there exists $G' \in \mathscr{A}$ such that $\sigma_{G'} > 0$, (ii) $\sum_{G' \in \mathscr{A}} \sigma_{G'} = 1/|aut\,\mathscr{A}|$. In this context, for all $n \geq 1$ and $k \geq 0$, define $\Psi^{n,k} \in \mathsf{S}(V)^{\otimes n}$ by the following recursion relation:*

$$\Psi^{1,0} := 1, \qquad \Psi^{1,k} := 0 \quad \text{if } k > 0, \tag{36}$$

$$\Psi^{n,k} := \frac{1}{k+n-1} \times \sum_{q=0}^{k} \sum_{p=2}^{n} \sum_{i=1}^{n-p+1} \left((q+p-1)\, \Phi_{i,i+1,\ldots,i+p-1}^{p,q} \cdot \Delta_i^{p-1} \left(\Psi^{n-p+1,k-q} \right) \right). \tag{37}$$

Then, $\Psi^{n,k} = \sum_{G \in V_X^{n,k}} \alpha_G S_{1,\ldots,n}^{G}$, where $\alpha_G \in \mathbb{Q}$ and $X := \cup_{p=2}^{n} \cup_{q=0}^{k} X^{p,q}$. Moreover, for any equivalence class $\mathscr{C} \in \mathscr{E}(V_X^{n,k})$, the following holds: (i) there exists $G \in \mathscr{C}$ such that $\alpha_G > 0$, (ii) $\sum_{G \in \mathscr{C}} \alpha_G = 1/|aut\,\mathscr{C}|$.

Now, given a nonempty set $Y \subseteq V_{\text{biconn}}^{p,q}$, we use the following notation:

$$V_Y^{\mu} := \left\{ G \in V_{\text{conn}}^{(p-1)\mu+1, q\mu} : \mathscr{E}(\mathscr{V}_G) \subseteq \mathscr{E}(Y), \operatorname{card}(\mathscr{V}_G) = \mu \right\}, \tag{38}$$

with the convention $V_Y^0 := \{(\{1\}, \emptyset)\}$. When we only consider graphs whose biconnected components are all in Y, the previous recurrence can be transformed into one on the number of biconnected components.

Theorem 8. *Let $p > 1$ and $q \geq 0$ be fixed integers. Let $Y \subseteq V_{biconn}^{p,q}$ be a non-empty set. Suppose that $\phi_{1,\ldots,p}^{p,q} := \sum_{G' \in Y} \sigma_{G'} B_{1,\ldots,p}^{G'} \in \mathscr{B}_{p,q} \subset \mathsf{S}(V)^{\otimes p}$ with $\sigma_{G'} \in \mathbb{Q}$, is such that for any equivalence class $\mathscr{A} \in \mathscr{E}(Y)$ the following holds: (i) there exists $G' \in \mathscr{A}$ such that $\sigma_{G'} > 0$, (ii) $\sum_{G' \in \mathscr{A}} \sigma_{G'} = 1/|aut\,(\mathscr{A})|$. In this context, for all $\mu \geq 0$, define $\psi^{\mu} \in \mathsf{S}(V)^{\otimes (p-1)\mu+1}$ by the following recursion relation:*

$$\psi^0 := 1, \qquad \psi^{\mu} := \frac{1}{\mu} \sum_{i=1}^{(p-1)(\mu-1)+1} \phi_{i,i+1,\ldots,i+p-1}^{p,q} \cdot \Delta_i^{p-1} \left(\psi^{\mu-1} \right). \tag{39}$$

Then, $\psi^{\mu} = \sum_{G \in V_Y^{\mu}} \alpha_G S_{1,\ldots,(p-1)\mu+1}^{G}$ with $\alpha_G \in \mathbb{Q}$. Moreover, for any equivalence class $\mathscr{C} \in \mathscr{E}(V_Y^{\mu})$, the following holds: (i) there exists $G \in \mathscr{C}$ such that $\alpha_G > 0$, (ii) $\sum_{G \in \mathscr{C}} \alpha_G = 1/|aut\,(\mathscr{C})|$.

Proof. In this case, $\Psi^{n,k} = 0$ unless $n = (p-1)\mu + 1$ and $k = q\mu$, where $\mu \geq 0$. Therefore, the recurrence of Theorem 7 can be easily converted into a recurrence on the number of biconnected components by setting $\psi^{\mu} := \Psi^{(p-1)\mu+1, q\mu}$. □

For $p = 2$ and $q = 0$, we recover the formula to generate trees of [7]. As in that paper and [8, 23], we may extend the result to obtain further interesting recursion relations.

Proposition 9. *For all* $\mu > 0$,

$$\psi^{\mu} = \frac{1}{\mu} \times \sum_{i_1+\cdots+i_p=\mu-1} \left(\sum_{a_1=1}^{(p-1)i_1+1} \sum_{a_2=(p-1)i_1+2}^{(p-1)(i_1+i_2)+2} \cdots \sum_{a_p=(p-1)(i_1+\cdots+i_{p-1})+p}^{(p-1)(i_1+\cdots+i_p)+p} \phi^{p,q}_{a_1,\ldots,a_p} \right) \cdot \left(\psi^{i_1} \otimes \cdots \otimes \psi^{i_p} \right), \tag{40}$$

where $i_j = 0, \ldots, \mu - 1$ *for all* $j = 1, \ldots, p$.

Proof. Equation (40) is proved by induction on the number of biconnected components μ. This is easily verified for $\mu = 1$:

$$\psi^1 = \phi^{p,q}_{1,\ldots p} \cdot (1 \otimes \cdots \otimes 1) = \phi^{p,q}_{1,\ldots p}. \tag{41}$$

We now assume the formula to hold for $\psi^{\mu-1}$. Then, formula (37) yields

$$\psi^{\mu} = \frac{1}{\mu} \sum_{i=1}^{(p-1)(\mu-1)+1} \phi^{p,q}_{i,i+1,\ldots,i+p-1} \cdot \Delta_i^{p-1} \left(\psi^{\mu-1} \right)$$

$$= \frac{1}{\mu(\mu-1)} \sum_{i=1}^{(p-1)(\mu-1)+1} \phi^{p,q}_{i,i+1,\ldots,i+p-1} \cdot \Delta_i^{p-1} \left(\sum_{i_1+\cdots+i_p=\mu-2} \left(\sum_{a_1=1}^{(p-1)i_1+1} \sum_{a_2=(p-1)i_1+2}^{(p-1)(i_1+i_2)+2} \cdots \sum_{a_p=(p-1)(i_1+\cdots+i_{p-1})+p}^{(p-1)(i_1+\cdots+i_p)+p} \phi^{p,q}_{a_1,\ldots,a_p} \right) \cdot \left(\psi^{i_1} \otimes \cdots \otimes \psi^{i_p} \right) \right)$$

$$= \frac{1}{\mu(\mu-1)} \times \sum_{i_1+\cdots+i_p=\mu-2} \sum_{i=1}^{(p-1)(\mu-1)+1} \left(\Delta_i^{p-1} \left(\sum_{a_1=1}^{(p-1)i_1+1} \sum_{a_2=(p-1)i_1+2}^{(p-1)(i_1+i_2)+2} \cdots \sum_{a_p=(p-1)(i_1+\cdots+i_{p-1})+p}^{(p-1)(i_1+\cdots+i_p)+p} \phi^{p,q}_{a_1,\ldots,a_p} \right) \cdot \phi^{p,q}_{i,i+1,\ldots,i+p-1} \cdot \Delta_i^{p-1} \left(\psi^{i_1} \otimes \cdots \otimes \psi^{i_p} \right) \right)$$

$$= \frac{1}{\mu(\mu-1)} \times \left(\sum_{i_1+\cdots+i_p=\mu-2} \left(\sum_{i=1}^{(p-1)i_1+1} \Delta_i^{p-1} \left(\sum_{a_1=1}^{(p-1)i_1+1} \sum_{a_2=(p-1)i_1+2}^{(p-1)(i_1+i_2)+2} \cdots \sum_{a_p=(p-1)(i_1+\cdots+i_{p-1})+p}^{(p-1)(i_1+\cdots+i_p)+p} \sum_{a_p=(p-1)(i_1+\cdots+i_{p-1})+p}^{(p-1)(i_1+\cdots+i_p)+p} \phi^{p,q}_{a_1,\ldots,a_p} \right) \cdot \phi^{p,q}_{i,i+1,\ldots,i+p-1} \cdot \left(\Delta_i^{p-1} \left(\psi^{i_1} \right) \otimes \cdots \otimes \psi^{i_p} \right) \right.\right.$$

$$+ \sum_{i=(p-1)i_1+2}^{(p-1)(i_1+i_2)+2} \Delta_i^{p-1} \left(\sum_{a_1=1}^{(p-1)i_1+1} \sum_{a_2=(p-1)i_1+2}^{(p-1)(i_1+i_2)+2} \cdots \sum_{a_p=(p-1)(i_1+\cdots+i_{p-1})+p}^{(p-1)(i_1+\cdots+i_p)+p} \right) \cdot \phi^{p,q}_{i,i+1,\ldots,i+p-1} \cdot \left(\psi^{i_1} \otimes \Delta_i^{p-1} \left(\psi^{i_2} \right) \otimes \cdots \otimes \psi^{i_p} \right)$$

$$+ \cdots + \sum_{i=(p-1)(i_1+i_2+\cdots+i_{p-1})+p}^{(p-1)(i+1+\cdots+ip)+p} \Delta_i^{p-1} \left(\sum_{a_1=1}^{(p-1)i_1+1} \sum_{a_2=(p-1)i_1+2}^{(p-1)(i_1+i_2)+2} \cdots \sum_{a_p=(p-1)(i_1+\cdots+i_{p-1})+p}^{(p-1)(i_1+\cdots+i_p)+p} \phi^{p,q}_{a_1,\ldots,a_p} \right) \cdot \phi^{p,q}_{i,i+1,\ldots,i+p-1}$$

$$\cdot \left(\psi^{i_1} \otimes \cdots \otimes \Delta_i^{p-1}\left(\psi^{i_p}\right)\right)\Bigg)\Bigg)$$

$$= \frac{1}{\mu(\mu-1)}$$

$$\times \Bigg(\sum_{i_1+\cdots+i_p=\mu-2} \Bigg((i_1+1)$$

$$\times \Bigg(\sum_{a_1=1}^{(p-1)i_1+p} \sum_{a_2=(p-1)i_1+p+1}^{(p-1)(i_1+i_2)+p+1} \cdots \sum_{a_p=(p-1)(i_1+\cdots+i_{p-1})+2p-1}^{(p-1)(i_1+\cdots+i_p)+2p-1} \phi^{p,q}_{a_1,\dots,a_p} \Bigg)$$

$$\cdot \left(\psi^{i_1+1} \otimes \cdots \otimes \psi^{i_p}\right) + (i_2+1)$$

$$\times \Bigg(\sum_{a_1=1}^{(p-1)i_1+1} \sum_{a_2=(p-1)i_1+2}^{(p-1)(i_1+i_2)+p+1} \cdots \sum_{a_p=(p-1)(i_1+\cdots+i_{p-1})+2p-1}^{(p-1)(i_1+\cdots+i_p)+2p-1} \phi^{p,q}_{a_1,\dots,a_p} \Bigg)$$

$$\cdot \phi^{p,q}_{i,i+1,\dots,i+p-1} \cdot \left(\psi^{i_1} \otimes \psi^{i_2+1} \otimes \cdots \otimes \psi^{i_p}\right)$$

$$+ \cdots + (i_p+1)$$

$$\times \Bigg(\sum_{a_1=1}^{(p-1)i_1+1} \sum_{a_2=(p-1)i_1+2}^{(p-1)(i_1+i_2)+2} \cdots \sum_{a_p=(p-1)(i_1+\cdots+i_{p-1})+2p-1}^{(p-1)(i_1+\cdots+i_p)+p} \phi^{p,q}_{a_1,\dots,a_p} \Bigg)$$

$$\cdot \phi^{p,q}_{i,i+1,\dots,i+p-1} \cdot \left(\psi^{i_1} \otimes \cdots \otimes \psi^{i_p+1}\right)\Bigg)\Bigg)$$

$$= \frac{1}{\mu(\mu-1)}$$

$$\times \Bigg(\sum_{i_1=0}^{\mu-1} \sum_{i_2+\cdots+i_p=\mu-1-i_1} i_1 + \sum_{i_2=0}^{\mu-1} \sum_{i_1+i_3+\cdots+i_p=\mu-1-i_2} i_2$$

$$+ \cdots + \sum_{i_p=0}^{\mu-1} \sum_{i_1+\cdots+i_{p-1}=\mu-1-i_p} i_p \Bigg)$$

$$\times \Bigg(\sum_{a_1=1}^{(p-1)i_1+1} \sum_{a_2=(p-1)i_1+2}^{(p-1)(i_1+i_2)+2} \cdots \sum_{a_p=(p-1)(i_1+\cdots+i_{p-1})+p}^{(p-1)(i_1+\cdots+i_p)+p} \phi^{p,q}_{a_1,\dots,a_p} \Bigg)$$

$$\cdot \left(\psi^{i_1} \otimes \cdots \otimes \psi^{i_p}\right)$$

$$= \frac{1}{\mu} \sum_{i_2+\cdots+i_p=\mu-1} \Bigg(\sum_{a_1=1}^{(p-1)i_1+1} \sum_{a_2=(p-1)i_1+2}^{(p-1)(i_1+i_2)+2} \cdots \sum_{a_p=(p-1)(i_1+\cdots+i_{p-1})+p}^{(p-1)(i_1+\cdots+i_p)+p} \phi^{p,q}_{a_1,\dots,a_p} \Bigg)$$

$$\cdot \left(\psi^{i_1} \otimes \cdots \otimes \psi^{i_p}\right). \tag{42}$$

□

Note that formula (40) states that the linear combination of all the equivalence classes of connected graphs in V_Y^{μ} is given by summing over all the p-tuples of connected graphs with total number of biconnected components equal to $\mu - 1$, gluing a vertex of each of them to distinct vertices of a graph in $\mathscr{E}(Y)$ in all the possible ways.

8. Cayley-Type Formulas

Let $Y \subseteq V^{p,q}_{\text{biconn}}$. Recall that V_Y^{μ} is the set of connected graphs on μ biconnected components, each of which has p vertices and cyclomatic number q and is isomorphic to a graph in Y. We proceed to use formula (40) to recover some enumerative results which are usually obtained via generating functions and the Lagrange inversion formula, see [3]. In particular, we extend to graphs in V_Y^{μ} the result that the sum of the inverses of the orders of the groups of automorphisms of all the pairwise nonisomorphic trees on n vertices equals $n^{n-2}/n!$ [12, page 209].

Proposition 10. *Let $p > 1$ and $q \geq 0$ be fixed integers. Let $Y \subseteq V^{p,q}_{biconn}$ be a non-empty set. For all $\mu \geq 0$,*

$$\sum_{\mathscr{C}\in\mathscr{E}(V_Y^{\mu})} \frac{1}{|aut\,\mathscr{C}|} = \frac{((p-1)\mu+1)^{\mu-2}}{\mu!} \left(p \sum_{\mathscr{A}\in\mathscr{E}(Y)} \frac{1}{|aut\,\mathscr{A}|} \right)^{\mu}. \tag{43}$$

Proof. We first recall the following well-known identities derived from Abel's binomial theorem [14]:

$$\sum_{i=1}^{n-1} \binom{n}{i} i^{i-1} (n-i)^{n-i-1} = 2(n-1)n^{n-2}, \tag{44}$$

$$\sum_{i=0}^{n} \binom{n}{i} (x+i)^{i-1} (y+(n-i))^{n-i-1} = \left(\frac{1}{x}+\frac{1}{y}\right)(x+y+n)^{n-1}, \tag{45}$$

where x and y are non-zero numbers. A proof may be found in [15] for instance. From (45) follows that for all $p > 1$ and non-zero numbers $x_1, \ldots, x_p$, the identity

$$\sum_{i_1+\cdots+i_p=n} \binom{n}{i_1,\ldots,i_p} (x_1+i_1)^{i_1-1} \cdots (x_p+i_p)^{i_p-1} = \frac{x_1+\cdots+x_p}{x_1\cdots x_p}(x_1+\cdots+x_p+n)^{n-1} \tag{46}$$

holds. The proof proceeds by induction. For $p = 2$ the identity specializes to (45). We assume the identity (46) to hold for $p-1$. Then,

$$\begin{aligned}
&\sum_{i_1+\cdots+i_p=n} \binom{n}{i_1,\ldots,i_p} (x_1+i_1)^{i_1-1} \cdots (x_p+i_p)^{i_p-1} \\
&= \sum_{i_1=0}^{n} \sum_{i_2+\cdots+i_p=n-i_1} \binom{n}{i_1}\binom{n-i_1}{i_2,\ldots,i_p} \\
&\quad \times (x_1+i_1)^{i_1-1} \cdots (x_p+i_p)^{i_p-1} \\
&= \frac{x_2+\cdots+x_p}{x_2\cdots x_p} \\
&\quad \times \sum_{i_1=0}^{n} \binom{n}{i_1} (x_1+i_1)^{i_1-1} (x_2+\cdots+x_p+n-i_1)^{n-i_1-1} \\
&= \frac{x_2+\cdots+x_p}{x_2\cdots x_p} \\
&\quad \times \left(\frac{1}{x_1} + \frac{1}{x_2+\cdots+x_p}\right)(x_1+\cdots+x_p+n)^{n-1} \\
&= \frac{x_1+\cdots+x_p}{x_1\cdots x_p}(x_1+\cdots+x_p+n)^{n-1}.
\end{aligned} \tag{47}$$

We turn to the proof of formula (43). Let $I(\mu) := \sum_{\mathscr{C}\in\mathscr{C}(V_Y^\mu)}(1/|\mathrm{aut}\,\mathscr{C}|)$. Formula (40) induces the following recurrence for $I(\mu)$:

$$\begin{aligned}
I(0) &= 1, \\
I(\mu) &= \frac{1}{\mu}\sum_{\mathscr{A}\in\mathscr{E}(Y)} \frac{1}{|\mathrm{aut}\,\mathscr{A}|} \\
&\quad \times \sum_{i_1+\cdots+i_p=\mu-1} ((p-1)i_1+1) \\
&\quad \cdots ((p-1)i_p+1) I(i_1)\cdots I(i_p).
\end{aligned} \tag{48}$$

We proceed to prove by induction that $I(\mu) = (((p-1)\mu+1)^{\mu-2}/\mu!)(p\sum_{\mathscr{A}\in\mathscr{E}(Y)}(1/|\mathrm{aut}\,\mathscr{A}|))^\mu$. The result holds for $\mu = 0, 1$:

$$\begin{aligned}
I(0) &= 1, \\
I(1) &= \sum_{\mathscr{A}\in\mathscr{E}(Y)} \frac{1}{|\mathrm{aut}\,\mathscr{A}|}.
\end{aligned} \tag{49}$$

We assume the result to hold for all $0 \le i \le \mu - 1$. Then,

$$\begin{aligned}
I(\mu) &= \frac{1}{\mu}\sum_{i_1+\cdots+i_p=\mu-1} ((p-1)i_1+1) \\
&\quad \cdots ((p-1)i_p+1) I(i_1)\cdots I(i_p) \sum_{\mathscr{A}\in\mathscr{E}(Y)} \frac{1}{|\mathrm{aut}\,\mathscr{A}|}
\end{aligned} \tag{50}$$

$$\begin{aligned}
&= \frac{p^{\mu-1}}{\mu}\sum_{i_1+\cdots+i_p=\mu-1} \frac{1}{i_1!\cdots i_p!}((p-1)i_1+1)^{i_1-1} \\
&\quad \cdots ((p-1)i_p+1)^{i_p-1}\left(\sum_{\mathscr{A}\in\mathscr{E}(Y)} \frac{1}{|\mathrm{aut}\,\mathscr{A}|}\right)^\mu
\end{aligned} \tag{51}$$

$$\begin{aligned}
&= \frac{p^{\mu-1}}{\mu!}(p-1)^{\mu-1-p} \\
&\quad \times \sum_{i_1+\cdots+i_p=\mu-1} \binom{\mu-1}{i_1,\ldots,i_p}\left(i_1+\frac{1}{p-1}\right)^{i_1-1} \\
&\quad \cdots \left(i_p+\frac{1}{p-1}\right)^{i_p-1}\left(\sum_{\mathscr{A}\in\mathscr{E}(Y)} \frac{1}{|\mathrm{aut}\,\mathscr{A}|}\right)^\mu
\end{aligned} \tag{52}$$

$$\begin{aligned}
&= \frac{1}{\mu!}(p-1)^{\mu-1-p}(p-1)^{p-1} \\
&\quad \times \left(\mu-1+\frac{p}{p-1}\right)^{\mu-2}\left(p\sum_{\mathscr{A}\in\mathscr{E}(Y)} \frac{1}{|\mathrm{aut}\,\mathscr{A}|}\right)^\mu
\end{aligned} \tag{53}$$

$$= \frac{1}{\mu!}((p-1)\mu+1)^{\mu-2}\left(p\sum_{\mathscr{A}\in\mathscr{E}(Y)} \frac{1}{|\mathrm{aut}\,\mathscr{A}|}\right)^\mu, \tag{54}$$

where we used formula (46) in going from (52) to (53). This completes the proof of Proposition 10. □

The corollary is now established.

Corollary 11. *Let $p > 1$ and $q \ge 0$ be fixed integers. Let $Y \subseteq V_{biconn}^{p,q}$ be a non-empty set. For all $\mu \ge 0$, $\{G \in V_Y^\mu \mid V(G) = \{1,\ldots,(p-1)\mu+1\}$ is a set of cardinality*

$$\frac{((p-1)\mu+1)!\,((p-1)\mu+1)^{\mu-2}}{\mu!}\left(p\sum_{\mathscr{A}\in\mathscr{E}(Y)} \frac{1}{|aut\,\mathscr{A}|}\right)^\mu. \tag{55}$$

Proof. The result is a straightforward application of Lagrange's theorem to the symmetric group on the set $\{1,\ldots,(p-1)\mu+1\}$ and each of its subgroups $\mathrm{aut}\,\mathscr{C}$, for all $\mathscr{C}\in\mathscr{E}(V_Y^\mu)$. □

For $p = 2$ and $q = 0$, formula (55) specializes to Cayley's formula [11]:

$$(\mu+1)^{\mu-1} = n^{n-2}, \tag{56}$$

where $n = \mu + 1$ is the number of vertices of all the trees with μ biconnected components. In this particular case, the recurrence given by formula (48) can be easily transformed into a recurrence on the number of vertices n:

$$J(1) = 1,$$
$$J(n) = \frac{1}{2(n-1)} \sum_{i=1}^{n-1} i(n-i) J(i) J(n-i), \tag{57}$$

which by (44) yields $J(n) = n^{n-2}/n!$. Now, for $T(n) = n! J(n)$, we obtain Dziobek's recurrence for Cayley's formula [10, 24]:

$$T(1) = 1,$$
$$T(n) = \frac{1}{2(n-1)} \sum_{i=1}^{n-1} \binom{n}{i} i(n-i) T(i) T(n-i). \tag{58}$$

Furthermore, for graphs whose biconnected components are all complete graphs on p vertices, formula (55) yields

$$\frac{((p-1)\mu+1)!((p-1)\mu+1)^{\mu-2}}{(p-1)!^{\mu}\mu!} \tag{59}$$

in agreement with Husimi's result for this particular case [25] (see also [26]). Also, when the biconnected components are all cycles of length p, we recover a particular case of Leroux's result [27]:

$$\frac{((p-1)\mu+1)!((p-1)\mu+1)^{\mu-2}}{2^{\mu}\mu!}. \tag{60}$$

Acknowledgment

The research was supported through the fellowship SFRH/BPD/48223/2008 provided by the Portuguese Foundation for Science and Technology (FCT).

References

[1] R. C. Read, "A survey of graph generation techniques," in *Combinatorial Mathematics 8*, vol. 884 of *Lecture Notes in Math.*, pp. 77–89, Springer, Berlin, Germany, 1981.

[2] F. Harary and E. M. Palmer, *Graphical Enumeration*, Academic Press, New York, NY, USA, 1973.

[3] F. Bergeron, G. Labelle, and P. Leroux, *Combinatorial Species and Tree-Like Structures*, vol. 67, Cambridge University Press, Cambridge, UK, 1998.

[4] C. Itzykson and J. B. Zuber, *Quantum Field Theory*, McGraw-Hill, New York, NY, USA, 1980.

[5] C. Jordan, "Sur les assemblages de lignes," *Journal für die Reine und Angewandte Mathematik*, vol. 70, pp. 185–190, 1869.

[6] R. C. Read, "Every one a winner or how to avoid isomorphism search when cataloguing combinatorial configurations. Algorithmic aspects of combinatorics (Conference in Vancouver Island, BC, Canada, 1976)," *Annals of Discrete Mathematics*, vol. 2, pp. 107–120, 1978.

[7] Â. Mestre and R. Oeckl, "Combinatorics of n-point functions via Hopf algebra in quantum field theory," *Journal of Mathematical Physics*, vol. 47, no. 5, p. 052301, 16, 2006.

[8] Â. Mestre and R. Oeckl, "Generating loop graphs via Hopf algebra in quantum field theory," *Journal of Mathematical Physics*, vol. 47, no. 12, Article ID 122302, p. 14, 2006.

[9] Â. Mestre, "Combinatorics of 1-particle irreducible n-point functions via coalgebra in quantum field theory," *Journal of Mathematical Physics*, vol. 51, no. 8, Article ID 082302, 2010.

[10] O. Dziobek, "Eine formel der substitutionstheorie," *Sitzungsberichte der Berliner Mathematischen Gesellschaft*, vol. 17, pp. 64–67, 1947.

[11] A. Cayley, "A theorem on trees," *Quarterly Journal of Pure and Applied Mathematics*, vol. 23, pp. 376–378, 1889.

[12] G. Pólya, "Kombinatorische Anzahlbestimmungen für Gruppen, Graphen und chemische Verbindungen," *Acta Mathematica*, vol. 68, pp. 145–254, 1937.

[13] R. Tarjan, "Depth-first search and linear graph algorithms," *SIAM Journal on Computing*, vol. 1, no. 2, pp. 146–160, 1972.

[14] N. Abel, "Beweis eines Ausdruckes, von welchem die Binomial-Formel ein einzelner Fall ist," *Journal für die Reine und Angewandte Mathematik*, vol. 1, pp. 159–160, 1826.

[15] L. Székely, "Abel's binomial theorem," http://www.math.sc.edu/~szekely/abel.pdf.

[16] R. Diestel, *Graph Theory*, vol. 173, Springer, Berlin, Germany, 5rd edition, 2005.

[17] M. J. Atallah and S. Fox, *Algorithms and Theory of Computation Handbook*, CRC Press, Boca Raton, Fla, USA, 1998.

[18] G. Kirchhoff, "Über die Auflösung der Gleichungen, auf welche man bei der Untersuchung der linearen Vertheilung galvanischer Ströme geführt wird," *Annual Review of Physical Chemistry*, vol. 72, pp. 497–508, 1847.

[19] Â. Mestre, "Generating connected and 2-edge connected graphs," *Journal of Graph Algorithms and Applications*, vol. 13, no. 2, pp. 251–281, 2009.

[20] J. Cuntz, R. Meyer, and J. M. Rosenberg, *Topological and Bivariant K-Theory*, vol. 36, Birkhäuser, Basel, Switzerland, 2007.

[21] C. Kassel, *Quantum Groups*, vol. 155, Springer, New York, NY, USA, 1995.

[22] J.-L. Loday, *Cyclic Homology*, vol. 301, Springer, Berlin, Germany, 1992.

[23] M. Livernet, "A rigidity theorem for pre-Lie algebras," *Journal of Pure and Applied Algebra*, vol. 207, no. 1, pp. 1–18, 2006.

[24] J. W. Moon, "Various proofs of Cayley's formula for counting trees," in *A Seminar on Graph Theory*, pp. 70–78, Holt, Rinehart & Winston, New York, NY, USA, 1967.

[25] K. Husimi, "Note on Mayers' theory of cluster integrals," *The Journal of Chemical Physics*, vol. 18, pp. 682–684, 1950.

[26] J. Mayer, "Equilibrium statistical mechanics," in *The International Encyclopedia of Physical Chemistry and Chemical Physics*, Pergamon Press, Oxford, UK, 1968.

[27] P. Leroux, "Enumerative problems inspired by Mayer's theory of cluster integrals," *Electronic Journal of Combinatorics*, vol. 11, no. 1, research paper 32, p. 28, 2004.

Initial Ideals of Tangent Cones to the Richardson Varieties in the Orthogonal Grassmannian

Shyamashree Upadhyay

School of Mathematics, Tata Institute of Fundamental Research, Homi Bhabha Road, Colaba, Mumbai 400005, India

Correspondence should be addressed to Shyamashree Upadhyay; shyamashree.upadhyay@gmail.com

Academic Editor: Nantel Bergeron

A Richardson variety X_α^γ in the Orthogonal Grassmannian is defined to be the intersection of a Schubert variety X^γ in the Orthogonal Grassmannian and an opposite Schubert variety X_α therein. We give an explicit description of the initial ideal (with respect to certain conveniently chosen term order) for the ideal of the tangent cone at any T-fixed point of X_α^γ, thus generalizing a result of Raghavan and Upadhyay (2009). Our proof is based on a generalization of the Robinson-Schensted-Knuth (RSK) correspondence, which we call the Orthogonal-bounded-RSK (OBRSK).

1. Introduction

The Orthogonal Grassmannian is defined in Section 2. A Richardson variety X_α^γ in the Orthogonal Grassmannian (The Richardson varieties in the ordinary Grassmannian are also studied by Stanley in [1], where these varieties are called *skew Schubert varieties*. Discussion of these varieties in the ordinary Grassmannian also appears in [2].) is defined to be the intersection of a Schubert variety X^γ in the Orthogonal Grassmannian with an opposite Schubert variety X_α therein. In particular, the Schubert and opposite Schubert varieties are special cases of the Richardson varieties. In this paper, we provide an explicit description of the initial ideal (with respect to certain conveniently chosen term order) for the ideal of the tangent cone at any T-fixed point e_β of X_α^γ. It should be noted that the local properties of the Schubert varieties at T-fixed points determine their local properties at all other points, because of the B action; but this does not extend to the Richardson varieties, since Richardson varieties only have a T-action.

In Raghavan and Upadhyay [3], an explicit description of the initial ideal (with respect to certain conveniently chosen term orders) for the ideal of the tangent cone at any T-fixed point of a Schubert variety in the Orthogonal Grassmannian has been obtained. In this paper, we generalize the result of [3] to the case of the Richardson varieties in the Orthogonal Grassmannian.

Sturmfels [4] and Herzog and Trung [5] proved results on a class of determinantal varieties which are equivalent to the results of [6–8] for the case of the Schubert varieties (in the ordinary Grassmannian) at the T-fixed point e_{id}. The key to their proofs was to use a version of the Robinson-Schensted-Knuth correspondence (which we will call the "ordinary" RSK) in order to establish a degree-preserving bijection between a set of monomials defined by an initial ideal and a "standard monomial basis." The difficulty in extending this method of proof to the case of the Schubert varieties (in the ordinary Grassmannian) at an arbitrary T-fixed point e_β lies in generalizing this bijection, which is done in the three papers [6–8]; the work done in [8] is slightly more general, since it applies to the Richardson varieties, and not just to the Schubert varieties. These three bijections, when restricted to the Schubert varieties in the ordinary Grassmannian, are in fact the same bijection (This supports the conviction of the authors in [6] that this bijection is natural and that it is in some sense the only natural bijection satisfying the required geometric conditions.), although this is not immediately apparent. In the work of Kreiman in [8], this "generalized bijection" has been viewed for the first time as an extended version of the "ordinary" RSK correspondence, which he calls the Bounded-RSK correspondence; this viewpoint was not present in [6, 7]. Although the formulations of the bijections in [6, 7] are similar to each other, the formulation of the bijection in [8] is in terms of different combinatorial indexing sets.

The work done in [3] by Raghavan and Upadhyay does not involve any version of the RSK correspondence, unlike the work done by Herzog and Trung in [5]. The work done in [3] relies on a degree-preserving bijection between a set of monomials defined by an initial ideal and a "standard monomial basis," and this bijection is proved by Raghavan and Upadhyay in [9]. It is mentioned in [9] that it will be nice if the bijection proved therein can be viewed as a kind of "Bounded-RSK" correspondence, as done by Kreiman in [8] for the case of the Richardson varieties in the ordinary Grassmannian. This paper fulfills the expectation of [9] that one should be able to view the bijection there as a generalized-bounded-RSK correspondence, which we call here the Orthogonal-bounded-RSK correspondence (OBRSK). Kreiman has mentioned in his paper [8] that he believes in the possibility of adapting the methods of his paper ([8]) to the Richardson varieties in the Symplectic and the Orthogonal Grassmannian as well; this paper supports Kreiman's belief for the Orthogonal Grassmannian case.

2. The Orthogonal Grassmannian and Richardson Varieties in It

Fix an algebraically closed field $\mathfrak{k}$ of characteristic not equal to 2. Fix a natural number d, a vector space V of dimension $2d$ over $\mathfrak{k}$, and a nondegenerate symmetric bilinear form $\langle\, , \rangle$ on V. For k an integer such that $1 \le k \le 2d$, set $k^* := 2d + 1 - k$. Fix a basis $e_1, \dots, e_{2d}$ of V such that $\langle e_i, e_k\rangle$ equals 1 if $i = k^*$ and is 0 otherwise. Denote by $SO(V)$ the group of linear automorphisms of V that preserve the bilinear form $\langle\, , \rangle$ and also the volume form. A linear subspace of V is said to be *isotropic*, if the bilinear form $\langle\, , \rangle$ vanishes identically on it. Denote by $\mathfrak{M}_d(V)'$ the closed subvariety of the Grassmannian of d-dimensional subspaces consisting of the points corresponding to maximal isotropic subspaces. The action of $SO(V)$ on V induces an action on $\mathfrak{M}_d(V)'$. There are two orbits for this action. These orbits are isomorphic: acting by a linear automorphism that preserves the form but not the volume form gives an isomorphism. We denote by $\mathfrak{M}_d(V)$ the orbit of the span of $e_1, \dots, e_d$ and call it the *(even) Orthogonal Grassmannian*. One can define the Orthogonal Grassmannian in the case when the dimension of V is not necessarily even. But it is enough to consider the case when the dimension of V is even, the reason being the following (see also Section 2.1 of [9] for the same): suppose that the dimension of V is odd; say dimension of $V = 2d + 1$. Let $\widetilde{n} := 2d + 2$ and $\widetilde{V}$ be a vector space of dimension $\widetilde{n}$ with a nondegenerate symmetric form. Let $\widetilde{e}_1, \dots, \widetilde{e}_{\widetilde{n}}$ be a basis of $\widetilde{V}$ such that $\langle \widetilde{e}_i, \widetilde{e}_k\rangle$ equals 1 if $i = k^*$ and is 0 otherwise. Put $e := \widetilde{e}_{d+1}$ and $f := \widetilde{e}_{d+2}$. Take λ to be an element of the field such that $\lambda^2 = 1/2$. We can take V to be the subspace of $\widetilde{V}$ spanned by the vectors $\widetilde{e}_1, \dots, \widetilde{e}_d, \lambda e + \lambda f, \widetilde{e}_{d+3}, \dots$, and $\widetilde{e}_{\widetilde{n}}$, and a basis of V to be these vectors in that order.

There is a natural map from $\mathfrak{M}_{d+1}(\widetilde{V})'$ to $\mathfrak{M}_d(V)$: intersecting with V an isotropic subspace of $\widetilde{V}$ of dimension $d + 1$ gives an isotropic subspace of V of dimension d, and we denote this map by $\cap$. The map $\cap$ is onto, for every isotropic subspace of $\widetilde{V}$ (and hence of V) is contained in an isotropic subspace of $\widetilde{V}$ of dimension $d + 1$. In fact, more is true: the map $\cap$ is two-to-one. The map $\cap$ being two-to-one, it is also elementary to see that the two points in any fiber lie one in each component of $\mathfrak{M}_{d+1}(\widetilde{V})'$. We therefore get a natural isomorphism between $\mathfrak{M}_{d+1}(\widetilde{V})$ and $\mathfrak{M}_d(V)$. Therefore, now onwards we call the (even) Orthogonal Grassmannian $\mathfrak{M}_d(V)$ (as defined above for a $2d$-dimensional vector space V) the *Orthogonal Grassmannian*.

Let $\mathfrak{M}_d(V) \subseteq G_d(V) \hookrightarrow \mathbb{P}(\wedge^d V)$ be the Plücker embedding (where $G_d(V)$ denotes the Grassmannian of all d-dimensional subspaces of V). Thus $\mathfrak{M}_d(V)$ is a closed subvariety of the projective variety $G_d(V)$, and hence $\mathfrak{M}_d(V)$ inherits the structure of a projective variety.

We take B (resp., B^-) to be the subgroup of $SO(V)$ consisting of those elements that are upper triangular (resp., lower triangular) with respect to the basis $e_1, \dots, e_{2d}$ and the subgroup T of $SO(V)$ consisting of those elements that are diagonal with respect to $e_1, \dots, e_{2d}$. It can be easily checked that T is a maximal torus of $SO(V)$; B and B^- are the Borel subgroups of $SO(V)$ which contain T. The group $SO(V)$ acts transitively on $\mathfrak{M}_d(V)$, and the T-fixed points of $\mathfrak{M}_d(V)$ under this action are easily seen to be of the form $\langle e_{i_1}, \dots, e_{i_d}\rangle$ for $\{i_1, \dots, i_d\}$ in $I(d)$, where $I(d)$ is the set of subsets of $\{1, \dots, 2d\}$ of cardinality d satisfying the following two conditions:

(i) for each k, $1 \le k \le 2d$, the subset contains exactly one of k, k^*, and

(ii) the number of elements in the subset that exceed d is even.

We write $I(d, 2d)$ for the set of all d-element subsets of $\{1, \dots, 2d\}$. There is a natural partial order on $I(d, 2d)$ and so also on $I(d)$: $v = (v_1 < \cdots < v_d) \le w = (w_1 < \cdots < w_d)$ if and only if $v_1 \le w_1, \dots, v_d \le w_d$. For $\mu = \{\mu_1, \dots, \mu_d\} \in I(d, 2d)$, $\mu_1 < \cdots < \mu_d$, define the *complement* of μ as $\{1, \dots, 2d\} \setminus \mu$, and denote it by $\overline{\mu}$.

The B-orbits (as well as B^--orbits) of $\mathfrak{M}_d(V)$ are naturally indexed by its T-fixed points: each B-orbit (as well as B^--orbit) contains one and only one such point. Let $\alpha \in I(d)$ be arbitrary, and let e_α denote the corresponding T-fixed point of $\mathfrak{M}_d(V)$. The Zariski closure of the B- (resp., B^--) orbit through e_α, with canonical reduced scheme structure, is called a *Schubert variety* (resp., *opposite Schubert variety*) and denoted by X^α (resp., X_α). For $\alpha, \gamma \in I(d)$, the scheme-theoretic intersection $X_\alpha^\gamma = X_\alpha \cap X^\gamma$ is called a *Richardson variety*. Each B-orbit (as well as B^--orbit) being irreducible and open in its closure, it follows that B-orbit closures (resp., B^--orbit closures) are indexed by the B-orbits (resp., B^--orbits). Thus the set $I(d)$ becomes an indexing set for the Schubert varieties in $\mathfrak{M}_d(V)$, and the set consisting of all pairs of elements of $I(d)$ becomes an indexing set for the

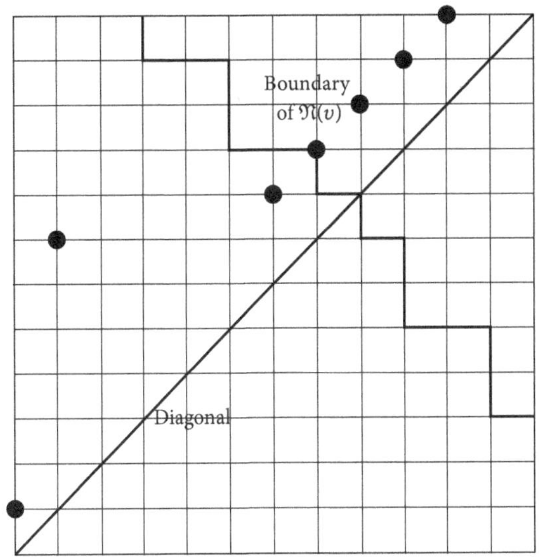

Figure 1

Richardson varieties in $\mathfrak{M}_d(V)$. It can be shown that X_α^γ is nonempty if and only if $\alpha \le \gamma$; that for $\beta \in I(d)$, $e_\beta \in X_\alpha^\gamma$ if and only if $\alpha \le \beta \le \gamma$.

3. Statement of the Problem and the Strategy of the Proof

3.1. Initial Statement of the Problem. The problem that is tackled in this paper is this: given a T-fixed point on a Richardson variety in $\mathfrak{M}_d(V)$, compute the initial ideal, with respect to some convenient term order, of the ideal of functions vanishing on the tangent cone to the Richardson variety at the given T-fixed point. The term order is specified in Section 3.4, and the answer is given in Theorem 7.

For the rest of this paper, α, β, and γ are arbitrarily fixed elements of $I(d)$ such that $\alpha \le \beta \le \gamma$. So, the problem tackled in this paper can be restated as follows. Given the Richardson variety X_α^γ in $\mathfrak{M}_d(V)$ and the T-fixed point e_β in it, find the initial ideal of the ideal of functions vanishing on the tangent cone at e_β to X_α^γ, with respect to some conveniently chosen term order. The tangent cone being a subvariety of the tangent space at e_β to $\mathfrak{M}_d(V)$, we first choose a convenient set of coordinates for the tangent space. But for that we need to fix some notation.

3.2. Basic Notation. For this subsection, let us fix an arbitrary element v of $I(d, 2d)$. We will be dealing extensively with ordered pairs (r, c), $1 \le r, c \le 2d$, such that r is not and c is an entry of v. Let $\mathfrak{R}(v)$ denote the set of all such ordered pairs, and set $\mathfrak{N}(v) := \{(r,c) \in \mathfrak{R}(v) \mid r > c\}$, $\mathfrak{O}\mathfrak{R}(v) := \{(r,c) \in \mathfrak{R}(v) \mid r < c^*\}$, $\mathfrak{O}\mathfrak{N}(v) := \{(r,c) \in \mathfrak{R}(v) \mid r > c, r < c^*\} = \mathfrak{O}\mathfrak{R}(v) \cap \mathfrak{N}(v)$, $\mathfrak{d}^v := \{(r,c) \in \mathfrak{R}(v) \mid r = c^*\}$, $\mathfrak{A}\mathfrak{R}(v) := \{(r,c) \in \mathfrak{R}(v) \mid r > c^*\}$, and $\mathfrak{A}\mathfrak{N}(v) := \{(r,c) \in \mathfrak{R}(v) \mid r > c, r > c^*\}$. We will refer to $\mathfrak{d}^v$ as the *diagonal*.

Figure 1 shows a drawing of $\mathfrak{R}(v)$. We think of r and c in (r, c) as row index and column index, respectively. The columns are indexed from left to right by the entries of v in ascending order and the rows from top to bottom by the entries of $\{1, \ldots, 2d\} \setminus v$ in ascending order. The points of $\mathfrak{d}^v$ are those on the diagonal, the points of $\mathfrak{O}\mathfrak{R}(v)$ are those that are (strictly) above the diagonal, and the points of $\mathfrak{N}(v)$ are those that are to the South-West of the polyline captioned "boundary of $\mathfrak{N}(v)$"—we draw the boundary so that points on the boundary belong to $\mathfrak{N}(v)$. The reader can readily verify that $d = 13$ and $v = (1, 2, 3, 4, 6, 7, 10, 11, 13, 15, 18, 19, \text{and } 22)$ for the particular picture drawn. The points of $\mathfrak{O}\mathfrak{R}(v)$ indicated by solid circles form an *extended* v-chain (see Figure 1); the definition of an extended v-chain is given later in Section 3.5.

We will be considering *monomials* in some of these sets. A *monomial*, as usual, is a subset with each member being allowed a multiplicity (taking values in the nonnegative integers). The *degree* of a monomial has also the usual sense: consider the underlying set of the monomial and look at the multiplicity with which each element of this underlying set appears in the monomial; the degree of the monomial is the sum of these multiplicities. The *intersection* of any two monomials in a set also has the natural meaning; it is the monomial whose underlying set is the intersection of the underlying sets of the given two monomials and the multiplicity of each element being the minimum of the multiplicities with which the element occurs in the two given monomials.

Given any two monomials A and B consisting of elements of $\mathfrak{R}(v)$, let set (A) and set (B) denote the underlying sets of A and B, respectively. We say that $B \subseteq A$ (as monomials) if set $(B) \subseteq \text{set}(A)$, and the multiplicity with which every element occurs in the monomial B is less than or equal to the multiplicity with which the same element occurs in the monomial A. Given two monomials A and B consisting of elements of $\mathfrak{R}(v)$ such that $B \subseteq A$ (as monomials), we can define a monomial called the *"monomial minus"* of B from A (denoted by $A\setminus_m B$) as follows. Take any element x of set(B). If the multiplicity with which x occurs in A is $m_x(A)$ and the multiplicity with which x occurs in B is $m_x(B)$, then the multiplicity with which x occurs in the monomial $A\setminus_m B$ is $m_x(A) - m_x(B)$. And any element in set$(A) \setminus$ set(B) occurs in the monomial $A\setminus_m B$ with the same multiplicity with which it occurs in A. This finishes the description of $A\setminus_m B$.

Remark 1. Note that in this subsection, and v was any element of $I(d, 2d)$, v was not necessarily in $I(d)$. In particular, all the above basic notations will hold true, if we take $v \in I(d)$ as well.

3.3. The Ideal of the Tangent Cone to X_α^γ at e_β. Let β be the element of $I(d)$ which was fixed at the beginning of this section. Consider the matrix of size $2d \times d$ whose columns are numbered by the entries of β and the rows by $\{1, \ldots, 2d\}$; the rows corresponding to the entries of β form the $d \times d$ identity matrix, and the remaining d rows form a skew symmetric matrix whose upper half entries are variables of the form $X_{(r,c)}$, where $(r,c) \in \mathfrak{O}\mathfrak{R}(\beta)$.

Let $\mathfrak{M}_d(V) \subseteq G_d(V) \hookrightarrow \mathbb{P}(\wedge^d V)$ be the Plücker embedding (where $G_d(V)$ denotes the Grassmannian of all d-dimensional subspaces of V). For θ in $I(d,2d)$, let p_θ denote the corresponding Plücker coordinate. Consider the affine patch $\mathbb{A}$ of $\mathbb{P}(\wedge^d V)$ given by $p_\beta \neq 0$. The affine patch $\mathbb{A}^\beta := \mathfrak{M}_d(V) \cap \mathbb{A}$ of the Orthogonal Grassmannian $\mathfrak{M}_d(V)$ is an affine space whose coordinate ring can be taken to be the polynomial ring in variables of the form $X_{(r,c)}$ with $(r,c) \in \mathfrak{O}\mathfrak{R}(\beta)$.

For $\theta \in I(d)$, consider the submatrix of the above mentioned matrix given by the rows numbered $\theta \setminus \beta$ and columns numbered $\beta \setminus \theta$. Such a submatrix is of even size and is skew-symmetric about antidiagonal, therefore its determinant is a square. The square root, which is determined up to sign, is called the *Pfaffian*. Let $f_{\theta,\beta}$ denote this Pfaffian. Set $Y_\alpha^\gamma(\beta) := X_\alpha^\gamma \cap \mathbb{A}^\beta$. From [10] we can deduce a set of generators for the ideal I of functions on $\mathbb{A}^\beta$ vanishing on $Y_\alpha^\gamma(\beta)$ (see Section 4.2 of [9] for the special case of the Schubert varieties). The following equation gives the generators:

$$I = \left(f_{\tau,\beta} \mid \tau \in I(d), \alpha \nleq \tau \text{ or } \tau \nleq \gamma \right). \quad (1)$$

We are interested in the tangent cone to X_α^γ at e_β or, what is the same, the tangent cone to $Y_\alpha^\gamma(\beta)$ at the origin. Observe that $f_{\tau,\beta}$ is a homogeneous polynomial. Because of this, $Y_\alpha^\gamma(\beta)$ itself is a cone and so is equal to its tangent cone at the origin. The ideal of the tangent cone to X_α^γ at e_β is therefore the ideal I in (1).

3.4. The Term Order. We now specify the term order $\rhd$ on monomials in the coordinate functions (of the tangent space to $\mathfrak{M}_d(V)$ at the torus fixed point e_β; it is easy to see that coordinate ring of the tangent space to $\mathfrak{M}_d(V)$ at e_β is the polynomial ring in variables of the form $X_{(r,c)}$, where $(r,c) \in \mathfrak{O}\mathfrak{R}(\beta)$) with respect to which the initial ideal of the ideal I of the tangent cone is to be taken.

Let $>$ be a total order on $\mathfrak{O}\mathfrak{R}(\beta)$ satisfying all of the following 6 conditions.

(i) $\mu > \nu$, if $\mu \in \mathfrak{O}\mathfrak{N}(\beta)$, $\nu \in \mathfrak{O}\mathfrak{R}(\beta) \setminus \mathfrak{O}\mathfrak{N}(\beta)$, and the row indices of μ and ν are equal.

(ii) $\mu > \nu$, if $\mu \in \mathfrak{O}\mathfrak{N}(\beta)$, $\nu \in \mathfrak{O}\mathfrak{N}(\beta)$, the row indices of μ and ν are equal, and the column index of μ exceeds that of ν.

(iii) $\mu > \nu$ if $\mu \in \mathfrak{O}\mathfrak{N}(\beta)$, $\nu \in \mathfrak{O}\mathfrak{R}(\beta)$, and the row index of μ is less than that of ν.

(iv) $\mu > \nu$, if $\mu \in \mathfrak{O}\mathfrak{R}(\beta) \setminus \mathfrak{O}\mathfrak{N}(\beta)$, $\nu \in \mathfrak{O}\mathfrak{N}(\beta)$, and the column indices of μ and ν are equal.

(v) $\mu > \nu$, if $\mu \in \mathfrak{O}\mathfrak{R}(\beta) \setminus \mathfrak{O}\mathfrak{N}(\beta)$, $\nu \in \mathfrak{O}\mathfrak{R}(\beta) \setminus \mathfrak{O}\mathfrak{N}(\beta)$, the column indices of μ and ν are equal, and the row index of μ exceeds that of ν.

(vi) $\mu > \nu$, if $\mu \in \mathfrak{O}\mathfrak{R}(\beta) \setminus \mathfrak{O}\mathfrak{N}(\beta)$, $\nu \in \mathfrak{O}\mathfrak{R}(\beta)$, and the column index of μ is less than that of ν.

Note here that the first 3 conditions above are the same as the conditions put on the total order $>_1$ as mentioned in Section 1.6 of [3]. Recall that in the paper [3], initial ideals of ideals of tangent cones at torus fixed points to the Schubert varieties in Orthogonal Grassmannians were computed, and the paper [3] did not deal with the Richardson varieties. The last 3 conditions above arise in this paper as an addition to the 3 conditions put on the total order $>_1$ (as mentioned in Section 1.6 of [3]), because here we are dealing with the Richardson varieties in $\mathfrak{M}_d(V)$.

Let $\rhd$ be the term order on monomials in $\mathfrak{O}\mathfrak{R}(\beta)$ given by the homogeneous lexicographic order with respect to $>$.

Remark 2. The total order $>$ on $\mathfrak{O}\mathfrak{R}(\beta)$ satisfying the 6 properties mentioned above can be realized as a concrete total order on $\mathfrak{O}\mathfrak{R}(\beta)$ if we put the following extra condition on it.

Let $r(\mu), r(\nu), c(\mu)$, and $c(\nu)$ denote the row index of μ, the row index of ν, the column index of μ, and the column index of ν, respectively. If $r(\mu) < r(\nu)$, $\mu \in \mathfrak{O}\mathfrak{R}(\beta) \setminus \mathfrak{O}\mathfrak{N}(\beta)$, $\nu \in \mathfrak{O}\mathfrak{N}(\beta)$, and $c(\nu) < c(\mu)$, then $\nu > \mu$ when $(r(\nu), c(\mu)) \notin \mathfrak{N}(\beta)$and $\mu > \nu$ when $(r(\nu), c(\mu)) \in \mathfrak{N}(\beta)$.

3.5. Extended ν-Chains and Associated Elements of $I(d)$. For this subsection, let ν be an arbitrarily fixed element of $I(d,2d)$ (not necessarily an element of $I(d)$, unless otherwise stated). For elements $\lambda = (R,C), \mu = (r,c)$ of $\mathfrak{R}(\nu)$, we write $\lambda > \mu$ if $R > r$ and $C < c$ (Note that these are strict inequalities.). A sequence $\lambda_1 > \cdots > \lambda_k$ of elements of $\mathfrak{O}\mathfrak{R}(\nu)$ is called an *extended* ν-chain. Note that an extended ν-chain can also be empty. Letting C to be an extended ν-chain, we define $C^+ := C \cap \mathfrak{O}\mathfrak{N}(\nu)$ and $C^- := C \cap (\mathfrak{O}\mathfrak{R}(\nu) \setminus \mathfrak{O}\mathfrak{N}(\nu))$. We call C^+ (resp., C^-) the *positive* (*resp.*, negative) parts of the extended ν-chain C. We call an extended ν-chain C *positive* (*resp.*, negative) if $C = C^+$ (*resp.*, $C = C^-$). The extended ν-chain C is called *nonempty* if it has at least one element in it. Note that if we specialize to the case when $\nu \in I(d)$, then whatever is called a ν-chain in Section 2.4.1 of [9] is a positive extended ν-chain over here. To every extended ν-chain C, we will now associate 2 subsets $\mathfrak{d}_C^\nu(+)$ and $\mathfrak{d}_C^\nu(-)$ of $\mathfrak{d}^\nu$ (each of even cardinality), but for that we first need to fix some notation and recall certain terminology from [3, 9].

Definition 3 (Pr and Pro). Given any subset D of $\mathfrak{O}\mathfrak{N}(\nu)$, let $\text{Pr}(D)$ denote the monomial whose underlying set is given by $\{\mu \mid \mu$ is a vertical or horizontal projection of some element of $D\}$ (both vertical and horizontal, as defined in Section 2.3 of [9]), and the cardinality of any element μ of this underlying set inside the monomial $\text{Pr}(D)$ equals the number of elements in D whose one of the projections (vertical or horizontal) is μ. For $\lambda = (r,c)$ in $\mathfrak{R}(\nu)$, define $\lambda^\# := (c^*, r^*)$. The involution $\lambda \mapsto \lambda^\#$ is just the reflection with respect to the diagonal $\mathfrak{d}^\nu$. For a subset $\mathfrak{E}$ of $\mathfrak{N}(\nu)$, the symbol $\mathfrak{E}^\#$ has the obvious meaning. One calls $\mathfrak{E}$ *symmetric* if $\mathfrak{E} = \mathfrak{E}^\#$. Given any symmetric subset E of $\mathfrak{N}(\nu)$, let one denote by $E(\text{top})$ the set $E \cap \mathfrak{O}\mathfrak{N}(\nu)$, by $E(\text{diag})$ the set $E \cap \mathfrak{d}^\nu$, and by $\text{Pro}(E)$ the monomial formed by taking the union of the subset $E(\text{diag})$ with the monomial $\text{Pr}(E(\text{top}))$. Let one make the definition of

Pro(E) more precise. The multiplicity with which any element occurs in the monomial Pro(E) is equal to the sum of the multiplicities with which the element occurs in the subset E(diag) and the monomial Pr(E(top)). So for any symmetric subset E of $\mathfrak{N}(v)$, Pro(E) is a monomial consisting of elements from the diagonal. Similarly for any subset D of $\mathfrak{ON}(v)$, Pr(D) is also a monomial consisting of elements from the diagonal.

Let $v \in I(d)$. Let $C : \alpha_1 = (r_1, c_1) > \alpha_2 = (r_2, c_2) > \cdots > \alpha_l = (r_l, c_l)$ be a positive extended v-chain in $\mathfrak{ON}(v)$. Two consecutive elements α_j and α_{j+1} of C are said to be *connected* if the following conditions are both satisfied:

(i) their legs are "intertwined", that is, $r_j^* \geq c_{j+1}$,

(ii) the point (r_{j+1}, r_j^*) belongs to $\mathfrak{N}(v)$.

Consider the coarsest equivalence relation on the elements of C generated by the above relation. The equivalence classes of C with respect to this equivalence relation are called the *connected components* of the v-chain C.

For any $\alpha = (r, c)$ in $\mathfrak{ON}(v)$, the elements $p_v(\alpha) := (c^*, c)$ and $p_h(\alpha) := (r, r^*)$ of the diagonal are called, respectively, the *vertical and horizontal projections of* α. Given the positive extended v-chain C as above, let $\mathfrak{S}_C$ be the monomial of $\mathfrak{N}(v)$ associated to C defined as follows.

First suppose that C is a connected v-chain in $\mathfrak{ON}(v)$. Observe that if there is at all an integer j $(1 \leq j \leq l)$ such that the horizontal projection $p_h(\alpha_j)$ does not belong to $\mathfrak{N}(v)$, then $j = l$. Define

$$\mathfrak{S}_C := \begin{cases} \{p_v(\alpha_1), \ldots, p_v(\alpha_l)\} & \text{if } l \text{ is even} \\ \{p_v(\alpha_1), \ldots, p_v(\alpha_l)\} \cup \{p_h(\alpha_l)\} & \\ \quad \text{if } l \text{ is odd, } p_h(\alpha_l) \in \mathfrak{N}(v) & \\ \{p_v(\alpha_1), \ldots, p_v(\alpha_{l-1})\} \cup \{\alpha_l, \alpha_l^\#\} & \\ \quad \text{if } l \text{ is odd, } p_h(\alpha_l) \notin \mathfrak{N}(v). & \end{cases} \tag{2}$$

For a v-chain C that is not necessarily connected, let $C = C_1 \cup C_2 \cup \cdots$ be the partition of C into its connected components, and set

$$\mathfrak{S}_C := \mathfrak{S}_{C_1} \cup \mathfrak{S}_{C_2} \cup \cdots. \tag{3}$$

Note that even if we take v to be in $I(d, 2d)$ (and not merely in $I(d)$) and define $\mathfrak{S}_C$ for any positive extended v-chain C exactly in the same way as we did above, there is no logical inconsistency. Hence we extend the definition of $\mathfrak{S}_C$ to any positive extended v-chain C, where $v \in I(d, 2d)$. Clearly $\mathfrak{S}_C$ is a symmetric subset of $\mathfrak{N}(v)$. Hence the monomial Pro($\mathfrak{S}_C$) is well defined for any positive extended v-chain C, where $v \in I(d, 2d)$.

Definition 4 (the flip map F). For any $v \in I(d, 2d)$ and any element $\lambda = (r, c) \in \mathfrak{R}(v)$, let $F(\lambda)$ be the element of $\mathfrak{R}(v^*)$ given by $F(\lambda) := (c, r)$. So F is an invertible map from $\mathfrak{R}(v)$ to $\mathfrak{R}(v^*)$ (note here that if $v \in I(d)$, then v^* needs not always to belong to $I(d)$); let one denote the inverse of F by F^{-1}. The map F naturally induces an invertible map from the set of all monomials in $\mathfrak{R}(v)$ to the set of all monomials in $\mathfrak{R}(v^*)$. One continues to call the induced map F and its inverse F^{-1}.

Definition 5 (the subsets $\mathfrak{d}_C^v(+)$ and $\mathfrak{d}_C^v(-)$ of $\mathfrak{d}^v$). Given any non-empty extended v-chain C, one will now associate 2 subsets $\mathfrak{d}_C^v(+)$ and $\mathfrak{d}_C^v(-)$ of $\mathfrak{d}^v$ (each of even cardinality) with it. Let

$$\mathfrak{d}_C^v(+) := \begin{cases} F^{-1}\left(\Pr(F(C)) \setminus_m \operatorname{Pro}\left(\mathfrak{S}_{F(C)}\right)\right) & \\ \qquad \text{if } C \text{ is negative} & \\ \operatorname{Pro}\left(\mathfrak{S}_{C^+}\right) \quad \text{if otherwise.} & \end{cases} \tag{4}$$

Similarly, let

$$\mathfrak{d}_C^v(-) := \begin{cases} \Pr(C) \setminus_m \operatorname{Pro}\left(\mathfrak{S}_C\right) & \text{if } C \text{ is positive} \\ F^{-1}\left(\operatorname{Pro}\left(\mathfrak{S}_{F(C^-)}\right)\right) & \text{if otherwise.} \end{cases} \tag{5}$$

It is an easy exercise to check that $\mathfrak{d}_C^v(+)$ and $\mathfrak{d}_C^v(-)$ thus defined are actually subsets (not monomials) of $\mathfrak{d}^v$ and that each of them has even cardinality.

Definition 6 (elements of $I(d)$ *associated with* $\mathfrak{d}_C^v(+)$ *and* $\mathfrak{d}_C^v(-)$). For this definition, we let v to be an arbitrary element in $I(d)$. Note that given any subset S of $\mathfrak{d}^v$ of even cardinality, one can naturally associate an element of $I(d)$ to it by removing those entries from v which appear as column indices in the elements of the set S and then adding to it the row indices of all the elements of S. It is easy to check that the resulting element actually belongs to $I(d)$. One denotes the resulting element by $I(d)(S)$. If S is empty, then $I(d)(S)$ is taken to be v itself.

Let $w_C^+(v) := I(d)(\mathfrak{d}_C^v(+))$ and $w_C^-(v) := I(d)(\mathfrak{d}_C^v(-))$. These are the two elements of $I(d)$ that we can naturally associate with the subsets $\mathfrak{d}_C^v(+)$ and $\mathfrak{d}_C^v(-)$ of $\mathfrak{d}^v$.

3.6. The Main Theorem and a Strategy of the Proof. Recall that the ideal of the tangent cone to X_α^γ at e_β is the ideal I given by (1). Let $\rhd$ be as in Section 3.4. For any element $f \in I$; let $\text{in}_\rhd f$ denote the initial term of f with respect to the term order $\rhd$. We define $\text{in}_\rhd I$ to be the ideal $\langle \text{in}_\rhd f \mid f \in I \rangle$ inside the polynomial ring $P := \mathfrak{k}[X_{(r,c)} \mid (r, c) \in \mathfrak{ON}(\beta)]$. For any monomial U in $\mathfrak{ON}(\beta)$, let us denote by X_U the product of all the elements $X_{(r,c)}$, where (r, c) runs over the elements in U with multiplicities.

Let $\text{Chains}_\alpha^\gamma(\beta)$ denote the set $\{X_C \mid C$ is a nonempty extended β-chain in $\mathfrak{ON}(\beta)$ such that either (i) or (ii) below holds$\}$.

(i) C^- is nonempty, and $\alpha \nleq w_{C^-}^-(\beta)$. (ii) C^+ is nonempty, and $w_{C^+}^+(\beta) \nleq \gamma$.

The main result of this paper is the following.

Theorem 7. $\text{in}_\rhd I = \langle \text{Chains}_\alpha^\gamma(\beta) \rangle$.

Example 8. Let $d = 5$, $\alpha = (1, 2, 3, 6, 7)$, $\beta = (1, 2, 4, 6, 8)$, and $\gamma = (1, 2, 5, 7, 8)$. Clearly then $\alpha \leq \beta \leq \gamma$. It can be verified that the set $\text{Chains}_\alpha^\gamma(\beta)$ consists of elements of the following 3 types.

(i) Any extended β-chain in $\mathfrak{ON}(\beta)$ which contains the element $(3, 6)$.

(ii) Any extended β-chain in $\mathfrak{ON}(\beta)$ which contains the element $(3, 4)$ and a nonempty positive part.

(iii) All positive extended β-chain in $\mathfrak{OR}(\beta)$ *except* the singleton-β-chain $\{(5,4)\}$.

One can therefore determine the corresponding initial ideal $\text{in}_{\triangleright}I$ using Theorem 7.

Remark 9. It follows from the statement of Theorem 7 that the set of all monomials in $\mathfrak{OR}(\beta)$ which contain at least one extended β-chain C such that $X_C \in \text{Chains}_\alpha^\gamma(\beta)$ form a vector space basis of the initial ideal $\text{in}_{\triangleright}I$ over the field $\mathfrak{k}$; see, for example, Example 8. In the special case when the Richardson variety is a Schubert variety, it is easy to see that the previous statement of this remark says exactly what has been said in the main theorem (Theorem 1.8.1) of [3].

Remark 10. Since $\text{in}_{\triangleright}I = \langle \text{Chains}_\alpha^\gamma(\beta) \rangle$, it follows that the dimension as a vector space of the graded piece of $P/\text{in}_{\triangleright}I$ of degree m equals the cardinality of the set of all monomials in $\mathfrak{OR}(\beta)$ of degree m which *do not* contain any extended β-chain C such that $X_C \in \text{Chains}_\alpha^\gamma(\beta)$.

Since P/I and $P/\text{in}_{\triangleright}I$ have the same Hilbert function, the same is true with $P/\text{in}_{\triangleright}I$ (in the previous statement) replaced by P/I. Hence if $R_\alpha^\gamma(\beta)$ denotes the coordinate ring of the tangent cone to X_α^γ at e_β, then the dimension as a vector space of the homogeneous piece of $R_\alpha^\gamma(\beta)$ of degree m *equals* the cardinality of the set of all monomials in $\mathfrak{OR}(\beta)$ of degree m which are such that for every extended β-chain C in the monomial, $\alpha \le w_{C^-}^-(\beta)$ and $w_{C^+}^+(\beta) \le \gamma$.

From Remark 10, the following corollary is immediate (the proof is similar to that of Corollary 2.5.2 of [9])

Corollary 11. *The multiplicity of X_α^γ at e_β equals the number of monomials in $\mathfrak{OR}(\beta)$ of maximal cardinality that are square-free and are such that, given any extended β-chain C in any of these monomials, one has: $\alpha \le w_{C^-}^-(\beta)$ and $w_{C^+}^+(\beta) \le \gamma$.*

We now briefly sketch the proof of Theorem 7 (omitting details, which can be found in Section 8). In order to introduce the main combinatorial objects of interest and outline a strategy of the proof, we will first need to prove that the set $\text{Chains}_\alpha^\gamma(\beta) \subseteq \text{in}_{\triangleright}I$, and this proof will follow from whatever is said in Remark 12.

Remark 12. Let C be a nonempty extended β-chain in $\mathfrak{OR}(\beta)$ such that $X_C \in \text{Chains}_\alpha^\gamma(\beta)$. If C^+ is nonempty and $w_{C^+}^+(\beta) \nleq \gamma$, then it can be proved that $X_{C^+} \in \text{in}_{\triangleright}I$, the proof being exactly the same as that in Section 4 of [3]. Then since $\text{in}_{\triangleright}I$ is an ideal and $X_C = X_{C^-}X_{C^+}$, it follows that $X_C \in \text{in}_{\triangleright}I$.

If C^- is nonempty and $\alpha \nleq w_{C^-}^-(\beta)$, look at $F(C^-)$, where F is the *flip* map as defined in Definition 4 from the set of all monomials in $\mathfrak{R}(\beta)$ to the set of all monomials in $\mathfrak{R}(\beta^*)$. Then $F(C^-)$ is a positive extended β^*-chain in $\mathfrak{OR}(\beta^*)$. We need to prove that $X_C \in \text{in}_{\triangleright}I$, for which it is enough to show that $X_{C^-} \in \text{in}_{\triangleright}I$. To prove that $X_{C^-} \in \text{in}_{\triangleright}I$, we will proceed in a way equivalent to the proof done in Section 4 of [3]. But there is a subtle difference between what is proved in Section 4 of [3] and what we are going to prove here; namely, Whatever was proved in Section 4 of [3] can be rephrased in the language of this paper as "Every positive extended β-chain D satisfying the property that $w_D^+(\beta) \nleq \gamma$ belongs to the initial ideal of the ideal of the tangent cone," but here we are going to prove that "Every negative extended β-chain D satisfying the property that $\alpha \nleq w_D^-(\beta)$ belongs to the initial ideal of the ideal of the tangent cone."

Because of this subtle difference, we need to construct certain gadgets for negative extended β-chains, which will play a role similar to the role of the objects like the *new forms*, Proj and Proj^e corresponding to positive extended β-chains (For positive extended β-chains, such objects are already defined in [3]). This construction is given in the following paragraph.

Consider the positive extended β^*-chain $F(C^-)$. We can construct *new forms*, Proj and Proj^e for $F(C^-)$ in the same way as they were constructed in [3], note here the fact that β^* may or may not belong to $I(d)$ does not really affect the construction of the new forms, Proj and Proj^e for $F(C^-)$. Then we apply the map F^{-1} to these objects constructed for $F(C^-)$, the resulting objects are the analogues of the *new forms*, Proj and Proj^e for the negative extended β-chain C^-. We apply a similar treatment to any other monomial related to $F(C^-)$ that we happen to encounter if we replace the "ν-chain A" in Section 4.2 of [3] with "the positive extended β^*-chain $F(C^-)$."

In Section 2.4 of [3], an element y_E of $I(d)$ corresponding to any ν-chain E (the notion of a ν-chain being as in Section 1.7 of [3]) has been defined. The analogous element of $I(d)$ for the negative extended β-chain C^- (we call it y_{C^-} here) can be obtained from $F^{-1}(\text{Proj}^e(F(C^-)))$ by following the natural process: *the column indices of elements of $F^{-1}(\text{Proj}^e(F(C^-)))$ occur as members of β; these are replaced by the row indices to obtain the desired element of $I(d)$ for C^-*. It is easy to check that y_{C^-} belongs to $I(d)$ and that $y_{C^-} \le w_{C^-}^-(\beta) \le \beta$. Since we already have that $\alpha \nleq w_{C^-}^-(\beta)$, it follows that $\alpha \nleq y_{C^-}$. These facts about y_{C^-} will be needed to produce an analogue of the main proof of [3] in our present case. To be more precise, these facts about y_{C^-} give the analogues of Propositions 2.4.1 and 2.4.2 of [3], and these two propositions had been used quite crucially inside the main proof of [3].

With all these analogues constructed for negative extended β-chains, we can proceed in an "equivalent" manner (Here, by "equivalent" we mean keeping track of the subtle difference as mentioned above and working accordingly.) as in the paper [3] and end up proving the desired fact, namely, $X_{C^-} \in \text{in}_{\triangleright}I$.

Since $\text{Chains}_\alpha^\gamma(\beta) \subseteq \text{in}_{\triangleright}I$, we have $\langle \text{Chains}_\alpha^\gamma(\beta) \rangle \subseteq \text{in}_{\triangleright}I$. To prove Theorem 7, we now need to show that $\langle \text{Chains}_\alpha^\gamma(\beta) \rangle \supseteq \text{in}_{\triangleright}I$. For this, it suffices to show that, in any degree, the number of monomials of $\langle \text{Chains}_\alpha^\gamma(\beta) \rangle$ is $\ge$ the number of monomials of $\text{in}_{\triangleright}I$. Or equivalently, it suffices to show that, in any degree, the number of monomials of $P/\langle \text{Chains}_\alpha^\gamma(\beta) \rangle$ is $\le$ the number of monomials of $P/\text{in}_{\triangleright}I$.

Consider the affine patch $Y_\alpha^\gamma(\beta)$ $(:= X_\alpha^\gamma \cap \mathbb{A}^\beta)$ of the Richardson variety X_α^γ. The following is a well known result.

Theorem 13. *$\mathfrak{k}[Y_\alpha^\gamma(\beta)] = P/I$, where $P = \mathfrak{k}[X_{(r,c)} \mid (r,c) \in \mathfrak{OR}(\beta)]$ and I is as in* (1).

TABLE 1: Two subsets of the ring $P = \mathfrak{k}[X_{(r,c)} | (r,c) \in \mathfrak{OR}(\beta)]$ and their indexing sets.

Set of elements in $P = \mathfrak{k}[X_{(r,c)} \| (r,c) \in \mathfrak{OR}(\beta)]$	Indexing set
Monomials of $P/\langle \mathrm{Chains}_\alpha^\gamma(\beta) \rangle$	Pairs of nonempty skew-symmetric lexicographic multisets on $\overline{\beta} \times \beta$ bounded by T_α, W_γ
Standard monomials on $Y_\alpha^\gamma(\beta)$	Nonvanishing skew-symmetric notched bitableaux on $\overline{\beta} \times \beta$ bounded by T_α, W_γ

Both the monomials of $P/\mathrm{in}_\triangleright I$ and the standard monomials on $Y_\alpha^\gamma(\beta)$ form a basis for P/I, and thus agree in cardinality in any degree. Therefore, to prove, that in any degree, the number of monomials of $P/\langle \mathrm{Chains}_\alpha^\gamma(\beta) \rangle$ is $\leq$ the number of monomials of $P/\mathrm{in}_\triangleright I$, it suffices to give a degree-preserving injection from the set of all monomials in $P/\langle \mathrm{Chains}_\alpha^\gamma(\beta) \rangle$ to the set of all standard monomials on $Y_\alpha^\gamma(\beta)$. We construct such an injection, the *Orthogonal-bounded*-RSK (OBRSK), from an indexing set of the former to an indexing set of the later. These indexing sets are given in Table 1.

4. Skew-Symmetric Notched Bitableaux

In this section onwards, the terminology and notation of Sections 4 and 5 of [8] will be in force. Recall the definition of a *semistandard* notched bitableau from Section 5 of [8].

Definition 14 (dual of an element with respect to a semistandard notched bitableau). Let (P, Q) be a semistandard notched bitableau. Let $p_{i,j}$ (resp., $q_{i,j}$) denote the entry in the ith row and jth column of P (resp., Q). For any row number i of P (or of Q), let k_i denote the total number of entries in the ith row of P (or Q). One calls the entry q_{i,k_i+1-j} of Q the *dual of* the entry $p_{i,j}$ of P *with respect to* (P, Q). Similarly, we call the entry p_{i,k_i+1-j} of P the *dual of* the entry $q_{i,j}$ of Q *with respect to* (P, Q).

Note that any entry of P or Q can be identified uniquely by specifying 4 coordinates, namely, the entry x, the tableau A ($A = P$ or Q) in which the entry lies, the row number i of the entry in the tableau A, and the column number j of the entry in the tableau A. Let $\mathrm{set}(P, Q)$ denote the set of all 4-tuples of the form (x, A, i, j). Given any 4-tuple $(x, A, i, j) \in \mathrm{set}(P, Q)$, let us denote by $D_{(P,Q)}(x, A, i, j)$ the 4-tuple which represents the dual of x with respect to (P, Q) (as defined in Definition 14). For $(x, A, i, j), (x', A', i', j') \in \mathrm{set}(P, Q)$, we say that $(x, A, i, j) \preceq (x', A', i', j')$ if $x \leq x'$. Similarly, we say that $(x, A, i, j) \prec (x', A', i', j')$ if $x < x'$ and $(x, A, i, j) \simeq (x', A', i', j')$ if $x = x'$.

A semistandard notched bitableau (P, Q) is said to be *skew-symmetric* if the following 2 conditions are satisfied simultaneously.

(i) The bitableau (P, Q) should be of *even* size; that is, the number of elements in each row of P and Q should be even.

(ii) For any $(x, A, i, j), (x', A', i', j') \in \mathrm{set}(P, Q)$, $(x, A, i, j) \preceq (x', A', i', j')$ *if and only if* $D_{(P,Q)}(x, A, i, j) \succeq D_{(P,Q)}(x', A', i', j')$. Moreover, $(x, A, i, j) \prec (x', A', i', j')$ *if and only if* $D_{(P,Q)}(x, A, i, j) \succ D_{(P,Q)}(x', A', i', j')$, and $(x, A, i, j) \simeq (x', A', i', j')$ *if and only if* $D_{(P,Q)}(x, A, i, j) \simeq D_{(P,Q)}(x', A', i', j')$.

Property (ii) will be henceforth referred to as *the duality property* associated to the skew-symmetric notched bitableau (P, Q). Note that a *skew-symmetric* notched bitableau is *a semistandard notched bitableau* by default. The *degree* of a skew-symmetric notched bitableau (P, Q) is the total number of boxes in P (or Q). The notions of *negative, positive,* and *nonvanishing* skew-symmetric notched bitableau remain the same as in Section 5 of [8]. The notion of a skew-symmetric notched bitableau (P, Q) being *bounded by* T, W (where T, W are subsets of $\mathbb{N}^2$), and the notion of *negative* and *positive parts* of a skew-symmetric notched bitableau (P, Q) remain the same as they were in Section 5 of [8].

If (P, Q) is a nonvanishing skew-symmetric notched bitableau, define $\iota(P, Q)$ to be the notched bitableau obtained by reversing the order of the rows of (Q, P). One checks that $\iota(P, Q)$ is a nonvanishing skew-symmetric notched bitableau. The map ι is an involution, and it maps negative skew-symmetric notched bitableaux to positive ones and vice versa. Thus ι gives a bijective pairing between the sets of negative and positive skew-symmetric notched bitableaux.

5. Pairs of Skew-Symmetric Lexicographic Multisets on $\mathbb{N}^2$

Recall the notions of a multiset on $\mathbb{N}^2$, finiteness of a multiset on $\mathbb{N}^2$ and degree of a multiset on $\mathbb{N}^2$, from Section 4 of [8]. By a *lexicographic multiset on* $\mathbb{N}^2$, we mean a finite multiset $\pi = \{(\alpha_1, \beta_1), \ldots, (\alpha_t, \beta_t)\}$ on $\mathbb{N}^2$ such that $\beta_k \geq \beta_{k+1}\ \forall k$, and if $\beta_k = \beta_{k+1}$, then $\alpha_k \geq \alpha_{k+1}, k = 1, \ldots, t-1$. Given a lexicographic multiset π on $\mathbb{N}^2$, let π^t denote the multiset on $\mathbb{N}^2$ (which is not necessarily lexicographic) obtained by switching the two coordinates of π. We call the multiset π^t on $\mathbb{N}^2$ the *transpose of the multiset* π. Consider a pair $\{\pi_1, \pi_2\}$ of multisets on $\mathbb{N}^2$ (not necessarily lexicographic), where both π_1 and π_2 are of the same *degree* (say, t). Let $\pi_1 = \{(a_1, b_1), \ldots, (a_t, b_t)\}$, and $\pi_2 = \{(d_1, c_1), \ldots, (d_t, c_t)\}$.

We call the first coordinate of the multiset π_1 the *a-cell,* the second coordinate of π_1 the *b-cell,* the first coordinate of π_2 the *d-cell,* and the second coordinate of π_2 the *c-cell.* Any entry in the pair $\{\pi_1, \pi_2\}$ of multisets on $\mathbb{N}^2$ can be identified uniquely by specifying 3 coordinates: the cell ∇ of $\{\pi_1, \pi_2\}$ in which the entry lies ($\nabla = a, b, c, or\ d$), the position i (counting from left to right) of the entry in the cell ∇, and the value $\sqsubseteq(\nabla, i)$ of the entry sitting in the ith position of the cell ∇.

Set $S_{\pi_1,\pi_2} := \{x \mid x = (\nabla, i, \sqsubseteq (\nabla, i)), \nabla \in \{a,b,c,d\}, i \in \{1,\ldots,t\}\}$. For any $x \in S_{\pi_1,\pi_2}$, let

$$D_{\pi_1,\pi_2}(x) := \begin{cases} (c, t+1-i, \sqsubseteq (c, t+1-i)) \\ \quad \text{if } x = (a, i, \sqsubseteq (a,i)) \\ (d, t+1-i, \sqsubseteq (d, t+1-i)) \\ \quad \text{if } x = (b, i, \sqsubseteq (b,i)) \\ (a, t+1-i, \sqsubseteq (a, t+1-i)) \\ \quad \text{if } x = (c, i, \sqsubseteq (c,i)) \\ (b, t+1-i, \sqsubseteq (b, t+1-i)) \\ \quad \text{if } x = (d, i, \sqsubseteq (d,i)). \end{cases} \quad \forall i \in \{1,\ldots,t\} \tag{6}$$

We call $D_{\pi_1,\pi_2}(x)$ the *dual of x with respect to the pair* $\{\pi_1,\pi_2\}$ *of multisets on* $\mathbb{N}^2$. Note that, for every $x \in S_{\pi_1,\pi_2}$, we have $D_{\pi_1,\pi_2}(x) \in S_{\pi_1,\pi_2}$. For any two elements $x, x' \in S_{\pi_1,\pi_2}$, where $x = (\nabla, i, \sqsubseteq (\nabla, i))$ and $x' = (\nabla', i', \sqsubseteq (\nabla', i'))$, we say that $x \precsim x'$ if $\sqsubseteq (\nabla, i) \le \sqsubseteq (\nabla', i')$. Similarly, we say that $x \precnsim x'$ if $\sqsubseteq (\nabla, i) < \sqsubseteq (\nabla', i')$ and $x \approx x'$ if $\sqsubseteq (\nabla, i) = \sqsubseteq (\nabla', i')$.

The above pair $\{\pi_1,\pi_2\}$ of multisets on $\mathbb{N}^2$ is said to be *skew-symmetric lexicographic* if the following conditions are satisfied simultaneously.

(i) π_1 is a lexicographic multiset on $\mathbb{N}^2$.

(ii) π_2^t is a lexicographic multiset on $\mathbb{N}^2$.

(iii) $a_i < d_{t+1-i}$ $\forall i \in \{1,\ldots,t\}$.

(iv) $b_i < c_{t+1-i}$ $\forall i \in \{1,\ldots,t\}$.

(v) For any $x, y \in S_{\pi_1,\pi_2}$, $x \precsim y$ *if and only if* $D_{\pi_1,\pi_2}(x) \succsim D_{\pi_1,\pi_2}(y)$. Moreover, $x \precnsim y$ *if and only if* $D_{\pi_1,\pi_2}(x) \succnsim D_{\pi_1,\pi_2}(y)$, and $x \approx y$ *if and only if* $D_{\pi_1,\pi_2}(x) \approx D_{\pi_1,\pi_2}(y)$ (→ this property is called the *duality property* associated to the pair $\{\pi_1,\pi_2\}$ of skew-symmetric lexicographic multisets on $\mathbb{N}^2$).

For any pair $\{\pi_1,\pi_2\}$ of skew-symmetric lexicographic multisets on $\mathbb{N}^2$, we define the *degree of the pair* to be 2 times the degree of π_1 (or of π_2, they are the same). A pair $\{\pi_1,\pi_2\}$ of skew-symmetric lexicographic multisets on $\mathbb{N}^2$ is said to be *negative* if $a_k < b_k, k = 1,\ldots,t$, *positive* if $a_k > b_k, k = 1,\ldots,t$, and *nonempty* if $a_k \ne b_k, k = 1,\ldots,t$, where $\pi_1 = \{(a_1,b_1),\ldots,(a_t,b_t)\}$ and $\pi_2 = \{(d_1,c_1),\ldots,(d_t,c_t)\}$. Note that condition (v) above will imply that if $a_k < b_k$ $\forall k = 1,\ldots,t$, then $d_{t+1-k} < c_{t+1-k}$ $\forall k = 1,\ldots,t$. Similarly, if $a_k > b_k$ $\forall k = 1,\ldots,t$, then $d_{t+1-k} > c_{t+1-k}$ $\forall k = 1,\ldots,t$, and if $a_k \ne b_k$ $\forall k = 1,\ldots,t$, then $d_{t+1-k} \ne c_{t+1-k}$ $\forall k = 1,\ldots,t$.

Let $\{\pi_1,\pi_2\}$ be a pair of nonempty skew-symmetric lexicographic multisets on $\mathbb{N}^2$ given by $\pi_1 = \{(a_1,b_1),\ldots,(a_t,b_t)\}$ and $\pi_2 = \{(d_1,c_1),\ldots,(d_t,c_t)\}$. Let us denote by π_1^- (resp., π_1^+) the lexicographic multiset on $\mathbb{N}^2$ consisting of those elements of π_1 such that $a_i < b_i$ (resp., $a_i > b_i$). Let us denote by π_2^- (resp., π_2^+) the lexicographic multiset on $\mathbb{N}^2$ consisting of those elements of π_2 such that $d_i < c_i$ (resp., $d_i > c_i$). We call $\{\pi_1^-,\pi_2^-\}$ and $\{\pi_1^+,\pi_2^+\}$ the *negative* and *positive parts* respectively of the pair $\{\pi_1,\pi_2\}$. Note here that because of condition (v) above, π_1^- and π_2^- will have the same degree, and the same holds true for π_1^+ and π_2^+. It is easy to see now that both the pairs $\{\pi_1^-,\pi_2^-\}$ and $\{\pi_1^+,\pi_2^+\}$ of multisets on $\mathbb{N}^2$ are skew-symmetric lexicographic in their own right.

Given a lexicographic multiset π on $\mathbb{N}^2$, define $l(\pi)$ to be the lexicographic multiset on $\mathbb{N}^2$ obtained by first switching the two coordinates of π and then rearranging the elements so that the new multiset is lexicographic. Let l^t be a map from the set of all lexicographic multisets on $\mathbb{N}^2$ to itself given by first switching the two coordinates of a given lexicographic multiset π and then rearranging the elements so that the transpose of the resulting multiset becomes lexicographic.

We now define an involution L on the set of all pairs of skew-symmetric lexicographic multisets on $\mathbb{N}^2$ by $L(\{\pi_1,\pi_2\}) := \{l(\pi_1), l^t(\pi_2)\}$. It is easy to check that this map is well defined, and it maps pairs of negative skew-symmetric lexicographic multisets on $\mathbb{N}^2$ to positive ones, and vice versa. Thus L gives a bijective pairing between the set of all pairs of negative skew-symmetric lexicographic multisets on $\mathbb{N}^2$ and the set of all pairs of positive skew-symmetric lexicographic multisets on $\mathbb{N}^2$.

6. The Orthogonal-Bounded-RSK Correspondence

We next define the *Orthogonal-bounded-RSK correspondence* (OBRSK) as a function which maps a pair of negative skew-symmetric lexicographic multisets on $\mathbb{N}^2$ to a negative skew-symmetric notched bitableau. Let $\{\pi_1,\pi_2\}$ be a pair of negative skew-symmetric lexicographic multisets on $\mathbb{N}^2$ whose entries are labeled as in Section 5. We inductively form a sequence of notched bitableaux $(P^{(0)},Q^{(0)})$, $(P^{(1)},Q^{(1)})$, $\ldots$, $(P^{(t)},Q^{(t)})$, such that each $(P^{(i)},Q^{(i)})$ is of even size, and $P^{(i)}$ is semistandard on b_i for every $i = 1,\ldots,t$ as follows.

Let $(P^{(0)},Q^{(0)}) = (\emptyset,\emptyset)$, and let $b_0 = b_1$. Assume inductively that we have formed $(P^{(i)},Q^{(i)})$, such that the notched bitableau $(P^{(i)},Q^{(i)})$ is of even size, $P^{(i)}$ is semistandard on b_i and thus on b_{i+1}, since $b_{i+1} \le b_i$.

Let us first fix some notation and terminology. Let $p_{kj}^{(i)}$ (resp., $q_{kj}^{(i)}$) denote the entry in the kth row and jth column of $P^{(i)}$ (resp., $Q^{(i)}$). Let $2l_k^{(i)}$ denote the total number of entries (note that it is always even) in the kth row of $P^{(i)}$ (or $Q^{(i)}$).

Given an arbitrary notched tableau P and any row number k of P, we call the entry in the jth box *(counting from left to right)* the *forward jth entry of the kth row of P*. Similarly, we call the entry in the jth box *(counting from right to left)* of P the *backward jth entry of the kth row of P*.

It is now easy to see that the backward jth entry of the kth row of $Q^{(i)}$ is actually equal to the forward $(2l_k^{(i)}+1-j)$th entry of $Q^{(i)}$. We now describe the OBRSK correspondence for the pair $\{\pi_1,\pi_2\}$ of negative skew-symmetric lexicographic multisets on $\mathbb{N}^2$ as mentioned above in Section 5.

Perform the bounded insertion process $P^{(i)} \overset{b_{i+1}}{\leftarrow} a_{i+1}$ as in [8]. In this finite-step process of bounded insertion, suppose that a_{i+1} had bumped the "forward j_1th entry" of the 1st row of $P^{(i)<b_{i+1}}$ (see Section 3 of [8] for the notation $P^{(i)<b_{i+1}}$, it

comes in the paragraph right after Example 3.1 of [8]); the "forward j_1th entry" of the 1st row of $P^{(i)<b_{i+1}}$ has bumped the "forward j_2th entry" of the 2nd row of $P^{(i)<b_{i+1}}, \ldots$ and so on until, at some point, a number is placed in a new box at the right end of some row of $P^{(i)<b_{i+1}}$, say this happens at the row number $K_{(i)}$ of $P^{(i)<b_{i+1}}$. Say that the entry of the new box (as mentioned in the previous statement) becomes the forward $j_{K_{(i)}}$th entry of the $K_{(i)}$th row of $P^{(i)} \overset{b_{i+1}}{\leftarrow} a_{i+1}$. We then construct a new notched tableau as follows.

We let c_{t-i} bump the "backward j_1th entry" of the 1st row of $Q^{(i)}$; then we let the "backward j_1th entry" of the 1st row of $Q^{(i)}$ bump the "backward j_2th entry" of the 2nd row of $Q^{(i)}, \ldots$ and so on until, at some point, a number is placed in a new box at the backward $j_{K_{(i)}}$th position of the $K_{(i)}$th row of $Q^{(i)}$, shifting all entries in the backward 1st… up to (and including) the backward $(j_{K_{(i)}} - 1)$th positions of the $K_{(i)}$th row of $Q^{(i)}$ to the right by one box. Essentially, whatever we did for the bounded insertion process producing $P^{(i)} \overset{b_{i+1}}{\leftarrow} a_{i+1}$, we do a dual version of the same process on $Q^{(i)}$ with the integer c_{t-i}. We denote the resulting notched tableau by $Q^{(i)} \overset{\text{dual}}{\leftarrow} c_{t-i}$.

Note here that the tableaux $P^{(i)} \overset{b_{i+1}}{\leftarrow} a_{i+1}$ and $Q^{(i)} \overset{\text{dual}}{\leftarrow} c_{t-i}$ so constructed are of the same shape, but there exists one row in both of them in which the total number of entries is *odd*. We wanted to construct a notched bitableau $(P^{(i+1)}, Q^{(i+1)})$ inductively from $(P^{(i)}, Q^{(i)})$ which should be of even size. We make it possible in the following way.

Let $K_{(i)}$ be the row number of $P^{(i)}$ (or of $Q^{(i)}$) at which the above-mentioned insertion algorithm had stopped. Place $d_{t+1-(i+1)}$ $(= d_{t-i})$ in a new box at the rightmost end of the $K_{(i)}$th row of $P^{(i)} \overset{b_{i+1}}{\leftarrow} a_{i+1}$. We denote the resulting notched tableau by $P^{(i+1)}$. It is an easy exercise to see that $P^{(i+1)}$ as constructed above will be semistandard on b_{i+1}. After this, we place b_{i+1} in a new box at the leftmost end of the $K_{(i)}$th row of $Q^{(i)} \overset{\text{dual}}{\leftarrow} c_{t-i}$, shifting all previously existing entries in the $K_{(i)}$th row of $Q^{(i)} \overset{\text{dual}}{\leftarrow} c_{t-i}$ to the right by one box. We denote the resulting notched tableau by $Q^{(i+1)}$. Clearly $P^{(i+1)}$ and $Q^{(i+1)}$ have the same shape. Now we have got hold of a notched bitableau $(P^{(i+1)}, Q^{(i+1)})$ which is of even size.

Then OBRSK $(\{\pi_1, \pi_2\})$ is defined to be $(P^{(t)}, Q^{(t)})$. In the process above, we write $(P^{(i+1)}, Q^{(i+1)}) = (P^{(i)}, Q^{(i)}) \overset{b_{i+1}, c_{t+1-(i+1)}}{\longleftarrow} a_{i+1}, d_{t+1-(i+1)}$. In terms of this notation, OBRSK $(\{\pi_1, \pi_2\}) = ((\emptyset, \emptyset) \overset{b_1, c_t}{\leftarrow} a_1, d_t) \cdots \overset{b_t, c_1}{\leftarrow} a_t, d_1$.

Lemma 15. *With notation as in the definition of the* OBRSK *correspondence mentioned above, $P^{(i)}$ is row strict for all $i \in \{1, \ldots, t\}$.*

Proof. We will prove the lemma by induction on i. The base case (i.e., when $i = 1$) of induction is easy to see. Now let $i \in \{1, \ldots, t-1\}$. Assume inductively that $P^{(i)}$ is row strict. We will now prove that $P^{(i+1)}$ is row strict. That $P^{(i)} \overset{b_{i+1}}{\leftarrow} a_{i+1}$ is row strict follows in the same way as in [8]. Note that $P^{(i+1)}$ is obtained from $P^{(i)} \overset{b_{i+1}}{\leftarrow} a_{i+1}$ by adding d_{t-i} at the rightmost end of some row of $P^{(i)} \overset{b_{i+1}}{\leftarrow} a_{i+1}$, say the kth row. It now suffices to ensure that d_{t-i} is strictly bigger than all entries in the kth row of $P^{(i)} \overset{b_{i+1}}{\leftarrow} a_{i+1}$. It follows from the defining properties of the pair of negative skew-symmetric lexicographic multisets $\{\pi_1, \pi_2\}$ on $\mathbb{N}^2$ that d_{t-i} is bigger than or equal to all entries of $P^{(i)} \overset{b_{i+1}}{\leftarrow} a_{i+1}$. But here we need to prove something sharper; namely, d_{t-i} is strictly bigger than all entries in the kth row of $P^{(i)} \overset{b_{i+1}}{\leftarrow} a_{i+1}$. We will prove this now.

Clearly all the entries of $P^{(i)} \overset{b_{i+1}}{\leftarrow} a_{i+1}$ are contained in $\{a_1, \ldots, a_{i+1}\} \dot\cup \{d_{t+1-i}, \ldots, d_t\}$ (see Section 4 of [8] for the notation $\dot\cup$). Also it is easy to observe that $a_j < d_{t-i}$ $\forall j \in \{1, \ldots, i+1\}$. So if the rightmost element of the kth row of $P^{(i)} \overset{b_{i+1}}{\leftarrow} a_{i+1}$ equals a_j for some $j \in \{1, \ldots, i+1\}$, then we are done. Otherwise, the element in the rightmost end of the kth row of $P^{(i)} \overset{b_{i+1}}{\leftarrow} a_{i+1}$ is d_j for some $j \in \{t+1-i, \ldots, t\}$ (say j_0). If $d_{j_0} < d_{t-i}$, then we are done. If not, then clearly $d_{j_0} = d_{t-i}$. It then follows from duality that $b_{t+1-j_0} = b_{i+1}$, and it is also clear that $t + 1 - j_0 < i + 1$.

But it is an easy exercise to check that if $l, l' \in \{1, \ldots, t\}$ are such that $l < l'$ and $b_l = b_{l'}$, then the number of the row in which $d_{t+1-l'}$ lies in $P^{(l')}$ is strictly bigger than the number of the row in which d_{t+1-l} lies in $P^{(l')}$ (Here, the row number is counted from top to bottom.). So d_{j_0} and d_{t-i} cannot lie in the same row of $P^{(i+1)}$, a contradiction, hence proved. □

The proof of the following lemma appears in Section 9.

Lemma 16. *If $\{\pi_1, \pi_2\}$ is a pair of negative skew-symmetric lexicographic multisets on $\mathbb{N}^2$, then* OBRSK$(\{\pi_1, \pi_2\})$ *is a negative skew-symmetric notched bitableau.*

Lemma 17. *The map* OBRSK *is a degree-preserving bijection from the set of all pairs of negative skew-symmetric lexicographic multisets on $\mathbb{N}^2$ to the set of all negative skew-symmetric notched bitableaux.*

Proof. That OBRSK is degree preserving is obvious. To show that OBRSK is a bijection, we define its inverse, which we call the *reverse* of OBRSK or ROBRSK.

Note that the entire procedure used to form $(P^{(i+1)}, Q^{(i+1)})$ from $(P^{(i)}, Q^{(i)})$, a_{i+1}, b_{i+1}, c_{t-i} and d_{t-i}, $i = 1, \ldots, t-1$, is reversible. In other words, by knowing only $(P^{(i+1)}, Q^{(i+1)})$, we can retrieve $(P^{(i)}, Q^{(i)})$, $a_{i+1}, b_{i+1}, c_{t-i}$, and d_{t-i}. First, we obtain b_{i+1}; it is the minimum entry of $Q^{(i+1)}$. Look at the lowest row in which b_{i+1} appears in $Q^{(i+1)}$, say it is row number s (counting from top to bottom). In the same row (row number s, counting from top to bottom) of $P^{(i+1)}$, look at the rightmost entry: this entry is precisely d_{t-i}. Remove this entry (which is d_{t-i}) from the sth row of $P^{(i+1)}$, that will give us the notched tableau $P^{(i)} \overset{b_{i+1}}{\leftarrow} a_{i+1}$. Similarly remove the leftmost entry (which is b_{i+1}) from the sth row of $Q^{(i+1)}$, and all other entries in this row of $Q^{(i+1)}$ should be moved one box to the left: this will give us the notched tableau $Q^{(i)} \overset{\text{dual}}{\leftarrow} c_{t-i}$.

Then, in the sth row of $P^{(i)} \overset{b_{i+1}}{\leftarrow} a_{i+1}$, select the greatest entry which is less than b_{i+1}. This entry was the new box of the bounded insertion. If we begin reverse bounded insertion with this entry, we retrieve $P^{(i)}$ and a_{i+1}. Look at the *path* in $P^{(i)} \overset{b_{i+1}}{\leftarrow} a_{i+1}$ starting from the sth row to the topmost row, along which this reverse bounded insertion had happened. Trace the *"dual path"* in $Q^{(i)} \overset{\text{dual}}{\leftarrow} c_{t-i}$ and do a *dual* of the reverse bounded insertion (which was done originally on $P^{(i)} \overset{b_{i+1}}{\leftarrow} a_{i+1}$ to retrieve $P^{(i)}$ and a_{i+1}) on $Q^{(i)} \overset{\text{dual}}{\leftarrow} c_{t-i}$: that will give us $Q^{(i)}$ and c_{t-i} out of $Q^{(i)} \overset{\text{dual}}{\leftarrow} c_{t-i}$.

We call this process of obtaining $(P^{(i)}, Q^{(i)})$, a_{i+1}, b_{i+1}, c_{t-i}, and d_{t-i} from $(P^{(i+1)}, Q^{(i+1)})$ described in the paragraphs above a *reverse step* and denote it by $(P^{(i)}, Q^{(i)}) = (P^{(i+1)}, Q^{(i+1)}) \xrightarrow{b_{i+1}, c_{t-i}} a_{i+1}, d_{t-i}$. We call the process of applying all the reverse steps sequentially to retrieve $\{\pi_1, \pi_2\}$ from $(P^{(t)}, Q^{(t)})$ the *reverse of* OBRSK or ROBRSK.

If $(P^{(t)}, Q^{(t)})$ is an arbitrary negative skew-symmetric notched bitableau (which we do not assume to be OBRSK $(\{\pi_1, \pi_2\})$ for some $\{\pi_1, \pi_2\}$), then we can still apply a sequence of reverse steps to $(P^{(t)}, Q^{(t)})$, to sequentially obtain $(P^{(i)}, Q^{(i)})$, a_{i+1}, b_{i+1}, c_{t-i}, and d_{t-i}, $i = t-1, \ldots, 1$. For this process to be well defined, however, it must first be checked that the successive $(P^{(i)}, Q^{(i)})$ are negative skew-symmetric notched bitableaux. For this, it suffices to prove a statement very similar to that proved in Lemma 16; namely, "If (P, Q) is a negative skew-symmetric notched bitableau, then $(P', Q') := (P, Q) \xrightarrow{b,c} a, d$ is a negative skew-symmetric notched bitableau, $a < b$, $d < c$, $a < d$, $b < c$ are positive integers, d is greater than or equal to all entries of P, and b is less than or equal to all entries of Q." That $a < b$, $d < c$, $a < d$, and $b < c$ are positive integers, d is greater than or equal to all entries of P, and b is less than or equal to all entries of Q follows immediately from the definition of a reverse step. That (P', Q') is a negative skew-symmetric notched bitableau follows in much the same manner as the proof of Lemma 16; we omit the details.

It remains to be shown that the pair of multisets on $\mathbb{N}^2$ produced by applying this sequence of reverse steps to the arbitrary skew-symmetric notched bitableau $(P^{(t)}, Q^{(t)})$ is skew-symmetric lexicographic. The proof of this uses the duality property of skew-symmetric notched bitableaux, the facts mentioned in the preceding paragraph regarding the integers a, b, c, and d, and the rest of the proof goes similarly as in the proof of Lemma 6.3 of [8].

At each step, OBRSK and the reverse of ROBRSK are inverse to each other. Thus they are inverse maps. □

The map OBRSK can be extended to all pairs of nonempty skew-symmetric lexicographic multisets on $\mathbb{N}^2$. If $\{\pi_1, \pi_2\}$ is a pair of positive skew-symmetric lexicographic multisets on $\mathbb{N}^2$, then define OBRSK$(\{\pi_1, \pi_2\})$ to be $\iota(\text{OBRSK}(L(\{\pi_1, \pi_2\})))$, where the definition of the map ι is given at the end of Section 4. If $\{\pi_1, \pi_2\}$ is a pair of nonempty skew-symmetric lexicographic multisets on $\mathbb{N}^2$, with negative and positive parts $\{\pi_1^-, \pi_2^-\}$ and $\{\pi_1^+, \pi_2^+\}$, then define OBRSK$(\{\pi_1, \pi_2\})$ to be the skew-symmetric notched bitableau whose negative and positive parts are OBRSK $(\{\pi_1^-, \pi_2^-\})$ and OBRSK $(\{\pi_1^+, \pi_2^+\})$.

Example 18. Let $\pi_1 = \{(4,17), (3,17), (3,14), (7,10), (4,9)\}$, and $\pi_2 = \{(20,25), (19,22), (15,26), (12,26), (12,25)\}$. See Figure 2.

Therefore.

$$\text{OBRSK}(\pi_1, \pi_2) = \left(\begin{array}{|c|c|c|c|} \hline 3 & 4 & 12 & 19 \\ \hline 3 & 7 & 12 & 20 \\ \hline 4 & 15 & & \\ \hline \end{array}, \begin{array}{|c|c|c|c|} \hline 10 & 17 & 25 & 26 \\ \hline 9 & 17 & 22 & 26 \\ \hline 14 & 25 & & \\ \hline \end{array} \right). \quad (7)$$

7. Restricting the OBRSK Correspondence

7.1. Restricting by β. Let $\beta \in I(d)$. We say that a *skew-symmetric notched bitableau (P, Q) is on $\overline{\beta} \times \beta$*, if all entries of P are in $\overline{\beta}$, all entries of Q are in β, and the sum of any entry in P (or in Q) with its dual (with respect to (P, Q)) is $2d + 1$.

Given any monomial U in $\mathfrak{DR}(\beta)$, we can define a monomial $U^{\#}$ in $\mathfrak{AR}(\beta)$ as follows: $U^{\#} := \{(c^*, r^*) \mid (r, c) \in U\}$. We say that a pair $\{V_1, V_2\}$ of skew-symmetric lexicographic multisets on $\mathbb{N}^2$ is a *pair of skew-symmetric lexicographic multisets on $\overline{\beta} \times \beta$*, if V_1 is a monomial in $\mathfrak{DR}(\beta)$, V_2 is a monomial in $\mathfrak{AR}(\beta)$, a number of elements (counting multiplicities) in V_1 and V_2 are the same, and $V_2 = V_1^{\#}$. It is easy to see from the construction of OBRSK that, if $\{U, U^{\#}\}$ is a pair of nonempty skew-symmetric lexicographic multisets on $\overline{\beta} \times \beta$, then OBRSK $(\{U, U^{\#}\})$ is a nonvanishing skew-symmetric notched bitableau on $\overline{\beta} \times \beta$, and vice versa. Thus we obtain the following.

Corollary 19. *The map* OBRSK *restricts to a degree-preserving bijection from the set of all pairs of nonempty (resp., negative, positive) skew-symmetric lexicographic multisets on $\overline{\beta} \times \beta$ to the set of all nonvanishing (resp., negative, positive) skew-symmetric notched bitableaux on $\overline{\beta} \times \beta$.*

7.2. Restricting by T and W. Let T and W be negative and positive subsets of $\mathbb{N}^2$, respectively, satisfying the condition that:

$$T_{(1)}, T_{(2)}, W_{(1)}, \text{and } W_{(2)} \text{ are subsets of } \mathbb{N}, \quad (8)$$

where the notation $T_{(1)}, T_{(2)}, W_{(1)}$, and $W_{(2)}$ is explained in Section 4 of [8]. In this subsection, we will define the boundedness of a nonempty pair of skew-symmetric lexicographic multisets on $\mathbb{N}^2$ by T, W, and this is given in Definition 26, after some definitions which are necessary to understand it.

Definition 20. A *dual pair of chains* in $\mathbb{N}^2$ is a pair of subsets $\{C_1 = \{(e_1, f_1), \ldots, (e_m, f_m)\}, C_2 = \{(g_1, h_1), \ldots, (g_m, h_m)\}\}$

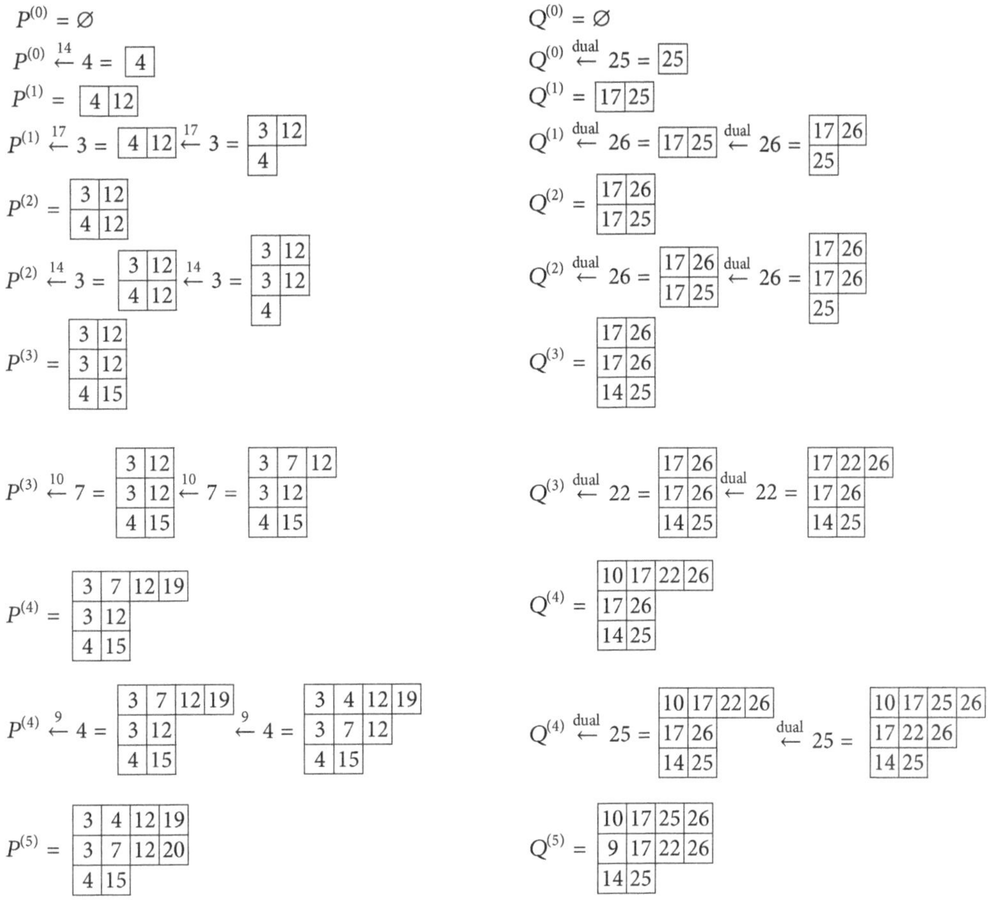

FIGURE 2

such that C_1 is a chain in $\mathbb{N}^2$ in the sense of Section 7 of [8], and $\{\sigma_1, \sigma_2\}$ is a pair of skew-symmetric lexicographic multisets on $\mathbb{N}^2$, where $\sigma_1 = \{(e_1, f_1), \ldots, (e_m, f_m)\}$ and $\sigma_2 = \{(g_1, h_1), \ldots, (g_m, h_m)\}$.

Definition 21. Let $\{U_1, U_2\}$ be a pair of skew-symmetric lexicographic multisets on $\mathbb{N}^2$. Let $\{C_1, C_2\}$ be a dual pair of chains in $\mathbb{N}^2$ such that C_i is contained in the underlying set of U_i for all $i = 1, 2$. Given any element in C_1 (say, the ith element counting from left to right), look at the first (counting from left to right) element in U_1 which is entrywise the same as the ith element of C_1. Call this element of U_1 the $i_{\min}$th element. Let t be the total number of elements in U_1. One calls the $(t + 1 - i_{\min})$th element of U_2 (counting from left to right) the dual element in U_2 corresponding to the ith element of C_1.

Definition 22. Given any pair $\{\pi_1, \pi_2\}$ of skew-symmetric lexicographic multisets on $\mathbb{N}^2$, where $\pi_1 = \{(a_1, b_1), \ldots, (a_t, b_t)\}$ and $\pi_2 = \{(d_1, c_1), \ldots, (d_t, c_t)\}$, one says that the elements (a_i, b_i) and (d_{t+1-i}, c_{t+1-i}) of π_1 and π_2, respectively, are dual to each other with respect to $\{\pi_1, \pi_2\}$.

Definition 23. Given a pair of $\{U_1, U_2\}$ of skew-symmetric lexicographic multisets on $\mathbb{N}^2$, one says that a dual pair $\{C_1, C_2\}$ of chains in $\mathbb{N}^2$ is a *dual pair of chains in* $\{U_1, U_2\}$, if the following two conditions are satisfied simultaneously: (i) C_i is contained in the underlying set of U_i for all $i = 1, 2$. (ii) Given any element of C_1 (say, the ith element counting from left to right), the dual element in U_2 corresponding to it is entrywise the same as the dual element of the ith element of C_1 with respect to $\{C_1, C_2\}$.

Definition 24. Given any row-strict notched bitableau (P, Q), we associate with it 2 subsets of $\mathbb{N}^2$ as follows. Let P_1 and Q_1 denote the topmost rows of P and Q, respectively. Let $p_{11} < \cdots < p_{1k_1}$ and $q_{11} < \cdots < q_{1k_1}$ denote the entries of P_1 and Q_1, respectively. We denote by $(P, Q)^{\text{up}}$ the subset of $\mathbb{N}^2$ given by $\{(p_{11}, q_{11}), \ldots, (p_{1k_1}, q_{1k_1})\}$. We denote by $(P, Q)^{\text{down}}$ the subset of $\mathbb{N}^2$ obtained similarly if we work with the lowermost rows of P and Q, instead of the topmost rows. See Example 25 for an illustration.

Example 25. Consider the row-strict notched bitableau

$$\left(\begin{array}{|c|c|c|c|} \hline H & J & T & Q \\ \hline G & I & S & P \\ \hline F & R \\ \cline{1-2} \end{array} \, , \, \begin{array}{|c|c|c|c|} \hline D & A & K & M \\ \hline E & B & L & N \\ \hline C & O \\ \cline{1-2} \end{array} \right) \qquad (9)$$

and call it $(\mathscr{P}, \mathscr{Q})$. Then $(\mathscr{P}, \mathscr{Q})^{\text{up}} = \{(H, D), (J, A), (T, K), (Q, M)\}$, and $(\mathscr{P}, \mathscr{Q})^{\text{down}} = \{(F, C), (R, O)\}$.

Definition 26. A nonempty pair of skew-symmetric lexicographic multisets $\{U_1, U_2\}$ on $\mathbb{N}^2$ is said to be *bounded by T, W* if for every dual pair $\{C_1, C_2\}$ of chains in $\{U_1, U_2\}$ one has

$$T \leq \left(P_{C_1^-,C_2^-}, Q_{C_1^-,C_2^-}\right)^{\text{up}}, \quad \left(P_{C_1^+,C_2^+}, Q_{C_1^+,C_2^+}\right)^{\text{down}} \leq W, \tag{10}$$

where one uses the order on multisets on $\mathbb{N}^2$ defined in Section 4 of [8]; $(P_{C_1^-,C_2^-}, Q_{C_1^-,C_2^-})$ is defined to be OBRSK $(\{C_1^-, C_2^-\})$, and $(P_{C_1^+,C_2^+}, Q_{C_1^+,C_2^+})$ is defined as OBRSK $(\{C_1^+, C_2^+\})$.

It is worthwhile to note that $(P_{C_1^-,C_2^-}, Q_{C_1^-,C_2^-})^{\text{up}}$, $(P_{C_1^+,C_2^+}, Q_{C_1^+,C_2^+})^{\text{down}}$, T, and W are subsets of $\mathbb{N}^2$, and they are *not* pairs of subsets!

With this definition, the OBRSK correspondence is a bounded function, in the sense that it maps bounded sets to bounded sets. More precisely, we have the following Lemma (to understand the statement of this lemma, we need to recall the notion of a semistandard notched bitableau being bounded by T, W from Section 5 of [8]), whose proof appears in Section 9.

Lemma 27. *If a pair $\{U_1, U_2\}$ of nonempty skew-symmetric lexicographic multisets on $\mathbb{N}^2$ is bounded by T, W, then* OBRSK$(\{U_1, U_2\})$ *is bounded by T, W.*

Definition 28. A dual pair $\{C_1, C_2\}$ of chains in $\mathbb{N}^2$ is called *a dual pair of chains in $\overline{\beta} \times \beta$* if $\{C_1, C_2\}$ is a pair of skew-symmetric lexicographic multisets on $\overline{\beta} \times \beta$.

Remark 29. A general dual pair of chains in $\overline{\beta} \times \beta$ will look like $\{C, C^\#\}$ for some extended β-chain C in $\mathfrak{D}\mathfrak{R}(\beta)$. Note also that if $\{U_1, U_2\}$ is a pair of skew-symmetric lexicographic multisets on $\overline{\beta} \times \beta$, that is, if $U_2 = U_1^\#$, then any dual pair of chains $\{C_1, C_2\}$ in $\{U_1, U_2\}$ must be a dual pair of chains in $\overline{\beta} \times \beta$; in other words, we must have $C_2 = C_1^\#$, where C_1 is an extended β-chain in $\mathfrak{D}\mathfrak{R}(\beta)$.

Remark 30. Let T and W be negative and positive subsets of $\mathbb{N}^2$, respectively, satisfying (8). A nonempty pair of skew-symmetric lexicographic multisets $\{U, U^\#\}$ on $\overline{\beta} \times \beta$ is said to be *bounded by T, W* if, for every dual pair of chains $\{C, C^\#\}$ in $\overline{\beta} \times \beta$ which is contained in the underlying set of $\{U, U^\#\}$,

$$T \leq \left(P_{C^-,C^{-\#}}, Q_{C^-,C^{-\#}}\right)^{\text{up}}, \quad \left(P_{C^+,C^{+\#}}, Q_{C^+,C^{+\#}}\right)^{\text{down}} \leq W, \tag{11}$$

where we use the order on multisets on $\mathbb{N}^2$ defined in Section 4 of [8], and $(P_{C^-,C^{-\#}}, Q_{C^-,C^{-\#}})$ (resp., $(P_{C^+,C^{+\#}}, Q_{C^+,C^{+\#}})$) is defined to be OBRSK $(\{C^-, C^{-\#}\})$ (resp., OBRSK $(\{C^+, C^{+\#}\})$). It is worthwhile to note that $(P_{C^-,C^{-\#}}, Q_{C^-,C^{-\#}})^{\text{up}}$, $(P_{C^+,C^{+\#}}, Q_{C^+,C^{+\#}})^{\text{down}}$, T, and W are subsets of $\mathbb{N}^2$; they are *not* pairs of subsets!

Let T and W be negative and positive subsets of $\overline{\beta} \times \beta$, respectively, satisfying (8). Combining Corollary 19 and Lemma 27, we obtain the following.

Corollary 31. *For any positive integer m, the number of pairs of nonempty skew-symmetric lexicographic multisets on $\overline{\beta} \times \beta$ bounded by T, W of degree $2m$ is less than or equal to the number of nonvanishing skew-symmetric notched bitableaux on $\overline{\beta} \times \beta$ bounded by T, W of degree $2m$.*

8. The Initial Ideal

In this section, we will prove the main result of the paper, that is, Theorem 7.

Proof of Theorem 7. We wish to show that $\text{in}_\triangleright I = \langle \text{Chains}_\alpha^\gamma(\beta) \rangle$. Since we already know from Remark 12 that *textChains*$_\alpha^\gamma(\beta) \subseteq \text{in}_\triangleright I$, it follows that $\langle \text{Chains}_\alpha^\gamma(\beta) \rangle \subseteq \text{in}_\triangleright I$. Hence it suffices to show that $\text{in}_\triangleright I \subseteq \langle \text{Chains}_\alpha^\gamma(\beta) \rangle$ or equivalently that, for any $m \geq 1$, cardinality of the set of all degree m monomials in $P/\langle \text{Chains}_\alpha^\gamma(\beta) \rangle$ is less than or equal to cardinality of the set of all degree m monomials in $P/\text{in}_\triangleright I$. Now since standard monomials on $Y_{\alpha,\beta}^\gamma$ (standard monomials on $Y_{\alpha,\beta}^\gamma$ is defined in Definition 32) and the monomials in $P/\text{in}_\triangleright I$ both induce homogeneous bases for P/I, it suffices to show that, for any $m \geq 1$, cardinality of the set of all degree m monomials in $P/\langle \text{Chains}_\alpha^\gamma(\beta) \rangle$ is less than or equal to cardinality of the set of all degree m standard monomials on $Y_{\alpha,\beta}^\gamma$.

We wish to give a different indexing set for the standard monomials on $Y_{\alpha,\beta}^\gamma$. Let I_β(skew-symm) denote the set of all pairs (R, S) such that all of the following conditions are satisfied.

(i) $R \subset \overline{\beta}$.

(ii) $S \subset \beta$.

(iii) $|R| = |S|$, and this cardinality is even.

(iv) If $R = \{r_1 < \cdots < r_{2l}\}$ and $S = \{s_1 < \cdots < s_{2l}\}$, then $r_i + s_{2l+1-i} = 2d + 1 \ \forall i \in \{1, \ldots, 2l\}$.

Defining $R - S := R \,\dot{\cup}\, (\beta \setminus S)$ (see Section 4 of [8]), we have the following fact, which is easily verified.

The map $(R, S) \mapsto R - S$ is a bijection from I_β (skew-symm) to $I(d)$, (Indeed, the inverse map is given by $\theta \mapsto (\theta \setminus \beta, \beta \setminus \theta)$.).

Note that under this bijection, $(\emptyset, \emptyset)$ maps to β. Let (R_α, S_α) and (R_γ, S_γ) be the preimages of the elements α and γ (of $I(d)$), respectively. Define T_α and W_γ to be any subsets of $\overline{\beta} \times \beta$ such that $(T_\alpha)_{(1)} = R_\alpha$, $(T_\alpha)_{(2)} = S_\alpha$, $(W_\gamma)_{(1)} = R_\gamma$, and $(W_\gamma)_{(2)} = S_\gamma$. Observe that T_α and W_γ satisfy (8).

Under this identification of I_β (skew-symm) with $I(d)$, the inequalities which define nonvanishing skew-symmetric notched bitableaux on $\overline{\beta} \times \beta$ bounded by T_α, W_γ are precisely the inequalities which define the standard monomials on $Y^{\gamma}_{\alpha,\beta}$ (the inequalities (12), (13), and (14) of this paper). Thus we obtain that the degree $2m$ nonvanishing skew-symmetric notched bitableaux on $\overline{\beta} \times \beta$ bounded by T_α, W_γ forms an indexing set for the degree m standard monomials on $Y^{\gamma}_{\alpha,\beta}$. Hence it suffices to show that, for any $m \geq 1$, cardinality of the set of all degree m monomials in $P/\langle \text{Chains}^{\gamma}_{\alpha}(\beta)\rangle$ is less than or equal to cardinality of the set of all nonvanishing skew-symmetric notched bitableaux on $\overline{\beta} \times \beta$ bounded by T_α, W_γ of degree $2m$.

On the other hand, it is easy to see that the set $\text{Chains}^{\gamma}_{\alpha}(\beta)$ can be rewritten in the following format: $\text{Chains}^{\gamma}_{\alpha}(\beta)$ is the set of all X_C, where C is any nonempty extended β-chain in $\mathfrak{OR}(\beta)$ such that either (i)$'$ or (ii)$'$ below holds.

(i)$'$ The chain C^- is nonempty, and $R_\alpha - S_\alpha \nleq (P_{C^-,C^{-\#}}, Q_{C^-,C^{-\#}})^{\text{up}}_{(1)} - (P_{C^-,C^{-\#}}, Q_{C^-,C^{-\#}})^{\text{up}}_{(2)}$.

(ii)$'$ The chain C^+ is nonempty, and $(P_{C^+,C^{+\#}}, Q_{C^+,C^{+\#}})^{\text{down}}_{(1)} - (P_{C^+,C^{+\#}}, Q_{C^+,C^{+\#}})^{\text{down}}_{(2)} \nleq R_\gamma - S_\gamma$.

It is now obvious that the pairs of nonempty skew-symmetric lexicographic multisets on $\overline{\beta} \times \beta$ bounded by T_α, W_γ of degree $2m$ form an indexing set for the degree m monomials of $P/\langle \text{Chains}^{\gamma}_{\alpha}(\beta)\rangle$. Hence we have that, for any $m \geq 1$, cardinality of the set of all degree m monomials in $P/\langle \text{Chains}^{\gamma}_{\alpha}(\beta)\rangle$ is equal to cardinality of the set of all pairs of nonempty skew-symmetric lexicographic multisets on $\overline{\beta} \times \beta$ bounded by T_α, W_γ of degree $2m$.

Again from Corollary 31, it follows that for any $m \geq 1$, cardinality of the set of all pairs of nonempty skew-symmetric lexicographic multisets on $\overline{\beta} \times \beta$ bounded by T_α, W_γ of degree $2m$ is less than or equal to cardinality of the set of all nonvanishing skew-symmetric notched bitableaux on $\overline{\beta} \times \beta$ bounded by T_α, W_γ of degree $2m$. Thus $\langle \text{Chains}^{\gamma}_{\alpha}(\beta)\rangle \supseteq \text{in}_{\rhd} I$. □

Definition 32. Let $P = \mathfrak{k}[X_{(r,c)} \mid (r,c) \in \mathfrak{OR}(\beta)]$. Recall the concept of a Pfaffian (denoted by $f_{\theta,\beta}$ for $\theta \in I(d)$) from Section 3.3 of this paper. One calls $f = f_{\theta_1,\beta} \cdots f_{\theta_r,\beta} \in P$ a *standard monomial* if $\theta_1, \ldots, \theta_r \in I(d)$,

$$\theta_1 \leq \cdots \leq \theta_r \tag{12}$$

and for each $i \in \{1, \ldots, r\}$, either

$$\theta_i < \beta \quad \text{or} \quad \theta_i > \beta. \tag{13}$$

If in addition, for $\alpha, \gamma \in I(d)$,

$$\alpha \leq \theta_1, \quad \theta_r \leq \gamma, \tag{14}$$

then we say that f is *standard on* $Y^{\gamma}_{\alpha,\beta}$. We define the *degree* of the standard monomial $f_{\theta_1,\beta} \cdots f_{\theta_r,\beta}$ to be the sum of the β-degrees of $\theta_1, \ldots, \theta_r$, where for any $\theta \in I(d)$; the β-degree is defined to be one-half the cardinality of $\theta \setminus \beta$.

Remark 33. In general, a standard monomial is not a monomial in the affine coordinates $X_{(r,c)}, (r,c) \in \mathfrak{OR}(\beta)$; rather, it is a polynomial. It is only a monomial in the $f_{\theta,\beta}$'s.

9. Proofs

In this section, we give proofs of Lemmas 16 and 27. To give a proof of Lemma 16, we need Lemma 34 mentioned below whose proof is left to the reader as an exercise.

Lemma 34. *If A and B are two finite multisets on $\mathbb{N}$ of the same degree such that $A \leq B$ (where the termwise order $\leq$ on multisets on $\mathbb{N}$ and the degree of a multiset on $\mathbb{N}$ are as defined in Section 4 of [8]) and p, q are natural numbers such that $p \leq q$, then $A \dot\cup \{p\} \leq B \dot\cup \{q\}$.*

9.1. Proof of Lemma 16

Proof. The proof is by induction. The base case of induction is easy to see, and hence we omit the details. Let $\{\pi_1^{(t-1)}, \pi_2^{(t-1)}\}$ be a pair of negative skew-symmetric lexicographic multisets on $\mathbb{N}^2$, where $\pi_1^{(t-1)} = \{(a_1,b_1), \ldots, (a_{t-1},b_{t-1})\}$ and $\pi_2^{(t-1)} = \{(d_2,c_2), \ldots, (d_t,c_t)\}$. Let $(P,Q) = \text{OBRSK}(\{\pi_1^{(t-1)}, \pi_2^{(t-1)}\})$. Assume inductively that (P,Q) is a negative skew-symmetric notched bitableau. Now let $\{\pi_1^{(t)}, \pi_2^{(t)}\}$ be a pair of negative skew-symmetric lexicographic multisets on $\mathbb{N}^2$ such that $\pi_1^{(t)} = \pi_1$ and $\pi_2^{(t)} = \pi_2$, where π_1 and π_2 are given by $\pi_1 = \{(a_1,b_1), \ldots, (a_t,b_t)\}$ and $\pi_2 = \{(d_1,c_1), \ldots, (d_t,c_t)\}$. Let $(P',Q') = \text{OBRSK}(\{\pi_1^{(t)}, \pi_2^{(t)}\})$. It suffices to show that (P',Q') is also a negative skew-symmetric notched bitableau.

We will first prove that (P',Q') is a negative semistandard notched bitableau. The fact that P' is row strict follows from Lemma 15. It then follows from duality that Q' is also row strict. Hence (P',Q') is row strict. Let r' be the total number of rows of P' (*or* Q'). Let P'_i, Q'_i, P_i, and Q_i denote the multiset consisting of all elements in the ith row of P', Q', P, and Q, respectively. In fact, these multisets (P'_i, Q'_i, P_i, Q_i) happen to be sets because P', Q', P, and Q are row strict. It needs to be shown that $P'_i - Q'_i \leq P'_{i+1} - Q'_{i+1}$ for $1 \leq i \leq r'-1$, and $P'_{r'} - Q'_{r'} \lessdot \emptyset$ (see Section 4 of [8] for the notation $\lessdot$). We will prove the former statement first. To prove that $P'_i - Q'_i \leq P'_{i+1} - Q'_{i+1}$ for $1 \leq i \leq r'-1$, there will be three nontrivial possibilities which will be registered below as Cases 1, 2, and 3, respectively.

Case 1. a_t and d_1 are added to the first row of P, and c_1 and b_t are added to the first row of Q. All rows other than the 1st row of P and Q remain unchanged.

In this case, it suffices to show that $P_1 \dot\cup Q_2 \dot\cup \{a_t, d_1\} \leq P_2 \dot\cup Q_1 \dot\cup \{c_1, b_t\}$. It follows from induction hypothesis that $P_1 \dot\cup Q_2 \leq P_2 \dot\cup Q_1$. Since $\{\pi_1^{(t)}, \pi_2^{(t)}\}$ is a pair of negative skew-symmetric lexicographic multisets on $\mathbb{N}^2$, we have $a_t < b_t$ and $d_1 < c_1$. Therefore $a_t \leq b_t$ and $d_1 \leq c_1$. These facts together with Lemma 34 (applied twice) imply easily that $P_1 \dot\cup Q_2 \dot\cup \{a_t, d_1\} \leq P_2 \dot\cup Q_1 \dot\cup \{c_1, b_t\}$. Hence we are done in this case.

Case 2. x_p bumps y_p from P_i, y_p bumps z_p from P_{i+1}, and the dual bumping happens on Q_i and Q_{i+1}. Let us express the dual bumping by saying that x_q bumps y_q from Q_i and y_q bumps z_q from Q_{i+1}.

Clearly then, $x_p \leq y_p \leq z_p < b_t$, and hence $x_q \geq y_q \geq z_q$ by the property (ii) in the definition of a skew-symmetric notched bitableau. Again since all entries in Q are $\geq b_t$, we get that $x_q \geq y_q \geq z_q \geq b_t$.

We need to show that $P_i' - Q_i' \leq P_{i+1}' - Q_{i+1}'$, or, in other words, $P_i' \dot{\cup} Q_{i+1}' \leq P_{i+1}' \dot{\cup} Q_i'$. Note that $P_i' \dot{\cup} Q_{i+1}' = [(P_i \dot{\cup} Q_{i+1}) \setminus \{y_p, z_q\}] \dot{\cup} \{x_p, y_q\}$ and $P_{i+1}' \dot{\cup} Q_i' = [(P_{i+1} \dot{\cup} Q_i) \setminus \{z_p, y_q\}] \dot{\cup} \{x_q, y_p\}$. It suffices to show that $(P_i \dot{\cup} Q_{i+1}) \setminus \{y_p, z_q\} \leq (P_{i+1} \dot{\cup} Q_i) \setminus \{z_p, y_q\}$ because if we can prove this much, then since $x_p \leq y_p$ and $x_q \geq y_q$ we can apply Lemma 34 and get done with the proof as in Case 1.

We will now prove that $(P_i \dot{\cup} Q_{i+1}) \setminus \{y_p, z_q\} \leq (P_{i+1} \dot{\cup} Q_i) \setminus \{z_p, y_q\}$. Since $z_p < b_t$ and all entries of Q_i are $\geq b_t$, therefore z_p is the smallest element in $P_{i+1} \dot{\cup} Q_i$ which is $\geq y_p$. It then follows from duality that z_q is the biggest element in $P_i \dot{\cup} Q_{i+1}$, that is, $\leq y_q$. Also since $y_p < b_t \leq z_q$, we have $y_p < z_q$, and hence by duality $y_q > z_p$. Also by induction hypothesis, $P_i \dot{\cup} Q_{i+1} \leq P_{i+1} \dot{\cup} Q_i$. All these facts put together prove the required thing easily, and we are done in this case.

Case 3. x_p bumps y_p from P_i, and y_p along with d_1 are added to P_{i+1}. The dual phenomenon happens with Q_i and Q_{i+1}; we express the dual phenomenon by saying that x_q bumps y_q from Q_i, and y_q along with b_t are added to Q_{i+1}.

Clearly then, $x_p \leq y_p < b_t$, and hence $x_q \geq y_q$ by the property (ii) in the definition of a Skew-symmetric notched bitableau. Again since all entries in Q are $\geq b_t$, we get that $x_q \geq y_q \geq b_t$. Also $y_p < d_1$ since the position of d_1 is at the rightmost end of that row of P_{i+1} in which y_p is placed. Hence by duality, we have $y_q > b_t$.

We need to show that $P_i' - Q_i' \leq P_{i+1}' - Q_{i+1}'$, or, in other words, $P_i' \dot{\cup} Q_{i+1}' \leq P_{i+1}' \dot{\cup} Q_i'$. Note that $P_i' \dot{\cup} Q_{i+1}' = (P_i \dot{\cup} Q_{i+1} \setminus \{y_p\}) \dot{\cup} \{b_t\} \dot{\cup} \{x_p, y_q\}$, and $P_{i+1}' \dot{\cup} Q_i' = (P_{i+1} \dot{\cup} Q_i \setminus \{y_q\}) \dot{\cup} \{d_1\} \dot{\cup} \{x_q, y_p\}$. Using Lemma 34, it is enough to prove that $(P_i \dot{\cup} Q_{i+1} \setminus \{y_p\}) \dot{\cup} \{b_t\} \leq (P_{i+1} \dot{\cup} Q_i \setminus \{y_q\}) \dot{\cup} \{d_1\}$, which we will do now.

Let $P_i \dot{\cup} Q_{i+1}$ be given by $X_1 \leq \cdots \leq X_{2n}$, and let $P_{i+1} \dot{\cup} Q_i$ be given by $Y_1 \leq \cdots \leq Y_{2n}$ (since the bitableau (P, Q) is of *even* size, the cardinality of $P_i \dot{\cup} Q_{i+1}$ or of $P_{i+1} \dot{\cup} Q_i$ is even). Note that there are no elements of P_{i+1} which are $\geq y_p$ and $< b_t$. Also since b_t is $\leq$ all elements of Q_i, therefore we can conclude that there are no elements of $P_{i+1} \dot{\cup} Q_i$ which are $\geq y_p$ and $< b_t$. It follows from duality that there are no elements of $P_i \dot{\cup} Q_{i+1}$ which are $\leq y_q$ as well as $> d_1$. Let $(P_i \dot{\cup} Q_{i+1} \setminus \{y_p\}) \dot{\cup} \{b_t\}$ be given by

$$\begin{aligned} X_1 \leq \cdots \leq X_{i-1} \leq X_{i+1} \leq \cdots \leq X_j \\ \leq b_t \leq X_{j+1} \leq \cdots \leq X_{2n} \end{aligned} \tag{15}$$

and let $(P_{i+1} \dot{\cup} Q_i \setminus \{y_q\}) \dot{\cup} \{d_1\}$ be given by

$$\begin{aligned} Y_1 \leq \cdots \leq Y_{2n-j} \leq d_1 \leq Y_{2n-j+1} \leq \cdots \leq Y_{2n-i} \\ \leq Y_{2n-i+2} \leq \cdots \leq Y_{2n}. \end{aligned} \tag{16}$$

This means that $y_p = X_i$ and its dual $y_q = Y_{2n-i+1}$ (the ith element counting from right to left in the ordered sequence $Y_1 \leq \cdots \leq Y_{2n}$). Also since $y_p = X_i$ and $y_p < b_t$, it follows that $i < j$. Observe that b_t is the jth element (counting from left to right) in the ordered sequence $X_1 \leq \cdots \leq X_{i-1} \leq X_{i+1} \leq \cdots \leq X_j \leq b_t \leq X_{j+1} \leq \cdots \leq X_{2n}$ of integers and d_1 (the dual of b_t with respect to the pair $\{\pi_1, \pi_2\}$ of multisets) is the jth element (counting from right to left) in the ordered sequence $Y_1 \leq \cdots \leq Y_{2n-j} \leq d_1 \leq Y_{2n-j+1} \leq \cdots \leq Y_{2n-i} \leq Y_{2n-i+2} \leq \cdots \leq Y_{2n}$ of integers. Consider any element in $(P_i \dot{\cup} Q_{i+1} \setminus \{y_p\}) \dot{\cup} \{b_t\}$. It can either be X_k for some $k \neq i$ or b_t. If it is X_k for some $k \neq i$, then we either need to show that $X_k \leq Y_k$, $X_k \leq d_1$, or $X_k \leq Y_{k-1}$ (these are the only 3 possibilities; the last possibility $X_k \leq Y_{k-1}$ happens if and only if $k \in \{2n - j + 2, \ldots, 2n - i + 1\}$). If it is b_t, then we need to show that $b_t \leq$ the jth element (counting from left to right) of $(P_{i+1} \dot{\cup} Q_i \setminus \{y_q\}) \dot{\cup} \{d_1\}$, which can either be Y_j, d_1, or Y_{j-1}.

After all these observation(s), the proof follows easily using the inequalities $i < j$, $x_p \leq y_p < b_t$, $x_q \geq y_q \geq b_t$, $y_p < d_1$, and $y_q > b_t$ and the facts that there are no elements of $P_{i+1} \dot{\cup} Q_i$ which lie in the interval $[y_p, b_t)$ and that there are no elements of $P_i \dot{\cup} Q_{i+1}$ which lie in the interval $(d_1, y_q]$.

We have now proved that (P', Q') is a semistandard notched bitableau. To prove that the semistandard notched bitableau (P', Q') is negative, it is enough to prove that $P_{r'}' - Q_{r'}' < \emptyset$, where r' denotes the total number of rows of P' (*or* Q'). If $P_{r'}' = P_{r'}$ and $Q_{r'}' = Q_{r'}$, then since (P, Q) is assumed to be a negative skew-symmetric notched bitableau, it follows immediately that $P_{r'}' - Q_{r'}' < \emptyset$. Otherwise, we do the following: let $P_{r'}'$ and $Q_{r'}'$ be given by $\lambda_1 < \cdots < \lambda_{s'}$ and $\delta_1 < \cdots < \delta_{s'}$, respectively. It follows easily from the way the OBRSK algorithm works that there exists at least one entry in $P_{r'}'$ which is $< b_t$. Let λ_l denote the largest entry in $P_{r'}'$ which is $< b_t$. Since $b_t \leq \delta_1$, it now follows easily that $\lambda_j < \delta_j \ \forall j \in \{1, \ldots, l\}$. Since λ_l is the largest entry in $P_{r'}'$ which is $< b_t$, therefore $l \geq 1$, and it follows from duality that $\delta_{s'+1-l} = \delta_{s'-(l-1)} > d_1$. Hence $\delta_{s'} \geq \delta_{s'-(l-1)} > d_1 = \lambda_{s'}$. It only remains to show that if $l + 1 < s'$, then $\lambda_j < \delta_j \ \forall j \in \{l + 1, \ldots, s' - 1\}$. Clearly since $l \geq 1$ and $l + 1 < s'$, it follows that, in this case, the last row of P (resp., Q) is the r'th row, namely $P_{r'}$ (resp., $Q_{r'}$), and $\delta_{s'-(l-1)}$ is the new box of the dual insertion in $Q_{r'}$. Since $\lambda_{s'} = d_1, \delta_{s'-(l-1)} > d_1, P_{r'} - Q_{r'} < \emptyset$ by induction hypothesis and $\lambda_j < \lambda_{s'} \ \forall j \in \{l + 1, \ldots, s' - 1\}$, it is now easy to see that $\lambda_j < \delta_j \ \forall j \in \{l + 1, \ldots, s' - 1\}$.

Hence we have proved that (P', Q') is a negative semistandard notched bitableau. The fact that (P', Q') is skew-symmetric is easy to see from the very construction of the OBRSK algorithm and from the fact that the pair $\{\pi_1^{(t)}, \pi_2^{(t)}\}$ of multisets is skew-symmetric, lexicographic. □

9.2. Proof of Lemma 27

Proof (proof of Lemma 27). Let T be a negative subset of $\mathbb{N}^2$ such that $T_{(1)}, T_{(2)}$ are subsets of $\mathbb{N}$. Suppose that $\{U_1 = \{(a_1, b_1), \ldots, (a_t, b_t)\}, U_2 = \{(d_1, c_1), \ldots, (d_t, c_t)\}\}$ is a pair of negative skew-symmetric lexicographic multisets on $\mathbb{N}^2$. For each $m = 1, \ldots, t$, let $U_1^{(m)} := \{(a_1, b_1), \ldots, (a_m, b_m)\}$ and $U_2^{(m)} := \{(d_{t+1-m}, c_{t+1-m}), \ldots, (d_t, c_t)\}$. Let $(P^{(m)}, Q^{(m)}) :=$

OBRSK($\{U_1^{(m)}, U_2^{(m)}\}$). For proving Lemma 27, it suffices to show that if $\{U_1, U_2\}$ is bounded by $T, \emptyset$, then $(P^{(m)}, Q^{(m)})$ is bounded by $T, \emptyset$ for all $m = 1, \dots, t$ because then the proof of Lemma 27 will follow similarly as the proofs of parts (iii), (iv), and (v) of Lemma 11.1 in [8].

Assume that $\{U_1, U_2\}$ is bounded by $T, \emptyset$. For every $m \in \{1, \dots, t\}$, let $P_1^{(m)}$ (resp., $Q_1^{(m)}$) denote the topmost row of $P^{(m)}$ (resp., $Q^{(m)}$). We need to show that $T_{(1)} - T_{(2)} \le P_1^{(m)} - Q_1^{(m)}$. Equivalently, we need to show that for all positive integers z, $|(T_{(1)} - T_{(2)})^{\le z}| \ge |(P_1^{(m)} - Q_1^{(m)})^{\le z}|$, where we use the definition $A - B := A \dot{\cup} (\mathbb{N} \setminus B)$ for any two subsets A and B of $\mathbb{N}$ (see Section 4 of [8]). We will prove this by induction on m. Let us first see the base case of induction. When $m = 1$, $U_1^{(m)} = \{(a_1, b_1)\}$ and $U_2^{(m)} = \{(d_t, c_t)\}$. Let $C_1 = \{(a_1, b_1)\}$ and $C_2 = \{(d_t, c_t)\}$. Since $\{U_1, U_2\}$ is bounded by $T, \emptyset$, it follows that for the dual pair $\{C_1, C_2\}$ of chains in $\{U_1, U_2\}$, we have $T \le (P_{C_1^-, C_2^-}, Q_{C_1^-, C_2^-})^{\text{up}} \le \emptyset$. But it can be easily seen that $(P_{C_1^-, C_2^-}, Q_{C_1^-, C_2^-})^{\text{up}} = \{(a_1, b_1), (d_t, c_t)\}$. This proves the base case of induction. Fix an integer $k \in \{1, \dots, t-1\}$, and assume that $(P^{(m)}, Q^{(m)})$ is bounded by $T, \emptyset$ for all $m \le k$. We will now show that $(P^{(k+1)}, Q^{(k+1)})$ is bounded by $T, \emptyset$.

Before we begin the proof, let us fix some notation for the rest of this subsection. Define $\{p_1^{(k)}, \dots, p_{2c_k}^{(k)}\}$ to be the topmost row of $P^{(k)}$ and $\{q_1^{(k)}, \dots, q_{2c_k}^{(k)}\}$ the topmost row of $Q^{(k)}$, both listed in increasing order. Let $(P, Q) = (P^{(k)}, Q^{(k)})$, $a = a_{k+1}$, $b = b_{k+1}$, $c = c_{t-k}$, $d = d_{t-k}$, $(P', Q') = (P^{(k+1)}, Q^{(k+1)})$, $\{p_1, \dots, p_{2\hat{c}}\} = \{p_1^{(k)}, \dots, p_{2c_k}^{(k)}\}$, and $\{q_1, \dots, q_{2\hat{c}}\} = \{q_1^{(k)}, \dots, q_{2c_k}^{(k)}\}$. Note that $\{p_1, \dots, p_{2\hat{c}}\} \subset \{a_1, \dots, a_k\} \dot{\cup} \{d_{t+1-k}, \dots, d_t\}$ and $\{q_1, \dots, q_{2\hat{c}}\} \subset \{b_1, \dots, b_k\} \dot{\cup} \{c_{t+1-k}, \dots, c_t\}$. Let P_1 (resp., Q_1) denote the topmost row of P (resp., Q). Similarly let P_1' (resp., Q_1') denote the topmost row of P' (resp., Q'). We consider two cases corresponding to the two ways in which (P_1', Q_1') can be obtained from (P_1, Q_1).

Case 1. P_1' is obtained by a bumping p_l in P_1, for some $1 \le l \le 2\hat{c}$, that is, $P_1' = (P_1 \setminus \{p_l\}) \dot{\cup} \{a\}$ and $Q_1' = (Q_1 \setminus \{q_{2\hat{c}+1-l}\}) \dot{\cup} \{c\}$.

Since b is less than or equal to all elements of $\{b_1, \dots, b_k\}$ and $b_i < c_{t+1-i}$ $\forall i \in \{1, \dots, t\}$, it follows that $b \le$ all elements of $\{b_1, \dots, b_k\} \dot{\cup} \{c_{t+1-k}, \dots, c_t\}$. Therefore $a < b \le q_1$, and hence by duality $c > d \ge p_{2\hat{c}}$. The fact that a bumps p_l implies that $a \le p_l$ and $p_l < b$. Hence $a \le p_l < b \le q_1$, and therefore by duality, we have $c \ge q_{2\hat{c}+1-l} > d \ge p_{2\hat{c}}$. For $z < a$, $p_l \le z < q_{2\hat{c}+1-l}$, or $z \ge c$,

$$\begin{aligned} \left|(T_{(1)} - T_{(2)})^{\le z}\right| &\ge \left|(P_1 - Q_1)^{\le z}\right| \\ &= \left|(P_1' - Q_1')^{\le z}\right|, \end{aligned} \tag{17}$$

where the first inequality follows from induction hypothesis, and second equality follows from the facts that $p_{l-1} < a \le p_l < b \le q_1 \le q_{2\hat{c}+1-l} \le c$ and $p_{2\hat{c}} \le d < c$. If $a = p_l$, then we are done. Thus we assume that $a < p_l$ (and hence by duality that $c > q_{2\hat{c}+1-l}$). We now need to consider only two possible positions of z, namely, $a \le z < p_l$ and $q_{2\hat{c}+1-l} \le z < c$. We claim that, for z such that $a \le z < p_l$ or $q_{2\hat{c}+1-l} \le z < c$, there exists a dual pair of chains $\{C_{k+1}^{(1)}, C_{k+1}^{(2)}\}$ in $\{U_1^{(k+1)}, U_2^{(k+1)}\}$ such that $|((P_{C_{k+1}^{(1)-}, C_{k+1}^{(2)-}}, Q_{C_{k+1}^{(1)-}, C_{k+1}^{(2)-}})_{(1)}^{\text{up}} - (P_{C_{k+1}^{(1)-}, C_{k+1}^{(2)-}}, Q_{C_{k+1}^{(1)-}, C_{k+1}^{(2)-}})_{(2)}^{\text{up}})^{\le z}| \ge |(P_1' - Q_1')^{\le z}|$. Assuming the claim and using the fact that $T \le (P_{C_{k+1}^{(1)-}, C_{k+1}^{(2)-}}, Q_{C_{k+1}^{(1)-}, C_{k+1}^{(2)-}})^{\text{up}}$ (which is because $\{U_1, U_2\}$ is bounded by $T, \emptyset$), we are done. The claim follows from Lemmas 41 and 42.

Case 2. P_1' is obtained by adding a to P_1 in position l from the left and adding d to P_1 at the rightmost end (after $p_{2\hat{c}}$); Q_1' is obtained from Q_1 by adding b to the leftmost end of Q_1 and adding c at the backward lth position of Q_1. That is, $P_1' = P_1 \dot{\cup} \{a, d\} = \{p_1 < \cdots < p_{l-1} < a < p_l < \cdots < p_{2\hat{c}} < d\}$, and $Q_1' = Q_1 \dot{\cup} \{b, c\} = \{b < q_1 < \cdots < q_{2\hat{c}+1-l} < c < q_{2\hat{c}+1-(l-1)} < \cdots < q_{2\hat{c}}\}$.

Note that $p_{l-1} < a < b < q_1$, $a < b \le p_l$ (since $b > p_l$ would require that a bumps p_l in the bounded insertion process), and $a < b < d < c$. For $z < a$,

$$\left|(T_{(1)} - T_{(2)})^{\le z}\right| \ge \left|(P_1 - Q_1)^{\le z}\right| = \left|(P_1' - Q_1')^{\le z}\right|, \tag{18}$$

where the first inequality follows from induction hypothesis and the second equality follows from the facts that $p_{l-1} < a < p_l$ and $a < b < q_1$. For z such that $b \le z < d$, note that

$$\begin{aligned} \left|(P_1' - Q_1')^{\le z}\right| &= \left|(P_1' \dot{\cup} (\mathbb{N} \setminus Q_1'))^{\le z}\right| \\ &= \left|(P_1')^{\le z}\right| + \left|(\mathbb{N} \setminus Q_1')^{\le z}\right| \\ &= \left(\left|(P_1)^{\le z}\right| + 1\right) + \left(\left|(\mathbb{N} \setminus Q_1)^{\le z}\right| - 1\right) \\ &= \left|(P_1)^{\le z}\right| + \left|(\mathbb{N} \setminus Q_1)^{\le z}\right| \\ &= \left|(P_1 - Q_1)^{\le z}\right|. \end{aligned} \tag{19}$$

Hence for $b \le z < d$, we have

$$\left|(T_{(1)} - T_{(2)})^{\le z}\right| \ge \left|(P_1 - Q_1)^{\le z}\right| = \left|(P_1' - Q_1')^{\le z}\right|. \tag{20}$$

For $z \ge c$,

$$\begin{aligned} \left|(P_1' - Q_1')^{\le z}\right| &= \left|(P_1' \dot{\cup} (\mathbb{N} \setminus Q_1'))^{\le z}\right| \\ &= \left|(P_1')^{\le z}\right| + \left|(\mathbb{N} \setminus Q_1')^{\le z}\right| \\ &= \left(\left|(P_1)^{\le z}\right| + 2\right) + \left(\left|(\mathbb{N} \setminus Q_1)^{\le z}\right| - 2\right) \\ &= \left|(P_1)^{\le z}\right| + \left|(\mathbb{N} \setminus Q_1)^{\le z}\right| \\ &= \left|(P_1 - Q_1)^{\le z}\right|. \end{aligned} \tag{21}$$

Hence for $z \ge c$, we have

$$\left|(T_{(1)} - T_{(2)})^{\le z}\right| \ge \left|(P_1 - Q_1)^{\le z}\right| = \left|(P_1' - Q_1')^{\le z}\right|. \tag{22}$$

It now remains to show that for z such that $a \le z < b$ or $d \le z < c$,

$$\left|(T_{(1)} - T_{(2)})^{\le z}\right| \ge \left|(P_1' - Q_1')^{\le z}\right|. \tag{23}$$

We claim that for z such that $a \le z < b$ or $d \le z < c$, there exists a dual pair of chains $\{S_{k+1}^{(1)}, S_{k+1}^{(2)}\}$ in $\{U_1^{(k+1)}, U_2^{(k+1)}\}$ such that

$$\Big|\Big(\Big\{\text{topmost row of } P_{S_{k+1}^{(1)}, S_{k+1}^{(2)}}\Big\} \dot{\cup} \Big(\mathbb{N} \setminus \Big\{\text{topmost row of } Q_{S_{k+1}^{(1)}, S_{k+1}^{(2)}}\Big\}\Big)\Big)^{\le z}\Big| \ge \left|(P_1' - Q_1')^{\le z}\right|. \tag{24}$$

Assuming the claim, we are done as in Case 1. But the proof of the claim is similar to the proof of the claim made in Case 1, so we omit it.

Hence we are done with the required proof modulo, the proof of the claim made in Case 1. As mentioned in Case 1, the claim follows from Lemmas 41 and 42. □

It now remains to prove Lemmas 41 and 42. But before we go to prove these two lemmas, let us have a close look at their statements and try to understand the reason behind generating such complicated lemmas for the purpose of proving the claim. It is clear that once we understand the reason behind the statement of Lemma 41 being so complicated, Lemma 42 is a natural thing to be proved for finishing the proof of the claim. Observe that Lemma 41 looks like a Generalization of Lemma 11.1 (i) of [8] in the present case of the Orthogonal grassmannian. In Lemma 11.1 (i) of [8], Kreiman says that, for $1 \le j \le m(k)$, there exists a chain $C_{k,j}$ in $U^{(k)}$ which has j elements, the last of which has first component $p_j^{(k)}$. We need to compare this statement of Kreiman with the statement of Lemma 41 here. The fact that we need a dual pair of chains here (in Lemma 41) instead of "a chain" (as mentioned in Lemma 11.1 (i) of [8]) is quite natural to believe. But it is not apparently clear as to why we need the dual pair of chains in Lemma 41 to satisfy a "strange" condition which is condition (##) (this condition is explained in Definition 40). We will now illustrate by an example (see Example 35 and the illustration associated to it below) that asking for the dual pair of chains in Lemma 41 to satisfy condition (##) is not at all strange, in fact it is the only natural thing to be asked for in the present case of the Orthogonal grassmannian. Another explanation for the same is provided in Remark 36.

Example 35. Let $d = 5$ and $\beta = (5, 7, 8, 9, 10)$. Consider the multiset $U = \{(1, 9), (3, 7)\}$ in $\mathbb{N}^2$ contained in the underlying set of $\overline{\beta} \times \beta$. Applying the map BRSK (which is defined in Section 6 of [8]) to the multiset U, we get BRSK (U) = $(\boxed{1}\boxed{3}, \boxed{7}\boxed{9})$. Set $C_{1,1} = C_{2,1} = \{(1, 9)\}$ and $C_{2,2} = U$. It is easy to see that the chains $C_{1,1}, C_{2,1}$, and $C_{2,2}$ satisfy the required conditions of lemma 11.1 (i) of [8].

On the other hand, let us take the pair of negative skew-symmetric lexicographic multisets on $\mathbb{N}^2$ given by U_1, U_2, where $U_1 = U$ and $U_2 = U^\# = \{(4, 8), (2, 10)\}$. Observe that OBRSK $(\{U, U^\#\})$ = $(\boxed{1}\boxed{2}\boxed{3}\boxed{4}, \boxed{7}\boxed{8}\boxed{9}\boxed{X})$, where $X = 10$ (we have denoted 10 by the alphabet X here because the young tableau package used here does not work with 2-digit integers). Set $C_{1,1}^{(1)} = \{(1, 9)\}, C_{1,1}^{(2)} = \{(2, 10)\}$, $C_{1,2}^{(1)} = \{(1, 9)\}, C_{1,2}^{(2)} = \{(2, 10)\}$, $C_{2,1}^{(1)} = \{(1, 9)\}, C_{2,1}^{(2)} = \{(2, 10)\}$, $C_{2,2}^{(1)} = \{(1, 9)\}, C_{2,2}^{(2)} = \{(2, 10)\}$, $C_{2,3}^{(1)} = U_1, C_{2,3}^{(2)} = U_2$, and $C_{2,4}^{(1)} = U_1, C_{2,4}^{(2)} = U_2$. It is easy to see that according to the notation of Lemma 41, we have $m(k) = 2$ for $k = 1$ and $m(k) = 4$ for $k = 2$. It can also be easily verified that, for each $k \in \{1, 2\}$ and each $j \in \{1, \ldots, m(k)\}$, the dual pair $\{C_{k,j}^{(1)}, C_{k,j}^{(2)}\}$ of chains (as mentioned above) satisfies the required condition (which is condition (##) with respect to the bitableau $(P^{(k)}, Q^{(k)})$ and the integer j) of Lemma 41. In fact, more is true; these dual pairs of chains satisfy condition (##)-strong (this condition is explained in Definition 39) with respect to the bitableau $(P^{(k)}, Q^{(k)})$ and the integer j for every $k \in \{1, 2\}$ and $j \in \{1, \ldots, m(k)\}$.

Illustration Associated to Example 35. A possible analog of Lemma 11.1 (i) of [8] in the present case of the Orthogonal grassmannian would be to demand for the dual pair of chains $\{C_{k,j}^{(1)}, C_{k,j}^{(2)}\}$ in Lemma 41 to have exactly j elements in $C_{k,j}^{(1)}$ (as well as in $C_{k,j}^{(2)}$) and the last element of $C_{k,j}^{(1)}$ to have first coordinate $p_j^{(k)}$. But it is easy to see from Example 35 that such a demand is meaningless. Instead, it would be natural and meaningful to demand for the dual pair $\{C_{k,j}^{(1)}, C_{k,j}^{(2)}\}$ of chains in Lemma 41 to satisfy condition (##)-strong (this is introduced in Definition 39) with respect to the bitableau $(P^{(k)}, Q^{(k)})$ and the integer j. But in Lemma 41, we demand the dual pair of chains $\{C_{k,j}^{(1)}, C_{k,j}^{(2)}\}$ to satisfy condition (##) (this is introduced in Definition 40) which is little weaker a condition to be satisfied than condition (##)-strong. The reason behind this is that if we demand for the dual pair $\{C_{k,j}^{(1)}, C_{k,j}^{(2)}\}$ of chains to satisfy condition (##)-strong instead of condition (##), then the proof of Lemma 41 becomes more difficult. This difficulty (in the proof of Lemma 41) arises at the point(s) where we need to show that the dual pair $\{C_{k+1,l}^{(1)}, C_{k+1,l}^{(2)}\}$ of chains in $\{U_1^{(k+1)}, U_2^{(k+1)}\}$ satisfies condition (##)-strong. It is easier to show that the dual pair $\{C_{k+1,l}^{(1)}, C_{k+1,l}^{(2)}\}$ of chains satisfies condition (##). This completes the explanation behind the statement of Lemma 41 being so complicated.

Remark 36. Note that the properties of the dual pair $\{C_{k,j}^{(1)}, C_{k,j}^{(2)}\}$ of chains in Lemma 41 are a Generalization of the properties of the chain $C_{k,j}$ of Lemma 11.1 of [8]. In fact, this Generalization precisely reveals the difference between the OBRSK algorithm (when applied to a dual pair of chains) in the present case of the Orthogonal Grassmannian and the BRSK algorithm of [8] (when applied to a chain) in the case of the ordinary Grassmannian. Equivalently, this Generalization reveals the difference between the concept of "$\mathfrak{D}$-domination of a chain by an element of $I(d)$" in the case of the Orthogonal Grassmannian (as in Section 2.4.4 of [9]) and the concept of "domination of a chain by an element of $I(d,2d)$" in the case of the ordinary Grassmannian as in [6].

Let us now prove Lemmas 41 and 42, but for the proofs we need some definitions which are the following.

Definition 37. Let (P,Q) and k_1 be as in Definition 24. Let $j \in \{1,\ldots,k_1\}$. With notation as in Definition 24, one calls (p_{1j}, q_{1j}) the jth element of $(P,Q)^{\text{up}}$, and we denote by $(P,Q)^{\text{up}}_{\leq j}$ the subset $\{(p_{11},q_{11}),\ldots,(p_{1j},q_{1j})\}$ of $(P,Q)^{\text{up}}$. See Example 38 for an illustration.

Example 38. Consider the row-strict notched bitableau $(\mathscr{P},\mathscr{Q})$ of Example 25. If we take $j=3$, then the jth element of $(\mathscr{P},\mathscr{Q})^{\text{up}}$ is (T,K) and $(\mathscr{P},\mathscr{Q})^{\text{up}}_{\leq j}$ equals $\{(H,D),(J,A),(T,K)\}$. We can work out similarly for other possible values of j.

Definition 39. Let (P,Q) be a negative skew-symmetric notched bitableau. Let $\{p_1,\ldots,p_{2c}\}$ and $\{q_1,\ldots,q_{2c}\}$ be the topmost row of P and Q, respectively, both listed in increasing order. Let $j \in \{1,\ldots,2c\}$. We say that a dual pair $\{C_1,C_2\}$ of chains in $\mathbb{N}^2$ satisfies condition (##)-strong with respect to the bitableau (P,Q) and the integer j if all the following conditions are satisfied simultaneously.

(i) $|C_1|, |C_2| \leq j$.

(ii) The first coordinate of the jth element of $(P_{C_1^-,C_2^-}, Q_{C_1^-,C_2^-})^{\text{up}}$ is p_j.

(iii) All the entries which occur as first coordinates of elements of C_1 form a subset of the set of all entries which occur as first coordinates of elements of $(P_{C_1^-,C_2^-}, Q_{C_1^-,C_2^-})^{\text{up}}_{\leq j}$.

Definition 40. Let (P,Q) be a negative skew-symmetric notched bitableau. Let $\{p_1,\ldots,p_{2c}\}$ and $\{q_1,\ldots,q_{2c}\}$ be the topmost row of P and Q, respectively, both listed in increasing order. Let $j \in \{1,\ldots,2c\}$. One says that a dual pair $\{C_1,C_2\}$ of chains in $\mathbb{N}^2$ satisfies condition (##) with respect to the bitableau (P,Q) and the integer j if all the following conditions are satisfied simultaneously.

(i) $|C_1|, |C_2| \leq j$.

(ii) There exists an integer χ_j which is $\geq j$ such that the first coordinate of the χ_jth element of $(P_{C_1^-,C_2^-}, Q_{C_1^-,C_2^-})^{\text{up}}$ is p_j.

(iii) All the entries which occur as first coordinates of elements of C_1 form a subset of the set of all entries which occur as first coordinates of elements of $(P_{C_1^-,C_2^-}, Q_{C_1^-,C_2^-})^{\text{up}}_{\leq \chi_j}$.

Lemma 41. *Let $\{U_1 = \{(a_1,b_1),\ldots,(a_t,b_t)\}, U_2 = \{(d_1,c_1),\ldots,(d_t,c_t)\}\}$ be a pair of negative skew-symmetric lexicographic multisets on $\mathbb{N}^2$. For $k = 1,\ldots,t$, let $U_1^{(k)} := \{(a_1,b_1),\ldots,(a_k,b_k)\}$ and $U_2^{(k)} := \{(d_{t+1-k},c_{t+1-k}),\ldots,(d_t,c_t)\}$. Let $(P^{(k)},Q^{(k)}) := \mathrm{OBRSK}(\{U_1^{(k)},U_2^{(k)}\})$. Define $\{p_1^{(k)},\ldots,p_{2c_k}^{(k)}\}$ to be the topmost row of $P^{(k)}$ and $\{q_1^{(k)},\ldots,q_{2c_k}^{(k)}\}$ the topmost row of $Q^{(k)}$, both listed in increasing order. Let $m(k) := \max\{m \in \{1,\ldots,2c_k\} \mid p_m^{(k)} < q_1^{(k)}\} = |(P_1^{(k)})^{<q_1^{(k)}}|$. Then for $1 \leq j \leq m(k)$, there exists a dual pair of chains $\{C_{k,j}^{(1)},C_{k,j}^{(2)}\}$ in $\{U_1^{(k)},U_2^{(k)}\}$ which satisfies condition (##) with respect to the bitableau $(P^{(k)},Q^{(k)})$ and the integer j. Let one denote by $\chi_{k,j}$ the integer corresponding to condition (##) (item number (ii), which is $\geq j$).*

Proof. We prove this by induction on k, with $k=1$ the starting point of induction. When $k=1$, $U_1^{(1)} = \{(a_1,b_1)\}$ and $U_2^{(1)} = \{(d_t,c_t)\}$. There are two possible cases, namely, when $m(k)=1$ and $m(k)=2$. In both the cases, for every $j \in \{1,\ldots,m(k)\}$, take $C_{1,j}^{(1)} = \{(a_1,b_1)\}$ and $C_{1,j}^{(2)} = \{(d_t,c_t)\}$. This proves the base case of induction.

Let $(P,Q) = (P^{(k)},Q^{(k)})$, $a = a_{k+1}$, $b = b_{k+1}$, $c = c_{t-k}$, $d = d_{t-k}$, $(P',Q') = (P^{(k+1)},Q^{(k+1)})$, $\{p_1,\ldots,p_{2\tilde{c}}\} = \{p_1^{(k)},\ldots,p_{2c_k}^{(k)}\}$, and $\{q_1,\ldots,q_{2\tilde{c}}\} = \{q_1^{(k)},\ldots,q_{2c_k}^{(k)}\}$. Note that $\{p_1,\ldots,p_{2\tilde{c}}\} \subset \{a_1,\ldots,a_k\} \dot\cup \{d_{t+1-k},\ldots,d_t\}$ and $\{q_1,\ldots,q_{2\tilde{c}}\} \subset \{b_1,\ldots,b_k\} \dot\cup \{c_{t+1-k},\ldots,c_t\}$. Let P_1 (resp., Q_1) denote the topmost row of P (resp., Q). Similarly let P_1' (resp., Q_1') denote the topmost row of P' (resp., Q').

Since b is less than or equal to all elements of $\{b_1,\ldots,b_k\}$ and $b_i < c_{t+1-i}$ $\forall i \in \{1,\ldots,t\}$, it follows that $b \leq$ all elements of $\{b_1,\ldots,b_k\} \dot\cup \{c_{t+1-k},\ldots,c_t\}$. Therefore $a < b \leq q_1$, and hence by duality $c > d \geq p_{2\tilde{c}}$. We consider two cases corresponding to the two ways in which (P_1',Q_1') can be obtained from (P_1,Q_1).

Case 1. P_1' is obtained by a bumping p_l in P_1, for some $1 \leq l \leq 2\tilde{c}$; that is, $P_1' = (P_1 \setminus \{p_l\}) \dot\cup \{a\}$ and $Q_1' = (Q_1 \setminus \{q_{2\tilde{c}+1-l}\}) \dot\cup \{c\}$.

The fact that a bumps p_l implies that $a \leq p_l$ and $p_l < b$. Hence $a \leq p_l < b \leq q_1$, and therefore, by duality, we have $c \geq q_{2\tilde{c}+1-l} > d \geq p_{2\tilde{c}}$. This implies that $m(k+1) = m(k)$. For $j \in \{1,\ldots,m(k)\} \setminus \{l\}$, set $C_{k+1,j}^{(1)} = C_{k,j}^{(1)}$ and $C_{k+1,j}^{(2)} = C_{k,j}^{(2)}$ (Note that in these cases $p_j^{(k)} = p_j^{(k+1)}$.). We now consider the case when $j = l$. If $l = 1$, then set $C_{k+1,l}^{(1)} = \{(a,b)\}$ and $C_{k+1,l}^{(2)} = \{(d,c)\}$. Otherwise consider the dual pair of chains $\{C_{k,l-1}^{(1)},C_{k,l-1}^{(2)}\}$ in $\{U_1^{(k)},U_2^{(k)}\}$. Let $C_{k+1,l}^{(1)} := C_{k,l-1}^{(1)} \cup \{(a,b)\}$ and $C_{k+1,l}^{(2)} := C_{k,l-1}^{(2)} \cup \{(d,c)\}$. Note that the dual pair $\{C_{k+1,l}^{(1)},C_{k+1,l}^{(2)}\}$ of chains in $\{U_1^{(k+1)},U_2^{(k+1)}\}$ satisfies the required conditions.

Case 2. P_1' is obtained by adding a to P_1 in position l from the left and adding d to P_1 at the rightmost end (after $p_{2\tilde{c}}$); Q_1' is obtained from Q_1 by adding b to the leftmost end of Q_1 and adding c at the backward lth position of Q_1. That is, $P_1' = P_1 \dot\cup \{a,d\} = \{p_1 < \cdots < p_{l-1} < a < p_l < \cdots < p_{2\tilde{c}} < d\}$, and $Q_1' \dot\cup \{b,c\} = \{b < q_1 < \cdots < q_{2\tilde{c}+1-l} < c < q_{2\tilde{c}+1-(l-1)} < \cdots < q_{2\tilde{c}}\}$.

Since $p_{l-1} < a < b < q_1$, it follows that $m(k) \geq (l-1)$. Note that $a < b \leq p_l$ (since $b > p_l$ would require that a bumps p_l in the bounded insertion process). Thus $m(k+1) = l$. For $j \in \{1,\ldots,l-1\}$, set $C_{k+1,j}^{(1)} = C_{k,j}^{(1)}$ and $C_{k+1,j}^{(2)} = C_{k,j}^{(2)}$.

Consider the dual pair $\{C^{(1)}_{k,l-1}, C^{(2)}_{k,l-1}\}$ of chains in $\{U^{(k)}_1, U^{(k)}_2\}$. Let $C^{(1)}_{k+1,l} := C^{(1)}_{k,l-1} \cup \{(a,b)\}$ and $C^{(2)}_{k+1,l} := C^{(2)}_{k,l-1} \cup \{(d,c)\}$. Note that the dual pair $\{C^{(1)}_{k+1,l}, C^{(2)}_{k+1,l}\}$ of chains in $\{U^{(k+1)}_1, U^{(k+1)}_2\}$ satisfies the required conditions. □

Lemma 42. *Let (P,Q) be a negative skew-symmetric notched bitableau. Let a, b, c, and d be integers such that $(P,Q) \overset{b,c}{\leftarrow} a,d$ is well defined. Let $(P',Q') := (P,Q) \overset{b,c}{\leftarrow} a,d$. Let P_1 (resp., Q_1) denote the topmost row of P (resp., Q). Similarly let P'_1 (resp., Q'_1) denote the topmost row of P' (resp., Q'). Let $\{p_1,\dots,p_{2\widetilde{c}}\}$ (resp., $\{q_1,\dots,q_{2\widetilde{c}}\}$) denote P_1 (resp., Q_1), both listed in increasing order. Suppose P'_1 is obtained by a bumping p_l in P_1, for some $1 \le l \le 2\widetilde{c}$; that is, suppose $P'_1 = (P_1 \setminus \{p_l\}) \dot\cup \{a\}$ and $Q'_1 = (Q_1 \setminus \{q_{2\widetilde{c}+1-l}\}) \dot\cup \{c\}$. Let $\{C_1, C_2\}$ be a dual pair of negative chains in $\mathbb{N}^2$ which satisfies condition (##) with respect to the bitableau (P',Q') and the integer l. Let $C_1 = \{(e_1,f_1),\dots,(e_r,f_r),(a,b)\}$ and $C_2 = \{(d,c),(g_r,h_r),\dots,(g_1,h_1)\}$, where $e_1 < \cdots < e_r < a < p_l < b < f_r < \cdots < f_1$ and $h_1 > \cdots > h_r > c > q_{2\widetilde{c}+1-l} > d > g_r > \cdots > g_1$. Then for z such that $a \le z < p_l$ or $q_{2\widetilde{c}+1-l} \le z < c$, one has*

$$\left|\left(\left(P_{C_1,C_2}, Q_{C_1,C_2}\right)^{\mathrm{up}}_{(1)} - \left(P_{C_1,C_2}, Q_{C_1,C_2}\right)^{\mathrm{up}}_{(2)}\right)^{\le z}\right| \ge \left|\left(P'_1 - Q'_1\right)^{\le z}\right|. \tag{25}$$

Proof. Note that for $a \le z < p_l$, we have $|((P_{C_1,C_2}, Q_{C_1,C_2})^{\mathrm{up}}_{(1)} - (P_{C_1,C_2}, Q_{C_1,C_2})^{\mathrm{up}}_{(2)})^{\le z}| = |(\{\text{topmost row of } P_{C_1,C_2}\} - \{\text{topmost row of } Q_{C_1,C_2}\})^{\le z}| = |(\{\text{topmost row of } P_{C_1,C_2}\} \dot\cup (\mathbb{N} \setminus \{\text{topmost row of } Q_{C_1,C_2}\}))^{\le z}| = |\{\text{topmost row of } P_{C_1,C_2}\}^{\le z} \dot\cup \{\mathbb{N}\}^{\le z}| \ge \chi_l + z \ge l + z$, where the last equality (not inequality!) is because $b \le$ all entries in Q_{C_1,C_2} and $a \le z < p_l < b$ (All the other inequalities and equalities being obvious.).

Also, $p_1 < \cdots < p_{l-1} < a < p_l < b \le q_1 < \cdots < q_{2\widetilde{c}}$, and $b < c$. Thus for $a \le z < p_l$, $|(P'_1 - Q'_1)^{\le z}| = |(P'_1 \dot\cup (\mathbb{N} \setminus Q'_1))^{\le z}| = |(P'_1 \dot\cup \mathbb{N})^{\le z}| = l + z$.

Hence we are done in the case $a \le z < p_l$. Now for z such that $q_{2\widetilde{c}+1-l} \le z < c$, we need to show that

$$\begin{aligned}&\left|\left(\left\{\text{topmost row of } P_{C_1,C_2}\right\}\right.\right.\\ &\left.\left.\quad \dot\cup \left(\mathbb{N} \setminus \left\{\text{topmost row of } Q_{C_1,C_2}\right\}\right)\right)^{\le z}\right| \\ &\quad \ge \left|\left(P'_1 \dot\cup \left(\mathbb{N} \setminus Q'_1\right)\right)^{\le z}\right|.\end{aligned} \tag{26}$$

Recall that $P'_1 = (P_1 \setminus \{p_l\}) \dot\cup \{a\}$ and $Q'_1 = (Q_1 \setminus \{q_{2\widetilde{c}+1-l}\}) \dot\cup \{c\}$. Since $(P',Q') = (P,Q) \overset{b,c}{\leftarrow} a,d$, therefore $d \ge$ all entries of P'. Hence $d \ge$ all entries of P'_1. On the other hand, since $a < p_l < b$, it follows from duality that $c > q_{2\widetilde{c}+1-l} > d$. So we have $p_1 < \cdots < p_{l-1} < a < p_{l+1} < \cdots < p_{2\widetilde{c}} \le d < q_{2\widetilde{c}+1-l} < c < q_{2\widetilde{c}+1-(l-1)} < \cdots < q_{2\widetilde{c}}$. It is now easy to observe that for z such that $q_{2\widetilde{c}+1-l} \le z < c$, the number of elements in Q'_1 which are $\le z$ is $2\widetilde{c} - l$. Hence the number of elements in $\mathbb{N} \setminus Q'_1$ which are $\le z$ will be $z - (2\widetilde{c} - l)$.

It is also clear that all the elements of P'_1 are $\le z$, and there are $2\widetilde{c}$ many elements in P'_1. Therefore,

$$\begin{aligned}\left|\left(P'_1 \dot\cup \left(\mathbb{N} \setminus Q'_1\right)\right)^{\le z}\right| &= 2\widetilde{c} + (z - (2\widetilde{c} - l)) \\ &= 2\widetilde{c} + z - 2\widetilde{c} + l = z + l.\end{aligned} \tag{27}$$

Let $\alpha_1 < \cdots < \alpha_{2\widetilde{c}}$ and $\beta_1 < \cdots < \beta_{2\widetilde{c}}$ denote the topmost rows of P_{C_1,C_2} and Q_{C_1,C_2}, respectively. It follows from the algorithm of OBRSK applied on $\{C_1, C_2\}$ that $d \ge \alpha_{2\widetilde{c}} > \cdots > \alpha_1$. On the other hand, since $a < p_l < b$, it follows from duality that $c > q_{2\widetilde{c}+1-l} > d$. Hence combining all these, we have $c > q_{2\widetilde{c}+1-l} > d \ge \alpha_{2\widetilde{c}} > \alpha_1$. So for z such that $q_{2\widetilde{c}+1-l} \le z < c$, the number of elements in the topmost row of P_{C_1,C_2} which are $\le z$ is $2\widetilde{c}$.

Since $\{C_1, C_2\}$ satisfies condition (##) with respect to the bitableau (P',Q') and the integer l, it follows that there exists an integer $\chi_l (\ge l)$ such that the χ_lth entry (counting from left to right) of the topmost row of P_{C_1,C_2} is a. Hence it follows from duality that the backward χ_lth entry (i.e., the χ_lth entry counting from right to left) of the topmost row of Q_{C_1,C_2} is c. Therefore for z such that $q_{2\widetilde{c}+1-l} \le z < c$, the number of elements in the topmost row of Q_{C_1,C_2} which are $\le z$ is equal to X_0, where X_0 is some nonnegative integer such that $X_0 \le 2\widetilde{c} - \chi_l$. But $\chi_l \ge l$, hence $-\chi_l \le -l$, and therefore $X_0 \le 2\widetilde{c} - \chi_l \le 2\widetilde{c} - l$.

Therefore, the number of elements in $(\mathbb{N} \setminus \{\text{topmost row of } Q_{C_1,C_2}\})$ which are $\le z$ is $z - X_0$. Hence $|(\{\text{topmost row of } P_{C_1,C_2}\} \dot\cup (\mathbb{N} \setminus \{\text{topmost row of } Q_{C_1,C_2}\}))^{\le z}| = 2\widetilde{c} + z - X_0 \ge 2\widetilde{c} + z - (2\widetilde{c} - l) = z + l = |(P'_1 \dot\cup (\mathbb{N} \setminus Q'_1))^{\le z}|$. □

References

[1] R. P. Stanley, "Some combinatorial aspects of the Schubert calculus," in *Combinatoire et représentation du groupe symétrique*, vol. 579 of *Lecture Notes in Mathematics*, Springer, Berlin, Germany, 1977.

[2] W. V. D. Hodge and D. Pedoe, *Methods of Algebraic Geometry*, vol. 1 of *Cambridge Mathematical Library*, Cambridge University Press, Cambridge, UK, 1994.

[3] K. N. Raghavan and S. Upadhyay, "Initial ideals of tangent cones to Schubert varieties in orthogonal Grassmannians," *Journal of Combinatorial Theory Series A*, vol. 116, no. 3, pp. 663–683, 2009.

[4] B. Sturmfels, "Gröbner bases and Stanley decompositions of determinantal rings," *Mathematische Zeitschrift*, vol. 205, no. 1, pp. 137–144, 1990.

[5] J. Herzog and N. V. Trung, "Gröbner bases and multiplicity of determinantal and Pfaffian ideals," *Advances in Mathematics*, vol. 96, no. 1, pp. 1–37, 1992.

[6] V. Kodiyalam and K. N. Raghavan, "Hilbert functions of points on Schubert varieties in Grassmannians," *Journal of Algebra*, vol. 270, no. 1, pp. 28–54, 2003.

[7] V. Kreiman and V. Lakshmibai, *Monomial bases and applications for Richardson and Schubert varieties in ordinary and affine Grassmannians [Ph.D. thesis]*, Northeastern University, 2003.

[8] V. Kreiman, "Local properties of Richardson varieties in the Grassmannian via a bounded Robinson-Schensted-Knuth correspondence," *Journal of Algebraic Combinatorics*, vol. 27, no. 3, pp. 351–382, 2008.

[9] K. N. Raghavan and S. Upadhyay, "Hilbert functions of points on Schubert varieties in orthogonal Grassmannians," *Journal of Algebraic Combinatorics*, vol. 31, no. 3, pp. 355–409, 2010.

[10] V. Lakshmibai and C. S. Seshadri, "Geometry of G/P—II. The work of de Concini and Procesi and the basic conjectures," *Proceedings of the Indian Academy of Sciences A*, vol. 87, no. 2, pp. 1–54, 1978.

Bounds for the Largest Laplacian Eigenvalue of Weighted Graphs

Sezer Sorgun

Department of Mathematics, Faculty of Sciences and Arts, Nevşehir University, 50300 Nevşehir, Turkey

Correspondence should be addressed to Sezer Sorgun; srgnrzs@gmail.com

Academic Editor: Jun-Ming Xu

Let G be weighted graphs, as the graphs where the edge weights are positive definite matrices. The Laplacian eigenvalues of a graph are the eigenvalues of Laplacian matrix of a graph G. We obtain two upper bounds for the largest Laplacian eigenvalue of weighted graphs and we compare these bounds with previously known bounds.

1. Introduction

Let $G = (V, E)$ be simple graphs, as graphs which have no loops or parallel edges such that V is a finite set of vertices and E is a set of edges.

A weighted graph is a graph each edge of which has been assigned to a square matrix called the weight of the edge. All the weight matrices are assumed to be of same order and to be positive matrix. In this paper, by "weighted graph" we mean "a weighted graph with each of its edges bearing a positive definite matrix as weight," unless otherwise stated.

The notations to be used in paper are given in the following.

Let G be a weighted graph on n vertices. Denote by $w_{i,j}$ the positive definite weight matrix of order p of the edge ij, and assume that $w_{ij} = w_{ji}$. We write $i \sim j$ if vertices i and j are adjacent. Let $w_i = \sum_{j:j\sim i} w_{ij}$. be the weight matrix of the vertex i.

The Laplacian matrix of a graph G is defined as $L(G) = (l_{ij})$, where

$$l_{i,j} = \begin{cases} w_i; & \text{if } i = j, \\ -w_{ij}; & \text{if } i \sim j, \\ 0; & \text{otherwise.} \end{cases} \quad (1)$$

The zero denotes the $p \times p$ zero matrix. Hence $L(G)$ is square matrix of order np. Let λ_1 denote the largest eigenvalue of $L(G)$. In this paper we also use to avoid the confusion that $\rho_1(w_{ij})$ is the spectral radius of w_{ij} matrix. If V is the disjoint union of two nonempty sets V_1 and V_2 such that every vertex i in V_1 has the same $\rho_1(w_i)$ and every vertex j in V_2 has the same $\rho_1(w_j)$, then G is called a weight-semiregular graph. If $\rho_1(w_i) = \rho_1(w_j)$ in weight semiregular graph, then G is called a weighted-regular graph.

Upper and lower bounds for the largest Laplacian eigenvalue for unweighted graphs have been investigated to a great extent in the literature. Also there are some studies about the bounds for the largest Laplacian eigenvalue of weighted graphs [1–3]. The main result of this paper, contained in Section 2, gives two upper bounds on the largest Laplacian for weighted graphs, where the edge weights are positive definite matrices. These upper bounds are attained by the same methods in [1–3]. We also compare the upper bounds with the known upper bounds in [1–3]. We also characterize graphs which achieve the upper bound. The results clearly generalize some known results for weighted and unweighted graphs.

2. The Known Upper Bounds for the Largest Laplacian Eigenvalue of Weighted Graphs

In this section, we present the upper bounds for the largest Laplacian eigenvalue of weighted graphs and very useful lemmas to prove theorems.

Theorem 1 (Horn and Johnson [4]). *Let $A \in M_n$ be Hermitian, and let the eigenvalues of A be ordered such that $\lambda_n \le \lambda_{n-1} \le \cdots \le \lambda_1$. Then,*

$$\lambda_n x^T x \le x^T A x \le \lambda_1 x^T x \tag{2}$$

$$\lambda_{\max} = \lambda_1 = \max_{x \ne 0} \frac{x^T A x}{x^T x} = \max_{x^T x = 1} x^T A x$$
$$\lambda_{\min} = \lambda_n = \min_{x \ne 0} \frac{x^T A x}{x^T x} = \min_{x^T x = 1} x^T A x \tag{3}$$

for all $x \in \mathbb{C}^n$.

Lemma 2 (Horn and Johnson [4]). *Let B be a Hermitian $n \times n$ matrix with eigenvalues $\lambda_1 \ge \lambda_2 \ge \cdots \ge \lambda_n$; then for any $\overline{x} \in R^n$ $(\overline{x} \ne \overline{0})$, $\overline{y} \in R^n$ $(\overline{y} \ne \overline{0})$,*

$$\left|\overline{x}^T B \overline{y}\right| \le \lambda_1 \sqrt{\overline{x}^T \overline{x}} \sqrt{\overline{y}^T \overline{y}}. \tag{4}$$

Equality holds if and only if $\overline{x}$ is an eigenvector of B corresponding to λ_1 and $\overline{y} = \alpha \overline{x}$ for some $\alpha \in R$.

Lemma 3 (see [1]). *Let G be a $(\rho_1(w_i), \rho_1(w_j))$-semiregular bipartite graph of order n such that the first l vertices of the same largest eigenvalue $\rho_1(w_i)$ and the remaining m vertices of the same largest eigenvalue $\rho_1(w_j)$. Also let $\overline{x}$ be a common eigenvector of w_{ij} corresponding to the largest eigenvalue $\rho_1(w_{ij})$ for all i, j, where $w_i = \sum_{k \in N_i} w_{ik}$ for all i. Then $\rho_1(w_i) + \rho_1(w_j)$ is the largest eigenvalue of $L(G)$ and the corresponding eigenvector is*

$$\Big(\underbrace{\rho_1(w_i)\overline{x}^T, \ldots, \rho_1(w_i)\overline{x}^T}_{l}, \underbrace{-\rho_1(w_j)\overline{x}^T, \ldots, -\rho_1(w_j)\overline{x}^T}_{m}\Big). \tag{5}$$

Theorem 4 (see [1]). *Let G be a simple connected weighted graph. Then*

$$\lambda_1 \le \max_{i \sim j} \left\{ \rho_1 \left(\sum_{k:k \sim i} w_{ik} \right) + \sum_{k:k \sim j} \rho_1 \left(w_{jk} \right) \right\}, \tag{6}$$

where w_{ij} is the positive definite weight matrix of order p of the edge ij. Moreover equality holds in (6) if and only if

(i) *G is a weight-semiregular bipartite graph,*

(ii) *w_{ij} have a common eigenvector corresponding to the largest eigenvalue $\rho_1(w_{ij})$ for all i, j.*

Theorem 5 (see [2]). *Let G be a simple connected weighted graph. Then*

$$\lambda_1 \le \max_{i \sim j} \left\{ \sqrt{ \begin{array}{l} \sum_{k:k \sim i} \rho_1 (w_{ik}) \left(\sum_{r:r \sim i} \rho_1 (w_{ir}) + \sum_{s:s \sim k} \rho_1 (w_{ks}) \right) \\ + \sum_{k:k \sim j} \rho_1 (w_{jk}) \left(\sum_{r:r \sim j} \rho_1 (w_{jr}) + \sum_{s:s \sim k} \rho_1 (w_{ks}) \right) \end{array} } \right\}, \tag{7}$$

where w_{ij} is the positive definite weight matrix of order p of the edge ij. Moreover equality holds in (7) if and only if

(i) *G is a bipartite semiregular graph;*

(ii) *w_{ij} have a common eigenvector corresponding to the largest eigenvalue $\rho_1(w_{ij})$ for all i, j.*

Corollary 6 (see [2]). *Let G be a simple connected weighted graph where each edge weight w_{ij} is a positive number. Then*

$$\lambda_1 \le \max_i \left\{ \sqrt{2 w_i \left(w_i + \overline{w}_i \right)} \right\}, \tag{8}$$

where $\overline{w}_i = (\sum_{k:k \sim i} w_{ik} w_k)/w_i$ and w_i is the weight of vertex i. Moreover equality holds if and only if G is a bipartite regular graph.

Corollary 7 (see [2]). *Let G be a simple connected weighted graph where each edge weight w_{ij} is a positive number. Then*

$$\lambda_1 \le \max_{i \sim j} \left\{ \sqrt{w_i \left(w_i + \overline{w}_i \right) + w_j \left(w_j + \overline{w}_j \right)} \right\}, \tag{9}$$

where $\overline{w}_i = (\sum_{k:k \sim i} w_{ik} w_k)/w_i$ and w_i is the weight of vertex i. Moreover equality holds if and only if G is a bipartite semiregular graph.

Theorem 8 (see [2]). *Let G be a simple connected weighted graph. Then*

$$\lambda_1 \le \max_{i \sim j} \left\{ \frac{\rho_1 (w_i) + \rho_1 (w_j) + \sqrt{\left(\rho_1 (w_i) - \rho_1 (w_j) \right)^2 + 4 \overline{y}_i \overline{y}_j}}{2} \right\}, \tag{10}$$

where $\overline{y}_i = (\sum_{k:k \sim i} \rho_1(w_{ik}) \rho_1(w_k))/\rho_1(w_i)$ and w_{ij} is the positive definite weight matrix of order p of the edge ij. Moreover equality holds in (10) if and only if

(i) *G is a weighted-regular graph or G is a weight-semiregular bipartite graph;*

(ii) *w_{ij} have a common eigenvector corresponding to the largest eigenvalue $\rho_1(w_{ij})$ for all i, j.*

Corollary 9 (see [2]). *Let G be a simple connected weighted graph where each edge weight w_{ij} is a positive number. Then*

$$\lambda_1 \le \max_i \left\{ w_i + \overline{w}_i \right\}, \tag{11}$$

where $\overline{w}_i = (\sum_{k:k \sim i} w_{ik} w_k)/w_i$ and w_i is the weight of vertex i. Moreover equality holds if and only if G is a bipartite semiregular graph or G is a bipartite regular graph.

Theorem 10 (see [3]). *Let G be a simple connected weighted graph. Then*

$$\lambda_1 \le \max_i \left\{ \sqrt{\begin{array}{c} \rho_1^2(w_i) + \sum_{k:k\sim i} \rho_1^2(w_{ik}) + \sum_{k:k\sim i} \rho_1(w_i w_{ik} + w_{ik} w_k) \\ + \sum_{1\le i,t\le n} \sum_{s\in N_i\cap N_t} \rho_1(w_{is} w_{st}) \end{array}} \right\}, \qquad (12)$$

where w_{ik} is the positive definite weight matrix of order p of the edge ik and $N_i \cap N_k$ is the set of common neighbours of i and k. Moreover equality holds in (12) *if and only if*

(i) *G is a weight-semiregular bipartite graph;*

(ii) *w_{ik} have a common eigenvector corresponding to the largest eigenvalue $\rho_1(w_{ik})$ for all i, k.*

Corollary 11 (see [3]). *Let G be a simple connected weighted graph where each edge weight w_{ij} is a positive number. Then*

$$\lambda_1 \le \max_i \left\{ \sqrt{\begin{array}{c} w_i^2 + \sum_{k:k\sim i} w_{ik}^2 + \sum_{k:k\sim i} (w_i w_{ik} + w_k w_{ik}) \\ + \sum_{1\le i,t\le n} \sum_{s\in N_i\cap N_t} w_{is} w_{st} \end{array}} \right\}. \qquad (13)$$

Moreover equality holds if and only if G is a bipartite semiregular graph.

3. Two Upper Bounds on the Largest Laplacian Eigenvalue of Weighted Graphs

In this section we present two upper bounds for the largest eigenvalue of weighted graphs and compare the bounds with some examples.

Theorem 12. *Let G be a simple connected weighted graph. Then*

$$\lambda_1 \le \max_{i\sim j} \left\{ \frac{\rho_1(w_i) + \rho_1(w_j) + \sqrt{(\rho_1(w_i) - \rho_1(w_j))^2 + 4\left(\sum_{k:k\sim i} \rho_1(w_{ik})\right)\left(\sum_{k:k\sim j} \rho_1(w_{jk})\right)}}{2} \right\}, \qquad (14)$$

where w_{ij} is the positive definite weight matrix of order p of the edge ij. Moreover equality holds in (14) *if and only if*

(i) *G is a weighted-regular graph or G is a weight-semiregular bipartite graph;*

(ii) *w_{ij} have a common eigenvector corresponding to the largest eigenvalue $\rho_1(w_{ij})$ for all i, j.*

Proof. Let $\overline{X} = (\overline{x}_1, \overline{x}_2, \ldots, \overline{x}_n)^T$ be an eigenvector corresponding to the largest eigenvalue λ_1 of $L(G)$. We assume that $\overline{x}_i$ is the vector component of $\overline{X}$ such that

$$\overline{x}_i^T \overline{x}_i = \max_{k\in V} \left\{ \overline{x}_k^T \overline{x}_k \right\}. \qquad (15)$$

Since $\overline{X}$ is nonzero, so is $\overline{x}_i$. Let

$$\overline{x}_j^T \overline{x}_j = \max_{k:k\sim i} \left\{ \overline{x}_k^T \overline{x}_k \right\} \qquad (16)$$

be. The (i, j)th element of $L(G)$ is

$$\begin{cases} w_i; & \text{if } i = j \\ -w_{i,j}; & \text{if } i \sim j \\ 0; & \text{otherwise.} \end{cases} \qquad (17)$$

We have

$$L(G)\overline{X} = \lambda_1 \overline{X}. \qquad (18)$$

From the ith equation of (18), we have

$$\left(\lambda_1 I_{p,p} - w_i\right) \overline{x}_i = -\sum_{k:k\sim i} w_{ik} \overline{x}_k, \qquad (19)$$

that is,

$$\overline{x}_i^T \left(\lambda_1 I_{p,p} - w_i\right) \overline{x}_i = -\sum_{k:k\sim i} \overline{x}_i^T w_{ik} \overline{x}_k \qquad (20)$$

$$\le \sum_{k:k\sim i} \left| \overline{x}_i^T w_{ik} \overline{x}_k \right| \qquad (21)$$

$$\le \sum_{k:k\sim i} \rho_1(w_{ik}) \sqrt{\overline{x}_i^T \overline{x}_i} \sqrt{\overline{x}_k^T \overline{x}_k} \quad \text{by (4)} \qquad (22)$$

$$\le \sum_{k:k\sim i} \rho_1(w_{ik}) \sqrt{\overline{x}_i^T \overline{x}_i} \sqrt{\overline{x}_j^T \overline{x}_j} \quad \text{by (16)}. \qquad (23)$$

From (23) we have

$$\begin{aligned} (\lambda_1 - \rho_1(w_i)) \overline{x}_i^T \overline{x}_i &\le \overline{x}_i^T \left(\lambda_1 I_{p,p} - w_i\right) \overline{x}_i \quad \text{by (2)} \\ &\le \sum_{k:k\sim i} \rho_1(w_{ik}) \sqrt{\overline{x}_i^T \overline{x}_i} \sqrt{\overline{x}_j^T \overline{x}_j}, \end{aligned} \qquad (24)$$

that is,

$$(\lambda_1 - \rho_1(w_i))\overline{x}_i^T\overline{x}_i \le \sum_{k:k\sim i} \rho_1(w_{ik})\sqrt{\overline{x}_i^T\overline{x}_i}\sqrt{\overline{x}_j^T\overline{x}_j}. \quad (25)$$

From the jth equation of (18), we get

$$(\lambda_1 I_{p,p} - w_j)\overline{x}_j = -\sum_{k:k\sim j} w_{jk}\overline{x}_k, \quad (26)$$

that is,

$$\overline{x}_j^T(\lambda_1 I_{p,p} - w_i)\overline{x}_j = -\sum_{k:k\sim j}\overline{x}_j^T w_{jk}\overline{x}_k \quad (27)$$

$$\le \sum_{k:k\sim j}\left|\overline{x}_j^T w_{jk}\overline{x}_k\right| \quad (28)$$

$$\le \sum_{k:k\sim j}\rho_1(w_{jk})\sqrt{\overline{x}_j^T\overline{x}_j}\sqrt{\overline{x}_k^T\overline{x}_k} \quad \text{by (4)} \quad (29)$$

$$\le \sum_{k:k\sim j}\rho_1(w_{jk})\sqrt{\overline{x}_i^T\overline{x}_i}\sqrt{\overline{x}_j^T\overline{x}_j} \quad \text{by (15)}. \quad (30)$$

Similarly, from (30) we get

$$\begin{aligned}(\lambda_1 - \rho_1(w_j))\overline{x}_j^T\overline{x}_j &\le \overline{x}_j^T(\lambda_1 I_{p,p} - w_j)\overline{x}_j \quad \text{by (2)}\\ &\le \sum_{k:k\sim j}\rho_1(w_{jk})\sqrt{\overline{x}_i^T\overline{x}_i}\sqrt{\overline{x}_j^T\overline{x}_j},\end{aligned} \quad (31)$$

that is,

$$(\lambda_1 - \rho_1(wj))\overline{x}_j^T\overline{x}_j \le \sum_{k:k\sim j}\rho_1(w_{jk})\sqrt{\overline{x}_i^T\overline{x}_i}\sqrt{\overline{x}_k^T\overline{x}_k}. \quad (32)$$

So, from (25) and (32) we have

$$\begin{aligned}&(\lambda_1 - \rho_1(w_j))(\lambda_1 - \rho_1(w_i))\overline{x}_j^T\overline{x}_j\overline{x}_i^T\overline{x}_i\\ &\le \left(\sum_{k:k\sim i}\rho_1(w_{ik})\right)\cdot\left(\sum_{k:k\sim j}\rho_1(w_{jk})\right)\overline{x}_j^T\overline{x}_j\overline{x}_i^T\overline{x}_i.\end{aligned} \quad (33)$$

Hence we get

$$\begin{aligned}&(\lambda_1 - \rho_1(w_j))(\lambda_1 - \rho_1(w_i))\\ &\le \left(\sum_{k:k\sim i}\rho_1(w_{ik})\right)\cdot\left(\sum_{k:k\sim j}\rho_1(w_{jk})\right),\end{aligned} \quad (34)$$

that is,

$$\begin{aligned}&\lambda_1^2 - \lambda_1(\rho_1(w_j) + \rho_1(w_i)) + \rho_1(w_i)\rho_1(w_j)\\ &-\left(\sum_{k:k\sim i}\rho_1(w_{ik})\right)\cdot\left(\sum_{k:k\sim j}\rho_1(w_{jk})\right) \le 0,\end{aligned} \quad (35)$$

that is,

$$\begin{aligned}\lambda_1 &\le \frac{\rho_1(w_i) + \rho_1(w_j)}{2}\\ &+ \frac{\sqrt{(\rho_1(w_i) - \rho_1(w_j))^2 + 4\left(\sum_{k:k\sim i}\rho_1(w_{ik})\right)\left(\sum_{k:k\sim j}\rho_1(w_{jk})\right)}}{2}\\ &\le \max_{i\sim j}\left\{\frac{\rho_1(w_i)+\rho_1(w_j)}{2} + \frac{\sqrt{(\rho_1(w_i)-\rho_1(w_j))^2+4\left(\sum_{k:k\sim i}\rho_1(w_{ik})\right)\left(\sum_{k:k\sim j}\rho_1(w_{jk})\right)}}{2}\right\}.\end{aligned} \quad (36)$$

This completes the proof of (14).

Now suppose that equality holds in (14). Then all inequalities in the previous argument must be equalities.

From equality in (23), we get

$$\overline{x}_i^T\overline{x}_i = \overline{x}_k^T\overline{x}_k \quad \text{for all } k, k\sim i. \quad (37)$$

Since $\overline{x}_i \neq 0$, we get that $\overline{x}_k \neq 0$ for all k, $k \sim i$. From equality in (22) and Lemma 2, we get that $\overline{x}_i$ is an eigenvector of w_{ik} for the largest eigenvalue $\rho_1(w_{ik})$. Hence we say that $\overline{x}_k = a\overline{x}_i$ for some a, for any k, $k \sim i$.

On the other hand, from (37) we get

$$a^2\overline{x}_i^T\overline{x}_i = \overline{x}_i^T\overline{x}_i, \quad (38)$$

that is,

$$a^2 = 1 \quad \text{as } \overline{x}_i^T\overline{x}_i > 0. \quad (39)$$

From equality in (21), we have

$$-\sum_{k:k\sim i}\overline{x}_i^T w_{ik}\overline{x}_k = \sum_{k:k\sim i}\left|\overline{x}_i^T w_{ik}\overline{x}_k\right|. \quad (40)$$

Since $\overline{x}_k = a\overline{x}_i$, from (40) we get

$$\begin{aligned}-\sum_{k:k\sim i}a\overline{x}_i^T w_{ik}\overline{x}_i &= \sum_{k:k\sim i}|a|\left|\overline{x}_i^T w_{ik}\overline{x}_k\right|\\ &= \sum_{k:k\sim i}|a|\,\overline{x}_i^T w_{ik}\overline{x}_k \quad \text{as } \overline{x}_i^T w_{ik}\overline{x}_k > 0.\end{aligned} \quad (41)$$

Hence we get

$$a = -1 \quad (42)$$

from equalities in (41). Therefore we have

$$\overline{x}_k = -\overline{x}_i \quad \text{for all } k, k\sim i. \quad (43)$$

Similarly from equality in (29), we get that $\overline{x}_j$ is an eigenvector of w_{jk} for the largest eigenvalue $\rho_1(w_{jk})$. Hence we say that $\overline{x}_k = b\overline{x}_j$ for some b, for any k, $k \sim j$. From equality in (16) we have

$$\overline{x}_j^T\overline{x}_j = \overline{x}_k^T\overline{x}_k \quad \text{for } k \sim i, \tag{44}$$

that is,

$$b^2\overline{x}_j^T\overline{x}_j = \overline{x}_j^T\overline{x}_j, \tag{45}$$

that is,

$$b^2 = 1 \quad \text{as } \overline{x}_j^T\overline{x}_j > 0. \tag{46}$$

Applying the same methods as previously, we get

$$b = -1. \tag{47}$$

Therefore we have

$$\overline{x}_k = -\overline{x}_j \quad \text{for all } k, k \sim j. \tag{48}$$

For $i \sim j$

$$\overline{x}_i = -\overline{x}_j. \tag{49}$$

Hence we take that $U = \{k : \overline{x}_k = \overline{x}_i\}$ and $W = \{k : \overline{x}_k = -\overline{x}_i\}$ from (43), (48), and (49). So, $N_j \subset U$ and $N_i \subset W$. Also, $U \neq W \neq \emptyset$ since $\overline{x}_i \neq 0$. Further, for any vertex $s \in N_{N_i}$ there exists a vertex $r \in N_i$ such that $r \sim j\ell r \sim s$, where N_{N_i} is the neighbor of neighbor set of vertex i. Therefore $\overline{x}_r = -\overline{x}_i$ and $\overline{x}_s = \overline{x}_i$. So $N_{N_i} \subset U$. By similar argument we can present that $N_{N_j} \subset W$. Continuing the procedure, it is easy to see, since G is connected, that $V = U \cup W$ and that the subgraphs induced by U and W, respectively, are empty graphs. Hence G is bipartite. Moreover, $\overline{x}_i$ is a common eigenvector of w_{ik} and w_i for the largest eigenvalue $\rho_1(w_{ik})$ and $\rho_1(w_i)$.

For $i, k \in U$

$$\lambda_1\overline{x}_i = w_i\overline{x}_i + \sum_{k:k\sim i} w_{ik}\overline{x}_i = w_k\overline{x}_i + \sum_{k:k\sim i} w_{ik}\overline{x}_i, \tag{50}$$

that is,

$$w_i\overline{x}_i = w_k\overline{x}_i. \tag{51}$$

Since $\overline{x}_i$ is an eigenvector of w_i corresponding to the largest eigenvalue of $\rho_1(w_i)$ for all i, we get

$$\rho_1(w_i)\overline{x}_i = \rho_1(w_k)\overline{x}_i, \tag{52}$$

that is,

$$(\rho_1(w_i) - \rho_1(w_k))\overline{x}_i = 0, \tag{53}$$

that is,

$$\rho_1(w_i) = \rho_1(w_k) \quad \text{as } \overline{x}_i \neq 0. \tag{54}$$

Therefore we get that $\rho_1(w_i)$ is constant for all $i \in U$. Similarly we can show that $\rho_1(w_j)$ is constant for all $j \in W$.

Hence G is a bipartite semiregular graph.

Conversely, suppose that conditions (i)-(ii) of the theorem hold for the graph G. Let G be $(\rho_1(w_i), \rho_1(w_j))$-semiregular bipartite graph. Let x be a common eigenvector of w_{ik} corresponding to the largest eigenvalue $\rho_1(w_{ik})$ for all i, k. Then we have

$$\begin{aligned} \rho_1(w_i) &= \sum_{k:k\sim i} \rho_1(w_{ik}), \\ \rho_1(w_j) &= \sum_{k:k\sim j} \rho_1(w_{jk}). \end{aligned} \tag{55}$$

By Lemma 3, we get

$$\lambda_1 = \rho_1(w_i) + \rho_1(w_j), \tag{56}$$

that is,

$$\lambda_1 = \frac{\rho_1(w_i) + \rho_1(w_j)}{2} + \frac{\sqrt{(\rho_1(w_i) - \rho_1(w_j))^2 + 4\left(\sum_{k:k\sim i}\rho_1(w_{ik})\right)\left(\sum_{k:k\sim j}\rho_1(w_{jk})\right)}}{2}. \tag{57}$$

□

Corollary 13 (see [1]). *Let G be a simple connected weighted graph where each edge weight $w_{i,j}$ is a positive number. Then*

$$\lambda_1 \leq \max_{i\sim j}\{w_i + w_j\}. \tag{58}$$

Moreover equality holds in (58) if and only if G is bipartite semiregular graph.

Proof. We have $\rho_1(w_i) = w_i$ and $\rho_1(w_{ij}) = w_{ij}$ for all i, j. From Theorem 12, we get the required result. □

Corollary 14 (see [5]). *Let G be a simple connected unweighted graph. Then*

$$\lambda_1 \leq \max_{i\sim j}\{d_i + d_j\}, \tag{59}$$

where d_i is the degree of vertex i. Moreover equality holds in (59) if and only if G is a bipartite regular graph or G is a bipartite semiregular graph.

Proof. For unweighted graph, $w_{i,j} = 1$ for $i \sim j$. Therefore $w_i = d_i$. Using Corollary 6, we get the required results. □

Theorem 15. *Let G be a simple connected weighted graph. Then*

$$\lambda_1 \leq \max_{i\sim j}\left\{\sqrt{\left(\rho_1(w_i) + \sum_{k:k\sim i}\rho_1(w_{ik})\right)\left(\rho_1(w_j) + \sum_{k:k\sim j}\rho_1(w_{jk})\right)}\right\}, \tag{60}$$

where w_{ij} is the positive definite weight matrix of order p of the edge ij. Moreover equality holds in (60) if and only if

(i) *G is a weighted-regular bipartite graph;*

(ii) *w_{ij} have a common eigenvector corresponding to the largest eigenvalue $\rho_1(w_{ij})$ for all i, j.*

Proof. Let $\overline{X} = (\overline{x}_1, \overline{x}_2, \ldots, \overline{x}_n)^T$ be an eigenvector corresponding to the largest eigenvalue λ_1 of $L(G)$. We assume that $\overline{x}_i$ is the vector component of $\overline{X}$ such that

$$\overline{x}_i^T \overline{x}_i = \max_{k \in V} \left\{ \overline{x}_k^T \overline{x}_k \right\}. \quad (61)$$

Since $\overline{X}$ is nonzero, so is $\overline{x}_i$. Let

$$\overline{x}_j^T \overline{x}_j = \max_{k:k\sim j} \left\{ \overline{x}_k^T \overline{x}_k \right\} \quad (62)$$

be. We have

$$L(G)\overline{X} = \lambda_1 \overline{X}. \quad (63)$$

From the ith equation of (43), we have

$$\lambda_1 \overline{x}_i = w_i \overline{x}_i - \sum_{k:k\sim i} w_{ik} \overline{x}_k, \quad (64)$$

that is,

$$\begin{aligned}
\lambda_1 \overline{x}_i^T \overline{x}_i &= \left| \overline{x}_i^T w_i \overline{x}_i - \sum_{k:k\sim i} \overline{x}_i^T w_{ik} \overline{x}_k \right| \\
&\le \left| \overline{x}_i^T w_i \overline{x}_i \right| + \sum_{k:k\sim i} \left| \overline{x}_i^T w_{ik} \overline{x}_k \right| \\
&\le \rho_1(w_i) \overline{x}_i^T \overline{x}_i + \sum_{k:k\sim i} \rho_1(w_{ik}) \sqrt{\overline{x}_i^T \overline{x}_i} \sqrt{\overline{x}_k^T \overline{x}_k} \quad \text{by (2)} \\
&\le \rho_1(w_i) \overline{x}_i^T \overline{x}_i + \sum_{k:k\sim i} \rho_1(w_{ik}) \overline{x}_i^T \overline{x}_i \quad \text{by (40)}.
\end{aligned} \quad (65)$$

Hence we get

$$\lambda_1 \le \rho_1(w_i) + \sum_{k:k\sim i} \rho_1(w_{ik}). \quad (66)$$

By the same method, from the jth equation of (43), we have

$$\lambda_1 \overline{x}_j = w_j \overline{x}_j - \sum_{k:k\sim j} w_{jk} \overline{x}_k, \quad (67)$$

that is,

$$\begin{aligned}
\lambda_1 \overline{x}_j^T \overline{x}_j &= \left| \overline{x}_j^T w_j \overline{x}_j - \sum_{k:k\sim j} \overline{x}_j^T w_{jk} \overline{x}_k \right| \\
&\le \left| \overline{x}_j^T w_j \overline{x}_j \right| + \sum_{k:k\sim j} \left| \overline{x}_j^T w_{jk} \overline{x}_k \right| \\
&\le \rho_1(w_j) \overline{x}_j^T \overline{x}_j + \sum_{k:k\sim j} \rho_1(w_{jk}) \sqrt{\overline{x}_j^T \overline{x}_j} \sqrt{\overline{x}_k^T \overline{x}_k} \quad \text{by (2)} \\
&\le \rho_1(w_j) \overline{x}_j^T \overline{x}_j + \sum_{k:k\sim j} \rho_1(w_{jk}) \overline{x}_j^T \overline{x}_j \quad \text{by (41)}.
\end{aligned} \quad (68)$$

Hence we get

$$\lambda_1 \le \rho_1(w_j) + \sum_{k:k\sim j} \rho_1(w_{jk}). \quad (69)$$

From (49) and (58), we have

$$\lambda_1^2 \le \left(\rho_1(w_i) + \sum_{k:k\sim i} \rho_1(w_{ik}) \right) \left(\rho_1(w_j) + \sum_{k:k\sim j} \rho_1(w_{jk}) \right), \quad (70)$$

that is,

$$\lambda_1 \le \max_{i\sim j} \left\{ \sqrt{\left(\rho_1(w_i) + \sum_{k:k\sim i} \rho_1(w_{ik}) \right) \left(\rho_1(w_j) + \sum_{k:k\sim j} \rho_1(w_{jk}) \right)} \right\}. \quad (71)$$

This completes the proof of (60).

Now we show the case of equality in (60). By similar method in Theorem 12. In the part of equalit, the necessary condition can show easily. So we will show the sufficient condition.

Suppose that conditions (i)-(ii) of Theorem hold for the graph G. We must prove that

$$\lambda_1 = \max_{i\sim j} \left\{ \sqrt{\left(\rho_1(w_i) + \sum_{k:k\sim i} \rho_1(w_{ik}) \right) \left(\rho_1(w_j) + \sum_{k:k\sim j} \rho_1(w_{jk}) \right)} \right\}. \quad (72)$$

Let G be regular bipartite graph. Therefore we have $\rho_1(w_i) = \alpha$ for $i \in U$ and $\rho_1(w_j) = \alpha$ for $j \in W$ such that $V = U \cup W$. Let $\overline{x}$ be a common eigenvector of w_{ik} corresponding to the largest eigenvalue $\rho_1(w_{ik})$ for all i, k. Hence we have

$$\rho_1(w_i) = \sum_{k:k\sim i} \rho_1(w_{ik}). \quad (73)$$

From (71) we get that

$$\lambda_1 \le 2\alpha. \quad (74)$$

On the other hand, the following equation can be easily verified:

$$(2\alpha)\begin{pmatrix}\overline{x}\\ \overline{x}\\ \vdots\\ \overline{x}\\ -\overline{x}\\ -\overline{x}\\ \vdots\\ -\overline{x}\end{pmatrix} = \begin{pmatrix} w_1 & \cdot & 0 & -w_{1,k+1} & \cdot & -w_{1,n}\\ 0 & \cdot & 0 & -w_{2,k+1} & \cdot & -w_{2,n}\\ \cdots & \cdot & \cdots & \cdots & \cdot & \cdots\\ 0 & \cdot & w_k & -w_{k,k+1} & \cdot & -w_{k,n}\\ -w_{k+1,1} & \cdot & -w_{k+1,k} & w_{k+1} & \cdot & 0\\ -w_{k+1,2} & \cdot & -w_{k+2,k} & 0 & \cdot & 0\\ \cdots & \cdot & \cdots & \cdots & \cdot & \cdots\\ -w_{n,1} & \cdot & -w_{n,k} & 0 & \cdot & w_n \end{pmatrix} \times \begin{pmatrix}\overline{x}\\ \overline{x}\\ \vdots\\ \overline{x}\\ -\overline{x}\\ -\overline{x}\\ \vdots\\ -\overline{x}\end{pmatrix}. \tag{75}$$

Thus 2α is an eigenvalue of $L(G)$. Since λ_1 is the largest eigenvalue of $L(G)$, we get

$$2\alpha \leq \lambda_1. \tag{76}$$

So from (74) and (76) we obtain

$$\lambda_1 = \max_{i\sim j}\left\{\sqrt{\left(\rho_1(w_i) + \sum_{k:k\sim i}\rho_1(w_{ik})\right)\left(\rho_1(w_j) + \sum_{k:k\sim j}\rho_1(w_{jk})\right)}\right\}. \tag{77}$$

□

Corollary 16. *Let G be a simple connected weighted graph where each edge weight $w_{i,j}$ is a positive number. Then*

$$\lambda_1 \leq \max_{i\sim j}\left\{2\sqrt{w_i w_j}\right\}. \tag{78}$$

Moreover equality holds in (78) if and only if G is bipartite semiregular graph.

Proof. We have $\rho_1(w_i) = w_i$ and $\rho_1(w_{ij}) = w_{ij}$ for all i, j. From Theorem 15 we get the required result. □

Corollary 17. *Let G be a simple connected unweighted graph. Then*

$$\lambda_1 \leq \max_{i\sim j}\left\{2\sqrt{d_i d_j}\right\}, \tag{79}$$

where d_i is the degree of vertex i. Moreover equality holds in (79) if and only if G is a bipartite regular graph or G is a bipartite semiregular graph.

Proof. For unweighted graph, $w_{i,j} = 1$ for $i \sim j$. Therefore $w_i = d_i$. Using Corollary 16, we get the required results. □

Example 18. Let $G_1 = (V_1, E_1)$ and $G_2 = (V_2, E_2)$ be a weighted graph where $V_1 = \{1,2,3,4\}$, $E_1 = \{\{1,3\},\{2,4\},\{3,4\}\}$ and each weight is the positive definite matrix of order three. Let $V_2 = \{1,2,3,4,5,6,7\}$, $E_2 = \left\{\begin{smallmatrix}\{1,4\},\{2,4\},\{3,4\},\\ \{4,5\},\{5,6\},\{5,7\}\end{smallmatrix}\right\}$ such that each weight is the positive definite matrix of order two. Assume that the following Laplacian matrices of G_1 and G_2 are as follows:

$$L(G_1) = \begin{bmatrix} 1 & 0 & 0 & 0 & 0 & 0 & -1 & 0 & 0 & 0 & 0 & 0\\ 0 & 5 & 3 & 0 & 0 & 0 & 0 & -5 & -3 & 0 & 0 & 0\\ 0 & 3 & 3 & 0 & 0 & 0 & 0 & -3 & -3 & 0 & 0 & 0\\ 0 & 0 & 0 & 2 & -1 & 0 & 0 & 0 & 0 & -2 & 1 & 0\\ 0 & 0 & 0 & -1 & 2 & -1 & 0 & 0 & 0 & 1 & -2 & 1\\ 0 & 0 & 0 & 0 & -1 & 2 & 0 & 0 & 0 & 0 & 1 & -2\\ -1 & 0 & 0 & 0 & 0 & 0 & 7 & 2 & -2 & 6 & 2 & -2\\ 0 & -5 & -3 & 0 & 0 & 0 & 2 & 11 & 1 & 2 & 6 & -2\\ 0 & -3 & -3 & 0 & 0 & 0 & -2 & 1 & 13 & -2 & -2 & 10\\ 0 & 0 & 0 & 2 & -1 & 0 & 6 & 2 & -2 & 8 & 1 & -2\\ 0 & 0 & 0 & -1 & 2 & -1 & 2 & 6 & -2 & -3 & 8 & -3\\ 0 & 0 & 0 & 0 & -1 & 2 & -2 & -2 & 10 & -2 & -2 & 12 \end{bmatrix},$$

$$L(G_2) = \begin{bmatrix} 1 & 1 & 0 & 0 & 0 & 0 & -1 & -1 & 0 & 0 & 0 & 0 & 0 & 0 \\ 1 & 2 & 0 & 0 & 0 & 0 & -1 & -2 & 0 & 0 & 0 & 0 & 0 & 0 \\ 0 & 0 & 1 & 1 & 0 & 0 & -1 & -1 & 0 & 0 & 0 & 0 & 0 & 0 \\ 0 & 0 & 1 & 3 & 0 & 0 & -1 & -3 & 0 & 0 & 0 & 0 & 0 & 0 \\ 0 & 0 & 0 & 0 & 1 & 1 & -1 & -1 & 0 & 0 & 0 & 0 & 0 & 0 \\ 0 & 0 & 0 & 0 & 1 & 4 & -1 & -4 & 0 & 0 & 0 & 0 & 0 & 0 \\ -1 & -1 & -1 & -1 & -1 & -1 & 4 & 4 & -1 & -1 & 0 & 0 & 0 & 0 \\ -1 & -2 & -1 & -3 & -1 & -4 & 4 & 14 & -1 & -5 & 0 & 0 & 0 & 0 \\ 0 & 0 & 0 & 0 & 0 & 0 & -1 & -1 & 3 & 3 & -1 & -1 & -1 & -1 \\ 0 & 0 & 0 & 0 & 0 & 0 & -1 & -5 & 3 & 18 & -1 & -6 & -1 & 7 \\ 0 & 0 & 0 & 0 & 0 & 0 & 0 & 0 & -1 & -1 & 1 & 1 & 0 & 0 \\ 0 & 0 & 0 & 0 & 0 & 0 & 0 & 0 & -1 & -6 & 1 & 6 & 0 & 0 \\ 0 & 0 & 0 & 0 & 0 & 0 & 0 & 0 & -1 & -1 & 0 & 0 & 1 & 1 \\ 0 & 0 & 0 & 0 & 0 & 0 & 0 & 0 & -1 & -7 & 0 & 0 & 1 & 7 \end{bmatrix}. \tag{80}$$

The largest eigenvalues of $L(G_1)$ and $L(G_2)$ are $\lambda_1 = 25,66$, $\lambda_2 = 26.16$ rounded two decimal places and the previously mentioned bounds give the following results:

	(6)	(7)	(10)	(12)	(14)	(60)
G_1	32.90	32.88	27.90	29.55	30.88	30.93
G_2	34.12	29.86	27.11	27.22	34.05	33.90

(81)

For G_1, we see that the upper bounds in (14) and (60) are better than upper bounds in (6) and (7). But they are not better than upper bounds in (10) and (12) from (81).

For also G_2, we see that upper bounds in (14) and (60) are only better than the upper bound in (6).

Consequently, we cannot exactly compare all the bounds for weighted graphs, where the weights are positive definite matrices. Modifications according to each weight of edges, especially for matrices can be shown.

References

[1] K. Ch. Das and R. B. Bapat, "A sharp upper bound on the largest Laplacian eigenvalue of weighted graphs," *Linear Algebra and Its Applications*, vol. 409, pp. 153–165, 2005.

[2] K. Ch. Das, "Extremal graph characterization from the upper bound of the Laplacian spectral radius of weighted graphs," *Linear Algebra and Its Applications*, vol. 427, no. 1, pp. 55–69, 2007.

[3] S. Sorgun and Ş. Büyükköse, "On the bounds for the largest Laplacian eigenvalues of weighted graphs," *Discrete Optimization*, vol. 9, no. 2, pp. 122–129, 2012.

[4] R. A. Horn and C. R. Johnson, *Matrix Analysis*, Cambridge University Press, Cambridge, UK, 1985.

[5] W. N. Anderson Jr. and T. D. Morley, "Eigenvalues of the Laplacian of a graph," *Linear and Multilinear Algebra*, vol. 18, no. 2, pp. 141–145, 1985.

On Bondage Numbers of Graphs: A Survey with Some Comments

Jun-Ming Xu

School of Mathematical Sciences, University of Science and Technology of China, Wentsun Wu Key Laboratory of CAS, Hefei, Anhui 230026, China

Correspondence should be addressed to Jun-Ming Xu; xujm@ustc.edu.cn

Academic Editor: Chính T. Hoang

The domination number of a graph G is the smallest number of vertices which dominate all remaining vertices by edges of G. The bondage number of a nonempty graph G is the smallest number of edges whose removal from G results in a graph with domination number greater than the domination number of G. The concept of the bondage number was formally introduced by Fink et al. in 1990. Since then, this topic has received considerable research attention and made some progress, variations, and generalizations. This paper gives a survey on the bondage number, including known results, conjectures, problems, and some comments, also selectively summarizes other types of bondage numbers.

1. Introduction

For terminology and notation on graph theory not given here, the reader is referred to Xu [1]. Let $G = (V, E)$ be a finite, undirected, and simple graph. We call $|V|$ and $|E|$ the order and size of G and denote them by $\upsilon = \upsilon(G)$ and $\varepsilon = \varepsilon(G)$, respectively, unless otherwise specified. Through this paper, the notations P_n, C_n, and K_n always denote a path, a cycle, and a complete graph of order n, respectively, the notation $K_{n_1,n_2,\ldots,n_t}$ denotes a complete t-partite graph with $n_1 \leqslant n_2 \leqslant \cdots \leqslant n_t$ and $n_t > 1$, $K_t(m) = K_{n_1,n_2,\ldots,n_t}$ with $n_1 = \cdots = n_t = m$, and $K_{1,n-1}$ is a star. For two vertices x and y in a connected graph G, we use $d_G(x, y)$ to denote the distance between x and y in G.

For a vertex x in G, let $N_G(x)$ be the *open set of neighbors* of x and $N_G[x] = N[x] = N_G(x) \cup \{x\}$ the *closed set of neighbors* of x. For a subset $X \subset V(G)$, $N_G(X) = (\cup_{x \in X} N_G(x)) \cap \overline{X}$, and $N_G[X] = N_G(X) \cup X$, where $\overline{X} = V(G) \setminus X$. Let E_x be the set of edges incident with x in G; that is, $E_x = \{xy \in E(G) : y \in N_G(x)\}$. We denote the degree of x by $d_G(x) = |E_x|$. The maximum and the minimum degrees of G are denoted by $\Delta(G)$ and $\delta(G)$, respectively. A vertex of degree zero is called an *isolated vertex*. An edge incident with a vertex of degree one is called a *pendant edge*.

The bondage number is an important parameter of graphs which is based upon the well-known domination number.

A subset $S \subseteq V(G)$ is called a *dominating set* of G if $N_G[S] = V(G)$; that is, every vertex x in $\overline{S}$ has at least one neighbor in S. The *domination number* of G, denoted by $\gamma(G)$, is the minimum cardinality among all dominating sets; that is,

$$\gamma(G) = \min\{|S| : S \subseteq V(G), N_G[S] = V(G)\}. \qquad (1)$$

A dominating set S is called a γ-set of G if $|S| = \gamma(G)$.

The domination is such an important and classic conception that it has become one of the most widely studied topics in graph theory and also is frequently used to study property of networks. The domination, with many variations and generalizations, is now well studied in graph and networks theory. The early vast literature on domination includes the bibliography compiled by Hedetniemi and Laskar [2] and a thorough study of domination appears in the books by Haynes et al. [3, 4]. However, the problem determining the domination number for general graphs was early proved to be NP-complete (see GT2 in Appendix in Garey and Johnson [5], 1979).

Among various problems related with the domination number, some focus on graph alterations and their effects on the domination number. Here, we are concerned with a particular graph alternation, the removal of edges from a graph.

Graphs with domination numbers changed upon the removal of an edge were first investigated by Walikar and Acharya [6] in 1979. A graph is called *edge-domination-critical graph* if $\gamma(G-e) > \gamma(G)$ for every edge e in G. The edge-domination-critical graph was characterized by Bauer et al. [7] in 1983; that is, a graph is edge-domination-critical if and only if it is the union of stars.

However, for lots of graphs, the domination number is out of the range of one-edge removal. It is immediate that $\gamma(H) \geqslant \gamma(G)$ for any spanning subgraph H of G. Every graph G has a spanning forest T with $\gamma(G) = \gamma(T)$, and so, in general, a graph has a nonempty subset $F \subseteq E(G)$ for which $\gamma(G-F) = \gamma(G)$.

Then it is natural for the alternation to be generalized to the removal of several edges, which is just enough to enlarge the domination number. That is the idea of the bondage number.

A measure of the efficiency of a domination in graphs was first given by Bauer et al. [7] in 1983, who called this measure as domination *line-stability*, defined as the minimum number of lines (i.e., edges) which when removed from G increases γ.

In 1990, Fink et al. [8] formally introduced the bondage number as a parameter for measuring the vulnerability of the interconnection network under link failure. The minimum dominating set of sites plays an important role in the network for it dominates the whole network with the minimum cost. So we must consider whether its function remains good when the network is attacked. Suppose that someone such as a saboteur does not know which sites in the network take part in the dominating role, but does know that the set of these special sites corresponds to a minimum dominating set in the related graph. Then how many links does he has to attack so that the cost cannot remains the same in order to dominate the whole network? That minimum number of links is just the bondage number.

The *bondage number* $b(G)$ of a nonempty undirected graph G is the minimum number of edges whose removal from G results in a graph with larger domination number. The precise definition of the bondage number is as follows:

$$b(G) = \min\{|B| : B \subseteq E(G), \gamma(G-B) > \gamma(G)\}. \quad (2)$$

Since the domination number of every spanning subgraph of a nonempty graph G is at least as great as $\gamma(G)$, the bondage number of a nonempty graph is well defined.

We call such an edge-set B with $\gamma(G-B) > \gamma(G)$ the *bondage set* and the minimum one the *minimum bondage set*. In fact, if B is a minimum bondage set, then $\gamma(G-B) = \gamma(G)+1$ because the removal of one single edge cannot increase the domination number by more than one. If $b(G)$ does not exist, for example, empty graphs, we define $b(G) = \infty$.

It is quite difficult to compute the exact value of the bondage number for general graphs since it strongly depends on the domination number of the graphs. Much work focused on the bounds on the bondage number as well as the restraints on some particular classes of graphs or networks. The purpose of this paper is to give a survey of results and research methods related to these topics for graphs and digraphs. For some results and research methods, we will make some comments to develop our further study.

The rest of the paper is organized as follows. Section 2 gives some preliminary results and complexity. Sections 3 and 4 survey some upper bounds and lower bounds, respectively. The results for some special classes of graphs and planar graphs are stated in Sections 5 and 6, respectively. In Section 7, we introduce some results on crossing number restraints. In Sections 8 and 9, we are concerned about other and generalized types of bondage numbers, respectively. In Section 10, we introduce some results for digraphs. In the last section, we introduce some results for vertex-transitive graphs by applying efficient dominating sets.

2. Simplicity and Complexity

As we have known from Introduction, the bondage number is an important parameter for measuring the stability or the vulnerability of a domination in a graph or a network. Our aim is to compute the bondage number for any given graphs or networks. One has determined the exact value of the bondage number for some graphs with simple structure. For arbitrarily given graph, however, it has been proved that determining its bondage number is NP-hard.

2.1. Exact Values for Ordinary Graphs. We begin our investigation of the bondage number by computing its value for several well-known classes of graphs with simple structure. In 1990, Fink et al. [8] proposed the concept of the bondage number and completely determined the exact values of bondage numbers of some ordinary graphs, such as complete graphs, paths, cycles, and complete multipartite graphs.

Theorem 1 (Fink et al. [8], 1990). *The exact values of bondage numbers of the following class of graphs are completely determined:*

$$b(K_n) = \left\lceil \frac{n}{2} \right\rceil;$$

$$b(P_n) = \begin{cases} 2 & \text{if } n \equiv 1 \pmod 3, \\ 1 & \text{otherwise}, \end{cases}$$

$$b(C_n) = \begin{cases} 3 & \text{if } n \equiv 1 \pmod 3, \\ 2 & \text{otherwise}, \end{cases}$$

$$b(K_{n_1,n_2,\ldots,n_t}) = \begin{cases} \left\lceil \frac{j}{2} \right\rceil & \text{if } n_j = 1 \text{ and } n_{j+1} \geqslant 2, \text{ for some } j, \ 1 \leqslant j < t, \\ 2t-1 & \text{if } n_1 = n_2 = \cdots = n_t = 2, \\ \sum_{i=1}^{t-1} n_i & \text{otherwise}. \end{cases} \quad (3)$$

The complete graph K_n is the unique $(n-1)$-regular graph of order $n \geqslant 2$, $b(K_n) = \lceil n/2 \rceil$ by Theorem 1. The t-partite graph $K_t(2)$ with $t = n/2$ is an $(n-2)$-regular graph G of order $n \geqslant 2$, and not unique, $b(G) = n - 1$ by Theorem 1 for an even integer $n \geqslant 4$. For an $(n - 3)$-regular graph G of order $n \geqslant 4$, Hu and Xu [9] obtained the following result.

Theorem 2 (Hu and Xu [9]). $b(G) = n - 3$ *for any* $(n-3)$-*regular graph* G *of order* $n \geqslant 4$.

Up to the present, no results have been known for k-regular graphs with $3 \leqslant k \leqslant n - 4$. For general graphs, there are the following two results.

Theorem 3 (Teschner [10], 1997). *If* G *is a nonempty graph with a unique minimum dominating set, then* $b(G) = 1$.

Theorem 4 (Bauer et al. [7], 1983). *If any vertex of a graph* G *is adjacent with two or more vertices of degree one, then* $b(G) = 1$.

Bauer et al. [7] observed that the star is the unique graph with the property that the bondage number is 1 and the deletion of any edge results in the domination number increasing. Motivated by this fact, Hartnell and Rall [11] proposed the concept of uniformly bonded graphs. A graph is called to be *uniformly bonded* if it has bondage number b and the deletion of any b edges results in a graph with increased domination number. Unfortunately, there are a few uniformly bonded graphs.

Theorem 5 (Hartnell and Rall [11], 1999). *The only uniformly bonded graphs with bondage number 2 are* C_3 *and* P_4. *The unique uniformly bonded graph with bondage number 3 is* C_4. *There are no such graphs for bondage number greater than 3.*

As we mentioned, to compute the exact value of bondage number for a graph strongly depends upon its domination number. In this sense, studying the bondage number can greatly inspire one's research concerned with dominations. However, determining the exact value of domination number for a given graph is quite difficult. In fact, even if the exact value of the domination number for some graph is determined, it is still very difficulty to compute the value of the bondage number for that graph. For example, for the hypercube Q_n, we have $\gamma(Q_n) = 2^{n-1}$, but we have not yet determined $b(Q_n)$ for any $n \geqslant 2$.

Perhaps Theorems 3 and 4 provide an approach to compute the exact value of bondage number for some graphs by establishing some sufficient conditions for $b(G) = b$. In fact, we will see later that Theorem 3 plays an important role in determining the exact values of the bondage numbers for some graphs. Thus, to study the bondage number, it is important to present various characterizations of graphs with a unique minimum dominating set.

2.2. Characterizations of Trees. For trees, Bauer et al. [7] in 1983 from the point of view of the domination line-stability, independently and Fink et al. [8] in 1990 from the point of view of the domination edge-vulnerability obtained the following result.

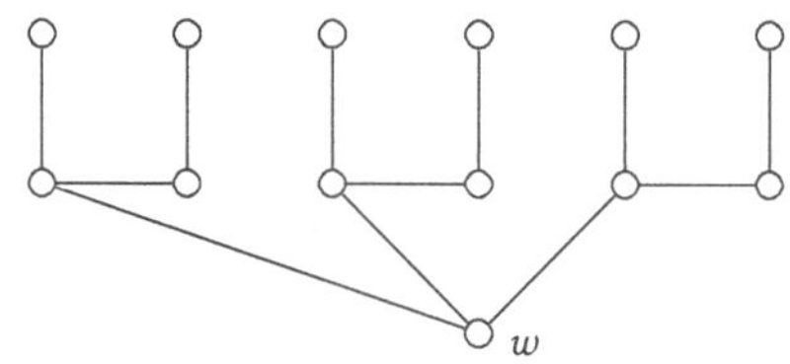

FIGURE 1: Tree F_t.

Theorem 6. *For any nontrivial tree* T, $b(T) \leqslant 2$.

By Theorem 6, it is natural to classify all trees according to their bondage numbers. Fink et al. [8] proved that a forbidden subgraph characterization to classify trees with different bondage numbers is impossible, since they proved that if F is a forest, then F is an induced subgraph of a tree T with $b(T) = 1$ and a tree T' with $b(T') = 2$. However, they pointed out that the complexity of calculating the bondage number of a tree of order n is at most $O(n^2)$ by methodically removing each pair of edges.

Even so, some characterizations, whether a tree has bondage number 1 or 2, have been found by several authors; see, for example, [10, 12, 13].

First we describe the method due to Hartnell and Rall [12], by which all trees with bondage number 2 can be constructed inductively. An important tree F_t in the construction is shown in Figure 1. To characterize this construction, we need some terminologies:

(1) attach a path P_n to a vertex x of a tree, which means to link x and one end-vertex of the P_n by an edge.

(2) attach F_t to a vertex x, which means to link x and a vertex y of F_t by an edge.

The following are four operations on a tree T:

Type 1: attach a P_2 to $x \in V(T)$, where $\gamma(T - x) = \gamma(T)$ and x belongs to at least one γ-set of T (such a vertex exists, say, one end-vertex of P_5).

Type 2: attach a P_3 to $x \in V(T)$, where $\gamma(T - x) < \gamma(T)$.

Type 3: attach F_1 to $x \in V(T)$, where x belongs to at least one γ-set of T.

Type 4: attach F_t, $t \geqslant 2$, to $x \in V(T)$, where x can be any vertex of T.

Let $\mathscr{C} = \{T : T$ is a tree, and $T = K_1$, $T = P_4$, and $T = F_t$ for some $t \geqslant 2$, or T can be obtained from P_4 or F_t $(t \geqslant 2)$ by a finite sequence of operations of Types $1, 2, 3, 4\}$.

Theorem 7 (Hartnell and Rall [12], 1992). *A tree has bondage number 2 if and only if it belongs to* $\mathscr{C}$.

Looking at different minimum dominating sets of a tree, Teschner [10] presented a totally different characterization of the set of trees having bondage number 1. They defined a vertex to be *universal* if it belongs to each minimum dominating set, and to be *idle* if it does not belong to any minimum dominating set.

Theorem 8 (Teschner [10], 1997). *A tree T has bondage number 1 if and only if T has a universal vertex or an edge xy satisfying*

(1) *x and y are neither universal nor idle; and*

(2) *all neighbors of x and y (except for x and y) are idle.*

For a positive integer k, a subset $I \subseteq V(G)$ is called a *k-independent set* (also called a *k-packing*) if $d_G(x, y) > k$ for any two distinct vertices x and y in I. When $k = 1$, 1-set is the normal independent set. The maximum cardinality among all k-independent sets is called the *k-independence number* (or *k-packing number*) of G, denoted by $\alpha_k(G)$. A k-independent set I is called an *α_k-set* if $|I| = \alpha_k(G)$. A graph G is said to be *α_k-stable* if $\alpha_k(G) = \alpha_k(G - e)$ for every edge e of G. There are two important results on k-independent sets.

Proposition 9 (Topp and Vestergaard [13], 2000). *A tree T is α_k-stable if and only if T has a unique α_k-set.*

Proposition 10 (Meir and Moon [14], 1975). *$\alpha_2(G) \leqslant \gamma(G)$ for any connected graph G with equality for any tree.*

Hartnell et al. [15], independently and Topp and Vestergaard [13] also gave a constructive characterization of trees with bondage number 2 and, applying Proposition 10, presented another characterization of those trees.

Theorem 11 (Hartnell et al. [15], 1998; Top and Vestergaard [13], 2000). *$b(T) = 2$ for a tree T if and only if T has a unique α_2-set.*

According to this characterization, Hartnell et al. [15] presented a linear algorithm for determining the bondage number of a tree.

In this subsection, we introduce three characterizations for trees with bondage number 1 or 2. The characterization in Theorem 7 is constructive, constructing all trees with bondage number 2, a natural and straightforward method, by a series of graph-operations. The characterization in Theorem 8 is a little advisable, by describing the inherent property of trees with bondage number 1. The characterization in Theorem 11 is wonderful, by using a strong graph-theoretic concept, α_k-set. In fact, this characterization is a byproduct of some results related to α_k-sets for trees. It is that this characterization closed the relation between two concepts, the bondage number and the k-independent set, and hence is of research value and important significance.

2.3. Complexity for General Graphs. As mentioned above, the bondage number of a tree can be determined by a linear time algorithm. According to this algorithm, we can determine within polynomial time the domination number of any tree by removing each edge and verifying whether the domination number is enlarged according to the known linear time algorithm for domination numbers of trees.

However, it is impossible to find a polynomial time algorithm for bondage numbers of general graphs. If such an algorithm A exists, then the domination number of any nonempty undirected graph G can be determined within polynomial time by repeatedly using A. Let $G_0 = G$ and $G_{i+1} = G_i - B_i$, where B_i is the minimum edge set of G_i found by A such that $\gamma(G_i - B_i) = \gamma(G_{i-1}) + 1$ for each $i = 0, 1, \ldots$; we can always find the minimum B_i whose removal from G_i enlarges the domination number, until $G_k = G_{k-1} - B_{k-1}$ is empty for some $k \geqslant 1$, though B_{k-1} is not empty. Then $\gamma(G) = \gamma(G_k) - k = \upsilon(G) - k$. As known to all, if NP $\neq P$, the minimum dominating set problem is NP-complete, and so polynomial time algorithms for the bondage number do not exist unless NP $= P$.

In fact, Hu and Xu [16] have recently shown that the problem determining the bondage number of general graphs is NP-hard.

Problem 1. Consider the decision problem:

Bondage Problem

Instance: a graph G and a positive integer b $(\leqslant \varepsilon(G))$.

Question: is $b(G) \leqslant b$?

Theorem 12 (Hu and Xu [16], 2012). *The bondage problem is NP-hard.*

The basic way of the proof is to follow Garey and Johnson's techniques for proving NP-hardness [5] by describing a polynomial transformation from the known NP-complete problem: 3-satisfiability problem.

Theorem 12 shows that we are unable to find a polynomial time algorithm to determine bondage numbers of general graphs unless NP $= P$. At the same time, this result also shows that the following study on the bondage number is of important significance.

(i) Find approximation polynomial algorithms with performance ratio as small as possible.

(ii) Find the lower and upper bounds with difference as small as possible.

(iii) Determine exact values for some graphs, specially well-known networks.

Unfortunately, we cannot prove whether or not determining the bondage number is NP-problem since for any subset $B \subset E(G)$ it is not clear that there is a polynomial algorithm to verify $\gamma(G - B) > \gamma(G)$. Since the problem of determining the domination number is NP-complete, we conjecture that it is not in NP. This is a worthwhile task to be studied further.

Motivated by the linear time algorithm of Hartnell et al. to compute the bondage number of a tree, we can make an attempt to consider whether there is a polynomial time algorithm to compute the bondage number for some special classes of graphs such as planar graphs, Cayley graphs, or graphs with some restrictions of graph-theoretical parameters such as degree, diameter, connectivity, and domination number.

3. Upper Bounds

By Theorem 12, since we cannot find a polynomial time algorithm for determining the exact values of bondage numbers of general graphs, it is weightily significative to establish some sharp bounds on the bondage number of a graph. In this section, we survey several known upper bounds on the bondage number in terms of some other graph-theoretical parameters.

3.1. Most Basic Upper Bounds. We start this subsection with a simple observation.

Observation 1 (Teschner [10], 1997). Let H be a spanning subgraph obtained from a graph G by removing k edges. Then $b(G) \leqslant b(H) + k$.

If we select a spanning subgraph H such that $b(H) = 1$, then Observation 1 yields some upper bounds on the bondage number of a graph G. For example, take $H = G - E_x + e$ and $k = d_G(x) - 1$, where $e \in E_x$, the following result can be obtained.

Theorem 13 (Bauer et al. [7], 1983). *If there exists at least one vertex x in a graph G such that $\gamma(G - x) \geqslant \gamma(G)$, then $b(G) \leqslant d_G(x) \leqslant \Delta(G)$.*

The following early result obtained can be derived from Observation 1 by taking $H = G - E_{xy}$ and $k = d(x) + d(y) - 2$.

Theorem 14 (Bauer et al. [7], 1983; Fink et al. [8], 1990). $b(G) \leqslant d(x) + d(y) - 1$ *for any two adjacent vertices x and y in a graph G; that is,*

$$b(G) \leqslant \min_{xy \in E(G)} \{d_G(x) + d_G(y) - 1\}. \tag{4}$$

This theorem gives a natural corollary obtained by several authors.

Corollary 15 (Bauer et al. [7], 1983; Fink et al. [8], 1990). *If G is a graph without isolated vertices, then $b(G) \leqslant \Delta(G) + \delta(G) - 1$.*

In 1999, Hartnell and Rall [11] extended Theorem 14 to the following more general case, which can be also derived from Observation 1 by taking $H = G - E_x - E_y + (x, z, y)$ if $d_G(x, y) = 2$, where (x, z, y) is a path of length 2 in G.

Theorem 16 (Hartnell and Rall [11], 1999). $b(G) \leqslant d(x) + d(y) - 1$ *for any distinct two vertices x and y in a graph G with $d_G(x, y) \leqslant 2$; that is,*

$$b(G) \leqslant \min_{d_G(x,y) \leqslant 2} \{d_G(x) + d_G(y) - 1\}. \tag{5}$$

Corollary 17 (Fink et al. [8], 1990). *If a vertex of a graph G is adjacent with two or more vertices of degree one, then $b(G) = 1$.*

We remark that the bounds stated in Corollary 15 and Theorem 16 are sharp. As indicated by Theorem 1, one class of graphs in which the bondage number achieves these bounds is the class of cycles whose orders are congruent to 1 modulo 3.

On the other hand, Hartnell and Rall [17] sharpened the upper bound in Theorem 14 as follows, which can be also derived from Observation 1.

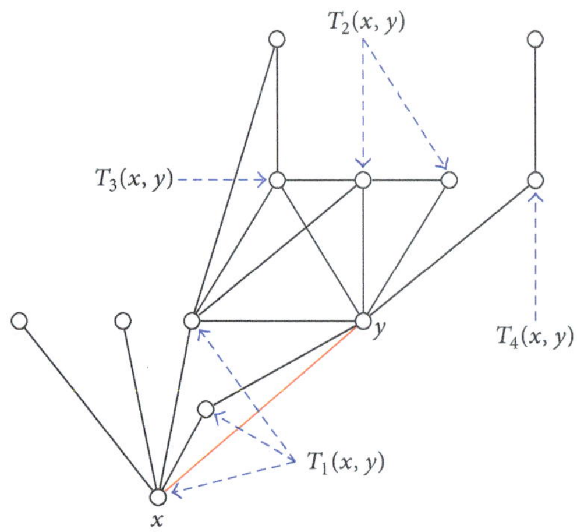

Figure 2: Illustration of $T_i(x, y)$ for $i = 1, 2, 3, 4$.

Theorem 18 (Hartnell and Rall [17], 1994). $b(G) \leqslant d_G(x) + d_G(y) - 1 - |N_G(x) \cap N_G(y)|$ *for any two adjacent vertices x and y in a graph G; that is,*

$$b(G) \leqslant \min_{xy \in E(G)} \{d_G(x) + d_G(y) - 1 - |N_G(x) \cap N_G(y)|\}. \tag{6}$$

These results give simple but important upper bounds on the bondage number of a graph, and is also the foundation of almost all results on bondage numbers upper bounds obtained till now.

By careful consideration of the nature of the edges from the neighbors of x and y, Wang [18] further refined the bound in Theorem 18. For any edge $xy \in E(G)$, $N_G(y)$ contains the following four subsets:

(1) $T_1(x, y) = N_G[x] \cap N_G(y)$;

(2) $T_2(x, y) = \{w \in N_G(y) : N_G(w) \subseteq N_G(y) - x\}$;

(3) $T_3(x, y) = \{w \in N_G(y) : N_G(w) \subseteq N_G(z) - x$ for some $z \in N_G(x) \cap N_G(y)\}$;

(4) $T_4(x, y) = N_G(y) \setminus (T_1(x, y) \cup T_2(x, y) \cup T_3(x, y))$.

The illustrations of $T_1(x, y)$, $T_2(x, y)$, $T_3(x, y)$, and $T_4(x, y)$ are shown in Figure 2 (corresponding vertices pointed by dashed arrows).

Theorem 19 (Wang [18], 1996). *For any nonempty graph G,*

$$b(G) \leqslant \min_{xy \in E(G)} \{d_G(x) + |T_4(x, y)|\}. \tag{7}$$

The graph in Figure 2 shows that the upper bound given in Theorem 19 is better than that in Theorems 16 and 18, for the upper bounds obtained from these two theorems are $d_G(x) + d_G(y) - 1 = 11$ and $d_G(x) + d_G(y) - |N_G(x) \cap N_G(y)| = 9$, respectively, while the upper bound given by Theorem 19 is $d_G(x) + |T_4(x, y)| = 6$.

The following result is also an improvement of Theorem 14, in which $t = 2$.

Theorem 20 (Teschner [10], 1997). *If G contains a complete subgraph K_t with $t \geqslant 2$, then $b(G) \leqslant \min_{xy\in E(K_t)}\{d_G(x)+d_G(y)-t+1\}$.*

Following Fricke et al. [19], a vertex x of a graph G is *γ-good* if x belongs to some γ-set of G and *γ-bad* if x belongs to no γ-set of G. Let $A(G)$ be the set of γ-good vertices, and let $B(G)$ be the set of γ-bad vertices in G. Clearly, $\{A(G), B(G)\}$ is a partition of $V(G)$. Note there exists some $x \in A$ such that $\gamma(G-x)=\gamma(G)$, say, one end-vertex of P_5. Samodivkin [20] presented some sharp upper bounds for $b(G)$ in terms of γ-good and γ-bad vertices of G.

Theorem 21 (Samodivkin [20], 2008). *Let G be a graph.*

(a) *Let $C(G)=\{x\in V(G):\gamma(G-x)\geqslant\gamma(G)\}$. If $C(G)\neq\emptyset$, then*

$$b(G)\leqslant \min\{d_G(x)+\gamma(G)-\gamma(G-x):x\in C(G)\}. \quad (8)$$

(b) *If $B\neq\emptyset$, then*

$$b(G)\leqslant \min\{|N_G(x)\cap A|:x\in B(G)\}. \quad (9)$$

Theorem 22 (Samodivkin [20], 2008). *Let G be a graph. If $A^+(G)=\{x\in A(G):d_G(x)\geqslant 1$ and $\gamma(G-x)<\gamma(G)\}\neq\emptyset$, then*

$$b(G)\leqslant \min_{x\in A^+(G),y\in B(G-x)}\{d_G(x)+|N_G(y)\cap A(G-x)|\}. \quad (10)$$

Proposition 23 (Samodivkin [20], 2008). *Under the notation of Theorem 19, if $x\in A^+(G)$, then $(T_1(x,y)-\{x\})\cup T_2(x,y)\cup T_3(x,y)\subseteq N_G(y)\setminus B(G-x)$.*

By Proposition 23, if $x\in A^+(G)$, then

$$\begin{aligned} d_G(x)&+\min_{y\in N_G(x)}\{|T_4(x,y)|\}\\ &\geqslant d_G(x)+\min_{y\in N_G(x)}\{|N_G(y)\cap A(G-x)|\}\\ &\geqslant d_G(x)+\min_{y\in B(G-x)}\{|N_G(y)\cap A(G-x)|\}. \end{aligned} \quad (11)$$

Hence, Theorem 19 can be seen to follow from Theorem 22. Any graph G with $b(G)$ achieving the upper bound in Theorem 19 can be used to show that the bound in Theorem 22 is sharp.

Let $t \geqslant 2$ be an integer. Samodivkin [20] constructed a very interesting graph G_t to show that the upper bound in Theorem 22 is better than the known bounds. Let $H_0,H_1,H_2,\ldots,H_{t+1}$ be mutually vertex-disjoint graphs such that $H_0\cong K_2$, $H_{t+1}\cong K_{t+3}$ and $H_i\cong K_{t+3}-e$ for each $i=1,2,\ldots,t$. Let $V(H_0)=\{x,y\}$, $x_{t+1}\in V(H_{t+1})$ and

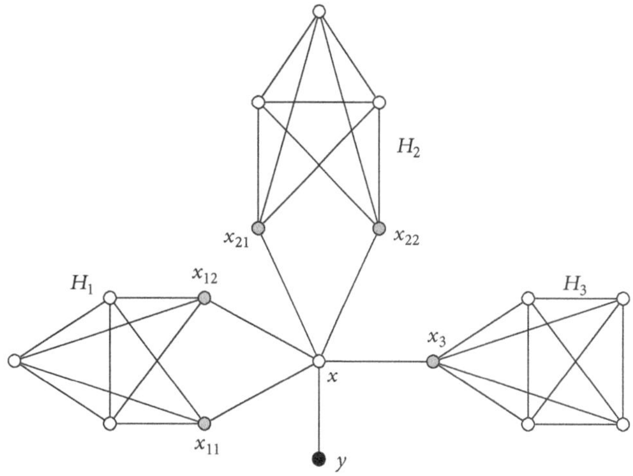

FIGURE 3: The graph G_2.

$x_{i1},x_{i2}\in V(H_i)$, $x_{i1}x_{i2}\notin E(H_i)$ for each $i=1,2,\ldots,t$. The graph G_t is defined as follows:

$$\begin{aligned} V(G_t)&=\bigcup_{k=0}^{t+1}V(H_k),\\ E(G_t)&=\left(\bigcup_{k=0}^{t+1}E(H_k)\right)\cup\left(\bigcup_{i=1}^{t}\{xx_{i1},xx_{i2}\}\right)\cup\{xx_{t+1}\}. \end{aligned} \quad (12)$$

Such a constructed graph G_t is shown in Figure 3 when $t=2$.

Observe that $\gamma(G_t)=t+2$, $A(G_t)=V(G_t)$, $d_{G_t}(x)=2t+2$, $d_{G_t}(x_{t+1})=t+3$, $d_{G_t}(y)=1$, and $d_{G_t}(z)=t+2$ for each $z\in V(G_t-\{x,y,x_{t+1}\})$. Moreover, $\gamma(G-y)<\gamma(G)$ and $\gamma(G_t-z)=\gamma(G_t)$ for any $z\in V(G_t)-\{y\}$. Hence, each of the bounds stated in Theorems 13–20 is greater than or equals $t+2$.

Consider the graph G_t-xy. Clearly, $\gamma(G_t-xy)=\gamma(G_t)$ and

$$\begin{aligned} B(G_t-xy)&=B(G_t-y)\\ &=\{x\}\cup V(H_{t+1}-x_{t+1})\cup\left(\bigcup_{k=1}^{t}\{x_{k_1},x_{k_2}\}\right). \end{aligned} \quad (13)$$

Therefore, $N_{G_t}(x)\cap G(G_t-y)=\{x_{t+1}\}$ which implies that the upper bound stated in Theorem 22 is equal to $d_{G_t}(y)+|\{x_{t+1}\}|=2$. Clearly, $b(G_t)=2$, and hence this bound is sharp for G_t.

From the graph G_t, we obtain the following statement immediately.

Proposition 24. *For every integer $t\geqslant 2$, there is a graph G such that the difference between any upper bound stated in Theorems 13–20 and the upper bound in Theorem 22 is equal to t.*

Although Theorem 22 supplies us with the upper bound that is closer to $b(G)$ for some graph G than what any one of

Theorems 13–20 provides, it is not easy to determine the sets $A^+(G)$ and $B(G)$ mentioned in Theorem 22 for an arbitrary graph G. Thus, the upper bound given in Theorem 22 is of theoretical importance, but not applied since, until now, we have not found a new class of graphs whose bondage numbers are determined by Theorem 22.

The above-mentioned upper bounds on the bondage number are involved in only degrees of two vertices. Hartnell and Rall [11] established an upper bound on $b(G)$ in terms of the numbers of vertices and edges of G with order n. For any connected graph G, let $\overline{\delta}(G)$ represent the average degree of vertices in G; that is, $\overline{\delta}(G) = (1/n)\sum_{x\in V} d_G(x)$. Hartnell and Rall first discovered the following proposition.

Proposition 25. *For any connected graph G, there exist two vertices x and y with distance at most two and with the property that* $d_G(x) + d_G(y) \leqslant 2\overline{\delta}(G)$.

Using Proposition 25 and Theorem 16, Hartnell and Rall gave the following bound.

Theorem 26 (Hartnell and Rall [11], 1999). $b(G) \leqslant 2\overline{\delta}(G) - 1$ *for any connected graph G.*

Note that $2\varepsilon = n\overline{\delta}(G)$ for any graph G with order n and size ε. Theorem 26 implies the following bound in terms of order n and size ε.

Corollary 27. $b(G) \leqslant (4\varepsilon/n) - 1$ *for any connected graph G with order n and size ε.*

A lower bound on ε in terms of order n and the bondage number is obtained from Corollary 28 immediately.

Corollary 28. $\varepsilon \geqslant (n/4)(b(G) + 1)$ *for any connected graph G with order n and size ε.*

Hartnell and Rall [11] gave some graphs G with order n to show that for each value of $b(G)$; the lower bound on $\varepsilon(G)$ given in the Corollary 28 is sharp for some values of n.

3.2. Bounds Implied by Connectivity. Use $\kappa(G)$ and $\lambda(G)$ to denote the vertex-connectivity and the edge-connectivity of a connected graph G, respectively, which are the minimum numbers of vertices and edges whose removal result in G disconnected. The famous Whitney inequality is stated as $\kappa(G) \leqslant \lambda(G) \leqslant \delta(G)$ for any graph or digraph G. Corollary 15 is improved by several authors as follows.

Theorem 29 (Hartnell and Rall [17], 1994; Teschner [10], 1997). *If G is a connected graph, then* $b(G) \leqslant \lambda(G) + \Delta(G) - 1$.

The upper bound given in Theorem 29 can be attained. For example, a cycle C_{3k+1} with $k \geqslant 1$, $b(C_{3k+1}) = 3$ by Theorem 1. Since $\kappa(C_{3k+1}) = \lambda(C_{3k+1}) = 2$, we have $\kappa(C_{3k+1}) + \lambda(C_{3k+1}) - 1 = 2 + 2 - 1 = 3 = b(C_{3k+1})$.

Motivated by Corollary 15, Theorems 29, and the Whitney inequality, Dunbar et al. [21] naturally proposed the following conjecture.

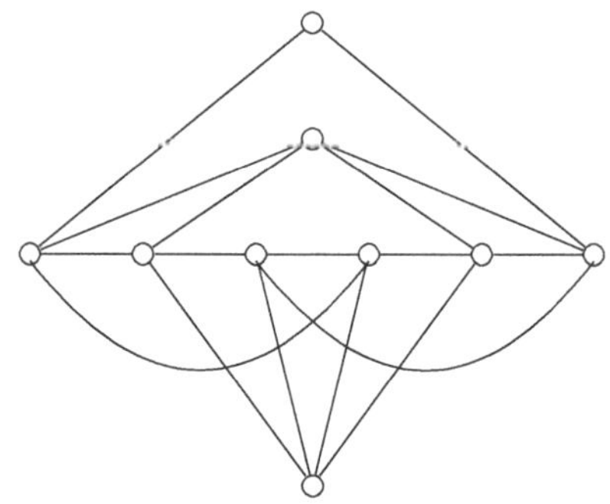

FIGURE 4: A graph H with $\gamma(H) = 3$ and $b(H) = 5$.

Conjecture 30. *If G is a connected graph, then* $b(G) \leqslant \Delta(G) + \kappa(G) - 1$.

However, Liu and Sun [22] presented a counterexample to this conjecture. They first constructed a graph H showed in Figure 4 with $\gamma(H) = 3$ and $b(H) = 5$. Then, let G be the disjoint union of two copies of H by identifying two vertices of degree two. They proved that $b(G) \geqslant 5$. Clearly, G is a 4-regular graph with $\kappa(G) = 1$ and $\lambda(G) = 2$, and so $b(G) \leqslant 5$ by Theorem 29. Thus, $b(G) = 5 > 4 = \Delta(G) + \kappa(G) - 1$.

With a suspicion of the relationship between the bondage number and the vertex-connectivity of a graph, the following conjecture is proposed.

Conjecture 31 (Liu and Sun [22], 2003). *For any positive integer r, there exists a connected graph G such that* $b(G) \geqslant \Delta(G) + \kappa(G) + r$.

To the knowledge of the author, until now no results have been known about this conjecture.

We conclude this subsection with the following remarks. From Theorem 29, if Conjecture 31 holds for some connected graph G, then $\lambda(G) > \kappa(G) + r$, which implies that G is of large edge-connectivity and small vertex-connectivity. Use $\xi(G)$ to denote the minimum edge-degree of G; that is,

$$\xi(G) = \min_{xy\in E(G)} \{d_G(x) + d_G(y) - 2\}. \tag{14}$$

Theorem 14 implies the following result.

Proposition 32. $b(G) \leqslant \xi(G) + 1$ *for any graph G.*

Use $\lambda'(G)$ to denote the restricted edge-connectivity of a connected graph G, which is the minimum number of edges whose removal result in G disconnected and no isolated vertices.

Proposition 33 (Esfahanian and Hakimi [23], 1988). *If G is neither* $K_{1,n}$ *nor* K_3, *then* $\lambda'(G) \leqslant \xi(G)$.

Combining Propositions 32 and 33, we propose a conjecture.

Conjecture 34. *If a connected G is neither* $K_{1,n}$ *nor* K_3, *then* $b(G) \leqslant \delta(G) + \lambda'(G) - 1$.

A cycle C_{3k+1} also satisfies $b(C_{3k+1}) = 3 = \delta(C_{3k+1}) + \lambda'(C_{3k+1}) - 1$ since $\lambda'(C_{3k+1}) = \delta(C_{3k+1}) = 2$ for any integer $k \geqslant 1$.

For the graph H shown in Figure 4, $\lambda'(H) = 4$ and $\delta(H) = 2$, and so $b(H) = 5 = \delta(H) + \lambda'(H) - 1$.

For the 4-regular graph G constructed by Liu and Sun [22] obtained from the disjoint union of two copies of the graph H showed in Figure 4 by identifying two vertices of degree two, we have $b(G) \geqslant 5$. Clearly, $\lambda'(G) = 2$. Thus, $b(G) = 5 = \delta(G) + \lambda'(G) - 1$.

For the 4-regular graph G_t constructed by Samodivkin [20], see Figure 3 for $t = 2$, $b(G_t) = 2$. Clearly, $\delta(G_t) = 1$ and $\lambda'(G) = 2$. Thus, $b(G) = 2 = \delta(G) + \lambda'(G) - 1$.

These examples show that if Conjecture 34 is true then the given upper bound is tight.

3.3. Bounds Implied by Degree Sequence. Now let us return to Theorem 16, from which Teschner [10] obtained some other bounds in terms of the degree sequences. The *degree sequence* $\pi(G)$ of a graph G with vertex-set $V = \{x_1, x_2, \ldots, x_n\}$ is the sequence $\pi = (d_1, d_2, \ldots, d_n)$ with $d_1 \leqslant d_2 \leqslant \cdots \leqslant d_n$, where $d_i = d_G(x_i)$ for each $i = 1, 2, \ldots, n$. The following result is essentially a corollary of Theorem 16.

Theorem 35 (Teschner [10], 1997). *Let G be a nonempty graph with degree sequence $\pi(G)$. If $\alpha_2(G) = t$, then $b(G) \leqslant d_t + d_{t+1} - 1$.*

Combining Theorem 35 with Proposition 10 (i.e., $\alpha_2(G) \leqslant \gamma(G)$), we have the following corollary.

Corollary 36 (Teschner [10], 1997). *Let G be a nonempty graph with the degree sequence $\pi(G)$. If $\gamma(G) = \gamma$, then $b(G) \leqslant d_\gamma + d_{\gamma+1} - 1$.*

In [10], Teschner showed that these two bounds are sharp for arbitrarily many graphs. Let $H = C_{3k+1} + \{x_1x_4, x_1x_{3k-1}\}$, where C_{3k+1} is a cycle $(x_1, x_2, \ldots, x_{3k+1}, x_1)$ for any integer $k \geqslant 2$. Then $\gamma(H) = k + 1$ and so $b(H) \leqslant 2 + 2 - 1 = 3$ by Corollary 36. Since C_{3k+1} is a spanning subgraph of H and $\gamma(G) = \gamma(H)$, Observation 1 yields that $b(H) \geqslant b(C_{3k+1}) = 3$, and so $b(H) = 3$.

Although various upper bounds have been established as the above, we find that the appearance of these bounds is essentially based upon the local structures of a graph, precisely speaking, the structures of the neighborhoods of two vertices within distance 2. Even if these bounds can be achieved by some special graphs, it is more often not the case. The reason lies essentially in the definition of the bondage number, which is the minimum value among all bondage sets, an integral property of a graph. While it easy to find upper bounds just by choosing some bondage set, the gap between the exact value of the bondage number and such a bound obtained only from local structures of a graph is often large. For example, a star $K_{1,\Delta}$, however large Δ is, $b(K_{1,\Delta}) = 1$. Therefore, one has been longing for better bounds upon some integral parameters. However, as what we will see below, it is difficult to establish such upper bounds.

3.4. Bounds in γ-Critical Graphs. A graph G is called a *vertex domination-critical graph* (*vc-graph* or *γ-critical* for short) if $\gamma(G - x) < \gamma(G)$ for any $x \in V(G)$, proposed by Brigham et al. [24] in 1988. Several families of graphs are known to be γ-critical. From definition, it is clear that if G is a γ-critical graph, then $\gamma(G) \geqslant 2$. The class of γ-critical graphs with $\gamma = 2$ is characterized as follows.

Proposition 37 (Brigham et al. [24], 1988). *A graph G with $\gamma(G) = 2$ is a γ-critical graph if and only if G is a complete graph K_{2t} ($t \geqslant 2$) with a perfect matching removed.*

A more interesting family is composed of the n-critical graphs $G_{m,n}$ defined for $m, n \geqslant 2$ by the circulant undirected graph $G(N, \pm S)$, where $N = (m+1)(n-1)+1$ and $S = \{1, 2, \ldots, \lfloor m/2 \rfloor\}$, in which $V(G(N, \pm S)) = N$ and $E(G(N, \pm S)) = \{ij : |j - i| \equiv s \pmod{N}, s \in S\}$.

The reason why γ-critical graphs are of special interest in this context is easy to see that they play an important role in study of the bondage number. For instance, it immediately follows from Theorem 13 that if $b(G) > \Delta(G)$ then G is a γ-critical graph. The γ-critical graphs are defined exactly in this way. In order to find graphs G with a high bondage number (i.e., higher than $\Delta(G)$) and beyond its general upper bounds for the bondage number, we, therefore, have to look at γ-critical graphs.

The bondage numbers of some γ-critical graphs have been examined by several authors; see, for example, [10, 20, 25, 26]. From Theorem 13, we know that the bondage number of a graph G is bounded from above by $\Delta(G)$ if G is not a γ-critical graph. For γ-critical graphs, it is more difficult to find an upper bound. We will see that the bondage numbers of γ-critical graphs in general are not even bounded from above by $\Delta + c$ for any fixed natural number c.

In this subsection, we introduce some upper bounds for the bondage number of a γ-critical graph. By Proposition 37, we easily obtain the following result.

Theorem 38. *If G is a γ-critical graph with $\gamma(G) = 2$, then $b(G) \leqslant \Delta + 1$.*

In Section 4, by Theorem 59, we will see that the equality in Theorem 38 holds; that is, $b(G) = \Delta + 1$ if G is a γ-critical graph with $\gamma(G) = 2$.

Theorem 39 (Teschner [10], 1997). *Let G be a γ-critical graph with degree sequence $\pi(G)$. Then $b(G) \leqslant \max\{\Delta(G) + 1, d_1 + d_2 + \cdots + d_{\gamma-1}\}$, where $\gamma = \gamma(G)$.*

As we mentioned above, if G is a γ-critical graph with $\gamma(G) = 2$, then $b(G) = \Delta + 1$, which shows that the bound given in Theorem 39 can be attained for $\gamma = 2$. However, we have not known whether this bound is tight for general $\gamma \geqslant 3$. Theorem 39 gives the following corollary immediately.

Corollary 40 (Teschner [10], 1997). *Let G be a γ-critical graph with degree sequence $\pi(G)$. If $\gamma(G) = 3$, then $b(G) \leqslant \max\{\Delta(G) + 1, \delta(G) + d_2\}$.*

From Theorem 38 and Corollary 40, we have $b(G) \leqslant 2\Delta(G)$ if G is a γ-critical graph with $\gamma(G) \leqslant 3$. The following result shows that this bound is not tight.

Theorem 41 (Teschner [26], 1995). *Let G be a γ-critical graph graph with $\gamma(G) \leqslant 3$. Then $b(G) \leqslant (3/2)\Delta(G)$.*

Until now, we have not known whether the bound given in Theorem 41 is tight or not. We state two conjectures on γ-critical graphs proposed by Samodivkin [20]. The first of them is motivated by Theorems 22 and 19.

Conjecture 42 (Samodivkin [20], 2008). *For every connected nontrivial γ-critical graph G,*

$$\min_{x \in A^+(G), y \in B(G-x)} \{d_G(x) + |N_G(y) \cap A(G-x)|\} \leqslant \frac{3}{2}\Delta(G). \quad (15)$$

To state the second conjecture, we need the following result on γ-critical graphs.

Proposition 43. *If G is a γ-critical graph, then $\upsilon(G) \leqslant (\Delta(G)+1)(\gamma(G)-1)+1$; moreover, if the equality holds, then G is regular.*

The upper bound in Proposition 43 is sharp in the sense that equality holds for the infinite class of γ-critical graphs $G_{m,n}$ defined in the beginning of this subsection. In Proposition 43, the first result is due to Brigham et al. [24] in 1988; the second is due to Fulman et al. [27] in 1995.

For a γ-critical graph G with $\gamma = 3$, by Proposition 43, $\upsilon(G) \leqslant 2\Delta(G) + 3$.

Theorem 44 (Teschner [26], 1995). *If G is a γ-critical graph with $\gamma = 3$ and $\upsilon(G) = 2\Delta(G) + 3$, then $b(G) \leqslant \Delta + 1$.*

We have not known yet whether the equality in Theorem 44 holds or not. However, Samodivkin proposed the following conjecture.

Conjecture 45 (Samodivkin [20], 2008). *If G is a γ-critical graph with $(\Delta(G)+1)(\gamma-1)+1$ vertices, then $b(G) = \Delta(G)+1$.*

In general, based on Theorem 41, Teschner proposed the following conjecture.

Conjecture 46 (Teschner [26], 1995). *If G is a γ-critical graph, then $b(G) \leqslant (3/2)\Delta(G)$ for $\gamma \geqslant 4$.*

We conclude this subsection with some remarks. Graphs which are minimal or critical with respect to a given property or parameter frequently play an important role in the investigation of that property or parameter. Not only are such graphs of considerable interest in their own right, but also a knowledge of their structure often aids in the development of the general theory. In particular, when investigating any finite structure, a great number of results are proven by induction. Consequently, it is desirable to learn as much as possible about those graphs that are critical with respect to a given property or parameter so as to aid and abet such investigations.

In this subsection, we survey some results on the bondage number for γ-critical graphs. Although these results are not very perfect, they provides a feasible method to approach the bondage number from different viewpoints. In particular, the methods given in Teschner [26] worth further exploration and development.

The following proposition is maybe useful for us to further investigate the bondage number of a γ-critical graph.

Proposition 47 (Brigham et al. [24], 1988). *If G has a nonisolated vertex x such that the subgraph induced by $N_G(x)$ is a complete graph, then G is not γ-critical.*

This simple fact shows that any γ-critical graph contains no vertices of degree one.

3.5. Bounds Implied by Domination. In the preceding subsection, we introduced some upper bounds on the bondage numbers for γ-critical graphs by consideration of dominations. In this subsection, we introduce related results for a general graph with given domination number.

Theorem 48 (Fink et al. [8], 1990). *For any connected graph G of order n ($\geqslant 2$),*

(a) $b(G) \leqslant n - 1$;

(b) $b(G) \leqslant \Delta(G) + 1$ *if* $\gamma(G) = 2$;

(c) $b(G) \leqslant n - \gamma(G) + 1$.

While the upper bound of $n-1$ on $b(G)$ is not particularly good for many classes of graphs (e.g., trees and cycles), it is an attainable bound. For example, if G is a complete t-partite graph $K_t(2)$ for $t = n/2$ and $n \leqslant 4$ an even integer, then the three bounds on $b(G)$ in Theorem 48 are sharp.

Teschner [26] consider a graph with $\gamma(G) = 1$ or $\gamma(G) = 2$. The next result is almost trivial but useful.

Lemma 49. *Let G be a graph with order n and $\gamma(G) = 1$, t the number of vertices of degree $n-1$. Then $b(G) = \lceil t/2 \rceil$.*

Since $t \leqslant \Delta(G)$ clearly, Lemma 49 yields the following result immediately.

Theorem 50 (Teschner [26], 1995). $b(G) \leqslant (1/2)\Delta + 1$ *for any graph G with* $\gamma(G) = 1$.

For a complete graph K_{2k+1}, $b(K_{2k+1}) = k+1 = (1/2)\Delta+1$, which shows that the upper bound given in Theorem 50 can be attained. For a graph G with $\gamma(G) = 2$, by Theorem 59 later, the upper bound given in Theorem 48 (b) can be also attained by a 2-critical graph (see Proposition 37).

3.6. Two Conjectures. In 1990, when Fink et al. [8] introduced the concept of the bondage number; they proposed the following conjecture.

Conjecture 51. $b(G) \leqslant \Delta(G) + 1$ *for any nonempty graph G.*

Although some results partially support Conjecture 51, Teschner [28] in 1993 found a counterexample to this conjecture, the cartesian product $K_3 \times K_3$, which shows that $b(G) = \Delta(G) + 2$.

If a graph G is a counterexample to Conjecture 51, it must be a γ-critical graph by Theorem 13. That is why the vertex domination-critical graphs are of special interest in the literature.

Now, we return to Conjecture 51. Hartnell and Rall [17] and Teschner [29], independently, proved that $b(G)$ can be much greater than $\Delta(G)$ by showing the following result.

Theorem 52 (Hartnell and Rall [17], 1994; Teschner [29]. 1996). *For an integer $n \geqslant 3$, let G_n be the cartesian product $K_n \times K_n$. Then $b(G_n) = 3(n-1) = (3/2)\Delta(G_n)$.*

This theorem shows that there exist no upper bounds of the form $b(G) \leqslant \Delta(G) + c$ for any integer c. Teschner [26] proved that $b(G) \leqslant (3/2)\Delta(G)$ for any graph G with $\gamma(G) \leqslant 2$ (see Theorems 50 and 48) and for γ-critical graphs with $\gamma(G) = 3$ and proposed the following conjecture.

Conjecture 53 (Teschner [26], 1995). *$b(G) \leqslant (3/2)\Delta(G)$ for any graph G.*

We believe that this conjecture is true, but so far no much work about it has been known.

4. Lower Bounds

Since the bondage number is defined as the smallest number of edges whose removal results in an increase of domination number, each constructive method that creates a concrete bondage set leads to an upper bound on the bondage number. For that reason, it is hard to find lower bounds. Nevertheless, there are still a few lower bounds.

4.1. Bounds Implied by Subgraphs. The first lower bound on the bondage number is gotten in terms of its spanning subgraph.

Theorem 54 (Teschner [10], 1997). *Let H be a spanning subgraph of a nonempty graph G. If $\gamma(H) = \gamma(G)$, then $b(H) \leqslant b(G)$.*

By Theorem 1, $b(C_n) = 3$ if $n \equiv 1 \pmod 3$ and $b(C_n) = 2$, otherwise $b(P_n) = 2$ if $n \equiv 1 \pmod 3$ and $b(P_n) = 1$ otherwise. From these results and Theorem 54, we get the following two corollaries.

Corollary 55. *If G is hamiltonian with order $n \geqslant 3$ and $\gamma(G) = \lceil n/3 \rceil$, then $b(G) \geqslant 2$ and in addition $b(G) \geqslant 3$ if $n \equiv 1 \pmod 3$.*

Corollary 56. *If G with order $n \geqslant 2$ has a hamiltonian path and $\gamma(G) = \lceil n/3 \rceil$, then $b(G) \geqslant 2$ if $n \equiv 1 \pmod 3$.*

4.2. Other Bounds. The *vertex covering number* $\beta(G)$ of G is the minimum number of vertices that are incident with all edges in G. If G has no isolated vertices, then $\gamma(G) \leqslant \beta(G)$ clearly. In [30], Volkmann gave a lot of graphs with $\beta = \gamma$.

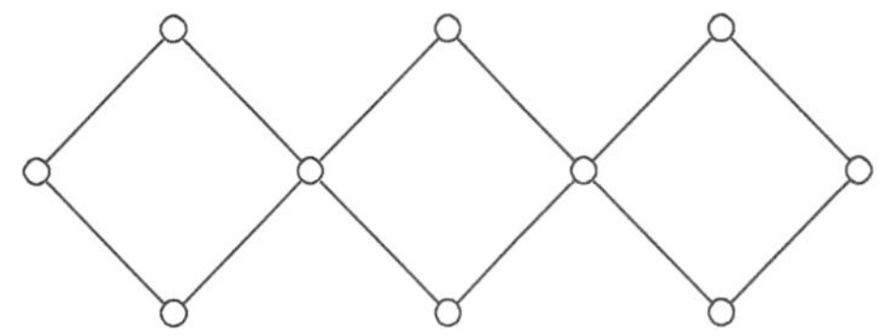

FIGURE 5: A γ-critical graph with $\beta = \gamma = 4$, $\delta = 2$, and $b = 3$.

Theorem 57 (Teschner [10], 1997). *Let G be a graph. If $\gamma(G) = \beta(G)$, then*

(a) $b(G) \geqslant \delta(G)$;

(b) $b(G) \geqslant \delta(G) + 1$ *if G is a γ-critical graph.*

The graph shown in Figure 5 shows that the bound given in Theorem 57(b) is sharp. For the graph G, it is a γ-critical graph and $\gamma(G) = \beta(G) = 4$. By Theorem 57, we have $b(G) \geqslant 3$. On the other hand, $b(G) \leqslant 3$ by Theorem 16. Thus, $b(G) = 3$.

Proposition 58 (Sanchis [31], 1991). *Let G be a graph G of order n ($\geqslant 6$) and size ε. If G has no isolated vertices and $3 \leqslant \gamma(G) \leqslant n/2$, then $\varepsilon \leqslant \binom{n-\gamma(G)+1}{2}$.*

Using the idea in the proof of Theorem 57, every upper bound for $\gamma(G)$ can lead to a lower bound for $b(G)$. In this way, Teschner [10] obtained another lower bound from Proposition 58.

Theorem 59 (Teschner [10], 1997). *Let G be a graph G of order n ($\geqslant 6$) and size ε. If $2 \leqslant \gamma(G) \leqslant n/2 - 1$, then*

(a) $b(G) \geqslant \min\{\delta(G), \varepsilon - \binom{n-\gamma(G)}{2}\}$;

(b) $b(G) \geqslant \min\{\delta(G) + 1, \varepsilon - \binom{n-\gamma(G)}{2}\}$ *if G is a γ-critical graph.*

The lower bound in Theorem 59(b) is sharp for a class of γ-critical graphs with domination number 2. Let G be the graph obtained from complete graph K_{2t} ($t \geqslant 2$) by removing a perfect matching. By Proposition 37, G is a γ-critical graph with $\gamma(G) = 2$. Then $b(G) \geqslant \delta(G) + 1 = \Delta(G) + 1$ by Theorem 59 and $b(G) \leqslant \Delta(G) + 1$ by Theorem 38, and so $b(G) = \Delta(G) + 1 = 2t - 1$.

As far as the author knows, there are no more lower bounds in the present literature. In view of applications of the bondage number, a network is vulnerable if its bondage number is small while it is stable if its bondage number is large. Therefore, better lower bounds let us learn better about the stability of networks from this point of view. Thus, it is of great significance to seek more lower bounds for various classes of graphs.

5. Results on Graph Operations

Generally speaking, it is quite difficult to determine the exact value of the bondage number for a given graph since it strongly depends on the dominating number of the graph. Thus, determining bondage numbers for some special graphs is interesting if the dominating numbers of those graphs are known or can be easily determined. In this section, we will introduce some known results on bondage numbers for some special classes of graphs obtained by some graph operations.

5.1. Cartesian Product Graphs. Let $G_1 = (V_1, E_1)$ and $G_2 = (V_2, E_2)$ be two graphs. The *Cartesian product* of G_1 and G_2 is an undirected graph, denoted by $G_1 \times G_2$, where $V(G_1 \times G_2) = V_1 \times V_2$; two distinct vertices x_1x_2 and y_1y_2, where $x_1, y_1 \in V(G_1)$ and $x_2, y_2 \in V(G_2)$, are linked by an edge in $G_1 \times G_2$ if and only if either $x_1 = y_1$ and $x_2y_2 \in E(G_2)$, or $x_2 = y_2$ and $x_1y_1 \in E(G_1)$. The Cartesian product is a very effective method for constructing a larger graph from several specified small graphs.

Theorem 60 (Dunbar et al. [21], 1998). *For $n \geqslant 3$,*

$$b(C_n \times K_2) = \begin{cases} 2 & if\ n \equiv 0\ or\ 1 \pmod 4, \\ 3 & if\ n \equiv 3 \pmod 4, \\ 4 & if\ n \equiv 2 \pmod 4. \end{cases} \tag{16}$$

For the Cartesian product $C_n \times C_m$ of two cycles C_n and C_m, where $n \geqslant 4$ and $m \geqslant 3$, Klavžar and Seifter [32] determined $\gamma(C_n \times C_m)$ for some n and m. For example, $\gamma(C_n \times C_4) = n$ for $n \geqslant 3$ and $\gamma(C_n \times C_3) = n - \lfloor n/4 \rfloor$ for $n \geqslant 4$. Kim [33] determined $b(C_3 \times C_3) = 6$ and $b(C_4 \times C_4) = 4$. For a general $n \geqslant 4$, the exact values of the bondage numbers of $C_n \times C_3$ and $C_n \times C_4$ are determined as follows.

Theorem 61 (Sohn et al. [34], 2007). *For $n \geqslant 4$,*

$$b(C_n \times C_3) = \begin{cases} 2 & if\ n \equiv 0 \pmod 4, \\ 4 & if\ n \equiv 1\ or\ 2 \pmod 4, \\ 5 & if\ n \equiv 3 \pmod 4. \end{cases} \tag{17}$$

Theorem 62 (Kang et al. [35], 2005). *$b(C_n \times C_4) = 4$ for $n \geqslant 4$.*

For larger m and n, Huang and Xu [36] obtained the following result; see Theorem 197 for more details.

Theorem 63 (Huang and Xu [36], 2008). *$b(C_{5i} \times C_{5j}) = 3$ for any positive integers i and j.*

Xiang et al. [37] determined that for $n \geqslant 5$,

$$b(C_n \times C_5) \begin{cases} = 3 & \text{if } n \equiv 0 \text{ or } 1 \pmod 5, \\ = 4 & \text{if } n \equiv 2 \text{ or } 4 \pmod 5, \\ \leqslant 7 & \text{if } n \equiv 3 \pmod 5. \end{cases} \tag{18}$$

For the Cartesian product $P_n \times P_m$ of two paths P_n and P_m, Jacobson and Kinch [38] determined $\gamma(P_n \times P_2) = \lceil (n+1)/2 \rceil$, $\gamma(P_n \times P_3) = n - \lfloor (n-1)/4 \rfloor$, and $\gamma(P_n \times P_4) = n + 1$ if $n = 1, 2, 3, 5, 6, 9$, and n otherwise. The bondage number of $P_n \times P_m$ for $n \geqslant 4$ and $2 \leqslant m \leqslant 4$ is determined as follows (Li [39] also determined $b(P_n \times P_2)$).

Theorem 64 (Hu and Xu [40], 2012). *For $n \geqslant 4$,*

$$b(P_n \times P_2) = \begin{cases} 1, & if\ n \equiv 1 \pmod 2 \\ 2, & if\ n \equiv 0 \pmod 2; \end{cases}$$

$$b(P_n \times P_3) = \begin{cases} 1, & if\ n \equiv 1\ or\ 2 \pmod 4 \\ 2, & if\ n \equiv 0\ or\ 3 \pmod 4, \end{cases} \tag{19}$$

$$b(P_n \times P_4) = 1 \quad for\ n \geqslant 14.$$

From the proof of Theorem 64, we find that if $P_n = \{0, 1, \ldots, n-1\}$ and $P_m = \{0, 1, \ldots, m-1\}$, then the removal of the vertex $(0, 0)$ in $P_n \times P_m$ does not change the domination number. If m increases, the effect of $(0, 0)$ for the domination number will be smaller and smaller from the probability. Therefore, we expect it is possible that $\gamma(P_n \times P_m - (0, 0)) = \gamma(P_n \times P_m)$ for $m \geqslant 5$ and give the following conjecture.

Conjecture 65. *$b(P_n \times P_m) \leqslant 2$ for $m \geqslant 5$ and $n \geqslant 4$.*

5.2. Block Graphs and Cactus Graphs. In this subsection, we introduce some results for corona graphs, block graphs, and cactus graphs.

The *corona* $G_1 \circ G_2$, proposed by Frucht and Harary [41], is the graph formed from a copy of G_1 and $\upsilon(G_1)$ copies of G_2 by joining the ith vertex of G_1 to the ith copy of G_2. In particular, we are concerned with the corona $H \circ K_1$, the graph formed by adding a new vertex v_i, and a new edge u_iv_i for every vertex u_i in H.

Theorem 66 (Carlson and Develin [42], 2006). *$b(H \circ K_1) = \delta(H) + 1$.*

This is a very important result, which is often used to construct a graph with required bondage number. In other words, this result implies that for any given positive integer b there is a graph G such that $b(G) = b$.

A *block graph* is a graph whose blocks are complete graphs. Each block in a *cactus graph* is either a cycle or a K_2. If each block of a graph is either a complete graph or a cycle, then we call this graph a *block-cactus graph.*

Theorem 67 (Teschner [43], 1997). *$b(G) \leq \Delta(G)$ for any block graph G.*

Teschner [43] characterized all block graphs with $b(G) = \Delta(G)$. In the same paper, Teschner found that γ-critical graphs are instrumental in determining bounds for the bondage number of cactus and block graphs and obtained the following result.

Theorem 68 (Teschner [43], 1997). *$b(G) \leqslant 3$ for any nontrivial cactus graph G.*

This bound can be achieved by $H \circ K_1$ where H is a nontrivial cactus graph without vertices of degree one by Theorem 66. In 1998, Dunbar et al. [21] proposed the following problem.

Problem 2. Characterize all cactus graphs with bondage number 3.

Some upper bounds for block-cactus graphs were also obtained.

Theorem 69 (Dunbar et al. [21], 1998). *Let G be a connected block-cactus graph with at least two blocks. Then $b(G) \leqslant \Delta(G)$.*

Theorem 70 (Dunbar et al. [21], 1998). *Let G be a connected block-cactus graph which is neither a cactus graph nor a block graph. Then $b(G) \leqslant \Delta(G) - 1$.*

5.3. Edge-Deletion and Edge-Contraction. In order to obtain further results of the bondage number, we may consider how the bondage number changes under some operations of a graph G, which remain the domination number unchanged or preserve some property of G, such as planarity. Two simple operations satisfying these requirements are the edge-deletion and the edge-contraction.

Theorem 71 (Huang and Xu [44], 2012). *Let G be any graph and $e \in E(G)$. Then $b(G-e) \geqslant b(G) - 1$. Moreover, $b(G-e) \leqslant b(G)$ if $\gamma(G-e) = \gamma(G)$.*

From Theorem 1, $b(C_n - e) = b(P_n) = b(C_n) - 1$ for any e in C_n, which shows that the lower bound on $b(G-e)$ is sharp. The upper bound can reach for any graph G with $b(G) \geqslant 2$.

Next, we consider the effect of the edge-contraction on the bondage number. Given a graph G, the contraction of G by the edge $e = xy$, denoted by G/xy, is the graph obtained from G by contracting two vertices x and y to a new vertex and then deleting all multiedges. It is easy to observe that for any graph G, $\gamma(G) - 1 \leqslant \gamma(G/xy) \leqslant \gamma(G)$ for any edge xy of G.

Theorem 72 (Huang and Xu [44], 2012). *Let G be any graph and xy be any edge in G. If $N_G(x) \cap N_G(y) = \emptyset$ and $\gamma(G/xy) = \gamma(G)$, then $b(G/xy) \geqslant b(G)$.*

We can use examples C_n and K_n to show that the conditions of Theorem 72 are necessary and the lower bound in Theorem 72 is tight. Clearly, $\gamma(C_n) = \lceil n/3 \rceil$ and $\gamma(K_n) = 1$; for any edge xy, $C_n/xy = C_{n-1}$ and $K_n/xy = K_{n-1}$. By Theorem 1, if $n \equiv 1 \pmod 3$, then $\gamma(C_n/xy) < \gamma(G)$ and $b(C_n/xy) = 2 < 3 = b(C_n)$. Thus, the result in Theorem 72 is generally invalid without the hypothesis $\gamma(G/xy) = \gamma(G)$. Furthermore, the condition $N_G(x) \cap N_G(y) = \emptyset$ cannot be omitted even if $\gamma(G/xy) = \gamma(G)$, since for odd n, $b(K_n/xy) = \lceil (n-1)/2 \rceil < \lceil n/2 \rceil = b(K_n)$, by Theorem 1.

On the other hand, $b(C_n/xy) = b(C_n) = 2$ if $n \equiv 0 \pmod 3$ and $b(K_n/xy) = b(K_n)$ if n is even, which shows that the equality in $b(G/xy) \geqslant b(G)$ may hold. Thus, the bound in Theorem 72 is tight. However, provided all the conditions, $b(G/xy)$ can be arbitrarily larger than $b(G)$. Given a graph H, let G be the graph formed from $H \circ K_1$ by adding a new vertex x and joining it to an vertex y of degree one in $H \circ K_1$. Then $G/xy = H \circ K_1$, $\gamma(G) = \gamma(G/xy)$, and $N_G(x) \cap N_G(y) = \emptyset$. But $b(G) = 1$ since $\gamma(G-xy) = \gamma(G/xy) + 1$, and $b(G/xy) = \delta(H) + 1$ by Theorem 66. The gap between $b(G)$ and $b(G/xy)$ is $\delta(H)$.

6. Results on Planar Graphs

From Section 2, we have seen that the bondage number for a tree has been completely solved. Moreover, a linear time algorithm to compute the bondage number of a tree is designed by Hartnell et al. [15]. It is quite natural to consider the bondage number for a planar graph. In this section, we state some results and problems on the bondage number for a planar graph.

6.1. Conjecture on Bondage Number. As mentioned in Section 3, the bondage number can be much larger than the maximum degree. But for a planar graph G, the bondage number $b(G)$ cannot exceed $\Delta(G)$ too much. It is clear that $b(G) \leqslant \Delta(G) + 4$ by Corollary 15 since $\delta(G) \leqslant 5$ for any planar graph G. In 1998, Dunbar et al. [21] posed the following conjecture.

Conjecture 73. *If G is a planar graph, then $b(G) \leqslant \Delta(G) + 1$.*

Because of its attraction, it immediately became the focus of attention when this conjecture is proposed. In fact, the main aim concerning the research of the bondage number for a planar graph is focused on this conjecture.

It has been mentioned in Theorems 1 and 60 that $b(C_{3k+1}) = 3 = \Delta + 1$, and $b(C_{4k+2} \times K_2) = 4 = \Delta + 1$. It is known that $b(K_6 - M) = 5 = \Delta + 1$, where M is a perfect matching of the complete graph K_6. These examples show that if Conjecture 73 is true, then the upper bound is sharp for $2 \leqslant \Delta \leqslant 4$.

Here, we note that it is sufficient to prove this conjecture for connected planar graphs, since the bondage number of a disconnected graph is simply the minimum of the bondage numbers of its components.

The first paper attacking this conjecture is due to Kang and Yuan [45], which confirmed the conjecture for every connected planar graph G with $\Delta \geqslant 7$. The proofs are mainly based on Theorems 16 and 18.

Theorem 74 (Kang and Yuan [45], 2000). *If G is a connected planar graph, then $b(G) \leqslant \min\{8, \Delta(G) + 2\}$.*

Obviously, in view of Theorem 74, Conjecture 73 is true for any connected planar graph with $\Delta \geqslant 7$.

6.2. Bounds Implied by Degree Conditions. As we have seen from Theorem 74, to attack Conjecture 73, we only need to consider connected planar graphs with maximum $\Delta \leqslant 6$. Thus, studying the bondage number of planar graphs by degree conditions is of significance. The first result on bounds implied by degree conditions is obtained by Kang and Yuan [45].

Theorem 75 (Kang and Yuan [45], 2000). *If G is a connected planar graph without vertices of degree 5, then $b(G) \leqslant 7$.*

Fischermann et al. [46] generalized Theorem 75 as follows.

Theorem 76 (Fischermann et al. [46], 2003). *Let G be a connected planar graph and X the set of vertices of degree 5 which have distance at least 3 to vertices of degrees 1, 2, and 3. If all vertices in X not adjacent with vertices of degree 4 are independent and not adjacent to vertices of degree 6, then $b(G) \leqslant 7$.*

Clearly, if G has no vertices of degree 5, then $X = \emptyset$. Thus, Theorem 76 yields Theorem 75.

Use n_i to denote the number of vertices of degree i in G for each $i = 1, 2, \ldots, \Delta(G)$. Using Theorem 18, Fischermann et al. obtained the following two theorems.

Theorem 77 (Fischermann et al. [46], 2003). *For any connected planar graph G,*

(1) *$b(G) \leqslant 7$ if $n_5 < 2n_2 + 3n_3 + 2n_4 + 12$;*

(2) *$b(G) \leqslant 6$ if G contains no vertices of degrees 4 and 5.*

Theorem 78 (Fischermann et al. [46], 2003). *For any connected planar graph G, $b(G) \leqslant \Delta(G) + 1$ if*

(1) *$\Delta(G) = 6$ and every edge $e = xy$ with $d_G(x) = 5$ and $d_G(y) = 6$ is contained in at most one triangle;*

(2) *$\Delta(G) = 5$ and no triangle contains an edge $e = xy$ with $d_G(x) = 5$ and $4 \leqslant d_G(y) \leqslant 5$.*

6.3. Bounds Implied by Girth Conditions. The girth $g(G)$ of a graph G is the length of the shortest cycle in G. If G has no cycles, we define $g(G) = \infty$.

Combining Theorem 74 with Theorem 78, we find that if a planar graph contains no triangles and has maximum degree $\Delta \geqslant 5$, then Conjecture 73 holds. This fact motivated Fischermann et al.'s [46] attempt to attack Conjecture 73 by girth restraints.

Theorem 79 (Fischermann et al. [46], 2003). *For any connected planar graph G,*

$$b(G) \leqslant \begin{cases} 6 & \text{if } g(G) \geqslant 4, \\ 5 & \text{if } g(G) \geqslant 5, \\ 4 & \text{if } g(G) \geqslant 6, \\ 3 & \text{if } g(G) \geqslant 8. \end{cases} \quad (20)$$

The first result in Theorem 79 shows that Conjecture 73 is valid for all connected planar graphs with $g(G) \geqslant 4$ and $\Delta(G) \geqslant 5$. It is easy to verify that the second result in Theorem 79 implies that Conjecture 73 is valid for all not 3-regular graphs of girth $g(G) \geqslant 5$, which is stated in the following corollary.

Corollary 80. *For any connected planar graph G, if G is not 3-regular and $g(G) \geqslant 5$, then $b(G) \leqslant \Delta(G) + 1$.*

The first result in Theorem 79 also implies that if G is a connected planar graph with no cycles of length 3, then $b(G) \leqslant 6$. Hou et al. [47] improved this result as follows.

Theorem 81 (Hou et al. [47], 2011). *For any connected planar graph G, if G contains no cycles of length i ($4 \leqslant i \leqslant 6$), then $b(G) \leqslant 6$.*

Since $b(C_{3k+1}) = 3$ for $k \geqslant 3$ (see Theorem 1), the last bound in Theorem 79 is tight. Whether other bounds in Theorem 79 are tight remains open. In 2003, Fischermann et al. [46] posed the following conjecture.

Conjecture 82. *For any connected planar graph G, $b(G) \leqslant 7$, and*

$$b(G) \leqslant \begin{cases} 5, & \text{if } g(G) \geqslant 4, \\ 4, & \text{if } g(G) \geqslant 5. \end{cases} \quad (21)$$

We conclude this subsection with a question on bondage numbers of planar graphs.

Question 1 (Fischermann et al. [46], 2003). Is there a planar graph G with $6 \leqslant b(G) \leqslant 8$?

In 2006, Carlson and Develin [42] showed that the corona $G = H \circ K_1$ for a planar graph H with $\delta(H) = 5$ has the bondage number $b(G) = \delta(H) + 1 = 6$ (see Theorem 66). Since the minimum degree of planar graphs is at most 5, then $b(G)$ can attach 6. If we take H as the graph of the icosahedron, then $G = H \circ K_1$ is such an example. The question for the existence of planar graphs with bondage number 7 or 8 remains open.

6.4. Comments on the Conjectures. Conjecture 73 is true for all connected planar graphs with minimum degree $\delta \leqslant 2$ by Theorem 16, or maximum degree $\Delta \geqslant 7$ by Theorem 74, or not γ-critical planar graphs by Theorem 13. Thus, to attack Conjecture 73, we only need to consider connected critical planar graphs with degree-restriction $3 \leqslant \delta \leqslant \Delta \leqslant 6$.

Recalling and anatomizing the proofs of all results mentioned in the preceding subsections on the bondage number for connected planar graphs, we find that the proofs of these results strongly depend upon Theorem 16 or Theorem 18. In other words, a basic way used in the proofs is to find two vertices x and y with distance at most two in a considered planar graph G such that

$$d_G(x) + d_G(y) \quad \text{or} \quad d_G(x) + d_G(y) - |N_G(x) \cap N_G(y)|, \quad (22)$$

which bounds $b(G)$, is as small as possible. Very recently, Huang and Xu [44] have considered the following parameter

$$B(G) = \min_{x,y \in V(G)} \{\{d_{xy} : 1 \leqslant d_G(x,y) \leqslant 2\} \cup \{d_{xy} - n_{xy} : d(x,y) = 1\}\}, \quad (23)$$

where $d_{xy} = d_G(x) + d_G(y) - 1$ and $n_{xy} = |N_G(x) \cap N_G(y)|$.

It follows from Theorems 16 and 18 that

$$b(G) \leqslant B(G). \qquad (24)$$

The proofs given in Theorems 74 and 79 indeed imply the following stronger results.

Theorem 83. *If G is a connected planar graph, then*

$$B(G) \leqslant \begin{cases} \min\{8, \Delta(G)+2\} \\ 6 & \text{if } g(G) \geqslant 4, \\ 5 & \text{if } g(G) \geqslant 5, \\ 4 & \text{if } g(G) \geqslant 6, \\ 3 & \text{if } g(G) \geqslant 8. \end{cases} \qquad (25)$$

Thus, using Theorem 16 or Theorem 18, if we can prove Conjectures 73 and 82, then we can prove the following statement.

Statement. If G is a connected planar graph, then

$$B(G) \leqslant \begin{cases} \Delta(G)+1, \\ 7, \\ 5, & \text{if } g(G) \geqslant 4, \\ 4, & \text{if } g(G) \geqslant 5. \end{cases} \qquad (26)$$

It follows from (24) that Statement implies Conjectures 73 and 82. However, Huang and Xu [44] gave examples to show that none of conclusions in Statement is true. As a result, they stated the following conclusion.

Theorem 84 (Huang and Xu [44], 2012). *It is not possible to prove Conjectures 73 and 82, if they are right, using Theorems 16 and 18.*

Therefore, a new method is needed to prove or disprove these conjectures. At the present time, one cannot prove these conjectures, and may consider to disprove them. If one of these conjectures is invalid, then there exists a *minimum counterexample* G with respect to $\upsilon(G)+\varepsilon(G)$. In order to obtain a minimum counterexample, one may consider two simple operations satisfying these requirements, the edge-deletion and the edge-contraction, which decrease $\upsilon(G)+\varepsilon(G)$ and preserve planarity. However, applying Theorems 71 and 72, Huang and Xu [44] presented the following results.

Theorem 85 (Huang and Xu [44], 2012). *It is not possible to construct minimum counterexamples to Conjectures 73 and 82 by the operation of an edge-deletion or an edge-contraction.*

7. Results on Crossing Number Restraints

It is quite natural to generalize the known results on the bondage number for planar graphs to for more general graphs in terms of graph-theoretical parameters. In this section, we consider graphs with crossing number restraints.

The *crossing number* $\mathrm{cr}(G)$ of G is the smallest number of pairwise intersections of its edges when G is drawn in the plane. If $\mathrm{cr}(G)=0$, then G is a planar graph.

7.1. General Methods. Use n_i to denote the number of vertices of degree i in G for $i=1,2,\ldots,\Delta(G)$. Huang and Xu obtained the following results.

Theorem 86 (Huang and Xu [48], 2007). *For any connected graph G,*

$$b(G) \leqslant \begin{cases} 6 & \text{if } g(G) \geqslant 4 \text{ and } 2\mathrm{cr}(G) < \frac{1}{2}a_1; \\ 5 & \text{if } g(G) \geqslant 5 \text{ and } 6\mathrm{cr}(G) < \frac{1}{6}a_2; \\ 4 & \text{if } g(G) \geqslant 6 \text{ and } 4\mathrm{cr}(G) < \frac{1}{4}a_3; \\ 3 & \text{if } g(G) \geqslant 8 \text{ and } 6\mathrm{cr}(G) < \frac{1}{6}a_4, \end{cases} \qquad (27)$$

where $a_1 = n_1 + 2n_2 + 2n_3 + \sum_{i=8}^{\Delta}(i-7)n_i + 8$; $a_2 = 3n_1 + 6n_2 + 5n_3 + \sum_{i=7}^{\Delta}(3i-18)n_i + 20$; $a_3 = n_1 + 2n_2 + \sum_{i=6}^{\Delta}(2i-10)n_i + 12$; *and* $a_4 = \sum_{i=5}^{\Delta}(3i-12)n_i + 16$.

Simplifying the conditions in Theorem 86, we obtain the following corollaries immediately.

Corollary 87 (Huang and Xu [48], 2007). *For any connected graph G,*

$$b(G) \leqslant \begin{cases} 6, & \text{if } g(G) \geqslant 4 \text{ and } \mathrm{cr}(G) \leqslant 3, \\ 5, & \text{if } g(G) \geqslant 5 \text{ and } \mathrm{cr}(G) \leqslant 4, \\ 4, & \text{if } g(G) \geqslant 6 \text{ and } \mathrm{cr}(G) \leqslant 2, \\ 3, & \text{if } g(G) \geqslant 8 \text{ and } \mathrm{cr}(G) \leqslant 2. \end{cases} \qquad (28)$$

Corollary 88 (Huang and Xu [48], 2007). *For any connected graph G,*

(a) $b(G) \leqslant 6$ *if G is not 4-regular,* $\mathrm{cr}(G) = 4$ *and* $g(G) \geqslant 4$;

(b) $b(G) \leqslant \Delta(G)+1$ *if G is not 3-regular,* $\mathrm{cr}(G) \leqslant 4$ *and* $g(G) \geqslant 5$;

(c) $b(G) \leqslant 4$ *if G is not 3-regular,* $\mathrm{cr}(G) = 3$ *and* $g(G) \geqslant 6$;

(d) $b(G) \leqslant 3$ *if* $\mathrm{cr}(G) = 3$, $g(G) \geqslant 8$ *and* $\Delta(G) \geqslant 5$.

These corollaries generalize some known results for planar graphs. For example, Corollary 87 contains Theorem 79; Corollary 88 (b) contains Corollary 80.

Theorem 89 (Huang and Xu [48], 2007). *For any connected graph G, if* $\mathrm{cr}(G) \leqslant n_3(G) + n_4(G) + 3$, *then* $b(G) \leqslant 8$.

Corollary 90 (Huang and Xu [48], 2007). *For any connected graph G, if* $\mathrm{cr}(G) \leqslant 3$, *then* $b(G) \leqslant 8$.

Perhaps being unaware of this result, in 2010, Ma et al. [49] proved that $b(G) \leqslant 12$ for any graph G with $\mathrm{cr}(G) = 1$.

Theorem 91 (Huang and Xu [48], 2007). *Let G be a connected graph, and* $I = \{x \in V(G) : d_G(x) = 5, d_G(x,y) \geqslant 3$ *if*

$d_G(y) \leqslant 3$, and $d_G(y) \neq 4$ for every $y \in N_G(x)\}$. If I is independent, no vertices adjacent to vertices of degree 6 and

$$\operatorname{cr}(G) < \max\left\{\frac{5n_3 + |I| - 2n_4 + 28}{11}, \frac{7n_3 + 40}{16}\right\}, \quad (29)$$

then $b(G) \leqslant 7$.

Corollary 92 (Huang and Xu [48], 2007). *Let G be a connected graph with $\operatorname{cr}(G) \leqslant 2$. If $I = \{x \in V(G) : d_G(x) = 5, d_G(y,x) \geqslant 3$ if $d_G(y) \leqslant 3$ and $d_G(y) \neq 4$ for every $y \in N_G(x)\}$ is independent, and no vertices adjacent to vertices of degree 6, then $b(G) \leqslant 7$.*

Theorem 93 (Huang and Xu [48], 2007). *Let G be a connected graph. If G satisfies*

(a) $5\operatorname{cr}(G) + n_5 < 2n_2 + 3n_3 + 2n_4 + 12$ *or*

(b) $7\operatorname{cr}(G) + 2n_5 < 3n_2 + 4n_4 + 24$,

then $b(G) \leqslant 7$.

Proposition 94 (Huang and Xu [48], 2007). *Let G be a connected graph with no vertices of degrees four and five. If $\operatorname{cr}(G) \leqslant 2$, then $b(G) \leqslant 6$.*

Theorem 95 (Huang and Xu [48], 2007). *If G is a connected graph with $\operatorname{cr}(G) \leqslant 4$ and not 4-regular when $\operatorname{cr}(G) = 4$, then $b(G) \leqslant \Delta(G) + 2$.*

The above results generalize some known results for planar graphs. For example, Corollary 90 and Theorem 95 contain Theorem 74; the first condition in Theorem 93 contains the second condition in Theorem 77.

From Corollary 88 and Theorem 95, we suggest the following questions.

Question 2. Is there a

(a) 4-regular graph G with $\operatorname{cr}(G) = 4$ such that $b(G) \geqslant 7$?

(b) 3-regular graph G with $\operatorname{cr} \leqslant 4$ and $g(G) \geqslant 5$ such that $b(G) = 5$?

(c) 3-regular graph G with $\operatorname{cr} = 3$ and $g(G) = 6$ or 7 such that $b(G) = 5$?

7.2. Carlson-Develin Methods. In this subsection, we introduce an elegant method presented by Carlson and Develin [42] to obtain some upper bounds for the bondage number of a graph.

Suppose that G is a connected graph. We say that G has *genus* ρ if G can be embedded in a surface S with ρ handles such that edges are pairwise disjoint except possibly for end-vertices. Let $\widetilde{G}$ be an embedding of G in surface S, and let $\phi(G)$ denote the number of regions in $\widetilde{G}$. The boundary of every region contains at least three edges and every edge is on the boundary of at most two regions (the two regions are identical when e is a cut-edge). For any edge e of G, let $r_G^1(e)$ and $r_G^2(e)$ be the numbers of edges comprising the regions in $\widetilde{G}$ which the edge e borders. It is clear that every $e = xy \in E(G)$,

$$\begin{aligned} r_G^2(e) \geqslant r_G^1(e) \geqslant 4 \quad & \text{if } |N_G(x) \cap N_G(y)| = 0, \\ r_G^2(e) \geqslant 4, \quad r_G^1(e) \geqslant 3 \quad & \text{if } |N_G(x) \cap N_G(y)| = 1, \\ r_G^2(e) \geqslant r_G^1(e) \geqslant 3 \quad & \text{if } |N_G(x) \cap N_G(y)| \geqslant 2. \end{aligned} \quad (30)$$

Following Carlson and Develin [42], for any edge $e = xy$ of G, we define

$$D_G(e) = \frac{1}{d_G(x)} + \frac{1}{d_G(y)} + \frac{1}{r_G^1(e)} + \frac{1}{r_G^2(e)} - 1. \quad (31)$$

By the well-known Euler's formula

$$\upsilon(G) - \varepsilon(G) + \phi(G) = 2 - 2\rho(G), \quad (32)$$

it is easy to see that

$$\sum_{e \in E(G)} D_G(e) = \upsilon(G) - \varepsilon(G) + \phi(G) = 2 - 2\rho(G). \quad (33)$$

If G is a planar graph, that is, $\rho(G) = 0$, then

$$\sum_{e \in E(G)} D_G(e) = \upsilon(G) - \varepsilon(G) + \phi(G) = 2. \quad (34)$$

Combining these formulas with Theorem 18, Carlson and Develin [42] gave a simple and intuitive proof of Theorem 74 and obtained the following result.

Theorem 96 (Carlson and Develin [42], 2006). *Let G be a connected graph which can be embedded on a torus. Then $b(G) \leqslant \Delta(G) + 3$.*

Recently, Hou and Liu [50] improved Theorem 96 as follows.

Theorem 97 (Hou and Liu [50], 2012). *Let G be a connected graph which can be embedded on a torus. Then $b(G) \leqslant 9$. Moreover, if $\Delta(G) \neq 6$, then $b(G) \leqslant 8$.*

By the way, Cao et al. [51] generalized the result in Theorem 79 to a connected graph G that can be embedded on a torus; that is,

$$b(G) \leq \begin{cases} 6, & \text{if } g(G) \geqslant 4 \text{ and } G \text{ is not 4-regular}, \\ 5, & \text{if } g(G) \geqslant 5, \\ 4, & \text{if } g(G) \geqslant 6 \text{ and } G \text{ is not 3-regular}, \\ 3, & \text{if } g(G) \geqslant 8. \end{cases} \quad (35)$$

Several authors used this method to obtain some results on the bondage number. For example, Fischermann et al. [46] used this method to prove the second conclusion in Theorem 78. Recently, Cao et al. [52] have used this method to deal with more general graphs with small crossing numbers. First, they found the following property.

Lemma 98. *$\delta(G) \leqslant 5$ for any graph G with $cr(G) \leqslant 5$.*

This result implies that $b(G) \leqslant \Delta(G) + 4$ by Corollary 15 for any graph G with $\mathrm{cr}(G) \leqslant 5$. Cao et al. established the following relation in terms of maximum planar subgraphs.

Lemma 99 (Cao et al. [52]). *Let G be a connected graph with crossing number* $\mathrm{cr}(G)$. *For any maximum planar subgraph H of G,*

$$\sum_{e\in E(H)} D_G(e) \geqslant 2 - \frac{2\mathrm{cr}(G)}{\delta(G)}. \quad (36)$$

Using Carlson-Develin method and combining Lemma 98 with Lemma 99, Cao et al. proved the following results.

Theorem 100 (Cao et al. [52]). *For any connected graph G,* $b(G) \leqslant \Delta(G) + 2$ *if G satisfies one of the following conditions:*

(a) $\mathrm{cr}(G) \leqslant 3$;

(b) $\mathrm{cr}(G) = 4$ *and G is not 4-regular;*

(c) $\mathrm{cr}(G) = 5$ *and G contains no vertices of degree 4.*

Theorem 101 (Cao et al. [52]). *Let G be a connected graph with* $\Delta(G) = 5$ *and* $\mathrm{cr}(G) \leqslant 4$. *If no triangles contain two vertices of degree 5, then* $b(G) \leqslant 6 = \Delta(G) + 1$.

Theorem 102 (Cao et al. [52]). *Let G be a connected graph with* $\Delta(G) \geqslant 6$ *and* $\mathrm{cr}(G) \leqslant 3$. *If* $\Delta(G) \geqslant 7$ *or if* $\Delta(G) = 6$, $\delta(G) \neq 3$ *and every edge* $e = xy$ *with* $d_G(x) = 5$ *and* $d_G(y) = 6$ *is contained in at most one triangle, then* $b(G) \leqslant \min\{8, \Delta(G)+1\}$.

Using Carlson-Develin method, it can be proved that $b(G) \leqslant \Delta(G) + 2$ if $\mathrm{cr}(G) \leqslant 3$ (see the first conclusion in Theorem 100), and not yet proved that $b(G) \leqslant 8$ if $\mathrm{cr}(G) \leqslant 3$, although it has been proved by using other method (see Corollary 90).

By using Carlson-Develin method, Samodivkin [25] obtained some results on the bondage number for graphs with some given properties.

Kim [53] showed that $b(G) \leqslant \Delta(G) + 2$ for a connected graph G with genus 1, $\Delta(G) \leqslant 5$ and having a toroidal embedding of which at least one region is not 4-sided. Recently, Gagarin and Zverovich [54] further have extended Carlson-Develin ideas to establish a nice upper bound for arbitrary graphs that can be embedded on orientable or nonorientable topological surface.

Theorem 103 (Gagarin and Zverovich [54], 2012). *Let G be a graph embeddable on an orientable surface of genus h and a nonorientable surface of genus k. Then* $b(G) \leqslant \min\{\Delta(G)+h+2, \Delta(G)+k+1\}$.

This result generalizes the corresponding upper bounds in Theorems 74 and 96 for any orientable or nonorientable topological surface. By investigating the proof of Theorem 103, Huang [55] found that the issue of the orientability can be avoided by using the Euler characteristic $\chi(= \upsilon(G) - \varepsilon(G) + \phi(G))$ instead of the genera h and k; the relations are $\chi = 2-2h$ and $\chi = 2-k$. To obtain the best result from Theorem 103, one wants h and k as small as possible; this is equivalent to having χ as large as possible.

According to Theorem 103, if G is planar ($h = 0, \chi = 2$) or can be embedded on the real projective plane ($k = 1, \chi = 1$), then $b(G) \leqslant \Delta(G) + 2$. In all other cases, Huang [55] had the following improvement for Theorem 103; the proof is based on the technique developed by Carlson-Develin and Gagarin-Zverovich and includes some elementary calculus as a new ingredient, mainly the intermediate-value theorem and the mean-value theorem.

Theorem 104 (Huang [55], 2012). *Let G be a graph embeddable on a surface whose Euler characteristic* χ *is as large as possible. If* $\chi \leqslant 0$, *then* $b(G) \leqslant \Delta(G)+\lfloor r\rfloor$, *where r is the largest real root of the following cubic equation in z:*

$$z^3 + 2z^2 + (6\chi - 7)z + 18\chi - 24 = 0. \quad (37)$$

In addition, if χ *decreases, then r increases.*

The following result is asymptotically equivalent to Theorem 104.

Theorem 105 (Huang [55], 2012). *Let G be a graph embeddable on a surface whose Euler characteristic* χ *is as large as possible. If* $\chi \leqslant 0$, *then* $b(G) < \Delta(G) + \sqrt{12-6\chi} + 1/2$, *or equivalently,* $b(G) \leqslant \Delta(G) + \lceil \sqrt{12-6\chi} - 1/2\rceil$.

Also, Gagarin and Zverovich [54] indicated that the upper bound in Theorem 103 can be improved for larger values of the genera h and k by adjusting the proofs and stated the following general conjecture.

Conjecture 106 (Gagarin and Zverovich [54], 2012). *For a connected graph G of orientable genus h and nonorientable genus k,*

$$b(G) \leqslant \min\{c_h, c_k', \Delta(G) + o(h), \Delta(G) + o(k)\}, \quad (38)$$

where c_h *and* c_k' *are constants depending, respectively, on the orientable and nonorientable genera of G.*

In the recent paper, Gagarin and Zverovich [56] provided constant upper bounds for the bondage number of graphs on topological surfaces, which can be used as the first estimation for the constants c_h and c_k' of Conjecture 106.

Theorem 107 (Gagarin and Zverovich [56], 2013). *Let G be a connected graph of order n. If G can be embedded on the surface of its orientable or nonorientable genus of the Euler characteristic* χ, *then*

(a) $\chi \geqslant 1$ *implies* $b(G) \leqslant 10$;

(b) $\chi \leqslant 0$ *and* $n > -12\chi$ *imply* $b(G) \leqslant 11$;

(c) $\chi \leqslant -1$ *and* $n \leqslant -12\chi$ *imply* $b(G) \leqslant 11 + 3\chi(\sqrt{17-8\chi} - 3)/(\chi - 1) = 11 + O(\sqrt{-\chi})$.

Theorem 79 shows that a connected planar triangle-free graph G has $b(G) \leqslant 6$. The following result generalizes to it all the other topological surfaces.

Theorem 108 (Gagarin and Zverovich [56], 2013). *Let G be a connected triangle-free graph of order n. If G can be embedded on the surface of its orientable or nonorientable genus of the Euler characteristic χ, then*

(a) $\chi \geqslant 1$ *implies* $b(G) \leqslant 6$;

(b) $\chi \leqslant 0$ *and* $n > -8\chi$ *imply* $b(G) \leqslant 7$;

(c) $\chi \leqslant -1$ *and* $n \leqslant -8\chi$ *imply* $b(G) \leqslant 7-4\chi/(1+\sqrt{2-\chi})$.

In [56], the authors also explicitly improved upper bounds in Theorem 103 and gave tight lower bounds for the number of vertices of graphs 2-cell embeddable on topological surfaces of a given genus. The interested reader is referred to the original paper for details. The following result is interesting although its proof is easy.

Theorem 109 (Samodivkin [57]). *Let G be a graph with order n and girth g. If G is embeddable on a surface whose Euler characteristic χ is as large as possible, then*

$$b(G) \leqslant 3 + \frac{8}{g-2} - \frac{4\chi g}{n(g-2)}. \tag{39}$$

If G is a planar graph with girth $g \geqslant 4 + i$ for any $i \in \{0, 1, 2\}$, then Theorem 109 leads to $b(G) \leqslant 6 - i$, which is the result in Theorem 79 obtained by Fischermann et al. [46]. Also, in many cases the bound stated in Theorem 109 is better than those given by Theorems 103 and 105.

8. Conditional Bondage Numbers

Since the concept of the bondage number is based upon the domination, all sorts of dominations, which are variations of the normal domination by adding restricted conditions to the normal dominating set, can develop a new "bondage number" as long as a variation of domination number is given. In this section, we survey results on the bondage number under some restricted conditions.

8.1. Total Bondage Numbers. A dominating set S of a graph G without isolated vertices is called *total* if the subgraph induced by S contains no isolated vertices. The minimum cardinality of a total dominating set is called the *total domination number* of G and denoted by $\gamma_T(G)$. It is clear that $\gamma(G) \leqslant \gamma_T(G) \leqslant 2\gamma(G)$ for any graph G without isolated vertices.

The total domination in graphs was introduced by Cockayne et al. [58] in 1980. Pfaff et al. [59, 60] in 1983 showed that the problem determining total domination number for general graphs is NP-complete, even for bipartite graphs and chordal graphs. Even now, total domination in graphs has been extensively studied in the literature. In 2009, Henning [61] gave a survey of selected recent results on total domination in graphs.

The *total bondage number* of G, denoted by $b_T(G)$, is the smallest cardinality of a subset $B \subseteq E(G)$ with the property that $G-B$ contains no isolated vertices and $\gamma_T(G-B) > \gamma_T(G)$.

From definition, $b_T(G)$ may not exist for some graphs, for example, $G = K_{1,n}$. We put $b_T(G) = \infty$ if $b_T(G)$ does not exist.

It is easy to see that $b_T(G)$ is finite for any connected graph G other than K_1, K_2, K_3, and $K_{1,n}$. In fact, for such a graph G, since there is a path of length 3 in G, we can find $B_1 \subseteq E(G)$ such that $G - B_1$ is a spanning tree T of G, containing a path of length 3. So $\gamma_T(T) = \gamma_T(G-B_1) \geqslant \gamma_T(G)$. For the tree T, we find $B_2 \subseteq E(T)$ such that $\gamma_T(T - B_2) > \gamma_T(T) \geqslant \gamma_T(G)$. Thus, we have $\gamma_T(G-B_2-B_1) > \gamma_T(G)$, and so $b_T(G) \leqslant |B_1|+|B_2|$. In the following discussion, we always assume that $b_T(G)$ exists when a graph G is mentioned.

In 1991, Kulli and Patwari [62] first studied the total bondage number of a graph and calculated the exact values of $b_T(G)$ for some standard graphs.

Theorem 110 (Kulli and Patwari [62], 1991). *For $n \geqslant 4$,*

$$b_T(C_n) = \begin{cases} 3 & \text{if } n \equiv 2 \pmod 4, \\ 2 & \text{otherwise}, \end{cases}$$
$$b_T(P_n) = \begin{cases} 2 & \text{if } n \equiv 2 \pmod 4, \\ 1 & \text{otherwise}, \end{cases}$$
$$b_T(K_{m,n}) = m \quad \text{with } 2 \leqslant m \leqslant n,$$
$$b_T(K_n) = \begin{cases} 4 & \text{for } n = 4, \\ 2n-5 & \text{for } n \geqslant 5. \end{cases} \tag{40}$$

Recently, Hu et al. [63] have obtained some results on the total bondage number of the Cartesian product $P_m \times P_n$ of two paths P_m and P_n.

Theorem 111 (Hu et al. [63], 2012). *For the Cartesian product $P_m \times P_n$ of two paths P_m and P_n,*

$$b_T(P_2 \times P_n) = \begin{cases} 1 & \text{if } n \equiv 0 \pmod 3, \\ 2 & \text{if } n \equiv 2 \pmod 3, \\ 3 & \text{if } n \equiv 1 \pmod 3, \end{cases}$$
$$b_T(P_3 \times P_n) = 1,$$
$$b_T(P_4 \times P_n) \begin{cases} = 1 & \text{if } n \equiv 1 \pmod 5, \\ = 2 & \text{if } n \equiv 4 \pmod 5, \\ \leqslant 3 & \text{if } n \equiv 2 \pmod 5, \\ \leqslant 4 & \text{if } n \equiv 0, 3 \pmod 5. \end{cases} \tag{41}$$

Generalized Petersen graphs are an important class of commonly used interconnection networks and have been studied recently. By constructing a family of minimum total dominating sets, Cao et al. [64] determined the total bondage number of the generalized Petersen graphs.

From Theorem 6, we know that $b(T) \leqslant 2$ for a nontrivial tree T. But given any positive integer k, Sridharan et al. [65] constructed a tree T for which $b_T(T) = k$. Let H_k be the tree obtained from a star $K_{1,k+1}$ by subdividing k edges twice. The tree H_7 is shown in Figure 6. It can be easily verified that $b_T(H_k) = k$. This fact can be stated as the following theorem.

Theorem 112 (Sridharan et al. [65], 2007). *For any positive integer k, there exists a tree T with $b_T(T) = k$.*

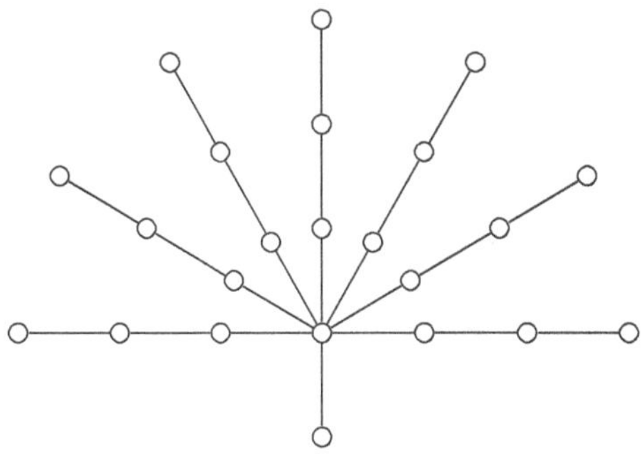

Figure 6: H_7.

Combining Theorem 1 with Theorem 112, we have what follows.

Corollary 113 (Sridharan et al. [65], 2007). *The bondage number and the total bondage number are unrelated, even for trees.*

However, Sridharan et al. [65] gave an upper bound on the total bondage number of a tree in terms of its maximum degree. For any tree T of order n, if $T \neq K_{1,n-1}$, then $b_T(T) = \min\{\Delta(T), (1/3)(n-1)\}$. Rad and Raczek [66] improved this upper bound and gave a constructive characterization of a certain class of trees attaching the upper bound.

Theorem 114 (Rad and Raczek [66], 2011). $b_T(T) \leqslant \Delta(T) - 1$ *for any tree T with maximum degree at least three.*

In general, the decision problem for $b_T(G)$ is NP-complete for any graph G. We state the decision problem for the total bondage as follows.

Problem 3. Consider the decision problem:

Total Bondage Problem

Instance: a graph G and a positive integer b.

Question: is $b_T(G) \leqslant b$?

Theorem 115 (Hu and Xu [16], 2012). *The total bondage problem is NP-complete.*

Consequently, in view of its computational hardness, it is significative to establish some sharp bounds on the total bondage number of a graph in terms of other graphic parameters. In 1991, Kulli and Patwari [62] showed that $b_T(G) \leqslant 2n - 5$ for any graph G with order $n \geqslant 5$. Sridharan et al. [65] improved this result as follows.

Theorem 116 (Sridharan et al. [65], 2007). *For any graph G with order n, if $n \geqslant 5$, then*

$$b_T(G) \leqslant \begin{cases} \frac{1}{3}(n-1) & \text{if } G \text{ contains no cycles,} \\ n-1 & \text{if } g(G) \geqslant 5, \\ n-2 & \text{if } g(G) = 4. \end{cases} \quad (42)$$

Rad and Raczek [66] also established some upper bounds on $b_T(G)$ for a general graph G. In particular, they gave an upper bound on $b_T(G)$ in terms of the girth of G.

Theorem 117 (Rad and Raczek [66], 2011). $b_T(G) \leqslant 4\Delta(G) - 5$ *for any graph G with $g(G) \geqslant 4$.*

8.2. Paired Bondage Numbers. A dominating set S of G is called to be *paired* if the subgraph induced by S contains a perfect matching. The *paired domination number* of G, denoted by $\gamma_P(G)$, is the minimum cardinality of a paired dominating set of G. Clearly, $\gamma_T(G) \leqslant \gamma_P(G)$ for every connected graph G with order at least two, where $\gamma_T(G)$ is the total domination number of G, and $\gamma_P(G) \leqslant 2\gamma(G)$ for any graph G without isolated vertices. Paired domination was introduced by Haynes and Slater [67, 68] and further studied in [69–72].

The *paired bondage number* of G with $\delta(G) \geqslant 1$, denoted by $b_P(G)$, is the minimum cardinality among all sets of edges $B \subseteq E$ such that $\delta(G - B) \geqslant 1$ and $\gamma_P(G - B) > \gamma_P(G)$.

The concept of the paired bondage number was first proposed by Raczek [73] in 2008. The following observations follow immediately from the definition of the paired bondage number.

Observation 2. Let G be a graph with $\delta(G) \geqslant 1$.

(a) If H is a subgraph of G such that $\delta(H) \geqslant 1$, then $b_P(H) \leqslant b_P(G)$.

(b) If H is a subgraph of G such that $b_P(H) = 1$ and k is the number of edges removed to form H, then $1 \leqslant b_P(G) \leqslant k + 1$.

(c) If $xy \in E(G)$ such that $d_G(x), d_G(y) > 1$, and xy belongs to each perfect matching of each minimum paired dominating set of G, then $b_P(G) = 1$.

Based on these observations, $b_P(P_n) \leqslant 2$. In fact, the paired bondage number of a path has been determined.

Theorem 118 (Raczek [73], 2008). *For any positive integer k, if $n \geqslant 2$, then*

$$b_P(P_n) = \begin{cases} 0 & \text{if } n = 2, 3, \text{ or } 5, \\ 1 & \text{if } n = 4k, 4k+3, \text{ or } 4k+6, \\ 2 & \text{otherwise.} \end{cases} \quad (43)$$

For a cycle C_n of order $n \geqslant 3$, since $\gamma_P(C_n) = \gamma_P(P_n)$, from Theorem 118 the following result holds immediately.

Corollary 119 (Raczek [73], 2008). *For any positive integer k, if $n \geqslant 3$, then*

$$b_P(C_n) = \begin{cases} 0 & \text{if } n = 3 \text{ or } 5, \\ 2 & \text{if } n = 4k, 4k+3, \text{ or } 4k+6, \\ 3 & \text{otherwise.} \end{cases} \quad (44)$$

A wheel W_n, where $n \geqslant 4$, is a graph with n vertices, formed by connecting a single vertex to all vertices of a cycle C_{n-1}. Of course $\gamma_P(W_n) = 2$.

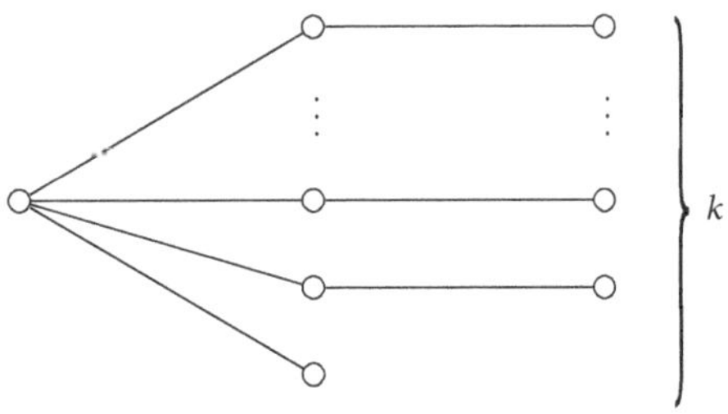

FIGURE 7: A tree T with $b_P(T) = k$.

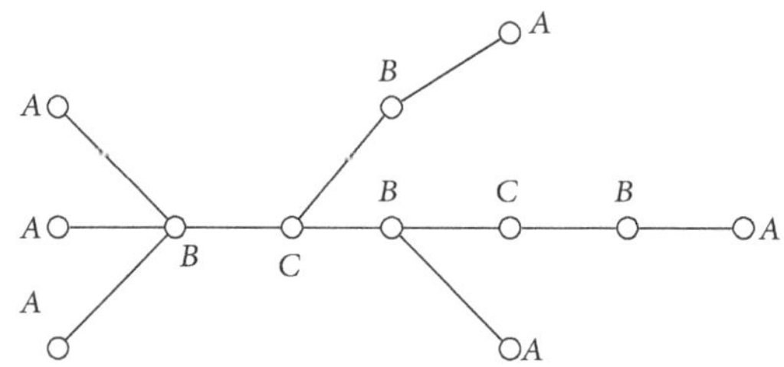

FIGURE 8: A tree T belong to the family $\mathscr{T}$.

Theorem 120 (Raczek [73], 2008). *For positive integers n and m,*

$$b_P(K_{m,n}) = m \quad \text{for } 1 < m \leqslant n,$$

$$b_P(W_n) = \begin{cases} 4 & \text{if } n = 4, \\ 3 & \text{if } n = 5, \\ 2 & \text{otherwise.} \end{cases} \tag{45}$$

Theorem 16 shows that if T is a nontrivial tree, then $b(T) \leqslant 2$. However, no similar result exists for paired bondage. For any nonnegative integer k, let T_k be a tree obtained by subdividing all but one edge of a star $K_{1,k+1}$ (as shown in Figure 7). It is easy to see that $b_P(T_k) = k$. We state this fact as the following theorem, which is similar to Theorem 112.

Theorem 121 (Raczek [73], 2008). *For any nonnegative integer k, there exists a tree T with $b_P(T) = k$.*

Consider the tree defined in Figure 5 for $k = 3$. Then $b(T_3) \leqslant 2$ and $b_P(T_3) = 3$. On the other hand, $b(P_4) = 2$ and $b_P(P_4) = 1$. Thus, we have what follows, which is similar to Corollary 113.

Corollary 122. *The bondage number and the paired bondage number are unrelated, even for trees.*

A constructive characterization of trees with $b(T) = 2$ is given by Hartnell and Rall in [12]. Raczek [73] provided a constructive characterization of trees with $b_P(T) = 0$. In order to state the characterization, we define a labeling and three simple operations on a tree T. Let $y \in V(T)$ and let $\ell(y)$ be the label assigned to y.

Operation $\mathscr{T}_1$. If $\ell(y) = B$, add a vertex x and the edge yx, and let $\ell(x) = A$.

Operation $\mathscr{T}_2$. If $\ell(y) = C$, add a path (x_1, x_2) and the edge yx_1, and let $\ell(x_1) = B$, $\ell(x_2) = A$.

Operation $\mathscr{T}_3$. If $\ell(y) = B$, add a path (x_1, x_2, x_3) and the edge yx_1, and let $\ell(x_1) = C$, $\ell(x_2) = B$, and $\ell(x_3) = A$.

Let $P_2 = (u, v)$ with $\ell(u) = A$ and $\ell(v) = B$. Let $\mathscr{T}$ be the class of all trees obtained from the labeled P_2 by a finite sequence of operations $\mathscr{T}_1$, $\mathscr{T}_2$, $\mathscr{T}_3$.

A tree T in Figure 8 belongs to the family $\mathscr{T}$.

Raczek [73] obtained the following characterization of all trees T with $b_P(T) = 0$.

Theorem 123 (Raczek [73], 2008). *For a tree T, $b_P(T) = 0$ if and only if T is in $\mathscr{T}$.*

We state the decision problem for the paired bondage as follows.

Problem 4. Consider the decision problem:

Paired Bondage Problem

Instance: a graph G and a positive integer b.

Question: is $b_P(G) \leqslant b$?

Conjecture 124. *The paired bondage problem is NP-complete.*

8.3. Restrained Bondage Numbers. A dominating set S of G is called to be *restrained* if the subgraph induced by $\overline{S}$ contains no isolated vertices, where $\overline{S} = V(G) \setminus S$. The *restrained domination number* of G, denoted by $\gamma_R(G)$, is the smallest cardinality of a restrained dominating set of G. It is clear that $\gamma_R(G)$ exists and $\gamma(G) \leqslant \gamma_R(G)$ for any nonempty graph G.

The concept of restrained domination was introduced by Telle and Proskurowski [74] in 1997, albeit indirectly, as a vertex partitioning problem. Concerning the study of the restrained domination numbers, the reader is referred to [74–79].

In 2008, Hattingh and Plummer [80] defined the *restrained bondage number* $b_R(G)$ of a nonempty graph G to be the minimum cardinality among all subsets $B \subseteq E(G)$ for which $\gamma_R(G - B) > \gamma_R(G)$.

For some simple graphs, their restrained domination numbers can be easily determined, and so restrained bondage numbers have been also determined. For example, it is clear that $\gamma_R(K_n) = 1$. Domke et al. [77] showed that $\gamma_R(C_n) = n - 2\lfloor n/3 \rfloor$ for $n \geqslant 3$, and $\gamma_R(P_n) = n - 2\lfloor (n-1)/3 \rfloor$ for $n \geqslant 1$. Using these results, Hattingh and Plummer [80] obtained the restricted bondage numbers for K_n, C_n, P_n, and $G = K_{n_1, n_2, \ldots, n_t}$ as follows.

Theorem 125 (Hattingh and Plummer [80], 2008). *For $n \geqslant 3$,*

$$b_R(K_n) = \begin{cases} 1 & \text{if } n = 3, \\ \left\lceil \dfrac{n}{2} \right\rceil & \text{otherwise,} \end{cases}$$

$$b_R(C_n) = \begin{cases} 1 & \text{if } n \equiv 0 \pmod 3, \\ 2 & \text{otherwise,} \end{cases}$$

$$b_R(P_n) = 1, \quad n \geqslant 4,$$

$$b_R(G) = \begin{cases} \left\lceil \frac{m}{2} \right\rceil & \text{if } n_m = 1 \text{ and } n_{m+1} \geqslant 2 \ (1 \leqslant m < t), \\ 2t-2 & \text{if } n_1 = n_2 = \cdots = n_t = 2 \ (t \geqslant 2), \\ 2 & \text{if } n_1 = 2 \text{ and } n_2 \geqslant 3 \ (t = 2), \\ \sum_{i=1}^{t-1} n_i - 1 & \text{otherwise}. \end{cases} \quad (46)$$

Theorem 126 (Hattingh and Plummer [80], 2008). *For a tree T of order n ($\geqslant 4$), $T \not\cong K_{1,n-1}$ if and only if $b_R(T) = 1$.*

Theorem 126 shows that the restrained bondage number of a tree can be computed in constant time. However, the decision problem for $b_R(G)$ is NP-complete even for bipartite graphs.

Problem 5. Consider the decision problem:

Restrained Bondage Problem

Instance: a graph G and a positive integer b.

Question: is $b_R(G) \leqslant b$?

Theorem 127 (Hattingh and Plummer [80], 2008). *The restrained bondage problem is NP-complete, even for bipartite graphs.*

Consequently, in view of its computational hardness, it is significative to establish some sharp bounds on the restrained bondage number of a graph in terms of other graphic parameters.

Theorem 128 (Hattingh and Plummer [80], 2008). *For any graph G with $\delta(G) \geqslant 2$,*

$$b_R(G) \leqslant \min_{xy \in E(G)} \{d_G(x) + d_G(y) - 2\}. \quad (47)$$

Corollary 129. *$b_R(G) \leqslant \Delta(G) + \delta(G) - 2$ for any graph G with $\delta(G) \geqslant 2$.*

Notice that the bounds stated in Theorem 128 and Corollary 129 are sharp. Indeed the class of cycles whose orders are congruent to 1, 2 (mod 3) have a restrained bondage number achieving these bounds (see Theorem 125).

Theorem 128 is an analogue of Theorem 14. A quite natural problem is whether or not, for restricted bondage number, there are analogues of Theorem 16, Theorem 18, Theorem 29, and so on. Indeed, we should consider all results for the bondage number whether or not there are analogues for restricted bondage number.

Theorem 130 (Hattingh and Plummer [80], 2008). *For any graph G with $\gamma_R(G) = 2$,*

$$b_R(G) \leqslant \Delta(G) + 1. \quad (48)$$

We now consider the relation between $b_R(G)$ and $b(G)$.

Theorem 131 (Hattingh and Plummer [80], 2008). *If $\gamma_R(G) = \gamma(G)$ for some graph G, then $b_R(G) \leqslant b(G)$.*

Corollary 129 and Theorem 131 provide a possibility to attack Conjecture 73 by characterizing planar graphs with $\gamma_R = \gamma$ and $b_R = b$ when $3 \leqslant \Delta \leqslant 6$.

However, we do not have $b_R(G) = b(G)$ for any graph G even if $\gamma_R(G) = \gamma(G)$. Observe that $\gamma_R(K_3) = \gamma(K_3)$, yet $b_R(K_3) = 1$ and $b(K_3) = 2$. We still may not claim that $b_R(G) = b(G)$ even in the case that every γ-set is a γ_R-set. The example K_3 again demonstrates this.

Theorem 132 (Hattingh and Plummer [80], 2008). *There is an infinite class of graphs in which each graph G satisfies $b(G) < b_R(G)$.*

In fact, such an infinite class of graphs can be obtained as follows. Let H be a connected graph, $\mathrm{BC}(H)$ a graph obtained from H by attaching ℓ ($\geqslant 2$) vertices of degree 1 to each vertex in H, and $\mathscr{B} = \{G : G = \mathrm{BC}(H)$ for some graph H such that $\delta(H) \geqslant 2\}$. By Theorem 3, we can easily check that $b(G) = 1 < 2 \leqslant \min\{\delta(H), \ell\} = b_R(G)$ for any $G \in \mathscr{B}$. Moreover, there exists a graph $G \in \mathscr{B}$ such that $b_R(G)$ can be much larger than $b(G)$, which is stated as the following theorem.

Theorem 133 (Hattingh and Plummer [80], 2008). *For each positive integer k, there is a graph G such that $k = b_R(G) - b(G)$.*

Combining Theorem 133 with Theorem 132, we have the following corollary.

Corollary 134. *The bondage number and the restrained bondage number are unrelated.*

Very recently, Jafari Rad et al. [81] have considered the total restrained bondage in graphs based on both total dominating sets and restrained dominating sets and obtained several properties, exact values, and bounds for the total restrained bondage number of a graph.

A restrained dominating set S of a graph G without isolated vertices is called to be a *total restrained dominating set* if S is a total dominating set. The *total restrained domination number* of G, denoted by $\gamma_{\mathrm{TR}}(G)$, is the minimum cardinality of a total restrained dominating set of G. For references on this domination in graphs, see [3, 61, 82–85]. The *total restrained bondage number* $b_{\mathrm{TR}}(G)$ of a graph G with no isolated vertex, is the cardinality of a smallest set of edges $B \subseteq E(G)$ for which $G - B$ has no isolated vertex and $\gamma_{\mathrm{TR}}(G - B) > \gamma_{\mathrm{TR}}(G)$. In the case that there is no such subset B, we define $b_{\mathrm{TR}}(G) = \infty$.

Ma et al. [84] determined that $\gamma_{\mathrm{TR}}(P_n) = n - 2\lfloor (n-2)/4 \rfloor$ for $n \geqslant 3$, $\gamma_{\mathrm{TR}}(C_n) = n - 2\lfloor n/4 \rfloor$ for $n \geqslant 3$, $\gamma_{\mathrm{TR}}(K_3) = 3$ and $\gamma_{\mathrm{TR}}(K_n) = 2$ for $n \neq 3$, and $\gamma_{\mathrm{TR}}(K_{1,n}) = 1 + n$ and $\gamma_{\mathrm{TR}}(K_{m,n}) = 2$ for $2 \leqslant m \leqslant n$. According to these results, Jafari Rad et al. [81] determined the exact values of total restrained bondage numbers for the corresponding graphs.

Theorem 135 (Jafari Rad et al. [81]). *For an integer n,*

$$b_{TR}(P_n) = \begin{cases} \infty & if\ 2 \leqslant n \leqslant 5, \\ 1, & if\ n \geqslant 5 \end{cases}$$

$$b_{TR}(C_n) = \begin{cases} \infty & if\ n = 3, \\ 1, & if\ 3 < n,\ n \equiv 0, 1 \pmod 4 \\ 2, & if\ 3 < n,\ n \equiv 2, 3 \pmod 4 \end{cases} \qquad (49)$$

$$b_{TR}(W_n) = 2 \quad for\ n \geqslant 6,$$

$$b_{TR}(K_n) = n - 1 \quad for\ n \geqslant 4,$$

$$b_{TR}(K_{m,n}) = m - 1 \quad for\ 2 \leqslant m \leqslant n.$$

$b_{TR}(K_n) = n-1$ shows the fact that for any integer $n \geqslant 3$ there exists a connected graph, namely, K_{n+1} such that $b_{TR}(K_{n+1}) = n$; that is, the total restrained bondage number is unbounded from the above.

A vertex is said to be a *support vertex* if it is adjacent to a vertex of degree one. Let $\mathscr{A}$ be the family of all connected graphs such that G belongs to $\mathscr{A}$ if and only if every edge of G is incident to a support vertex or G is a cycle of length three.

Proposition 136 (Cyman and Raczek [82], 2006). *For a connected graph G of order n, $\gamma_{TR}(G) = n$ if and only if $G \in \mathscr{A}$.*

Hence, if $G \in \mathscr{A}$, then the removal of any subset of edges of G cannot increase the total restrained domination number of G and thus $b_{TR}(G) = \infty$. When $G \notin \mathscr{A}$, a result similar to Theorem 6 is obtained.

Theorem 137 (Jafari Rad et al. [81]). *For any tree $T \notin \mathscr{A}$, $b_{TR}(T) \leqslant 2$. This bound is sharp.*

In the same paper, the authors provided a characterization for trees $T \notin \mathscr{A}$ with $b_{TR}(T) = 1$ or 2. The following result is similar to Observation 1.

Observation 3 (Jafari Rad et al. [81]). If k edges can be removed from a graph G to obtain a graph H without an isolated vertex and with $b_{TR}(H) = t$, then $b_{TR}(G) \leqslant k + t$.

Applying Theorem 137 and Observation 3, Jafari Rad et al. characterized all connected graphs G with $b_{TR}(G) = \infty$.

Theorem 138 (Jafari Rad et al. [81]). *For a connected graph G, $b_{TR}(G) = \infty$ if and only if $G \in \mathscr{A}$.*

8.4. Other Conditional Bondage Numbers. A subset $I \subseteq V(G)$ is called an *independent set* if no two vertices in I are adjacent in G. The maximum cardinality among all independent sets is called the *independence number* of G, denoted by $\alpha(G)$.

A dominating set S of a graph G is called to be *independent* if S is an independent set of G. The minimum cardinality among all independent dominating set is called the *independence domination number* of G and denoted by $\gamma_I(G)$.

Since an independent dominating set is not only a dominating set but also an independent set, $\gamma(G) \leqslant \gamma_I(G) \leqslant \alpha(G)$ for any graph G.

It is clear that a maximal independent set is certainly a dominating set. Thus, an independent set is maximal if and only if it is an independent dominating set, and so $\gamma_I(G)$ is the minimum cardinality among all maximal independent sets of G. This graph-theoretical invariant has been well studied in the literature; see, for example, Haynes et al. [3] for early results, Goddard and Henning [86] for recent results on independent domination in graphs.

In 2003, Zhang et al. [87] defined the *independence bondage number* $b_I(G)$ of a nonempty graph G to be the minimum cardinality among all subsets $B \subseteq E(G)$ for which $\gamma_I(G-B) > \gamma_I(G)$. For some ordinary graphs, their independence domination numbers can be easily determined, and so independence bondage numbers have been also determined. Clearly, $b_I(K_n) = 1$ if $n \geqslant 2$.

Theorem 139 (Zhang et al. [87], 2003). *For any integer $n \geqslant 4$,*

$$b_I(C_n) = \begin{cases} 1 & if\ n \equiv 1 \pmod 2, \\ 2 & if\ n \equiv 0 \pmod 2, \end{cases}$$

$$b_I(P_n) = \begin{cases} 1 & if\ n \equiv 0 \pmod 2, \\ 2 & if\ n \equiv 1 \pmod 2, \end{cases} \qquad (50)$$

$$b_I(K_{m,n}) = n \quad for\ m \leqslant n.$$

Apart from the above-mentioned results, as far as we find no other results on the independence bondage number in the present literature, we never knew much of any result on this parameter for a tree.

A subset S of $V(G)$ is called an *equitable dominating set* if for every $y \in V(G-S)$ there exists a vertex $x \in S$ such that $xy \in E(G)$ and $|d_G(x) - d_G(y)| \leqslant 1$ proposed by Dharmalingam [88]. The minimum cardinality of such a dominating set is called the *equitable domination number* and is denoted by $\gamma_E(G)$. Deepak et al. [89] defined the *equitable bondage number* $b_E(G)$ of a graph G to be the cardinality of a smallest set $B \subseteq E(G)$ of edges for which $\gamma_E(G-B) > \gamma_E(G)$ and determined $b_E(G)$ for some special graphs.

Theorem 140 (Deepak et al. [89], 2011). *For any integer $n \geqslant 2$,*

$$b_E(K_n) = \left\lceil \frac{n}{2} \right\rceil,$$

$$b_E(K_{m,n}) = m \quad for\ 0 \leqslant n - m \leqslant 1,$$

$$b_E(C_n) = \begin{cases} 3 & if\ n \equiv 1 \pmod 3 \\ 2 & otherwise, \end{cases}$$

$$b_E(P_n) = \begin{cases} 2 & if\ n \equiv 1 \pmod 3 \\ 1 & otherwise, \end{cases}$$

$$b_E(T) \leqslant 2 \quad for\ any\ nontrivial\ tree\ T,$$

$$b_E(G) \leqslant n - 1 \quad for\ any\ connected\ graph\ G\ of\ order\ n. \qquad (51)$$

As far as we know, there are no more results on the equitable bondage number of graphs in the present literature.

There are other variations of the normal domination by adding restricted conditions to the normal dominating set. For example, the *connected domination set* of a graph G, proposed by Sampathkumar and Walikar [90], is a dominating set S such that the subgraph induced by S is connected. The problem of finding a minimum connected domination set in a graph is equivalent to the problem of finding a spanning tree with maximum number of vertices of degree one [91], and has some important applications in wireless networks [92]; thus, it has received much research attention. However, for such an important domination, there is no result on its bondage number. We believe that the topic is of significance for future research.

9. Generalized Bondage Numbers

There are various generalizations of the classical domination, such as p-domination, distance domination, fractional domination, Roman domination, and rainbow domination. Every such a generalization can lead to a corresponding bondage. In this section, we introduce some of them.

9.1. p-Bondage Numbers. In 1985, Fink and Jacobson [93] introduced the concept of p-domination. Let p be a positive integer. A subset S of $V(G)$ is a *p-dominating set* of G if $|S \cap N_G(y)| \geqslant p$ for every $y \in \overline{S}$. The *p-domination number* $\gamma_p(G)$ is the minimum cardinality among all p-dominating sets of G. Any p-dominating set of G with cardinality $\gamma_p(G)$ is called a γ_p-set of G. Note that a γ_1-set is a classic minimum dominating set. Notice that every graph has a p-dominating set since the vertex-set $V(G)$ is such a set. We also note that a 1-dominating set is a dominating set, and so $\gamma(G) = \gamma_1(G)$. The p-domination number has received much research attention; see a state-of-the-art survey article by Chellali et al. [94].

It is clear from definition that every p-dominating set of a graph certainly contains all vertices of degree at most $p-1$. By this simple observation, to avoid the trivial case occurrence, we always assume that $\Delta(G) \geqslant p$. For $p \geqslant 2$, Lu et al. [95] gave a constructive characterization of trees with unique minimum p-dominating sets.

Recently, Lu and Xu [96] have introduced the concept to the *p-bondage number* of G, denoted by $b_p(G)$, as the minimum cardinality among all sets of edges $B \subseteq E(G)$ such that $\gamma_p(G-B) > \gamma_p(G)$. Clearly, $b_1(G) = b(G)$.

Lu and Xu [96] established a lower bound and an upper bound on $b_p(T)$ for any integer $p \geqslant 2$ and any tree T with $\Delta(T) \geqslant p$ and characterized all trees achieving the lower bound and the upper bound, respectively.

Theorem 141 (Lu and Xu [96], 2011). *For any integer $p \geqslant 2$ and any tree T with $\Delta(T) \geqslant p$,*

$$1 \leqslant b_p(T) \leqslant \Delta(T) - p + 1. \tag{52}$$

Let S be a given subset of $V(G)$ and, for any $x \in S$, let

$$N_p(x, S, G) = \{y \in \overline{S} \cap N_G(x) : |N_G(y) \cap S| = p\}. \tag{53}$$

Then, all trees achieving the lower bound 1 can be characterized as follows.

Theorem 142 (Lu and Xu [96], 2011). *Let T be a tree with $\Delta(T) \geqslant p \geqslant 2$. Then $b_p(T) = 1$ if and only if for any γ_p-set S of T there exists an edge $xy \in (S, \overline{S})$ such that $y \in N_p(x, S, T)$*

The notation $S(a,b)$ denotes a *double star* obtained by adding an edge between the central vertices of two stars $K_{1,a-1}$ and $K_{1,b-1}$. And the vertex with degree a (resp., b) in $S(a,b)$ is called the *L-central vertex* (resp., *R-central vertex*) of $S(a,b)$:

To characterize all trees attaining the upper bound given in Theorem 141, we define three types of operations on a tree T with $\Delta(T) = \Delta \geqslant p + 1$:

Type 1: attach a pendant edge to a vertex y with $d_T(y) \leqslant p-2$ in T.

Type 2: attach a star $K_{1,\Delta-1}$ to a vertex y of T by joining its central vertex to y, where y in a γ_p-set of T and $d_T(y) \leqslant \Delta - 1$.

Type 3: attach a double star $S(p, \Delta-1)$ to a pendant vertex y of T by coinciding its R-central vertex with y, where the unique neighbor of y is in a γ_p-set of T.

Let $\mathscr{B}$ = {T: a tree obtained from $K_{1,\Delta}$ or $S(p,\Delta)$ by a finite sequence of operations of Types 1, 2, and 3}.

Theorem 143 (Lu and Xu [96], 2011). *A tree with the maximum $\Delta \geqslant p+1$ has p-bondage number $\Delta - p + 1$ if and only if it belongs to $\mathscr{B}$.*

Theorem 144 (Lu and Xu [97], 2012). *Let $G_{m,n} = P_m \times P_n$. Then*

$$b_2(G_{2,n}) = 1 \quad \text{for } n \geqslant 2,$$
$$b_2(G_{3,n}) = \begin{cases} 2 & \text{if } n \equiv 1 \pmod 3, \\ 1 & \text{otherwise.} \end{cases} \quad \text{for } n \geqslant 2, \tag{54}$$
$$b_2(G_{4,n}) = \begin{cases} 1 & \text{if } n \equiv 3 \pmod 4, \\ 2 & \text{otherwise.} \end{cases} \quad \text{for } n \geqslant 7.$$

Recently, Krzywkowski [98] has proposed the concept of the nonisolating 2-bondage of a graph. The *nonisolating 2-bondage number* of a graph G, denoted by $b_2'(G)$, is the minimum cardinality among all sets of edges $B \subseteq E(G)$ such that $G - B$ has no isolated vertices and $\gamma_2(G-B) > \gamma_2(G)$. If for every $B \subseteq E(G)$, either $\gamma_2(G-B) = \gamma_2(G)$ or $G-B$ has isolated vertices, then define $b_2'(G) = 0$, and say that G is a 2-nonisolatedly strongly stable graph. Krzywkowski presented some basic properties of nonisolating 2-bondage in graphs, showed that for every nonnegative integer k there exists a tree T such $b_2'(T) = k$, characterized all 2-non-isolatedly strongly stable trees, and determined $b_2'(G)$ for several special graphs.

Theorem 145 (Krzywkowski [98], 2012). *For any positive integers n and m with $m \leqslant n$,*

$$b_2'(K_n) = \begin{cases} 0 & \text{if } n = 1,2,3 \\ \left\lfloor \frac{2n}{3} \right\rfloor & \text{otherwise,} \end{cases}$$

$$b_2'(P_n) = \begin{cases} 0 & \text{if } n = 1,2,3 \\ 1 & \text{otherwise,} \end{cases}$$

$$b_2'(C_n) = \begin{cases} 0 & \text{if } n = 3 \\ 1 & \text{if } n \text{ is even} \\ 2 & \text{if } n \text{ is odd,} \end{cases} \quad (55)$$

$$b_2'(K_{m,n}) = \begin{cases} 3 & \text{if } m = n = 3 \\ 1 & \text{if } m = n = 4 \\ 2 & \text{otherwise.} \end{cases}$$

9.2. Distance Bondage Numbers. A subset S of vertices of a graph G is said to be a *distance k-dominating set* for G if every vertex in G not in S is at distance at most k from some vertex of S. The minimum cardinality of all distance k-dominating sets is called the *distance k-domination number* of G and denoted by $\gamma_k(G)$ (do not confuse with the above-mentioned $\gamma_p(G)$!). When $k = 1$, a distance 1-dominating set is a normal dominating set, and so $\gamma_1(G) = \gamma(G)$ for any graph G. Thus, the distance k-domination is a generalization of the classical domination.

The relation between γ_k and α_k (the k-independence number, defined in Section 2.2) for a tree obtained by Meir and Moon [14], who proved that $\gamma_k(T) = \alpha_{2k}(T)$ for any tree T (a special case for $k = 1$ is stated in Proposition 10). The further research results can be found in Henning et al. [99], Tian and Xu [100–103], and Liu et al. [104].

In 1998, Hartnell et al. [15] defined the *distance k-bondage number* of G, denoted by $b_k(G)$, to be the cardinality of a smallest subset B of edges of G with the property that $\gamma_k(G - B) > \gamma_k(G)$. From Theorem 6, it is clear that if T is a nontrivial tree, then $1 \leqslant b_1(T) \leqslant 2$. Hartnell et al. [15] generalized this result to any integer $k \geqslant 1$.

Theorem 146 (Hartnell et al. [15], 1998). *For every nontrivial tree T and positive integer k, $1 \leqslant b_k(T) \leqslant 2$.*

Hartnell et al. [15] and Topp and Vestergaard [13] also characterized the trees having distance k-bondage number 2. In particular, the class of trees for which $b_1(T) = 2$ are just those which have a unique maximum 2-independent set (see Theorem 11).

Since, when $k = 1$, the distance 1-bondage number $b_1(G)$ is the classical bondage number $b(G)$, Theorem 12 gives the NP-hardness of deciding the distance k-bondage number of general graphs.

Theorem 147. *Given a nonempty undirected graph G and positive integers k and b with $b \leqslant \varepsilon(G)$, determining wether or not $b_k(G) \leqslant b$ is NP-hard.*

For a vertex x in G, the *open k-neighborhood* $N_k(x)$ of x is defined as $N_k(x) = \{y \in V(G) : 1 \leqslant d_G(x,y) \leq k\}$. The *closed k-neighborhood* $N_k[x]$ of x in G is defined as $N_k(x) \cup \{x\}$. Let

$$\Delta_k(G) = \max\{|N_k(x)| : \text{for any } x \in V(G)\}. \quad (56)$$

Clearly, $\Delta_1(G) = \Delta(G)$. The kth *power* of a graph G is the graph G^k with vertex-set $V(G^k) = V(G)$ and edge-set $E(G^k) = \{xy : 1 \leqslant d_G(x,y) \leqslant k\}$. The following lemma holds directly from the definition of G^k.

Proposition 148 (Tian and Xu [101]). *$\Delta(G^k) = \Delta_k(G)$ and $\gamma(G^k) = \gamma_k(G)$ for any graph G and each $k \geqslant 1$.*

A graph G is *k-distance domination-critical*, or *γ_k-critical* for short, if $\gamma_k(G - x) < \gamma_k(G)$ for every vertex x in G, proposed by Henning et al. [105]

Proposition 149 (Tian and Xu [101]). *For each $k \geqslant 1$, a graph G is γ_k-critical if and only if G^k is γ_k-critical.*

By the above facts, we suggest the following worthwhile problem.

Problem 6. Can we generalize the results on the bondage for G to G^k? In particular, do the following propositions hold?

(a) $b(G^k) = b_k(G)$ for any graph G and each $k \geqslant 1$.

(b) $b(G^k) \leqslant \Delta_k(G)$ if G is not γ_k-critical.

Let k and p be positive integers. A subset S of $V(G)$ is defined to be a *(k, p)-dominating set* of G if, for any vertex $x \in \overline{S}$, $|N_k(x) \cap S| \geqslant p$. The *(k, p)-domination number* of G, denoted by $\gamma_{k,p}(G)$, is the minimum cardinality among all (k,p)-dominating sets of G. Clearly, for a graph G, a $(1,1)$-dominating set is a classical dominating set, a $(k,1)$-dominating set is a distance k-dominating set, and a $(1,p)$-dominating set is the above-mentioned p-dominating set. That is, $\gamma_{1,1}(G) = \gamma(G)$, $\gamma_{k,1}(G) = \gamma_k(G)$, and $\gamma_{1,p}(G) = \gamma_p(G)$.

The concept of (k,p)-domination in a graph G is a generalized domination which combines distance k-domination and p-domination in G. So, the investigation of (k,p)-domination of G is more interesting and has received the attention of many researchers; see, for example, [106–109].

More general, Li and Xu [110] proposed the concept of the (ℓ, w)-domination number of a w-connected graph. A subset S of $V(G)$ is defined to be an *(ℓ, w)-dominating set* of G if, for any vertex $x \in \overline{S}$, there are w internally disjoint paths of length at most ℓ from x to some vertex in S. The *(ℓ, w)-domination number* of G, denoted by $\gamma_{\ell,w}(G)$, is the minimum cardinality among all (ℓ, w)-dominating sets of G. Clearly, $\gamma_{k,1}(G) = \gamma_k(G)$ and $\gamma_{1,p}(G) = \gamma_p(G)$.

It is quite natural to propose the concept of bondage number for (k,p)-domination or (ℓ, w)-domination. However, as far as we know, none has proposed this concept until today. This is a worthwhile topic for further research.

9.3. Fractional Bondage Numbers. If σ is a function mapping the vertex-set V into some set of real numbers, then for any subset $S \subseteq V$; let $\sigma(S) = \sum_{x \in S} \sigma(x)$. Also let $|\sigma| = \sigma(V)$.

A real-value function $\sigma : V \to [0,1]$ is a dominating function of a graph G if for every $x \in V(G)$, $\sigma(N_G[x]) \geqslant 1$. Thus, if S is a dominating set of G, σ is a function, where

$$\sigma(x) = \begin{cases} 1 & \text{if } x \in S, \\ 0 & \text{otherwise,} \end{cases} \tag{57}$$

then σ is a dominating function of G. The fractional domination number of G, denoted by $\gamma_F(G)$, is defined as follows.

$$\gamma_F(G) = \min\{|\sigma| : \sigma \text{ is a domination function of } G\}. \tag{58}$$

The γ_F-bondage number of G, denoted by $b_F(G)$, is defined as the minimum cardinality of a subset $B \subseteq E$ whose removal results in $\gamma_F(G - B) > \gamma_F(G)$.

Hedetniemi et al. [111] are the first to study fractional domination although Farber [112] introduced the idea indirectly. The concept of the fractional domination number was proposed by Domke and Laskar [113] in 1997. The fractional domination numbers for some ordinary graphs are determined.

Proposition 150 (Domke et al. [114], 1998; Domke and Laskar [113], 1997). *(a) If G is a k-regular graph with order n, then $\gamma_F(G) = n/k + 1$;*

(b) $\gamma_F(C_n) = n/3$, $\gamma_F(P_n) = \lceil n/3 \rceil$, $\gamma_F(K_n) = 1$;

(c) $\gamma_F(G) = 1$ if and only if $\Delta(G) = n - 1$;

(d) $\gamma_F(T) = \gamma(T)$ for any tree T;

(e) $\gamma_F(K_{n,m}) = (n(m+1) + m(n+1))/(nm-1)$, where $\max\{n,m\} > 1$.

The assertions (a)–(d) are due to Domke et al. [114] and the assertion (e) is due to Domke and Laskar [113]. According to these results, Domke and Laskar [113] determined the γ_F-bondage numbers for these graphs.

Theorem 151 (Domke and Laskar [113], 1997). *$b_F(K_n) = \lceil n/2 \rceil$; $b_F(K_{n,m}) = 1$ where $\max\{n,m\} > 1$,*

$$b_F(C_n) = \begin{cases} 2 & \text{if } n \equiv 0 \pmod 3, \\ 1 & \text{otherwise,} \end{cases}$$
$$b_F(P_n) = \begin{cases} 2 & \text{if } n \equiv 1 \pmod 3, \\ 1 & \text{otherwise,} \end{cases} \tag{59}$$

It is easy to see that for any tree T, $b_F(T) \leqslant 2$. In fact, since $\gamma_F(T) = \gamma(T)$ for any tree T, $\gamma_F(T') = \gamma(T')$ for any subgraph T' of T. It follows that

$$\begin{aligned} b_F(T) &= \min\{B \subseteq E(T) : \gamma_F(T - B) > \gamma_F(T)\} \\ &= \min\{B \subseteq E(T) : \gamma(T - B) > \gamma(T)\} \\ &= b(T) \leqslant 2. \end{aligned} \tag{60}$$

Except for these, no results on $b_F(G)$ have been known as yet.

9.4. Roman Bondage Numbers. A *Roman dominating function* on a graph G is a labeling $f : V \to \{0,1,2\}$ such that every vertex with label 0 has at least one neighbor with label 2. The weight of a Roman dominating function f on G is the value

$$f(G) = \sum_{u \in V(G)} f(u). \tag{61}$$

The minimum weight of a Roman dominating function on a graph G is called the *Roman domination number* of G, denoted by $\gamma_{RM}(G)$.

A Roman dominating function $f : V \to \{0,1,2\}$ can be represented by the ordered partition (V_0, V_1, V_2) (or (V_0^f, V_1^f, V_2^f) to refer to f) of V, where $V_i = \{v \in V \mid f(v) = i\}$. In this representation, its weight $f(G) = |V_1| + 2|V_2|$. It is clear that $V_1^f \cup V_2^f$ is a dominating set of G, called *the Roman dominating set*, denoted by $D_R^f = (V_1, V_2)$. Since $V_1^f \cup V_2^f$ is a dominating set when f is a Roman dominating function, and since placing weight 2 at the vertices of a dominating set yields a Roman dominating function, in [115], it is observed that

$$\gamma(G) \leqslant \gamma_{RM}(G) \leqslant 2\gamma(G). \tag{62}$$

A graph G is called to be *Roman* if $\gamma_{RM}(G) = 2\gamma(G)$.

The definition of the Roman domination function is proposed implicitly by Stewart [116] and ReVelle and Rosing [117]. Roman dominating numbers have been deeply studied. In particular, Bahremandpour et al. [118] showed that the problem determining the Roman domination number is NP-complete even for bipartite graphs.

Let G be a graph with maximum degree at least two. The *Roman bondage number* $b_{RM}(G)$ of G is the minimum cardinality of all sets $E' \subseteq E$ for which $\gamma_{RM}(G - E') > \gamma_{RM}(G)$. Since in study of Roman bondage number the assumption $\Delta(G) \geqslant 2$ is necessary, we always assume that when we discuss $b_{RM}(G)$, all graphs involved satisfy $\Delta(G) \geqslant 2$. The Roman bondage number $b_{RM}(G)$ is introduced by Jafari Rad and Volkmann in [119].

Recently, Bahremandpour et al. [118] have shown that the problem determining the Roman bondage number is NP-hard even for bipartite graphs.

Problem 7. Consider the decision problem:

Roman Bondage Problem

Instance: a nonempty bipartite graph G and a positive integer k.

Question: is $b_{RM}(G) \leqslant k$?

Theorem 152 (Bahremandpour et al. [118], 2013). *The Roman bondage problem is NP-hard even for bipartite graphs.*

The exact value of $b_{RM}(G)$ is known only for a few family of graphs including the complete graphs, cycles, and paths.

Theorem 153 (Jafari Rad and Volkmann [119], 2011). *For any positive integers n and m with $m \leqslant n$,*

$$b_{\mathrm{RM}}(P_n) = \begin{cases} 2, & if\ n \equiv 2 \pmod 3, \\ 1, & otherwise, \end{cases}$$
$$b_{\mathrm{RM}}(C_n) = \begin{cases} 3, & if\ n = 2 \pmod 3, \\ 2, & otherwise, \end{cases}$$
$$b_{\mathrm{RM}}(K_n) = \left\lceil \frac{n}{2} \right\rceil \quad for\ n \geqslant 3, \tag{63}$$
$$b_{\mathrm{RM}}(K_{m,n}) = \begin{cases} 1 & if\ m = 1\ and\ n \neq 1, \\ 4 & if\ m = n = 3, \\ m & otherwise. \end{cases}$$

Lemma 154 (Cockayne et al. [115], 2004). *If G is a graph of order n and contains vertices of degree $n-1$, then $\gamma_{\mathrm{RM}}(G) = 2$.*

Using Lemma 154, the third conclusion in Theorem 153 can be generalized to more general case, which is similar to Lemma 49.

Theorem 155 (Jafari Rad and Volkmann [119], 2011). *Let G be a graph with order $n \geq 3$ and t the number of vertices of degree $n-1$ in G. If $t \geqslant 1$, then $b_{\mathrm{RM}}(G) = \lceil t/2 \rceil$.*

Ebadi and PushpaLatha [120] conjectured that $b_{\mathrm{RM}}(G) \leqslant n-1$ for any graph G of order $n \geqslant 3$. Dehgardi et al. [121], Akbari and Qajar [122], independently, showed that this conjecture is true.

Theorem 156. $b_{RM}(G) \leqslant n-1$ *for any connected graph G of order $n \geqslant 3$.*

For a complete t-partite graph, we have the following result.

Theorem 157 (Hu and Xu [123], 2011). *Let $G = K_{m_1,m_2,\ldots,m_t}$ be a complete t-partite graph with $m_1 = \cdots = m_i < m_{i+1} \leq \cdots \leqslant m_t$, $t \geqslant 2$ and $n = \sum_{j=1}^{t} m_j$. Then*

$$b_{\mathrm{RM}}(G) = \begin{cases} \left\lceil \frac{i}{2} \right\rceil & if\ m_i = 1\ and\ n \geqslant 3, \\ 2 & if\ m_i = 2\ and\ i = 1, \\ i & if\ m_i = 2\ and\ i \geqslant 2, \\ 4 & if\ m_i = 3\ and\ i = t = 2, \\ n-1 & if\ m_i = 3\ and\ i = t \geqslant 3, \\ n - m_t & if\ m_i \geqslant 3\ and\ m_t \geqslant 4. \end{cases} \tag{64}$$

Consider a complete t-partite graph $K_t(3)$ for $t \geqslant 3$, $b_{\mathrm{RM}}(K_t(3)) = n-1$ by Theorem 157, where $n = 3t$, which shows that this upper bound of $n-1$ on $b_{\mathrm{RM}}(G)$ in Theorem 156 is sharp. This fact and $\gamma_R(K_t(3)) = 4$ give a negative answer to the following two problems posed by Dehgardi et al. [121].

Prove or Disprove: for any connected graph G of order $n \geq 3$, (i) $b_{\mathrm{RM}}(G) = n-1$ if and only if $G \cong K_3$; (ii) $b_{\mathrm{RM}}(G) \leqslant n - \gamma_R(G) + 1$.

Note that a complete t-partite graph $K_t(3)$ is $(n-3)$-regular and $b_{\mathrm{RM}}(K_t(3)) = n-1$ for $t \geqslant 3$, where $n = 3t$. Hu and Xu further determined that $b_R(G) = n-2$ for any $(n-3)$-regular graph G of order $n \leqslant 5$ except $K_t(3)$.

Theorem 158 (Hu and Xu [123], 2011). *For any $(n-3)$-regular graph G of order n other than $K_t(3)$, $b_{RM}(G) = n-2$ for $n \geqslant 5$.*

For a tree T with order $n \geqslant 3$, Ebadi and PushpaLatha [120], and Jafari Rad and Volkmann [119], independently, obtained an upper bound on $b_{\mathrm{RM}}(T)$.

Theorem 159. $b_{RM}(T) \leqslant 3$ *for any tree T with order $n \geqslant 3$.*

Theorem 160 (Bahremandpour et al. [118], 2013). $b_{\mathrm{RM}}(P_2 \times P_n) = 2$ *for $n \geqslant 2$.*

Theorem 161 (Jafari Rad and Volkmann [119], 2011). *Let G be a graph with order at least three.*

(a) *If (x, y, z) is a path of length 2 in G, then $b_{\mathrm{RM}}(G) \leqslant d_G(x) + d_G(y) + d_G(z) - 3 - |N_G(x) \cap N_G(y)|$.*

(b) *If G is connected, then $b_{\mathrm{RM}}(G) \leqslant \lambda(G) + 2\Delta(G) - 3$.*

Theorem 161 (b) implies $b_{\mathrm{RM}}(G) \leqslant \delta(G) + 2\Delta(G) - 3$ for a connected graph G. Note that for a planar graph G, $\delta(G) \leqslant 5$; moreover, $\delta(G) \leqslant 3$ if the girth at least 4, and $\delta(G) \leqslant 2$ if the girth at least 6. These two facts show that $b_{\mathrm{RM}}(G) \leqslant 2\Delta(G) + 2$ for a connected planar graph G. Jafari Rad and Volkmann [124] improved this bound.

Theorem 162 (Jafari Rad and Volkmann [124], 2011). *For any connected planar graph G of order $n \geqslant 3$ with girth $g(G)$,*

$$b_{\mathrm{RM}}(G) \leqslant \begin{cases} 2\Delta(G), & \\ \Delta(G) + 6, & \\ \Delta(G) + 5 & if\ G\ contains\ no\ vertices\ of\ degree\ five, \\ \Delta(G) + 4 & if\ g(G) \geqslant 4, \\ \Delta(G) + 3 & if\ g(G) \geqslant 5, \\ \Delta(G) + 2 & if\ g(G) \geqslant 6, \\ \Delta(G) + 1 & if\ g(G) \geqslant 8. \end{cases} \tag{65}$$

According to Theorem 157, $b_{\mathrm{RM}}(C_n) = 3 = \Delta(C_n)+1$ for a cycle C_n of length $n \geqslant 8$ with $n \equiv 2 \pmod 3$, and therefore the last upper bound in Theorem 162 is sharp, at least for $\Delta = 2$.

Combining the fact that every planar graph G with minimum degree 5 contains an edge xy with $d_G(x) = 5$ and $d_G(y) \in \{5, 6\}$ with Theorem 161 (a), Akbari et al. [125] obtained the following result.

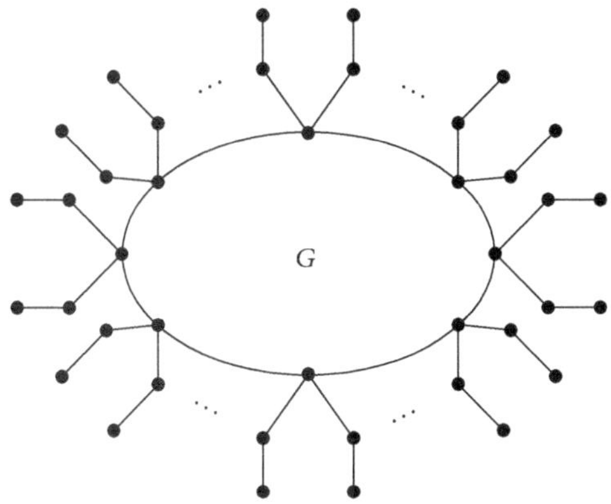

Figure 9: The graph $\widehat{G}$ is constructed from G.

Theorem 163 (Akbari et al. [125], 2012). *$b_{RM}(G) \leqslant 15$ for every planar graph G.*

It remains open to show whether the bound in Theorem 163 is sharp or not. Though finding a planar graph G with $b_{RM}(G) = 15$ seems to be difficult, Akbari et al. [125] constructed an infinite family of planar graphs with Roman bondage number equal to 7 by proving the following result.

Theorem 164 (Akbari et al. [125], 2012). *Let G be a graph of order n and $\widehat{G}$ is the graph of order $5n$ obtained from G by attaching the central vertex of a copy of a path P_5 to each vertex of G (see Figure 9). Then $\gamma_{RM}(\widehat{G}) = 4n$ and $b_{RM}(\widehat{G}) = \delta(G) + 2$.*

By Theorem 164, infinite many planar graphs with Roman bondage number 7 can be obtained by considering any planar graph G with $\delta(G) = 5$ (e.g., the icosahedron graph).

Conjecture 165 (Akbari et al. [125], 2012). *The Roman bondage number of every planar graph is at most 7.*

For general bounds, the following observation is directly obtained, which is similar to Observation 1.

Observation 4. Let G be a graph of order n with maximum degree at least two. Assume that H is a spanning subgraph obtained from G by removing k edges. If $\gamma_{RM}(H) = \gamma_{RM}(G)$, then $b_R(H) \leqslant b_{RM}(G) \leqslant b_{RM}(H) + k$.

Theorem 166 (Bahremandpour et al. [118], 2013). *For any connected graph G of order n, if $n \geqslant 3$ and $\gamma_{RM}(G) = \gamma(G) + 1$, then*

$$b_{RM}(G) \leqslant \min\{b(G), n_\Delta\}, \tag{66}$$

where n_Δ is the number of vertices with maximum degree Δ in G.

Observation 5. For any graph G, if $\gamma(G) = \gamma_{RM}(G)$, then $b_{RM}(G) \leqslant b(G)$.

Theorem 167 (Bahremandpour et al. [118], 2013). *For every Roman graph G,*

$$b_{RM}(G) \geqslant b(G). \tag{67}$$

The bound is sharp for cycles on n vertices where $n \equiv 0 \pmod 3$.

The strict inequality in Theorem 167 can hold, for example, $b(C_{3k+2}) = 2 < 3 = b_{RM}(C_{3k+2})$ by Theorem 157.

A graph G is called to be *vertex Roman domination-critical* if $\gamma_{RM}(G - x) < \gamma_{RM}(G)$ for every vertex x in G. If G has no isolated vertices, then $\gamma_{RM}(G) \leqslant 2\gamma(G) \leqslant 2\beta(G)$. If $\gamma_{RM}(G) = 2\beta(G)$, then $\gamma_{RM}(G) = 2\gamma(G)$ and hence G is a Roman graph. In [30], Volkmann gave a lot of graphs with $\gamma(G) = \beta(G)$. The following result is similar to Theorem 57.

Theorem 168 (Bahremandpour et al. [118], 2013). *For any graph G, if $\gamma_{RM}(G) = 2\beta(G)$, then*

(a) *$b_{RM}(G) \geqslant \delta(G)$;*

(b) *$b_{RM}(G) \geqslant \delta(G) + 1$ if G is a vertex Roman domination-critical graph.*

If $\gamma_{RM}(G) = 2$, then obviously $b_{RM}(G) \leqslant \delta(G)$. So we assume $\gamma_{RM}(G) \geqslant 3$. Dehgardi et al. [121] proved $b_{RM}(G) \leqslant (\gamma_{RM}(G) - 2)\Delta(G) + 1$ and posed the following problem: if G is a connected graph of order $n \geqslant 4$ with $\gamma_{RM}(G) \geqslant 3$, then

$$b_{RM}(G) \leqslant (\gamma_{RM}(G) - 2)\Delta(G). \tag{68}$$

Theorem 161 (a) shows that the inequality (68) holds if $\gamma_{RM}(G) \geqslant 5$. Thus, the bound in (68) is of interest only when $\gamma_{RM}(G)$ is 3 or 4.

Lemma 169 (Bahremandpour et al. [118], 2013). *For any nonempty graph G of order $n \geqslant 3$, $\gamma_{RM}(G) = 3$ if and only if $\Delta(G) = n - 2$.*

The following result shows that (68) holds for all graphs G of order $n \geqslant 4$ with $\gamma_{RM}(G) = 3$ or 4, which improves Theorem 161 (b).

Theorem 170 (Bahremandpour et al. [118], 2013). *For any connected graph G of order $n \geqslant 4$,*

$$b_{RM}(G) \leq \begin{cases} \Delta(G) = n - 2 & \text{if } \gamma_{RM}(G) = 3, \\ \Delta(G) + \delta(G) - 1 & \text{if } \gamma_{RM}(G) = 4 \end{cases} \tag{69}$$

with the first equality if and only if $G \cong C_4$.

Recently, Akbari and Qajar [122] have proved the following result.

Theorem 171 (Akbari and Qajar [122]). *For any connected graph G of order $n \geqslant 3$,*

$$b_{RM}(G) \leqslant n - \gamma_{RM}(G) + 5. \tag{70}$$

We conclude this section with the following problems.

Problem 8. Characterize all connected graphs G of order $n \geqslant 3$, for which $b_{\mathrm{RM}}(G) = n - 1$.

Problem 9. Prove or disprove: if G is a connected graph of order $n \geqslant 3$, then

$$b_{\mathrm{RM}}(G) \leqslant n - \gamma_{\mathrm{RM}}(G) + 3. \quad (71)$$

Note that $\gamma_R(K_t(3)) = 4$ and $b_{\mathrm{RM}}(K_t(3)) = n-1$, where $n = 3t$ for $t \geqslant 3$. The upper bound on $b_{\mathrm{RM}}(G)$ given in Problem 9 is sharp if Problem 9 has positive answer.

9.5. Rainbow Bondage Numbers. For a positive integer k, a *k-rainbow dominating function* of a graph G is a function f from the vertex-set $V(G)$ to the set of all subsets of the set $\{1, 2, \ldots, k\}$ such that for any vertex x in G with $f(x) = \emptyset$ the condition

$$\bigcup_{y \in N_G(x)} f(y) = \{1, 2, \ldots, k\} \quad (72)$$

is fulfilled. The weight of a k-rainbow dominating function f on G is the value

$$f(G) = \sum_{x \in V(G)} |f(x)|. \quad (73)$$

The *k-rainbow domination number* of a graph G, denoted by $\gamma_{rk}(G)$, is the minimum weight of a k-rainbow dominating function f of G.

Note that $\gamma_{r1}(G)$ is the classical domination number $\gamma(G)$. The k-rainbow domination number is introduced by Brešar et al. [126]. Rainbow domination of a graph G coincides with ordinary domination of the Cartesian product of G with the complete graph, in particular, $\gamma_{rk}(G) = \gamma(G \times K_k)$ for any positive integer k and any graph G [126]. Hartnell and Rall [127] proved that $\gamma(T) < \gamma_{r2}(T)$ for any tree T, no graph has $\gamma = \gamma_{rk}$ for $k \geqslant 3$, for any $k \geqslant 2$, and any graph G of order n:

$$\min\{n, \gamma(G) + k - 2\} \leqslant \gamma_{rk}(G) \leqslant k\gamma(G). \quad (74)$$

The introduction of rainbow domination is motivated by the study of paired domination in Cartesian products of graphs, where certain upper bounds can be expressed in terms of rainbow domination; that is, for any graph G and any graph H, if $V(H)$ can be partitioned into $k\gamma_P$-sets, then $\gamma_P(G \times H) \leqslant (1/k)\upsilon(H)\gamma_{rk}(G)$, proved by Brešar et al. [126].

Dehgardi et al. [128] introduced the concept of k-rainbow bondage number of graphs. Let G be a graph with maximum degree at least two. The k-rainbow bondage number $b_{rk}(G)$ of G is the minimum cardinality of all sets $B \subseteq E(G)$ for which $\gamma_{rk}(G - B) > \gamma_{rk}(G)$. Since in the study of k-rainbow bondage number the assumption $\Delta(G) \geqslant 2$ is necessary, we always assume that when we discuss $b_{rk}(G)$, all graphs involved satisfy $\Delta(G) \geqslant 2$.

The 2-rainbow bondage number of some special families of graphs has been determined.

Theorem 172 (Dehgardi et al. [128]). *For any integer $n \geqslant 3$,*

$$b_{r2}(P_n) = 1 \quad \text{for } n \geqslant 2,$$
$$b_{r2}(C_n) = \begin{cases} 1 & \text{if } n \equiv 0 \pmod 4, \\ 2 & \text{otherwise}, \end{cases}$$
$$b_{r2}(K_{n,n}) = \begin{cases} 1 & \text{if } n = 2, \\ 3 & \text{if } n = 3, \\ 6 & \text{if } n = 4, \\ m & \text{if } n \geqslant 5. \end{cases} \quad (75)$$

Theorem 173 (Dehgardi et al. [128]). *For any connected graph G of order $n \geqslant 3$, $b_{r2}(G) \leqslant \lambda(G) + 2\Delta(G) - 3$.*

Theorem 174 (Dehgardi et al. [128]). *For any tree T of order $n \geqslant 3$, $b_{r2}(T) \leqslant 2$.*

Theorem 175 (Dehgardi et al. [128]). *If G is a planar graph with maximum degree at least two, then $b_{r2}(G) \leqslant 15$. Moreover, if the girth $g(G) \geqslant 4$, then $b_{r2}(G) \leqslant 11$ and if $g(G) \geqslant 6$, then $b_{r2}(G) \leqslant 9$.*

Theorem 176 (Dehgardi et al. [128]). *For a complete bipartite graph $K_{p,q}$, if $2k + 1 \leqslant p \leqslant q$, then $b_{rk}(K_{p,q}) = p$.*

Theorem 177 (Dehgardi et al. [128]). *For a complete graph K_n, if $n \geqslant k + 1$, then $b_{rk}(K_n) = \lceil kn/(k+1) \rceil$.*

The special case $k = 1$ of Theorem 177 can be found in Theorem 1. The following is a simple observation similar to Observation 1.

Observation 6 (Dehgardi et al. [128]). Let G be a graph with maximum degree at least two, H a spanning subgraph obtained from G by removing k edges. If $\gamma_{rk}(H) = \gamma_{rk}(G)$, then $b_{rk}(H) \leqslant b_{rk}(G) \leqslant b_{rk}(H) + k$.

Theorem 178 (Dehgardi et al. [128]). *For every graph G with $\gamma_{rk}(G) = k\gamma(G)$, $b_{rk}(G) \geqslant b(G)$.*

A graph G is said to be *vertex k-rainbow domination-critical* if $\gamma_{rk}(G - x) < \gamma_{rk}(G)$ for every vertex x in G. It is clear that if G has no isolated vertices, then $\gamma_{rk}(G) \leqslant k\gamma(G) \leqslant k\beta(G)$, where $\beta(G)$ is the vertex-covering number of G. Thus, if $\gamma_{rk}(G) = k\beta(G)$, then $\gamma_{rk}(G) = k\gamma(G)$.

Theorem 179 (Dehgardi et al. [128]). *For any graph G, if $\gamma_{rk}(G) = k\beta(G)$, then*

(a) *$b_{rk}(G) \geqslant \delta(G)$;*

(b) *$b_{rk}(G) \geqslant \delta(G) + 1$ if G is vertex k-rainbow domination-critical.*

As far as we know, there are no more results on the k-rainbow bondage number of graphs in the literature.

10. Results on Digraphs

Although domination has been extensively studied in undirected graphs, it is natural to think of a dominating set as a one-way relationship between vertices of the graph. Indeed, among the earliest literatures on this subject, von Neumann and Morgenstern [129] used what is now called domination in digraphs to find solution (or kernels, which are independent dominating sets) for cooperative n-person games. Most likely, the first formulation of domination by Berge [130], in 1958, was given in the context of digraphs and, only some years latter by Ore [131], in 1962, for undirected graphs. Despite this history, examination of domination and its variants in digraphs has been essentially overlooked (see [4] for an overview of the domination literature). Thus, there are few, if any, such results on domination for digraphs in the literature.

The bondage number and its related topics for undirected graph have become one of major areas both in theoretical and applied researches. However, until recently, Carlson and Develin [42], Shan and Kang [132], and Huang and Xu [133, 134] studied the bondage number for digraphs, independently. In this section, we introduce their results for general digraphs. Results for some special digraphs such as vertex-transitive digraphs are introduced in the next section.

10.1. Upper Bounds for General Digraphs. Let $G = (V, E)$ be a digraph without loops and parallel edges. A digraph G is called to be *asymmetric* if whenever $(x, y) \in E(G)$ then $(y, x) \notin E(G)$ and to be *symmetric* if $(x, y) \in E(G)$ implies $(y, x) \in E(G)$. For a vertex x of $V(G)$, the sets of out-neighbors and in-neighbors of x are, respectively, defined as $N_G^+(x) = \{y \in V(G) : (x, y) \in E(G)\}$ and $N_G^-(x) = \{x \in V(G) : (x, y) \in E(G)\}$, the out-degree and the in-degree of x are, respectively, defined as $d_G^+(x) = |N_G^+(x)|$ and $d_G^-(x) = |N_G^+(x)|$. Denote the maximum and the minimum out-degree (resp., in-degree) of G by $\Delta^+(G)$ and $\delta^+(G)$ (resp., $\Delta^-(G)$ and $\delta^-(G)$). The degree $d_G(x)$ of x is defined as $d_G^+(x)+d_G^-(x)$, and the maximum and the minimum degrees of G are denoted by $\Delta(G)$ and $\delta(G)$, respectively; that is, $\Delta(G) = \max\{d_G(x) : x \in V(G)\}$ and $\delta(G) = \min\{d_G(x) : x \in V(G)\}$. Note that the definitions here are different from the ones in the textbook on digraphs; see for example, Xu [1].

A subset S of $V(G)$ is called a *dominating set* if $V(G) = S \cup N_G^+(S)$, where $N_G^+(S)$ is the set of out-neighbors of S. Then, just as for undirected graphs, $\gamma(G)$ is the minimum cardinality of a dominating set, and the *bondage number* $b(G)$ is the smallest cardinality of a set B of edges such that $\gamma(G - B) > \gamma(G)$ if such a subset $B \subseteq E(G)$ exists. Otherwise, we put $b(G) = \infty$.

Some basic results for undirected graphs stated in Section 3 can be generalized to digraphs. For example, Theorem 18 is generalized by Carlson and Develin [42], Huang and Xu [133], and Shan and Kang [132], independently, as follows.

Theorem 180. *Let G be a digraph and $(x, y) \in E(G)$. Then $b(G) \leqslant d_G(y) + d_G^-(x) - |N_G^-(x) \cap N_G^-(y)|$.*

Corollary 181. *For a digraph G, $b(G) \leqslant \delta^-(G) + \Delta(G)$.*

Since $\delta^-(G) \leqslant (1/2)\Delta(G)$, from Corollary 181, Conjecture 53 is valid for digraphs.

Corollary 182. *For a digraph G, $b(G) \leqslant (3/2)\Delta(G)$.*

In the case of undirected graphs, the bondage number of $G_n = K_n \times K_n$ achieves this bound. However, it was shown by Carlson and Develin in [42] that if we take the symmetric digraph G_n, we have $\Delta(G_n) = 4(n-1)$, $\gamma(G_n) = n$, and $b(G_n) \leqslant 4(n - 1) + 1 = \Delta(G_n) + 1$. So this family of digraphs cannot show that the bound in Corollary 182 is tight.

Corresponding to Conjecture 51, which is discredited for undirected graphs and is valid for digraphs, the same conjecture for digraphs can be proposed as follows.

Conjecture 183 (Carlson and Develin [42], 2006). *$b(G) \leqslant \Delta(G) + 1$ for any digraph G.*

If Conjecture 183 is true, then the following results shows that this upper bound is tight since $b(K_n \times K_n) = \Delta(K_n \times K_n)+1$ for a complete digraph K_n.

In 2007, Shan and Kang [132] gave some tight upper bounds on the bondage numbers for some asymmetric digraphs. For example, $b(T) \leqslant \Delta(T)$ for any asymmetric directed tree T; $b(G) \leqslant \Delta(G)$ for any asymmetric digraph G with order at least 4 and $\gamma(G) \leqslant 2$. For planar digraphs, they obtained the following results.

Theorem 184 (Shan and Kang [132], 2007). *Let G be a asymmetric planar digraph. Then $b(G) \leqslant \Delta(G)+2$; and $b(G) \leqslant \Delta(G) + 1$ if $\Delta(G) \geqslant 5$ and $d_G^-(x) \geqslant 3$ for every vertex x with $d_G(x) \geqslant 4$.*

10.2. Results for Some Special Digraphs. The exact values and bounds on $b(G)$ for some standard digraphs are determined.

Theorem 185 (Huang and Xu [133], 2006). *For a directed cycle C_n and a directed path P_n,*

$$b(C_n) = \begin{cases} 3 & \text{if } n \text{ is odd}, \\ 2 & \text{if } n \text{ is even}, \end{cases} \qquad b(P_n) = \begin{cases} 2 & \text{if } n \text{ is odd}, \\ 1 & \text{if } n \text{ is even}. \end{cases} \tag{76}$$

For the de Bruijn digraph $B(d, n)$ and the Kautz digraph $K(d, n)$,

$$b(B(d,n)) = d \quad \text{if } n \text{ is odd},$$
$$d \leqslant b(B(d,n)) \leqslant 2d \quad \text{if } n \text{ is even}, \tag{77}$$
$$b(K(d,n)) = d + 1.$$

Like undirected graphs, we can define the total domination number and the total bondage number. On the total bondage numbers for some special digraphs, the known results are as follows.

Theorem 186 (Huang and Xu [134], 2007). *For a directed cycle C_n and a directed path P_n, $b_T(P_n)$ and $b_T(C_n)$ all do not exist. For a complete digraph K_n,*

$$\begin{gathered} b_T(K_n) = \infty \quad \text{if } n = 1,2, \\ b_T(K_n) = 3 \quad \text{if } n = 3, \\ n \leqslant b_T(K_n) \leqslant 2n-3 \quad \text{if } n \geqslant 4. \end{gathered} \tag{78}$$

The extended de Bruijn digraph $EB(d,n;q_1,\dots,q_p)$ and the extended Kautz digraph $EK(d,n;q_1,\dots,q_p)$ are introduced by Shibata and Gonda [135]. If $p = 1$, then they are the de Bruijn digraph $B(d,n)$ and the Kautz digraph $K(d,n)$, respectively. Huang and Xu [134] determined their total domination numbers. In particular, their total bondage numbers for general cases are determined as follows.

Theorem 187 (Huang and Xu [134], 2007). *If $d \geqslant 2$ and $q_i \geqslant 2$ for each $i = 1,2,\dots,p$, then*

$$\begin{gathered} b_T\left(EB\left(d,n;q_1,\dots,q_p\right)\right) = d^p - 1, \\ b_T\left(EK\left(d,n;q_1,\dots,q_p\right)\right) = d^p. \end{gathered} \tag{79}$$

In particular, for the de Bruijn digraph $B(d,n)$ and the Kautz digraph $K(d,n)$,

$$b_T(B(d,n)) = d-1, \qquad b_T(K(d,n)) = d. \tag{80}$$

Zhang et al. [136] determined the bondage number in complete t-partite digraphs.

Theorem 188 (Zhang et al. [136], 2009). *For a complete t-partite digraph $K_{n_1,n_2,\dots,n_t}$, where $n_1 \leqslant n_2 \leqslant \cdots \leqslant n_t$,*

$$b\left(K_{n_1,n_2,\dots,n_t}\right) = \begin{cases} m & \text{if } n_m = 1,\ n_{m+1} \geqslant 2 \text{ for some, } m\,(1 \leqslant m < t) \\ 4t-3 & \text{if } n_1 = n_2 = \cdots = n_t = 2, \\ \sum_{i=1}^{t-1} n_i + 2(t-1) & \text{otherwise.} \end{cases} \tag{81}$$

Since an undirected graph can be thought of as a symmetric digraph, any result for digraphs has an analogy for undirected graphs in general. In view of this point, studying the bondage number for digraphs is more significant than for undirected graphs. Thus, we should further study the bondage number of digraphs and try to generalize known results on the bondage number and related variants for undirected graphs to digraphs, prove or disprove Conjecture 183. In particular, determine the exact values of $b(B(d,n))$ for an even n and $b_T(K_n)$ for $n \geqslant 4$.

11. Efficient Dominating Sets

A dominating set S of a graph G is called to be *efficient* if for every vertex x in G, $|N_G[x] \cap S| = 1$ if G is a undirected graph or $|N_G^-[x] \cap S| = 1$ if G is a directed graph. The concept of an efficient set was proposed by Bange et al. [137] in 1988. In the literature, an efficient set is also called *a perfect dominating set* (see Livingston and Stout [138] for an explanation).

From definition, if S is an efficient dominating set of a graph G, then S is certainly an independent set and every vertex not in S is adjacent to exactly one vertex in S.

It is also clear from definition that a dominating set S is efficient if and only if $\mathcal{N}_G[S] = \{N_G[x] : x \in S\}$ for the undirected graph G or $\mathcal{N}_G^+[S] = \{N_G^+[x] : x \in S\}$ for the digraph G is a partition of $V(G)$, where the induced subgraph by $N_G[x]$ or $N_G^+[x]$ is a star or an out-star with the root x.

The efficient domination has important applications in many areas, such as error-correcting codes, and receives much attention in the late years.

The concept of efficient dominating sets is a measure of the efficiency of domination in graphs. Unfortunately, as shown in [137], not every graph has an efficient dominating set and, moreover, it is an NP-complete problem to determine whether a given graph has an efficient dominating set. In addition, it was shown by Clark [139] in 1993 that for a wide range of p, almost every random undirected graph $G \in \mathcal{G}(\upsilon, p)$ has no efficient dominating sets. This means that undirected graphs possessing an efficient dominating set are rare. However, it is easy to show that every undirected graph has an orientation with an efficient dominating set (see Bange et al. [140]).

In 1993, Barkauskas and Host [141] showed that determining whether an arbitrary oriented graph has an efficient dominating set is NP-complete. Even so, the existence of efficient dominating sets for some special classes of graphs has been examined; see, for example, Dejter and Serra [142] and Lee [143] for Cayley graph, Gu et al. [144] for meshes and tori, Obradović et al. [145], Huang and Xu [36], Reji Kumar and MacGillivray [146] for circulant graphs, Harary graphs and tori, and Van Wieren et al. [147] for cube-connected cycles.

In this section, we introduce results of the bondage number for some graphs with efficient dominating sets, due to Huang and Xu [36].

11.1. Results for General Graphs. In this subsection, we introduce some results on bondage numbers obtained by applying efficient dominating sets. We first state the two following lemmas.

Lemma 189. *For any k-regular graph or digraph G of order n, $\gamma(G) \geqslant n/(k+1)$, with equality if and only if G has an efficient dominating set. In addition, if G has an efficient dominating set, then every efficient dominating set is certainly a γ-set, and vice versa.*

Let e be an edge and S a dominating set in G. We say that *e supports S* if $e \in (S,\overline{S})$, where $(S,\overline{S}) = \{(x,y) \in E(G) : x \in S, y \in \overline{S}\}$. Denote by $s(G)$ the minimum number of edges which support all γ-sets in G.

Lemma 190. *For any graph or digraph G, $b(G) \geqslant s(G)$, with equality if G is regular and has an efficient dominating set.*

A graph G is called to be *vertex-transitive* if its automorphism group $\mathrm{Aut}(G)$ acts transitively on its vertex-set $V(G)$. A vertex-transitive graph is regular. Applying Lemmas 189 and 190, Huang and Xu obtained some results on bondage numbers for vertex-transitive graphs or digraphs.

Theorem 191 (Huang and Xu [36], 2008). *For any vertex-transitive graph or digraph G of order n,*

$$b(G) \geqslant \begin{cases} \left\lceil \dfrac{n}{2\gamma(G)} \right\rceil & \text{if } G \text{ is undirected}, \\ \left\lceil \dfrac{n}{\gamma(G)} \right\rceil & \text{if } G \text{ is directed}. \end{cases} \quad (82)$$

Theorem 192 (Huang and Xu [36], 2008). *Let G be a k-regular graph of order n. Then*

$b(G) \leqslant k$ if G is undirected and $n \geqslant \gamma(G)(k+1)-k+1$;

$b(G) \leqslant k+1+\ell$ if G is directed and $n \geqslant \gamma(G)(k+1)-\ell$ with $0 \leqslant \ell \leqslant k-1$.

Next, we introduce a better upper bound on $b(G)$. To this aim, we need the following concept, which is a generalization of the concept of the edge-covering of a graph G. For $V' \subseteq V(G)$ and $E' \subseteq E(G)$, we say that E' covers V' and call E' an *edge-covering* for V' if there exists an edge $(x, y) \in E'$ for any vertex $x \in V'$. For $y \in V(G)$, let $\beta'[y]$ be the minimum cardinality over all edge-coverings for $N_G^-[y]$.

Theorem 193 (Huang and Xu [36], 2008). *If G is a k-regular graph with order $\gamma(G)(k+1)$, then $b(G) \leqslant \beta'[y]$ for any $y \in V(G)$.*

The upper bound on $b(G)$ given in Theorem 193 is tight in view of $b(C_n)$ for a cycle or a directed cycle C_n (see Theorems 1 and 185, resp.).

It is easy to see that for a k-regular graph G, $\lceil (k+1)/2 \rceil \leqslant \beta'[y] \leqslant k$ when G is undirected and $\beta'[y] = k+1$ when G is directed. By this fact and Lemma 189, the following theorem is merely a simple combination of Theorems 191 and 193.

Theorem 194 (Huang and Xu [36], 2008). *Let G be a vertex-transitive graph of degree k. If G has an efficient dominating set, then*

$$\left\lceil \frac{k+1}{2} \right\rceil \leqslant b(G) \leqslant k \quad \text{if } G \text{ is undirected}, \qquad b(G) = k+1 \quad \text{if } G \text{ is directed}. \quad (83)$$

Theorem 195 (Huang and Xu [36], 2008). *If G is an undirected vertex-transitive cubic graph with order $4\gamma(G)$ and girth $g(G) \leqslant 5$, then $b(G) = 2$.*

The proof of Theorem 195 leads to a byproduct. If $g = 5$, let $(u_1, u_2, u_3, u_4, u_5)$ be a cycle, and let $\mathscr{D}_i$ be the family of efficient dominating sets containing u_i for each $i \in \{1, 2, 3, 4, 5\}$. Then $\mathscr{D}_i \cap \mathscr{D}_j = \emptyset$ for $i = 1, 2, \ldots, 5$. Then G has at least $5s$ efficient dominating sets, where $s = s(G)$. But there are only $4s$ $(= ns/\gamma)$ distinct efficient dominating sets in G. This contradiction implies that an undirected vertex-transitive cubic graph with girth five has no efficient dominating sets. But a similar argument for $g(G) = 3, 4$ or $g(G) \geqslant 6$ could not give any contradiction. This is consistent with the result that $\mathrm{CCC}(n)$, a vertex-transitive cubic graph with girth n if $3 \leqslant n \leqslant 8$, or girth 8 if $n \geqslant 9$, has efficient dominating sets for all $n \geqslant 3$ except $n = 5$ (see Theorem 199 in the following subsection).

11.2. Results for Cayley Graphs. In this subsection, we use Theorem 194 to determine the exact values or approximative values of bondage numbers for some special vertex-transitive graphs by characterizing the existence of efficient dominating sets in these graphs.

Let Γ be a nontrivial finite group, S a nonempty subset of Γ without the identity element of Γ. A digraph G is defined as follows:

$$V(G) = \Gamma, \qquad (x, y) \in E(G) \Longleftrightarrow x^{-1}y \in S \quad \text{for any } x, y \in \Gamma \quad (84)$$

is called a *Cayley digraph* of the group Γ with respect to S, denoted by $C_\Gamma(S)$. If $S^{-1} = \{s^{-1} : s \in S\} = S$; then $C_\Gamma(S)$ is symmetric and is called a Cayley undirected graph, a Cayley graph for short. Cayley graphs or digraphs are certainly vertex-transitive (see Xu [1]).

A circulant graph $G(n; S)$ of order n is a Cayley graph $C_{Z_n}(S)$, where $Z_n = \{0, \ldots, n-1\}$ is the addition group of order n and S is a nonempty subset of Z_n without the identity element and, hence, is a vertex-transitive digraph of degree $|S|$. If $S^{-1} = S$, then $G(n; S)$ is an undirected graph. If $S = \{1, k\}$, where $2 \leqslant k \leqslant n-2$, we write $G(n; 1, k)$ for $G(n; \{1, k\})$ or $G(n; \{\pm 1, \pm k\})$ and call it a double-loop circulant graph.

Huang and Xu [36] showed that, for directed $G = G(n; 1, k)$, $\lceil n/3 \rceil \leqslant \gamma(G) \leqslant \lceil n/2 \rceil$ and G has an efficient dominating set if and only if $3 \mid n$ and $k \equiv 2 \pmod 3$; for directed $G = G(n; 1, k)$ and $k \neq n/2$, $\lceil n/5 \rceil \leqslant \gamma(G) \leqslant \lceil n/3 \rceil$ and G has an efficient dominating set if and only if $5 \mid n$ and $k \equiv \pm 2 \pmod 5$. By Theorem 194, we obtain the bondage number of a double loop circulant graph if it has an efficient dominating set.

Theorem 196 (Huang and Xu [36], 2008). *Let G be a double-loop circulant graph $G(n; 1, k)$. If G is directed with $3 \mid n$ and $k \equiv 2 \pmod 3$, or G is undirected with $5 \mid n$ and $k \equiv \pm 2 \pmod 5$, then $b(G) = 3$.*

The $m \times n$ torus is the Cartesian product $C_m \times C_n$ of two cycles and is a Cayley graph $C_{Z_m \times Z_n}(S)$, where $S = \{(0,1), (1,0)\}$ for directed cycles and $S = \{(0, \pm 1), (\pm 1, 0)\}$ for undirected cycles and, hence, is vertex-transitive. Gu et al. [144] showed that the undirected torus $C_m \times C_n$ has an efficient dominating set if and only if both m and n are multiples of 5. Huang and Xu [36] showed that the directed torus $C_m \times C_n$ has an efficient dominating set if and only if both m and n are multiples of 3. Moreover, they found a necessary condition for a dominating set containing the vertex $(0, 0)$ in $C_m \times C_n$ to be efficient and obtained the following result.

Theorem 197 (Huang and Xu [36], 2008). *Let $G = C_m \times C_n$. If G is undirected and both m and n are multiples of* 5 *or if G is directed and both m and n are multiples of* 3*, then $b(G) = 3$.*

The hypercube Q_n is the Cayley graph $C_\Gamma(S)$, where $\Gamma = Z_2 \times \cdots \times Z_2 = (Z_2)^n$ and $S = \{100\cdots 0, 010\cdots 0, \ldots, 00\cdots 01\}$. Lee [143] showed that Q_n has an efficient dominating set if and only if $n = 2^m - 1$ for a positive integer m. By Theorem 194, the following result is obtained immediately.

Theorem 198 (Huang and Xu [36], 2008). *If $n = 2^m - 1$ for a positive integer m, then $2^{m-1} \leqslant b(Q_n) \leqslant 2^m - 1$.*

Problem 10. Determine the exact value of $b(Q_n)$ for $n = 2^m - 1$.

The n-dimensional cube-connected cycle, denoted by CCC(n), is constructed from the n-dimensional hypercube Q_n by replacing each vertex x in Q_n with an undirected cycle C_n of length n and linking the ith vertex of the C_n to the ith neighbor of x. It has been proved that CCC(n) is a Cayley graph and, hence, is a vertex-transitive graph with degree 3. Van Wieren et al. [147] proved that CCC(n) has an efficient dominating set if and only if $n \neq 5$. From Theorems 194 and 195, the following result can be derived.

Theorem 199 (Huang and Xu [36], 2008). *Let $G = CCC(n)$ be the n-dimensional cube-connected cycles with $n \geqslant 3$ and $n \neq 5$. Then $\gamma(G) = n2^{n-2}$ and $2 \leqslant b(G) \leqslant 3$. In addition, $b(CCC(3)) = b(CCC(4)) = 2$.*

Problem 11. Determine the exact value of $b(CCC(n))$ for $n \geqslant 5$.

Acknowledgments

This work was supported by NNSF of China (Grants nos. 11071233, 61272008). The author would like to express his gratitude to the anonymous referees for their kind suggestions and comments on the original paper, to Professor Sheikholeslami and Professor Jafari Rad for sending the references [128] and [81], which resulted in this version.

References

[1] J.-M. Xu, *Theory and Application of Graphs*, vol. 10 of *Network Theory and Applications*, Kluwer Academic, Dordrecht, The Netherlands, 2003.

[2] S. T. Hedetniemi and R. C. Laskar, "Bibliography on domination in graphs and some basic definitions of domination parameters," *Discrete Mathematics*, vol. 86, no. 1–3, pp. 257–277, 1990.

[3] T. W. Haynes, S. T. Hedetniemi, and P. J. Slater, *Fundamentals of Domination in Graphs*, vol. 208 of *Monographs and Textbooks in Pure and Applied Mathematics*, Marcel Dekker, New York, NY, USA, 1998.

[4] T. W. Haynes, S. T. Hedetniemi, and P. J. Slater, Eds., *Domination in Graphs*, vol. 209 of *Monographs and Textbooks in Pure and Applied Mathematics*, Marcel Dekker, New York, NY, USA, 1998.

[5] M. R. Garey and D. S. Johnson, *Computers and Intractability*, W. H. Freeman and Company, San Francisco, Calif, USA, 1979.

[6] H. B. Walikar and B. D. Acharya, "Domination critical graphs," *National Academy Science Letters*, vol. 2, pp. 70–72, 1979.

[7] D. Bauer, F. Harary, J. Nieminen, and C. L. Suffel, "Domination alteration sets in graphs," *Discrete Mathematics*, vol. 47, no. 2-3, pp. 153–161, 1983.

[8] J. F. Fink, M. S. Jacobson, L. F. Kinch, and J. Roberts, "The bondage number of a graph," *Discrete Mathematics*, vol. 86, no. 1–3, pp. 47–57, 1990.

[9] F.-T. Hu and J.-M. Xu, "The bondage number of $(n-3)$-regular graph G of order n," *Ars Combinatoria*. to appear.

[10] U. Teschner, "New results about the bondage number of a graph," *Discrete Mathematics*, vol. 171, no. 1–3, pp. 249–259, 1997.

[11] B. L. Hartnell and D. F. Rall, "A bound on the size of a graph with given order and bondage number," *Discrete Mathematics*, vol. 197-198, pp. 409–413, 1999, 16th British Combinatorial Conference (London, 1997).

[12] B. L. Hartnell and D. F. Rall, "A characterization of trees in which no edge is essential to the domination number," *Ars Combinatoria*, vol. 33, pp. 65–76, 1992.

[13] J. Topp and P. D. Vestergaard, "α_k- and γ_k-stable graphs," *Discrete Mathematics*, vol. 212, no. 1-2, pp. 149–160, 2000.

[14] A. Meir and J. W. Moon, "Relations between packing and covering numbers of a tree," *Pacific Journal of Mathematics*, vol. 61, no. 1, pp. 225–233, 1975.

[15] B. L. Hartnell, L. K. Jørgensen, P. D. Vestergaard, and C. A. Whitehead, "Edge stability of the k-domination number of trees," *Bulletin of the Institute of Combinatorics and Its Applications*, vol. 22, pp. 31–40, 1998.

[16] F.-T. Hu and J.-M. Xu, "On the complexity of the bondage and reinforcement problems," *Journal of Complexity*, vol. 28, no. 2, pp. 192–201, 2012.

[17] B. L. Hartnell and D. F. Rall, "Bounds on the bondage number of a graph," *Discrete Mathematics*, vol. 128, no. 1–3, pp. 173–177, 1994.

[18] Y.-L. Wang, "On the bondage number of a graph," *Discrete Mathematics*, vol. 159, no. 1–3, pp. 291–294, 1996.

[19] G. H. Fricke, T. W. Haynes, S. M. Hedetniemi, S. T. Hedetniemi, and R. C. Laskar, "Excellent trees," *Bulletin of the Institute of Combinatorics and Its Applications*, vol. 34, pp. 27–38, 2002.

[20] V. Samodivkin, "The bondage number of graphs: good and bad vertices," *Discussiones Mathematicae. Graph Theory*, vol. 28, no. 3, pp. 453–462, 2008.

[21] J. E. Dunbar, T. W. Haynes, U. Teschner, and L. Volkmann, "Bondage, insensitivity, and reinforcement," in *Domination in Graphs*, vol. 209 of *Monographs and Textbooks in Pure and Applied Mathematics*, pp. 471–489, Dekker, New York, NY, USA, 1998.

[22] H. Liu and L. Sun, "The bondage and connectivity of a graph," *Discrete Mathematics*, vol. 263, no. 1–3, pp. 289–293, 2003.

[23] A.-H. Esfahanian and S. L. Hakimi, "On computing a conditional edge-connectivity of a graph," *Information Processing Letters*, vol. 27, no. 4, pp. 195–199, 1988.

[24] R. C. Brigham, P. Z. Chinn, and R. D. Dutton, "Vertex domination-critical graphs," *Networks*, vol. 18, no. 3, pp. 173–179, 1988.

[25] V. Samodivkin, "Domination with respect to nondegenerate properties: bondage number," *The Australasian Journal of Combinatorics*, vol. 45, pp. 217–226, 2009.

[26] U. Teschner, "A new upper bound for the bondage number of graphs with small domination number," *The Australasian Journal of Combinatorics*, vol. 12, pp. 27–35, 1995.

[27] J. Fulman, D. Hanson, and G. MacGillivray, "Vertex domination-critical graphs," *Networks*, vol. 25, no. 2, pp. 41–43, 1995.

[28] U. Teschner, "A counterexample to a conjecture on the bondage number of a graph," *Discrete Mathematics*, vol. 122, no. 1–3, pp. 393–395, 1993.

[29] U. Teschner, "The bondage number of a graph G can be much greater than $\Delta(G)$," *Ars Combinatoria*, vol. 43, pp. 81–87, 1996.

[30] L. Volkmann, "On graphs with equal domination and covering numbers," *Discrete Applied Mathematics*, vol. 51, no. 1-2, pp. 211–217, 1994.

[31] L. A. Sanchis, "Maximum number of edges in connected graphs with a given domination number," *Discrete Mathematics*, vol. 87, no. 1, pp. 65–72, 1991.

[32] S. Klavžar and N. Seifter, "Dominating Cartesian products of cycles," *Discrete Applied Mathematics*, vol. 59, no. 2, pp. 129–136, 1995.

[33] H. K. Kim, "A note on Vizing's conjecture about the bondage number," *Korean Journal of Mathematics*, vol. 10, no. 2, pp. 39–42, 2003.

[34] M. Y. Sohn, X. Yuan, and H. S. Jeong, "The bondage number of $C_3 \times C_n$," *Journal of the Korean Mathematical Society*, vol. 44, no. 6, pp. 1213–1231, 2007.

[35] L. Kang, M. Y. Sohn, and H. K. Kim, "Bondage number of the discrete torus $C_n \times C_4$," *Discrete Mathematics*, vol. 303, no. 1–3, pp. 80–86, 2005.

[36] J. Huang and J.-M. Xu, "The bondage numbers and efficient dominations of vertex-transitive graphs," *Discrete Mathematics*, vol. 308, no. 4, pp. 571–582, 2008.

[37] C. J. Xiang, X. Yuan, and M. Y. Sohn, "Domination and bondage number of $C_3 \times C_n$," *Ars Combinatoria*, vol. 97, pp. 299–310, 2010.

[38] M. S. Jacobson and L. F. Kinch, "On the domination number of products of graphs. I," *Ars Combinatoria*, vol. 18, pp. 33–44, 1984.

[39] Y.-Q. Li, "A note on the bondage number of a graph," *Chinese Quarterly Journal of Mathematics*, vol. 9, no. 4, pp. 1–3, 1994.

[40] F.-T. Hu and J.-M. Xu, "Bondage number of mesh networks," *Frontiers of Mathematics in China*, vol. 7, no. 5, pp. 813–826, 2012.

[41] R. Frucht and F. Harary, "On the corona of two graphs," *Aequationes Mathematicae*, vol. 4, pp. 322–325, 1970.

[42] K. Carlson and M. Develin, "On the bondage number of planar and directed graphs," *Discrete Mathematics*, vol. 306, no. 8-9, pp. 820–826, 2006.

[43] U. Teschner, "On the bondage number of block graphs," *Ars Combinatoria*, vol. 46, pp. 25–32, 1997.

[44] J. Huang and J.-M. Xu, "Note on conjectures of bondage numbers of planar graphs," *Applied Mathematical Sciences*, vol. 6, no. 65–68, pp. 3277–3287, 2012.

[45] L. Kang and J. Yuan, "Bondage number of planar graphs," *Discrete Mathematics*, vol. 222, no. 1–3, pp. 191–198, 2000.

[46] M. Fischermann, D. Rautenbach, and L. Volkmann, "Remarks on the bondage number of planar graphs," *Discrete Mathematics*, vol. 260, no. 1–3, pp. 57–67, 2003.

[47] J. Hou, G. Liu, and J. Wu, "Bondage number of planar graphs without small cycles," *Utilitas Mathematica*, vol. 84, pp. 189–199, 2011.

[48] J. Huang and J.-M. Xu, "The bondage numbers of graphs with small crossing numbers," *Discrete Mathematics*, vol. 307, no. 15, pp. 1881–1897, 2007.

[49] Q. Ma, S. Zhang, and J. Wang, "Bondage number of 1-planar graph," *Applied Mathematics*, vol. 1, no. 2, pp. 101–103, 2010.

[50] J. Hou and G. Liu, "A bound on the bondage number of toroidal graphs," *Discrete Mathematics, Algorithms and Applications*, vol. 4, no. 3, Article ID 1250046, 5 pages, 2012.

[51] Y.-C. Cao, J.-M. Xu, and X.-R. Xu, "On bondage number of toroidal graphs," *Journal of University of Science and Technology of China*, vol. 39, no. 3, pp. 225–228, 2009.

[52] Y.-C. Cao, J. Huang, and J.-M. Xu, "The bondage number of graphs with crossing number less than four," *Ars Combinatoria*. to appear.

[53] H. K. Kim, "On the bondage numbers of some graphs," *Korean Journal of Mathematics*, vol. 11, no. 2, pp. 11–15, 2004.

[54] A. Gagarin and V. Zverovich, "Upper bounds for the bondage number of graphs on topological surfaces," *Discrete Mathematics*, 2012.

[55] J. Huang, "An improved upper bound for the bondage number of graphs on surfaces," *Discrete Mathematics*, vol. 312, no. 18, pp. 2776–2781, 2012.

[56] A. Gagarin and V. Zverovich, "The bondage number of graphs on topological surfaces and Teschner's conjecture," *Discrete Mathematics*, vol. 313, no. 6, pp. 796–808, 2013.

[57] V. Samodivkin, "Note on the bondage number of graphs on topological surfaces," http://arxiv.org/abs/1208.6203 .

[58] E. J. Cockayne, R. M. Dawes, and S. T. Hedetniemi, "Total domination in graphs," *Networks*, vol. 10, no. 3, pp. 211–219, 1980.

[59] R. Laskar, J. Pfaff, S. M. Hedetniemi, and S. T. Hedetniemi, "On the algorithmic complexity of total domination," *Society for Industrial and Applied Mathematics*, vol. 5, no. 3, pp. 420–425, 1984.

[60] J. Pfaff, R. C. Laskar, and S. T. Hedetniemi, "NP-completeness of total and connected domination and irredundance for bipartite graphs," Tech. Rep. 428, Department of Mathematical Sciences, Clemson University, 1983.

[61] M. A. Henning, "A survey of selected recent results on total domination in graphs," *Discrete Mathematics*, vol. 309, no. 1, pp. 32–63, 2009.

[62] V. R. Kulli and D. K. Patwari, "The total bondage number of a graph," in *Advances in Graph Theory*, pp. 227–235, Vishwa, Gulbarga, India, 1991.

[63] F.-T. Hu, Y. Lu, and J.-M. Xu, "The total bondage number of grid graphs," *Discrete Applied Mathematics*, vol. 160, no. 16-17, pp. 2408–2418, 2012.

[64] J. Cao, M. Shi, and B. Wu, "Research on the total bondage number of a spectial network," in *Proceedings of the 3rd International Joint Conference on Computational Sciences and Optimization (CSO '10)*, pp. 456–458, May 2010.

[65] N. Sridharan, MD. Elias, and V. S. A. Subramanian, "Total bondage number of a graph," *AKCE International Journal of Graphs and Combinatorics*, vol. 4, no. 2, pp. 203–209, 2007.

[66] N. Jafari Rad and J. Raczek, "Some progress on total bondage in graphs," *Graphs and Combinatorics*, 2011.

[67] T. W. Haynes and P. J. Slater, "Paired-domination and the paired-domatic number," *Congressus Numerantium*, vol. 109, pp. 65–72, 1995.

[68] T. W. Haynes and P. J. Slater, "Paired-domination in graphs," *Networks*, vol. 32, no. 3, pp. 199–206, 1998.

[69] X. Hou, "A characterization of $2\gamma, \gamma_P$-trees," *Discrete Mathematics*, vol. 308, no. 15, pp. 3420–3426, 2008.

[70] K. E. Proffitt, T. W. Haynes, and P. J. Slater, "Paired-domination in grid graphs," *Congressus Numerantium*, vol. 150, pp. 161–172, 2001.

[71] H. Qiao, L. Kang, M. Cardei, and D.-Z. Du, "Paired-domination of trees," *Journal of Global Optimization*, vol. 25, no. 1, pp. 43–54, 2003.

[72] E. Shan, L. Kang, and M. A. Henning, "A characterization of trees with equal total domination and paired-domination numbers," *The Australasian Journal of Combinatorics*, vol. 30, pp. 31–39, 2004.

[73] J. Raczek, "Paired bondage in trees," *Discrete Mathematics*, vol. 308, no. 23, pp. 5570–5575, 2008.

[74] J. A. Telle and A. Proskurowski, "Algorithms for vertex partitioning problems on partial *k*-trees," *SIAM Journal on Discrete Mathematics*, vol. 10, no. 4, pp. 529–550, 1997.

[75] G. S. Domke, J. H. Hattingh, M. A. Henning, and L. R. Markus, "Restrained domination in trees," *Discrete Mathematics*, vol. 211, no. 1–3, pp. 1–9, 2000.

[76] G. S. Domke, J. H. Hattingh, M. A. Henning, and L. R. Markus, "Restrained domination in graphs with minimum degree two," *Journal of Combinatorial Mathematics and Combinatorial Computing*, vol. 35, pp. 239–254, 2000.

[77] G. S. Domke, J. H. Hattingh, S. T. Hedetniemi, R. C. Laskar, and L. R. Markus, "Restrained domination in graphs," *Discrete Mathematics*, vol. 203, no. 1–3, pp. 61–69, 1999.

[78] J. H. Hattingh and E. J. Joubert, "An upper bound for the restrained domination number of a graph with minimum degree at least two in terms of order and minimum degree," *Discrete Applied Mathematics*, vol. 157, no. 13, pp. 2846–2858, 2009.

[79] M. A. Henning, "Graphs with large restrained domination number," *Discrete Mathematics*, vol. 197-198, pp. 415–429, 1999, 16th British Combinatorial Conference (London, 1997).

[80] J. H. Hattingh and A. R. Plummer, "Restrained bondage in graphs," *Discrete Mathematics*, vol. 308, no. 23, pp. 5446–5453, 2008.

[81] N. Jafari Rad, R. Hasni, J. Raczek, and L. Volkmann, "Total restrained bondage in graphs," *Acta Mathematica Sinica*, vol. 29, no. 6, pp. 1033–1042, 2013.

[82] J. Cyman and J. Raczek, "On the total restrained domination number of a graph," *The Australasian Journal of Combinatorics*, vol. 36, pp. 91–100, 2006.

[83] M. A. Henning, "Graphs with large total domination number," *Journal of Graph Theory*, vol. 35, no. 1, pp. 21–45, 2000.

[84] D.-X. Ma, X.-G. Chen, and L. Sun, "On total restrained domination in graphs," *Czechoslovak Mathematical Journal*, vol. 55, no. 1, pp. 165–173, 2005.

[85] B. Zelinka, "Remarks on restrained domination and total restrained domination in graphs," *Czechoslovak Mathematical Journal*, vol. 55, no. 2, pp. 393–396, 2005.

[86] W. Goddard and M. A. Henning, "Independent domination in graphs: a survey and recent results," *Discrete Mathematics*, vol. 313, no. 7, pp. 839–854, 2013.

[87] J. H. Zhang, H. L. Liu, and L. Sun, "Independence bondage number and reinforcement number of some graphs," *Transactions of Beijing Institute of Technology*, vol. 23, no. 2, pp. 140–142, 2003.

[88] K. D. Dharmalingam, *Studies in Graph Theory—equitable domination and bottleneck domination [Ph.D. thesis]*, Madurai Kamaraj University, Madurai, India, 2006.

[89] G. Deepak, N. D. Soner, and A. Alwardi, "The equitable bondage number of a graph," *Research Journal of Pure Algebra*, vol. 1, no. 9, pp. 209–212, 2011.

[90] E. Sampathkumar and H. B. Walikar, "The connected domination number of a graph," *Journal of Mathematical and Physical Sciences*, vol. 13, no. 6, pp. 607–613, 1979.

[91] K. White, M. Farber, and W. Pulleyblank, "Steiner trees, connected domination and strongly chordal graphs," *Networks*, vol. 15, no. 1, pp. 109–124, 1985.

[92] M. Cadei, M. X. Cheng, X. Cheng, and D.-Z. Du, "Connected domination in Ad Hoc wireless networks," in *Proceedings of the 6th International Conference on Computer Science and Informatics (CS&I '02)*, 2002.

[93] J. F. Fink and M. S. Jacobson, "n-domination in graphs," in *Graph Theory with Applications to Algorithms and Computer Science*, Wiley-Interscience Publications, pp. 283–300, Wiley, New York, NY, USA, 1985.

[94] M. Chellali, O. Favaron, A. Hansberg, and L. Volkmann, "k-domination and k-independence in graphs: a survey," *Graphs and Combinatorics*, vol. 28, no. 1, pp. 1–55, 2012.

[95] Y. Lu, X. Hou, J.-M. Xu, and N. Li, "Trees with unique minimum *p*-dominating sets," *Utilitas Mathematica*, vol. 86, pp. 193–205, 2011.

[96] Y. Lu and J.-M. Xu, "The *p*-bondage number of trees," *Graphs and Combinatorics*, vol. 27, no. 1, pp. 129–141, 2011.

[97] Y. Lu and J.-M. Xu, "The 2-domination and 2-bondage numbers of grid graphs. A manuscript," http://arxiv.org/abs/1204.4514 .

[98] M. Krzywkowski, "Non-isolating 2-bondage in graphs," *The Journal of the Mathematical Society of Japan*, vol. 64, no. 4, 2012.

[99] M. A. Henning, O. R. Oellermann, and H. C. Swart, "Bounds on distance domination parameters," *Journal of Combinatorics, Information & System Sciences*, vol. 16, no. 1, pp. 11–18, 1991.

[100] F. Tian and J.-M. Xu, "Bounds for distance domination numbers of graphs," *Journal of University of Science and Technology of China*, vol. 34, no. 5, pp. 529–534, 2004.

[101] F. Tian and J.-M. Xu, "Distance domination-critical graphs," *Applied Mathematics Letters*, vol. 21, no. 4, pp. 416–420, 2008.

[102] F. Tian and J.-M. Xu, "A note on distance domination numbers of graphs," *The Australasian Journal of Combinatorics*, vol. 43, pp. 181–190, 2009.

[103] F. Tian and J.-M. Xu, "Average distances and distance domination numbers," *Discrete Applied Mathematics*, vol. 157, no. 5, pp. 1113–1127, 2009.

[104] Z.-L. Liu, F. Tian, and J.-M. Xu, "Probabilistic analysis of upper bounds for 2-connected distance *k*-dominating sets in graphs," *Theoretical Computer Science*, vol. 410, no. 38–40, pp. 3804–3813, 2009.

[105] M. A. Henning, O. R. Oellermann, and H. C. Swart, "Distance domination critical graphs," *Journal of Combinatorial Mathematics and Combinatorial Computing*, vol. 44, pp. 33–45, 2003.

[106] T. J. Bean, M. A. Henning, and H. C. Swart, "On the integrity of distance domination in graphs," *The Australasian Journal of Combinatorics*, vol. 10, pp. 29–43, 1994.

[107] M. Fischermann and L. Volkmann, "A remark on a conjecture for the (*k*,*p*)-domination number," *Utilitas Mathematica*, vol. 67, pp. 223–227, 2005.

[108] T. Korneffel, D. Meierling, and L. Volkmann, "A remark on the (2,2)-domination number," *Discussiones Mathematicae. Graph Theory*, vol. 28, no. 2, pp. 361–366, 2008.

[109] Y. Lu, X. Hou, and J.-M. Xu, "On the (2,2)-domination number of trees," *Discussiones Mathematicae. Graph Theory*, vol. 30, no. 2, pp. 185–199, 2010.

[110] H. Li and J.-M. Xu, "(d,m)-domination numbers of m-connected graphs," Rapports de Recherche 1130, LRI, Université Paris-Sud, 1997.

[111] S. M. Hedetniemi, S. T. Hedetniemi, and T. V. Wimer, "Linear time resource allocation algorithms for trees," Tech. Rep. URI-014, Department of Mathematical Sciences, Clemson University, 1987.

[112] M. Farber, "Domination, independent domination, and duality in strongly chordal graphs," *Discrete Applied Mathematics*, vol. 7, no. 2, pp. 115–130, 1984.

[113] G. S. Domke and R. C. Laskar, "The bondage and reinforcement numbers of γ_f for some graphs," *Discrete Mathematics*, vol. 167-168, pp. 249–259, 1997.

[114] G. S. Domke, S. T. Hedetniemi, and R. C. Laskar, "Fractional packings, coverings, and irredundance in graphs," vol. 66, pp. 227–238, 1988.

[115] E. J. Cockayne, P. A. Dreyer, Jr., S. M. Hedetniemi, and S. T. Hedetniemi, "Roman domination in graphs," *Discrete Mathematics*, vol. 278, no. 1–3, pp. 11–22, 2004.

[116] I. Stewart, "Defend the roman empire," *Scientific American*, vol. 281, pp. 136–139, 1999.

[117] C. S. ReVelle and K. E. Rosing, "Defendens imperium romanum: a classical problem in military strategy," *The American Mathematical Monthly*, vol. 107, no. 7, pp. 585–594, 2000.

[118] A. Bahremandpour, F.-T. Hu, S. M. Sheikholeslami, and J.-M. Xu, "On the Roman bondage number of a graph," *Discrete Mathematics, Algorithms and Applications*, vol. 5, no. 1, Article ID 1350001, 15 pages, 2013.

[119] N. Jafari Rad and L. Volkmann, "Roman bondage in graphs," *Discussiones Mathematicae. Graph Theory*, vol. 31, no. 4, pp. 763–773, 2011.

[120] K. Ebadi and L. PushpaLatha, "Roman bondage and Roman reinforcement numbers of a graph," *International Journal of Contemporary Mathematical Sciences*, vol. 5, pp. 1487–1497, 2010.

[121] N. Dehgardi, S. M. Sheikholeslami, and L. Volkman, "On the Roman k-bondage number of a graph," *AKCE International Journal of Graphs and Combinatorics*, vol. 8, pp. 169–180, 2011.

[122] S. Akbari and S. Qajar, "A note on Roman bondage number of graphs," *Ars Combinatoria*. to Appear.

[123] F.-T. Hu and J.-M. Xu, "Roman bondage numbers of some graphs," http://arxiv.org/abs/1109.3933 .

[124] N. Jafari Rad and L. Volkmann, "On the Roman bondage number of planar graphs," *Graphs and Combinatorics*, vol. 27, no. 4, pp. 531–538, 2011.

[125] S. Akbari, M. Khatirinejad, and S. Qajar, "A note on the roman bondage number of planar graphs," *Graphs and Combinatorics*, 2012.

[126] B. Brešar, M. A. Henning, and D. F. Rall, "Rainbow domination in graphs," *Taiwanese Journal of Mathematics*, vol. 12, no. 1, pp. 213–225, 2008.

[127] B. L. Hartnell and D. F. Rall, "On dominating the Cartesian product of a graph and αk," *Discussiones Mathematicae. Graph Theory*, vol. 24, no. 3, pp. 389–402, 2004.

[128] N. Dehgardi, S. M. Sheikholeslami, and L. Volkman, "The k-rainbow bondage number of a graph," to appear in Discrete Applied Mathematics.

[129] J. von Neumann and O. Morgenstern, *Theory of Games and Economic Behavior*, Princeton University Press, Princeton, NJ, USA, 1944.

[130] C. Berge, *Theorie des Graphes et ses Applications*, 1958.

[131] O. Ore, *Theory of Graphs*, vol. 38, American Mathematical Society Colloquium Publications, 1962.

[132] E. Shan and L. Kang, "Bondage number in oriented graphs," *Ars Combinatoria*, vol. 84, pp. 319–331, 2007.

[133] J. Huang and J.-M. Xu, "The bondage numbers of extended de Bruijn and Kautz digraphs," *Computers & Mathematics with Applications*, vol. 51, no. 6-7, pp. 1137–1147, 2006.

[134] J. Huang and J.-M. Xu, "The total domination and total bondage numbers of extended de Bruijn and Kautz digraphs," *Computers & Mathematics with Applications*, vol. 53, no. 8, pp. 1206–1213, 2007.

[135] Y. Shibata and Y. Gonda, "Extension of de Bruijn graph and Kautz graph," *Computers & Mathematics with Applications*, vol. 30, no. 9, pp. 51–61, 1995.

[136] X. Zhang, J. Liu, and J. Meng, "The bondage number in complete t-partite digraphs," *Information Processing Letters*, vol. 109, no. 17, pp. 997–1000, 2009.

[137] D. W. Bange, A. E. Barkauskas, and P. J. Slater, "Efficient dominating sets in graphs," in *Applications of Discrete Mathematics*, pp. 189–199, SIAM, Philadelphia, Pa, USA, 1988.

[138] M. Livingston and Q. F. Stout, "Perfect dominating sets," *Congressus Numerantium*, vol. 79, pp. 187–203, 1990.

[139] L. Clark, "Perfect domination in random graphs," *Journal of Combinatorial Mathematics and Combinatorial Computing*, vol. 14, pp. 173–182, 1993.

[140] D. W. Bange, A. E. Barkauskas, L. H. Host, and L. H. Clark, "Efficient domination of the orientations of a graph," *Discrete Mathematics*, vol. 178, no. 1–3, pp. 1–14, 1998.

[141] A. E. Barkauskas and L. H. Host, "Finding efficient dominating sets in oriented graphs," *Congressus Numerantium*, vol. 98, pp. 27–32, 1993.

[142] I. J. Dejter and O. Serra, "Efficient dominating sets in Cayley graphs," *Discrete Applied Mathematics*, vol. 129, no. 2-3, pp. 319–328, 2003.

[143] J. Lee, "Independent perfect domination sets in Cayley graphs," *Journal of Graph Theory*, vol. 37, no. 4, pp. 213–219, 2001.

[144] W. Gu, X. Jia, and J. Shen, "Independent perfect domination sets in meshes, tori and trees," Unpublished.

[145] N. Obradović, J. Peters, and G. Ružić, "Efficient domination in circulant graphs with two chord lengths," *Information Processing Letters*, vol. 102, no. 6, pp. 253–258, 2007.

[146] K. Reji Kumar and G. MacGillivray, "Efficient domination in circulant graphs," *Discrete Mathematics*, vol. 313, no. 6, pp. 767–771, 2013.

[147] D. Van Wieren, M. Livingston, and Q. F. Stout, "Perfect dominating sets on cube-connected cycles," *Congressus Numerantium*, vol. 97, pp. 51–70, 1993.

5

Graphs with no $K_{3,3}$ Minor Containing a Fixed Edge

Donald K. Wagner

Mathematical, Computer, and Information Sciences Division, Office of Naval Research, Arlington, VA 22203, USA

Correspondence should be addressed to Donald K. Wagner; don.wagner@navy.mil

Academic Editor: Chính T. Hoang

It is well known that every cycle of a graph must intersect every cut in an even number of edges. For planar graphs, Ford and Fulkerson proved that, for any edge e, there exists a cycle containing e that intersects every minimal cut containing e in exactly two edges. The main result of this paper generalizes this result to any nonplanar graph G provided G does not have a $K_{3,3}$ minor containing the given edge e. Ford and Fulkerson used their result to provide an efficient algorithm for solving the maximum-flow problem on planar graphs. As a corollary to the main result of this paper, it is shown that the Ford-Fulkerson algorithm naturally extends to this more general class of graphs.

1. Introduction

This paper examines the structure of paths and cuts in a graph relative to a fixed edge. In particular, let G be a graph, and let e be an edge of G. Define an *e-path* of G to be a path P such that $P \cup \{e\}$ is a cycle of G. Define an *e-cut* of G to be a cut of G that contains e (in this paper, paths and cycles do not have repeated nodes and are equated with their edge sets. Also, cuts are minimal; i.e., no cut properly contains another.) Ford and Fulkerson [1] showed that if G is planar, then there exists an e-path that intersects every e-cut in exactly one edge. This Ford-Fulkerson property does not hold for graphs in general. Specifically, take $G = K_{3,3}$. Then, for any edge e of G and any e-path P, one can always find an e-cut that intersects P in more than one edge. The Ford-Fulkerson property, however, is not confined solely to planar graphs; in particular, if $G = K_5$, then it is easy to find an e-path, for any choice of e, that intersects every e-cut in exactly one edge.

One of the main goals of this paper is to extend the Ford-Fulkerson result to a larger class of graphs. Motivated by the $K_{3,3}$ example above, it is shown that if $K_{3,3}$ is excluded in the proper way, then this goal can be achieved. Below is the main result of the paper. Throughout the paper, n denotes the number of nodes of a graph, and m the number of edges.

Theorem 1. *Let G be a graph, and let e be an edge of G. If $G \setminus e$ is connected, and G does not have a $K_{3,3}$ minor containing e, then there exists an e-path of G that intersects every e-cut of G in exactly one edge. Moreover, such an e-path can be found in $O(m)$ time.*

Theorem 1 is then used to provide a very simple $O(n^2)$-time algorithm for the maximum-flow problem for graphs in this class. This is within a logarithmic factor of the fastest maximum-flow algorithm, namely, the recent algorithm due to Orlin [2].

The remainder of the paper is outlined as follows. The next section introduces a graph decomposition, which serves as a key ingredient for the proof of Theorem 1. Section 3 contains the proof of Theorem 1, and Section 4 applies Theorem 1 to the maximum-flow problem.

2. Graph Decomposition

This section describes a connectivity-based decomposition for graphs that do not have a $K_{3,3}$ minor containing a fixed edge. This decomposition was introduced in Wagner [3]. For the sake of completeness, the key results are presented and proved here.

The notion of connectivity used here is that of Tutte [4]. A *k-separation*, for a positive integer k, of a connected graph G is a partition $\{E_1, E_2\}$ of the edge set of G such that $|E_1| \geq$

$k \le |E_2|$ and the edge-induced subgraphs $G[E_1]$ and $G[E_2]$ have at most k nodes in common. A connected graph G is *k-connected*, for $k \ge 2$, if it does not have a k'-separation for any $k' < k$. A k-separation $\{E_1, E_2\}$ of a k-connected graph G is an *internal* k-separation if $|E_1| \ge k + 1 \le |E_2|$.

The next theorem is a well-known result of Wagner [5].

Theorem 2. *Let G be a 3-connected graph. Then, G does not have a $K_{3,3}$ minor if and only if G is planar or isomorphic to K_5.*

It is sometimes more convenient to work with subdivisions rather than minors. A graph H is a *subdivision* of a graph K if it can be obtained from K by a sequence of the following operation: replace an edge xy by edges xz and yz, where z is a new node. If a graph G has a subgraph H that is a subdivision of a graph K, then G is said to have a *K subdivision*. It is well known and easy to prove that a 3-connected graph has a $K_{3,3}$ subdivision if and only if it has a $K_{3,3}$ minor.

If a graph H is a subdivision of a graph K, and x and y are nonadjacent nodes of K, then x and y are *independent* in H. The next lemma is due to Širáň [6].

Lemma 3. *Let G be a 3-connected graph, and let $e = xy$ be an edge of G. If e is not contained in any $K_{3,3}$ minor of G, then for any $K_{3,3}$ subdivision H of G, x and y are independent degree-three nodes of H.*

In this paper, 2- and 3-separations play a crucial role, as do the related notions of 2- and 3-sums. First, consider a 2-separation $\{E_1, E_2\}$ of a 2-connected graph G. Let $\{p, q\} := V(G[E_1]) \cap V(G[E_2])$, and let $\{t\}$ be a set disjoint from $E(G)$. For $i \in \{1, 2\}$, define G_i to be the graph obtained from $G[E_i]$ by adding t as an edge joining p and q. Then, $\{G_1, G_2\}$ is a *2-sum decomposition* of the graph G, and t is the *connecting* edge.

Now, let $\{E_1, E_2\}$ be an internal 3-separation of a 3-connected graph G. Let $\{x, y, z\} := V(G[E_1]) \cap V(G[E_2])$, and let S be the set of edges of G that have both end nodes in $\{x, y, z\}$. Let R be a set disjoint from $E(G)$ such that $|R \cup S| = 3$. For $i \in \{1, 2\}$, construct a graph G_i from $G[E_i \cup S]$ by adding the members of R in such a way that $T := R \cup S$ is a triangle of G_i and such that each edge of R has the same ends in G_1 as it does in G_2. Then, $\{G_1, G_2\}$ is a *3-sum decomposition* of G, and T is the *connecting* triangle.

It is well known that if $\{G_1, G_2\}$ is a k-sum decomposition of a k-connected graph G, for $k \in \{2, 3\}$ then both G_1 and G_2 are k-connected and are isomorphic to proper minors of G.

Two special kinds of internal 3-separations are needed. Both are defined for a given 3-connected graph G relative to a fixed edge e.

First, let $\{E_1, E_2\}$ be a internal 3-separation of G. If both ends of e are in $V(G[E_1]) \cap V(G[E_2])$, then the 3-separation $\{E_1, E_2\}$ is said to be *straddled* by e. Observe that in this case, e is in the connecting triangle of the corresponding 3-sum decomposition. The notion of a straddling edge can be found in the work of Tseng and Truemper [7]. It is also related to the concept of "contractibility," which traces back to the work of Tutte [8].Specifically, an edge of a 3-connected graph is *contractible* if its contraction results in a 3-connected graph. It is easy to see that an edge is not contractible if and only if it straddles an internal 3-separation.

The second special internal 3-separation is as follows. Let $\{E_1, E_2\}$ be an internal 3-separation of G, and suppose E_1 has exactly seven edges, say $e, f_1, \ldots, f_6$. Suppose further that $\{e, f_1, f_2\}$, $\{e, f_3, f_4\}$, and $\{e, f_5, f_6\}$ are triangles of G such that no two of $\{f_1, \ldots, f_6\}$ are parallel. Then, $G[E_1]$ is a *crown*, and $\{E_1, E_2\}$ is a *crown* 3-separation of G with respect to e. Observe that the crown $G[E_1]$ has three nodes of degree two, which, by the 3-connectivity of G, constitute the set $V(G[E_1]) \cap V(G[E_2])$. It also has two nodes of degree four, which are the ends of e.

Let $\{E_1, E_2\}$ be a *crown* 3-separation of G with respect to e, and let $\{G_1, G_2\}$ be the corresponding 3-sum decomposition. If G_2 is planar, then G is said to be *crown-planar* with respect to e. Crown-planar graphs show up in the decomposition established in Theorem 5. In the context of Theorem 5, crown-planar graphs can alternatively be described as being obtained from a 3-connected planar graph by duplicating a degree-three node x, where the fixed edge e joins x to its twin.

Let G be a graph, and H a subgraph of G. Let P be a path of G, the end nodes of which are nodes of H and the internal nodes of which are not nodes of H. Then, the subgraph $H \cup P$ of G is said to be obtained from H by *adjoining* P, and P is an *adjoinable* path of G with respect to H.

Let G be a graph, and e an edge of G. Let H be a $K_{3,3}$ subdivision of G, and suppose that e joins two independent degree-three nodes of H. Since $K_{3,3}$ has nine edges, the graph H consists of nine paths, each of which is a subdivision of an edge of $K_{3,3}$. The six such paths that share an end with e are called the *principal* paths of H with respect to e; the remaining three paths are the *support* paths of H. The $K_{3,3}$ subdivision H of G is *good* (resp., *bad*) with respect to e if all six (resp., at most five) of the principal paths with respect to e consist of a single edge.

Lemma 4. *Let G be a 3-connected graph, and let e be an edge of G. Then, either (i) G has a $K_{3,3}$ minor that contains e, (ii) G has an internal 3-separation that is straddled by e, or (iii) every $K_{3,3}$ subdivision of G is good with respect to e.*

Proof. Let $e = xy$. Suppose that neither (i) nor (iii) holds. If G is planar or isomorphic to K_5, then (iii) holds vacuously, and so, Theorem 2 implies that G has a $K_{3,3}$ minor, and thus a $K_{3,3}$ subdivision. By Lemma 3, x and y are independent degree-three nodes in every $K_{3,3}$ subdivision of G. Since (iii) does not hold, there exists a $K_{3,3}$ subdivision of G, say H, in which some principal path with respect to e, say Q_1, has at least two edges. Let u denote the end node of Q_1 not in $\{x, y\}$. Let Q_2 denote the other principal path that has u as an end node, and let S_1 denote the support path that has u as an end node. Denote the other end node of S_1 by z. Consistent with the above, assume H and Q_1 are chosen so that the number of edges in S_1 is as small as possible.

Claim. If an adjoinable path of G with respect to H has one end that is an internal node of either Q_1 or Q_2, then the other end of the path is a node of $V(Q_1 \cup Q_2 \cup S_1)$.

Proof of Claim. If the other end of the path is not in $V(Q_1 \cup Q_2 \cup S_1)$, then it is easy to check that adjoining the path to H results in a graph that has a $K_{3,3}$ minor that contains e, a contradiction. *End of Claim.*

Observe that $\{Q_1 \cup Q_2 \cup \{e\}, E(H) - (Q_1 \cup Q_2 \cup \{e\})\}$ is an internal 3-separation of H straddled by e. Thus, either (ii) holds or there exists an adjoinable path R_1 of G, one end of which, say r_1, is a internal node of Q_1 (say) and the other end of which, say t_1, is not in $V(Q_1 \cup Q_2)$. By the Claim, t_1 is a node of S_1; if it is an internal node of S_1, then a contradiction to the choice of H is obtained by adjoining R_1 to H and deleting the internal nodes of the ur_1-subpath of subpath of Q_1. Thus, $t_1 = z$.

Observe that $\{Q_1 \cup Q_2 \cup S_1, E(H) - (Q_1 \cup Q_2 \cup S_1)\}$ is an internal 3-separation of H straddled by e. Thus, either (ii) holds or there exists an adjoinable path R_2 of G with respect to H, one end of which, say r_2, is in $V(Q_1 \cup Q_2 \cup S_1)$, the other end of which, say t_2, is not in $V(Q_1 \cup Q_2 \cup S_1)$, and neither end of which is in $\{x, y, z\}$. By the Claim, r_2 is a node of S_1 and t_2 is a node of P, where P is one of the principal or support paths of H not in $\{Q_1, Q_2, S_1\}$. Moreover, r_2 must equal u, for otherwise a contradiction to the choice of H is obtained by adjoining R_2 to H and deleting the internal nodes of the zt_2-subpath of P. By the Claim, R_1 and R_2 are node disjoint. Now, $H \cup R_1 \cup R_2$ has a $K_{3,3}$ minor that contains e, a contradiction. □

Theorem 5 below is the main result of the section.

Theorem 5. *Let G be a 3-connected graph, and let e be an edge of G. Then, either (i) G is planar, (ii) G is isomorphic to* K_5, *(iii) G has a* $K_{3,3}$ *minor that contains e, (iv) G has an internal 3-separation that is straddled by e, or (v) G is crown-planar with respect to e.*

Proof. Lemma 4 and Theorem 2 together imply that either one of (i)–(iv) holds, or every $K_{3,3}$ subdivision of G is good with respect to e. Assume that none of (i)–(iv) hold and let H denote a $K_{3,3}$ subdivision of G that is good with respect to e. Let $e = xy$, and let z denote the common end node of the three support paths of H. Let u, v, and w denote the remaining degree-three nodes of H. Let S_1, S_2, and S_3 denote the three support paths of H with respect to e, and without loss of generality, assume that the ends of S_1 are u and z.

Observe that $\{\{S_1, ux, uy, e\}, E(H) - \{S_1, ux, uy, e\}\}$ is an internal 3-separation of H straddled by e. Since (iv) does not hold, there exists an adjoinable path R_1 of G with respect to H, one end of which is in $V(S_1)$, the other end of which is in $V(S_2)$ (say), and neither end of which is equal to z. Similarly, there exists an adjoinable path R_2 of G with respect to H, one end of which is $V(S_3)$, the other end of which is in $V(S_1)$ (say), and neither end of which is equal to z.

Observe that $\{\{e, ux, uy, vx, vy, wx, wy\}, S_1 \cup S_2 \cup S_3 \cup R_1 \cup R_2\}$ is a crown 3-separation with respect to e of $H \cup R_1 \cup R_2$. Thus, either G has a crown 3-separation with respect to e or there exists an adjoinable path R_3 of G with respect to $H \cup R_1 \cup R_2$, one end of which is in $\{x, y\}$ and the other end of which, call it t, is in $V(S_1 \cup S_2 \cup S_3 \cup R_1 \cup R_2)$. If $t \neq z$, then observe that $H \cup R_1 \cup R_3 \cup R_3$ contains a bad $K_{3,3}$ subdivision with respect to e, a contradiction (note, if $t \in \{u, v, w\}$, then by the 3-connectivity of G, R_3 has at least two edges). Thus, $t = z$. It can now be checked that $H \cup R_1 \cup R_2 \cup R_3$ contains a $K_{3,3}$ minor containing e, a contradiction.

Finally, it needs to be shown that if G has a crown 3-separation with respect to e, and none of (i)–(iv) hold, then G is crown-planar with respect to e. To see this, let $\{G_1, G_2\}$ be the 3-sum decomposition of G corresponding to the crown 3-separation with respect to e, where $e \in E(G_1)$. Then, it suffices to show that G_2 is planar. If this is not the case, then by Theorem 2, G_2 is either isomorphic to K_5 or has a $K_{3,3}$ subdivision. In either case, it is straightforward to see that G has a $K_{3,3}$ subdivision for which e does not join two independent nodes, contradicting Lemma 3. □

The next result is from Wagner [3]. It shows that if G is a simple 2-connected graph having an edge e that is not contained in a $K_{3,3}$ minor, then the number of edges of G is bounded $5n - 12$. The proof is a straightforward induction using 2- and 3-sum decompositions, together with Theorem 5 and the well-known fact that any planar graph has at most $3n - 6$ edges.

Lemma 6. *Let G be a simple 2-connected graph having at least three nodes. If, for some edge e, G does not have a* $K_{3,3}$ *minor containing e, then G has at most* $5n - 12$ *edges.*

3. Admissible Paths

This section presents a proof of Theorem 1. This section begins with two lemmas that relate an e-cut of a graph to that of a member of a k-sum decomposition of the graph.

Lemma 7. *Let G be a 2-connected graph, and let e be an edge of G. Suppose that* $\{E_1, E_2\}$ *is a 2-separation of G with* $e \in E_1$. *Let* $\{G_1, G_2\}$ *be the corresponding 2-sum decomposition, and let t denote the connecting edge. Let D be an e-cut of G. Then, either D or* $D - E_2 \cup \{t\}$ *is an e-cut of* G_1. *Moreover, in the latter case,* $D - E_1 \cup \{t\}$ *is a t-cut of* G_2.

Proof. Let p and q denote the nodes common to $G[E_1]$ and $G[E_2]$. Let $\{X, Y\}$ denote the node partition of $V(G)$ corresponding to the e-cut D of G.

First, it is shown that $D \cap E_2 \neq \emptyset$ if and only if $p \in X$ (say) and $q \in Y$. To this end, suppose that $p \in X$ and $q \in Y$. Observe, there exists a pq-path in $G[E_2]$, and any such path must contain an edge from D. Thus, $D \cap E_2 \neq \emptyset$. Now, suppose that $D \cap E_2 \neq \emptyset$, and let $f \in D \cap E_2$. Since $\{e, f\} \subseteq D$, there exists two paths, say P and Q, each of which joins an end node of f to an end node of e and such that $V(P) \subseteq X$ (say) and $V(Q) \subseteq Y$. Since $e \in E_1$ and $f \in E_2$, P must go through p (say) and Q through q. Thus, $p \in X$ and $q \in Y$.

Now, define $X_1 := X \cap V(G_1)$ and $Y_1 := Y \cap V(G_1)$. Then, X_1 and Y_1 are both nonempty since they each contain an end of e. Also, they are disjoint and their union equals $V(G_1)$. By the previous paragraph, it can be seen that the set of edges of G_1 that have exactly one end in X_1 is D (if $\{p, q\} \subseteq X_1$ (say)) or $D - E_2 \cup \{t\}$ (if $p \in X_1$ (say) and $q \in Y_1$). Thus, the conclusion that either D or $D - E_2 \cup \{t\}$ is an e-cut of G_1 follows provided the subgraphs $G_1[X_1]$ and $G_1[Y_1]$ are connected. To this end, let u and v be nodes of X_1, and let R be a uv-path in $G[X]$. If $R \subseteq E_1$, then R is a path of $G_1[X_1]$. Suppose that this is not the case. Then, $\{p, q\} \subseteq V(R)$. Moreover, the edges of $R \cap E_2$ constitute a pq-subpath S of R. Thus, replacing, in R, the subpath S by the edge t yields a uv-path in G_1. Thus, $G_1[X_1]$ is connected. Similarly, $G_1[Y_1]$ is connected.

Now, suppose that $D - E_2 \cup \{t\}$ is an e-cut of G_1. Showing that $D - E_1 \cup \{t\}$ is a t-cut of G_2 is done in a manner similar to the above. Specifically, define $X_2 := X \cap V(G_2)$ and $Y_2 := Y \cap V(G_2)$. Then, X_2 and Y_2 are nonempty since $p \in X_2$ and $q \in Y_2$. Also, they are disjoint and their union equals $V(G_2)$. Moreover, the set of edges of G_2 that have exactly one end in X_2 is precisely $D - E_1 \cup \{t\}$. Thus, the conclusion that $D - E_1 \cup \{t\}$ is a t-cut of G_2 follows provided the subgraphs $G_2[X_2]$ and $G_2[Y_2]$ are connected. To this end, let u and v be nodes of X_2, and let R be a uv-path in $G[X]$. Since $q \in Y_2$, $q \notin V(R)$. Therefore, $R \subseteq E_2$. Thus, $G_2[X_2]$ is connected. Similarly, $G_2[Y_2]$ is connected. □

Lemma 8. *Let G be a 3-connected graph, and let e be an edge of G. Suppose that $\{E_1, E_2\}$ is a 3-separation of G that is straddled by e. Let $\{G_1, G_2\}$ be the corresponding 3-sum decomposition, and let T denote the connecting triangle. Let D be an e-cut of G. Then, $D - E_2 \cup \{t, e\}$ is an e-cut of G_1 for some $t \in T - \{e\}$.*

Proof. Let x, y, and z denote the nodes common to $G[E_1]$ and $G[E_2]$. Let $T := \{e, t, t'\}$ with $e = xy$ and $t = yz$. Let $\{X, Y\}$ denote the node partition of $V(G)$ corresponding to the e-cut D of G. Without loss of generality, assume $\{x, z\} \subseteq X$ and $y \in Y$. Let $X_1 := X \cap V(G_1)$ and $Y_1 := Y \cap V(G_1)$. Then, X_1 and Y_1 are both nonempty since they each contain an end of e. Also, they are disjoint and their union equals $V(G_1)$. Moreover, the set of edges of G_1 that have exactly one end in X_1 is precisely $D - E_2 \cup \{t, e\}$. The result now follows provided the subgraphs $G_1[X_1]$ and $G_1[Y_1]$ are connected. To this end, let u and v be nodes of X_1, and let R be a uv-path in $G[X]$. If $R \subseteq E_1$, then R is a path of $G_1[X_1]$. Suppose that this is not the case. Then, $\{x, z\} \subseteq V(R)$. Moreover, the edges of $R \cap E_2$ constitute an xz-subpath S of P. Thus, replacing, in R, the subpath S by the edge t' yields a uv-path in G_1. Thus $G[X_1]$ is connected. Similarly, $G[Y_1]$ is connected. □

Let G be a graph, and let e be an edge of G. An e-path of G is called *admissible* if it intersects every e-cut in exactly one edge. The above two lemmas are now used to show how admissible paths relate to k-sums.

Lemma 9. *Let G be a 2-connected graph, and let e be an edge of G. Suppose that G has a 2-separation $\{E_1, E_2\}$ with $e \in E_1$. Let $\{G_1, G_2\}$ be the corresponding 2-sum, and let t be the connecting edge. Suppose that P_1 is an admissible e-path of G_1, and P_2 is an admissible t-path of G_2. If $t \notin P_1$, P_1 is an admissible e-path of G, and if $t \in P_1$, then $P_1 - \{t\} \cup P_2$ is an admissible e-path of G.*

Proof. Let D be an e-cut of G. By Lemma 7, either D or $D - E_2 \cup \{t\}$ is an e-cut of G_1.

First, consider the case that D is an e-cut of G_1. Since D is an e-cut of both G and G_1, it must be that $D \subseteq E_1$. In particular, $t \notin D$. Since P_1 is an admissible e-path of G_1, it follows that $|D \cap (P_1 - \{t\})| = 1$. Thus, if $t \notin P_1$, then P_1 is an admissible e-path of G, and if $t \in P_1$, then $P_1 - \{t\} \cup P_2$ is an admissible e-path of G.

Now, assume that $D - E_2 \cup \{t\}$ is an e-cut of G_1. Since P_1 is an admissible e-path of G_1, $|(D - E_2 \cup \{t\}) \cap P_1| = 1$. If $t \notin P_1$, then $P_1 \subseteq E_1$, from which it follows $|D \cap P_1| = 1$, implying that P_1 is an admissible e-path of G. Thus, assume $t \in P_1$. Since P_1 is an admissible e-path of G_1, $(D - E_2 \cup \{t\}) \cap P_1 = \{t\}$. Therefore, $(D \cap E_1) \cap (P_1 - \{t\}) = \emptyset$. Since P_2 is an admissible t-path of G, by Lemma 7, $|(D - E_1 \cup \{t\}) \cap P_2| = 1$. It follows that $|(D \cap P_1 - \{t\}) \cap P_2| = 1$, implying that $P_1 - \{t\} \cap P_2$ is an admissible e-path of G. □

Lemma 10. *Let G be a 3-connected graph, and let e be an edge of G. Suppose that G has an internal 3-separation straddled by e. Let $\{G_1, G_2\}$ be the corresponding 3-sum, and let T be the connecting triangle. Let P be an admissible e-path of G_1 that is edge disjoint from $T - \{e\}$. Then, P is an admissible e-path of G.*

Proof. Since P is edge disjoint from $T - \{e\}$, P is a path of G. Let D be an e-cut of G. By Lemma 8, $D - E_2 \cup \{t, e\}$ is an e-cut of G_1 for some $t \in T - \{e\}$. By assumption, $|(D - E_2 \cup \{t, e\}) \cap P| = 1$, which implies $|D \cap P| = 1$, as required. □

Below is the re-statement of Theorem 1 in terms of admissible paths.

Theorem 11. *Let G be a graph, and let e be an edge of G. If $G\backslash e$ is connected, and G does not have a $K_{3,3}$ minor containing e, then G has an admissible e-path. Moreover, such an e-path can be found in $O(m)$ time in general and in $O(n)$ time if G is 2-connected and simple.*

Proof. The proof is by a series of reductions.

(I) *Reduction to the Simple 2-Connected Case.* Since $G\backslash e$ is connected, every e-path of G is contained in the same block of G as e. Therefore, the search for an admissible e-path of G can be restricted to this block. Computing the blocks of G and identifying the one containing e can be done in $O(m)$ time; see Tarjan [9]. Also, since any e-path uses at most one edge from any parallel class, any parallel edges can be deleted, which requires $O(m)$ time. Now, by Lemma 6, m is $O(n)$.

(II) *Reduction to the 3-Connected Case.* The next step is to show that the admissible-path problem on G can be reduced in linear time to solve a sequence of admissible-path problems, where each problem in the sequence is defined on a graph that is 3-connected and does not contain a $K_{3,3}$ minor using a specified edge, and such that the total size of this sequence is linear in the size of G.

(IIa) *Admissible Paths and 2-Sum Decompositions.* The first step in defining this sequence is to examine the relationship between an instance of the admissible-path problem and a 2-sum decomposition in the underlying graph. So, suppose that G is not 3-connected, and let $\{E_1, E_2\}$ be a 2-separation of G with $e \in E_1$. Let $\{G_1, G_2\}$ be the corresponding 2-sum decomposition, and let $t = pq$ be the connecting edge. It is straightforward to verify that G_1 (resp., G_2) is 2-connected and does not have a $K_{3,3}$ minor containing e (resp., t). By Lemma 9, if P_1 is an admissible e-path of G_1, and P_2 is an admissible t-path of G_2, then an admissible e-path P of G is equal to either P_1, if $t \notin P_1$, and $P_1 - \{t\} \cup P_2$, otherwise.

(IIb) *The Reduction Procedure.* To turn the above relationship between admissible paths and 2-sum decomposition into a computationally efficient algorithm requires two straightforward ideas. First, one chooses the 2-sum decomposition judiciously, and second one applies this judicious choice recursively. Tutte [4] and Hopcroft and Tarjan [10] showed that one can always find a 2-separation $\{E_1, E_2\}$ of G such that $e \in E_1$, and in the resulting 2-sum decomposition $\{G_1, G_2\}$, G_2 is either 3-connected, a cycle on three or more edges, or a bond (i.e., the planar dual of a cycle) on three or more edges. Applying this choice of 2-sum decomposition recursively reduces the admissible-path problem on G to solve a sequence of admissible-path problems on a collection of graphs, say $\{H_1, \ldots, H_j\}$, every member of which is either 3-connected, a cycle, or a bond. Moreover, Hopcroft and Tarjan [10] showed that the sequence of 2-separations necessary to generate $\{H_1, \ldots, H_j\}$ can be found in $O(n)$ time and that the size of the collection, that is, $\sum_{i=1}^{j} |V(H_i)|$, is $O(n)$. Observe that an admissible path on a cycle or bond can be trivially found in linear time (in the size of the cycle or bond). Thus, in $O(n)$ time, the admissible-path problem on G can be reduced to solve a sequence of admissible-path problems, each of which is on a graph that is 3-connected and does not have a $K_{3,3}$ minor using a specified edge. Moreover, the total size of the graphs in the sequence is $O(n)$. Thus, it suffices to prove the theorem assuming G is 3-connected.

(III) *Reduction to the Planar Case.* Assume that G is 3-connected. By Theorem 5, G either is planar, isomorphic to K_5, crown-planar with respect to e or has an internal 3-separation straddled by e. These cases are considered one at a time. As a first step, it is shown that one can recognize which case is applicable in $O(n)$ time. Clearly, recognizing if G is isomorphic to K_5 can be done in constant time. Also, it is well known that planarity can be recognized in $O(n)$ time; see, for example, Hopcroft and Tarjan [11]. Determining whether G has an internal 3-separation straddled by e can be done by simply first deleting the ends of e and then determining if the resulting graph has a cut vertex; the latter can be done in $O(n)$ time using algorithm of Tarjan [9]. The only other possibility for G is that it is crown-planar with respect to e.

(IIIa) *The Base Cases.* This subcase considers these cases when G is either planar, isomorphic to K_5, or crown-planar with respect to e. For each of these three cases, it is shown how to find an admissible path of G in $O(n)$ time.

First, assume that G is planar. Then, as observed by Ford and Fulkerson [1], it is straightforward to find an admissible e-path of G. This is done as follows. Embed G in the plane, so that e is on the infinite face, and then delete e. Then, there exist two e-paths of G the union of which defines the outer face of $G \setminus e$. Ford and Fulkerson [1] showed that each of these paths is an admissible e-path of G (this is also easily seen by planar duality). Observe that if $G \setminus e$ is 2-connected, these two admissible e-paths are internally node disjoint, a fact that will be used later. Finding an embedding of a planar graph can be done in $O(n)$ time [11, 12], and so for planar graphs, it follows that the two admissible e-paths can be found in $O(n)$ time.

Second, consider the case that G is isomorphic to K_5 or is crown-planar with respect to e. In each of these cases, there exist two internally node-disjoint e-paths in G, each of which contains exactly two edges. Since, in general, every e-cut must intersect every e-path in an odd number of edges, it must be that each of these e-paths is admissible. Thus, finding these two admissible e-paths can be done in $O(n)$ time.

(IIIb) 3-*Sum Decompositions.* The final step is to analyze when G has an internal 3-separation straddled by e. Consider the graph $G \setminus \{x, y\}$, where x and y are the ends of e. Since G has an internal 3-separation, $G \setminus \{x, y\}$ has a cut node. Conversely, each cut node of $G \setminus \{x, y\}$ corresponds to an internal 3-separation of G straddled by e. Now, choose a block of $G \setminus \{x, y\}$ that has exactly one cut node in common with rest of $G \setminus \{x, y\}$, and let $\{E_1, E_2\}$ be the corresponding internal 3-separation of G. Let $\{G_1, G_2\}$ be the associated 3-sum decomposition of G, and let T be the connecting triangle. Then, by the choice the cut node, G_1 (say) does not have an internal 3-separation. Thus, by Theorem 5, G_1 is planar, isomorphic to K_5, or crown planar with respect to e. Moreover, since G_1 is 3-connected, $G_1 \setminus e$ is 2-connected. Thus, by Case (IIIa), G_1 has two internally node-disjoint admissible e-paths. Since these two paths are internally node disjoint, one of them, call it P, must be edge disjoint from $T - \{e\}$. Thus, by Lemma 10, P is an admissible e-path of G. Finding the appropriate internal 3-separation easily reduces to finding the cut nodes of $G \setminus \{x, y\}$ and so requires $O(n)$ time using Tarjan [9]. Once the appropriate internal 3-separation is identified, finding the admissible path requires $O(n)$ time by Case (IIIa). □

Theorem 11 can be strengthened as follows: if $G \setminus e$ is 2-connected (resp., 2-edge-connected), then there exist two internally node disjoint (resp., edge-disjoint) admissible e-paths. The proof of this requires a bit more work but follows the same line of reasoning.

Theorem 11 provides a sufficient condition for the existence of an admissible e-path. It is, however, not a necessary condition. For example, take the graph $K_{3,3}$, and let e be any edge. Now, add a new edge that creates a triangle T containing e. Then, $T - \{e\}$ is an e-path that intersects every e-cut in exactly one edge. It would be interesting to determine if one excludes triangles containing e, whether the condition is also necessary.

4. The Ford-Fulkerson Algorithm

Let G be a graph, and let x and y be distinguished nodes of G. Assume G has the edge $e = xy$, and consider an instance of the minimum e-cut problem (equivalently, the maximum-flow problem) defined on G. Ford and Fulkerson [1] provided a very simple algorithm for solving this problem provided that G is planar. This section shows that the Ford-Fulkerson algorithm extends virtually unchanged provided that G does not have a $K_{3,3}$ minor containing e; the running time of the algorithm is shown to be $O(n^2)$. This is within a logarithmic

```
Algorithm ford-fulkerson;
begin
    H := G;
    D := {e};
    while D does not contain an e-cut of H do
        begin
            set P to be an admissible e-path of H;
            ε := min{c_f | f ∈ P};
            c_f := c_f − ε for f ∈ P;
            choose f ∈ P such that c_f = 0 and set H ← H \ f;
            D ← D ∪ {f};
        end;
end;
```

ALGORITHM 1

factor of the fastest maximum-flow algorithm, namely the recent (and more complicated) algorithm due Orlin [2]. The only faster algorithm for graphs in this class is the $O(n)$-time algorithm in Wagner [3].

Consider the following property for the graph G.

Property A. For any subset X of edges with $e \in X$, if $G \setminus X$ has an e-path, then it has an admissible e-path.

Algorithm 1 is the Ford-Fulkerson algorithm, stated in terms of solving the minimum e-cut problem. An instance of the minimum e-cut problem is specified by (G, e, c), where c is a vector of nonnegative edge capacities.

The next theorem shows that the algorithm is correct provided Property A holds. In particular, Theorem 11, then implies that the algorithm works for any instance (G, e, c) provided G does not have a $K_{3,3}$ minor containing e.

Theorem 12. *Assume that G satisfies Property A. Then, Algorithm 1 correctly computes a minimum e-cut of (G, e, u). In particular, at termination, the set D contains a unique e-cut of G, and it is a minimum e-cut of G.*

Proof. Each execution of the *while* loop chooses an admissible e-path. Since, by assumption, G satisfies Property A, this step is well defined. Each execution of the *while* loop adds exactly one edge to D, and therefore eventually, D will contain an e-cut of G, and so the algorithm will terminate.

Assume that the *while* loop executes t times, and index the instantiations of H, D, P, f, c, and ϵ by $1, \ldots, t$. So, H_t denotes the instantiation of H at termination of the algorithm.

It is first shown that D_t contains a unique e-cut of G. This is equivalent to showing that H_t has exactly two components. Suppose this is not the case; that is, suppose H_t has at least three components. Since H_t was obtained from H_{t-1} by deleting exactly one edge, H_{t-1} must have at least two components. Let j denote the least index such that H_j has at least two components. Observe that, in H_j, the ends of e are in the same component, for otherwise the algorithm would have terminated at $j < t$, contradicting the definition of t. By definition, H_{j-1} is connected. Since H_j is obtained from H_{j-1} by deleting the edge f_{j-1}, this edge must be a cut edge of H_{j-1}. From the algorithm, f_{j-1} is contained in the e-path P_{j-1}. But this is impossible since no path, starting and ending in the same component of a graph, can contain a cut edge of the graph. This D_t contains a unique e-cut of G; denote this e-cut by D^*.

Now, consider (H_2, e, c^2), that is, the minimum e-cut problem that results after one execution of the *while* loop. The graph H_2 has one less edge than H_1. Applying Algorithm 1 to (H_2, e, c^2) will produce the set $D_t - \{f_1\}$, which, by induction, will contain a unique e-cut of H_2 that is also a minimum e-cut of (H_2, e, c^2); denote this e-cut by D^{**}. Observe that (G, e, u) is obtained from (H_2, e, u^2) first adding the edge f_1 with a capacity of zero and then adding ϵ_1 to the capacity of every edge of P_1. Consider these steps individually. Adding an edge of capacity zero effectively leaves the minimum e-cut problem unchanged. In particular, either $D^{**} \cup \{f_1\}$ or D^{**} is minimum e-cut for the resulting minimum e-cut problem, depending on whether the ends of f_1 are in different components of $H_2 \setminus D^{**}$ or not. In either case, the capacity of the resultant minimum e-cut is unchanged. The second step also effectively does not change the minimum e-cut problem. In particular, by the algorithm, P_1 is an admissible e-path. Therefore, adding ϵ_1 to each edge of P_1 adds ϵ_1 to the capacity of each e-cut. Consequently, either $D^{**} \cup \{f_1\}$ or D^{**} (again, depending on whether the ends of f_1 are in different components of $H_2 \setminus D^{**}$ or not) is a minimum e-cut of (G, e, u). In either case, this minimum e-cut of (G, e, u) is contained in D_t and, thus, by the uniqueness demonstrated in the previous paragraph, is equal to D^*, as required. □

Corollary. *If G does not have a $K_{3,3}$ minor containing e, then the complexity of Algorithm 1 is $O(n^2)$.*

Proof. By Theorem 11, finding the first admissible e-path requires $O(m)$ time. Also, note that the initial step in this first admissible-path computation reduces the graph to a simple 2-connected graph, and so by Lemma 6, the resulting number of edges is $O(n)$. Thus, each subsequent admissible-path computation requires $O(n)$ time. Since the algorithm deletes one edge every iteration, it requires at most $O(n)$ iterations. Thus, the algorithm requires $O(n^2)$ time. □

As a final note, Algorithm 1 can be dualized, using planar path-cut duality, to an algorithm for finding a shortest e-path in a graph. This dual version of Algorithm 1 can, in fact, be shown to be equivalent to Dijkstra's shortest-path algorithm.

References

[1] L. R. Ford and D. R. Fulkerson, "Maximal flow through a network," *Canadian Journal of Mathematics*, vol. 8, pp. 399–404, 1956.

[2] J. B. Orlin, "Max flows in $O(nm)$ time, or better," Working paper, 2012.

[3] D. K. Wagner, "$K_{3,3}$ minors and the maximum-flow problem," *Algorithmic Operations Research*, vol. 3, no. 1, pp. 30–42, 2008.

[4] W. T. Tutte, *Graph Theory*, Addison-Wesley, Reading, Mass, USA, 1984.

[5] K. Wagner, "Über eine Erweiterung eines Satzes von Kuratowski," *Deutsche Mathematik*, vol. 2, pp. 280–285, 1937.

[6] J. Širáň, "Edges and Kuratowski subgraphs of nonplanar graphs," *Mathematische Nachrichten*, vol. 113, pp. 187–190, 1983.

[7] F. T. Tseng and K. Truemper, "A decomposition of the matroids with the max-flow min-cut property," *Discrete Applied Mathematics*, vol. 15, no. 2-3, pp. 329–364, 1986.

[8] W. T. Tutte, "A theory of 3-connected graphs," *Proceedings of the Koninklijke Nederlandse Akademie van Wetenschappen A*, vol. 64, pp. 441–455, 1961.

[9] R. E. Tarjan, "Depth-first search and linear graph algorithms," *SIAM Journal on Computing*, vol. 1, no. 2, pp. 146–160, 1972.

[10] J. E. Hopcroft and R. E. Tarjan, "Dividing a graph into triconnected components," *SIAM Journal on Computing*, vol. 2, pp. 135–158, 1973.

[11] J. E. Hopcroft and R. E. Tarjan, "Efficient planarity testing," *Journal of the Association for Computing Machinery*, vol. 21, pp. 549–568, 1974.

[12] K. Mehlhorn and P. Mutzel, "On the embedding phase of the Hopcroft and Tarjan planarity testing algorithm," *Algorithmica*, vol. 16, no. 2, pp. 233–242, 1996.

6

Some New Classes of Open Distance-Pattern Uniform Graphs

Bibin K. Jose

Department of Mathematics, S.D. College Alappuzha, University of Kerala, Kerala 688003, India

Correspondence should be addressed to Bibin K. Jose; bibinkjose2002@yahoo.com

Academic Editor: Toufik Mansour

Given an arbitrary nonempty subset M of vertices in a graph $G = (V, E)$, each vertex u in G is associated with the set $f_M^o(u) = \{d(u, v) : v \in M, u \neq v\}$ and called its open M-distance-pattern. The graph G is called open distance-pattern uniform (odpu-) graph if there exists a subset M of $V(G)$ such that $f_M^o(u) = f_M^o(v)$ for all $u, v \in V(G)$, and M is called an open distance-pattern uniform (odpu-) set of G. The minimum cardinality of an odpu-set in G, if it exists, is called the odpu-number of G and is denoted by od(G). Given some property P, we establish characterization of odpu-graph with property P. In this paper, we characterize odpu-chordal graphs, and thereby characterize interval graphs, split graphs, strongly chordal graphs, maximal outerplanar graphs, and ptolemaic graphs that are odpu-graphs. We also characterize odpu-self-complementary graphs, odpu-distance-hereditary graphs, and odpu-cographs. We prove that the odpu-number of cographs is even and establish that any graph G can be embedded into a self-complementary odpu-graph H, such that G and $\overline{G}$ are induced subgraphs of H. We also prove that the odpu-number of a maximal outerplanar graph is either 2 or 5.

1. Introduction

All graphs considered in this paper are finite, simple, undirected, and connected. For graph theoretic terminology, we refer to Harary [1].

The concept of open distance-pattern and open distance-pattern uniform graphs was studied in [2]. Given an arbitrary nonempty subset M of vertices in a graph $G = (V, E)$, the open M-distance-pattern of a vertex u in G is defined to be the set $f_M^o(u) = \{d(u, v) : v \in M, u \neq v\}$, where $d(x, y)$ denotes the distance between the vertices x and y in G. If there exists a nonempty set $M \subseteq V(G)$ such that $f_M^o(u)$ is independent of the choice of u, then G is called open distance-pattern uniform (odpu-) graph, and the set M is called an open distance-pattern uniform (odpu-) set. The minimum cardinality of an odpu-set in G, if it exists, is the odpu-number of G and is denoted by od(G). In this paper, we characterize several classes of odpu graph such as odpu-chordal graphs, interval graphs, split graphs, strongly chordal graphs, maximal outerplanar graphs, ptolemaic graphs, self-complementary graphs, odpu-distance-hereditary graphs, and odpu-cographs. We need the following definitions and previous results.

For a vertex v in a connected graph G, the eccentricity $e(v)$ of v is the distance to a vertex farthest from v. The minimum eccentricity among the vertices of a connected graph G is the radius of G, denoted by $r(G)$, and the maximum eccentricity is its diameter, $d(G)$. A vertex v in a connected graph G is called a central vertex if $e(v) = r(G)$. The collection of all central vertices is called the center of G denoted by $Z(G)$.

In paper [2], it is proved that a graph G with radius $r(G)$ is an odpu graph if and only if the open distance-pattern of any vertex u in G is $f_M^o(u) = \{1, 2, \ldots, r(G)\}$, and a graph is an odpu-graph if and only if its centre $Z(G)$ is an odpu-set, thereby characterizing odpu-graphs, which in fact suggests an easy method to check the existence of an odpu-set for a given graph. The central subgraph $\langle Z(G)\rangle$ of a graph is the subgraph induced by the center.

Proposition 1 (see [2]). *For any graph G, $od(G) = 2$ if and only if there exist at least two vertices $x, y \in V(G)$ such that $deg(x) = deg(y) = |V(G)| - 1$, where $deg(x)$ denote the degree of the vertex x in G.*

Proposition 2 (see [2]). *There is no graph having odpu-number three.*

Proposition 3 (see [2]). *A graph G is an odpu graph if and only if its centre $Z(G)$ is an odpu set, and hence $|Z(G)| \geq 2$.*

Proposition 4 (see [2]). *All self-centered graphs are odpu graphs.*

Theorem 5 (see [2]). *The shadow graph of any complete graph K_n, $n \geq 3$ is an odpu-graph with odpu-number $n+2$ (the shadow graph $S(G)$ of a graph G is obtained from G by adding for each vertex v of G a new vertex v', called the shadow vertex of v, and joining v' to all the neighbors of v in G).*

Theorem 6 (see [2]). *Every odpu-graph G satisfies $r(G) \leq d(G) \leq r(G)+1$, where $r(G)$ and $d(G)$ denote the radius and diameter of G, respectively.*

The complement $\overline{G}$ of a graph G has the same vertices as G, and every pair of vertices are joined by an edge in $\overline{G}$ if and only if they are not joined in G. A self-complementary graph G is one that is isomorphic to its complement $\overline{G}$.

Proposition 7 (see [3]). *Let G be a nontrivial self-complementary graph. Then*

(i) *G has radius 2 and diameter 2 or 3,*

(ii) *G has diameter 3 if and only if it contains a dominating edge (an edge $e \in E$ is said to be a dominating edge if every edge in $E \setminus \{e\}$ is adjacent to e),*

(iii) *the number of vertices of eccentricity 3 is never greater than the number of vertices of eccentricity 2.*

Cographs (or complement reducible graphs) are defined as the class of graphs formed from a single vertex under the closure of the operations of union and complement. Cographs have the following two remarkable properties. Firstly, they are exactly the P_4-restricted graphs, and secondly, a cograph has a unique tree representation, called a cotree. In the cotree representation, leaves of the cotree represent vertices of the graph. Internal nodes of the cotree are labelled using 0 or 1 in such a way that (0) nodes and (1) nodes alternate along every path starting from the root which is always a (1) node. Each nonleaf vertex of T has at least two children. The root will have only one (0) node child if and only if the represented cograph is disconnected. Also, the cotree for a particular cograph is unique up to a permutation of the children of the internal nodes. Two vertices x and y of the graph are adjacent if and only if the unique path from x to the root of the tree meets the unique path from y to the root at a (1) node. In order to establish various properties about cographs, we label each internal node of a cotree as follows: the root is labelled (1), the children of a node with label (1) are labelled (0), and children of a node labelled (0) are labelled (1). Henceforth, we assume all cotrees to be labelled as such, and we will refer to the internal nodes of cotrees as (0)-nodes and (1)-nodes (cf. [4–7]).

Theorem 8 (see [4]). *A graph G has a cotree if and only if G is a cograph. In that case, the cotree representation is unique.*

A graph G is said to be chordal if every cycle of length at least 4 has a chord, that is, an edge joining nonconsecutive vertices of the cycle. A subset $S \subseteq V(G)$ is called a vertex separator of G for nonadjacent vertices a and b (or $a-b$ separator of G) if, in $G-S$ (the graph obtained from G by the removal of the vertices of S and incident edges), the vertices a and b belong to distinct connected components. Let $S(a,b,G)$ be the set of all $a-b$ separators of G. If no proper subset of $S \in S(a,b,G)$ belongs to $S(a,b,G)$, then S is called a minimal $a-b$ separator of G. Let $S_0(a,b,G)$ be the set of all minimal $a-b$ separators of G. The following propositions (cf. [8]) are needed in proving our main results.

Proposition 9 (see [8]). *If G is a chordal graph and x, y are distinct nonadjacent vertices in $\langle Z(G)\rangle$, the central graph of G, and $S_0 \in S_0(x,y,G)$, then the following conditions hold:*

(1) $S_0 \subseteq Z(G)$,

(2) *there are at least two distinct vertices $z_1, z_2 \in S_0$ such that for every $i = 1,2$ either $d_G(x,z_i) = 1$ or $d_G(y,z_i) = 1$. In particular, $|S_0| \geq 2$.*

Proposition 10 (see [8]). *If G is a chordal graph and x, y are distinct nonadjacent vertices in $\langle Z(G)\rangle$, the central subgraph of G, then $S_0(x,y,G) = S_0(x,y,\langle Z(G)\rangle)$.*

Proposition 11 (see [8]). *If G is a chordal graph, then $\langle Z(G)\rangle$ is connected.*

Proposition 12 (see [8]). *If G is a chordal graph, then $d(\langle Z(G)\rangle) \leq 3$.*

Proposition 13 (see [8]). *If G is a chordal graph with $(r,d) = (2,3)$, then $d(\langle Z(G)\rangle) \leq 2$.*

Proposition 14 (see [8]). *There are no self-centered chordal graph G with $(r,d) = (3,3)$. Consequently, for a self-centered chordal graph G, $r(G) \leq 2$.*

A distance-hereditary graph (cf. [9]) is a connected graph, which preserves the distance function for induced subgraphs. That is, the distance between any two nonadjacent vertices of any connected induced subgraph of such a graph is the same as the distance between these two vertices in the original graph.

Proposition 15 (see [9]). *Let G be a distance-hereditary graph, and let u be a vertex of G. Then, for each positive integer $k \leq \max d(u,v) : v \in V$ and vertices v, w of the same connected component of $N_k(u)$, one has $N(v) \cup N_{k-1}(u) = N(w) \cup N_{k-1}(u)$, where the ith neighborhood of the vertex u, $N_i(u) = \{v : d(u,v) = i\}$.*

A graph G is an interval graph if and only if there is a one-to-one correspondence between its vertices and a set of intervals on the real line, such that two vertices are adjacent if and only if the corresponding intervals have an intersection. It is also well known that a graph G is an interval graph if and only if G is chordal and asteroidal triple-free, where asteroidal triple is a set of three distinct vertices (v_1, v_2, v_3) such that there exists a path connecting v_i and v_j that contains no neighbor of v_k $(i \neq k \neq j)$, for every combination of $1 \leq i, j, k \leq 3$ (cf. [10]).

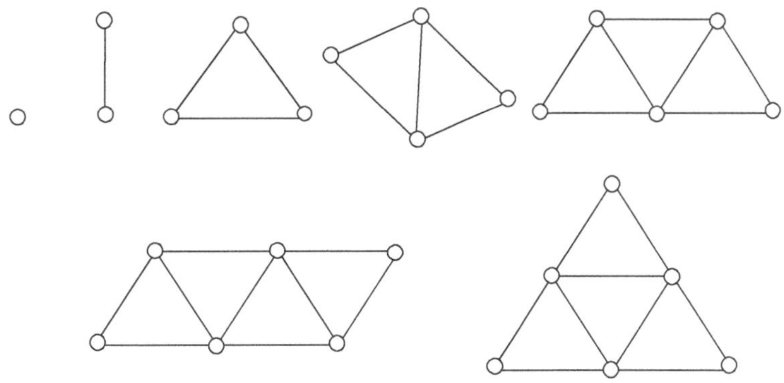

Figure 1

A graph is a split graph if and only if its vertices can be partitioned into an independent set, and vertices which induces a clique. For simplicity, given a split graph G, we call a vertex partition of $V(G) = (V_I, V_C)$, such that V_I is an independent set and vertices in V_C induce a clique, as an I-C-decomposition of G. It is known that a chordal graph whose complement is also a chordal graph is equivalent to a split graph (cf. [10]).

A graph G is strongly chordal if and only if G is chordal, and every even cycle of length six or more contains a chord splitting the cycle into two odd length paths (cf. [10]).

Ptolemaic graphs are exactly those graphs that are both chordal and distance-hereditary (cf. [11]).

A graph is outerplanar if it can be drawn in the plane with all nodes in the exterior boundary. It is called maximal outerplanar if no edge can be added without destroying its outerplanar property. Every maximal outerplanar graphs is chordal (cf. [12]).

Proposition 16 (see [12]). *If G is a maximal outerplanar graph, then its central subgraph $\langle Z(G)\rangle$ is isomorphic to one of the seven graphs in Figure 1.*

2. Main Results

The following theorem gives a complete characterization for the odpu-self-complementary graphs. Further, given any positive even integer $n \neq 2$, there exists an odpu, self-complementary graphs with odpu-number n. Also, we prove that it is possible to embed any graph G into a self-complementary, odpu-graph H with G and $\overline{G}$ being induced subgraphs of the graph H. Recall that a universal vertex means a vertex which is adjacent to all other vertices of the graph.

Theorem 17. *A self-complementary graph G is an odpu graph if and only if G has no universal vertex in $\langle Z(G)\rangle$.*

Proof. Assume that G is a self-complementary, odpu-graph. If v is a universal vertex in $\langle Z(G)\rangle$, then $f_M^o(v) = \{1\}$, where $M = Z(G)$. Hence, all the vertices of G must be adjacent to v in G. Hence, v is an isolated vertex in $\overline{G}$, a contradiction to the hypothesis that G is a self-complementary graph.

Conversely, let the self-complementary graph G be without universal vertex in $\langle Z(G)\rangle$. By Proposition 7, the possible radius r and diameter d of G are $r = d = 2$ or $r = 2$ and $d = 3$. If $r = d = 2$, then the graph G is self-centered, and hence it is an odpu-graph.

If $r = 2$ and $d = 3$, then by Proposition 7, G has a dominating edge uv. In this case, the end points of the dominating edge lie in the centre, and any vertex outside the centre is adjacent to exactly one end vertex of any dominating edge.

Let x be any arbitrary vertex in $Z(G)$. By hypothesis, x is not a universal vertex, and hence, there exists a vertex $y \in Z(G)$ such that $d(x, y) = 2$. Therefore, $2 \in f_M^o(x)$. Also since x is adjacent to a dominating edge, $1 \in f_M^o(x)$. Hence, $f_M^o(x) = \{1, 2\}$ for all $x \in Z(G)$.

Now, let v be an arbitrary vertex in $G - Z(G)$. Then, v is adjacent to exactly one vertex of a dominating edge xy. Hence, either $d(x, v) = 1$ and $d(y, v) = 2$ or $d(x, v) = 2$ and $d(y, v) = 1$. In both cases, $f_M^o(v) = \{1, 2\}$ for all $v \in V(G) - Z(G)$, and hence G is an odpu-graph. □

By Theorem 17 and Proposition 7, the following Corollary holds immediately.

Corollary 18. *A self-complementary graph G is an odpu-graph if and only if $r(G) = r(\langle Z(G)\rangle) = 2$.*

Theorem 19. *Given any even integer $k \geq 4$, there exists a self-complementary, odpu-graph with odpu-number k.*

Proof. First, we take a path P_4 and replace the end vertices of the path P_4 by copies of nP_2 and the interior vertices by copies of the complete n-partite graph $K_{2,2,\ldots,2}$; $n \geq 2$. Where two vertices of P_4 were joined by an edge, the corresponding graphs are now joined by all possible edges between them. Let the resulting graph be G_1. Clearly, G_1 is a self-complementary graph of $8n$ vertices with diameter 3. Moreover, if we add a $\overline{K_2}$ and join it to all the vertices of the copies of $K_{2,2,\ldots,2}$, we get a self-complementary graph G_2 of order $8n + 2$ and diameter 3 (see Figure 2).

Now, we claim that the odpu-number of G_1 is $4n$ and that of G_2 is $4n+2$. Since the eccentricities of the vertices in $K_{2,2,\ldots,2}$ are 2 and the eccentricities of the vertices of nP_2 are 3 in G_1, the center $Z(G_1)$ is the collection of all vertices of both

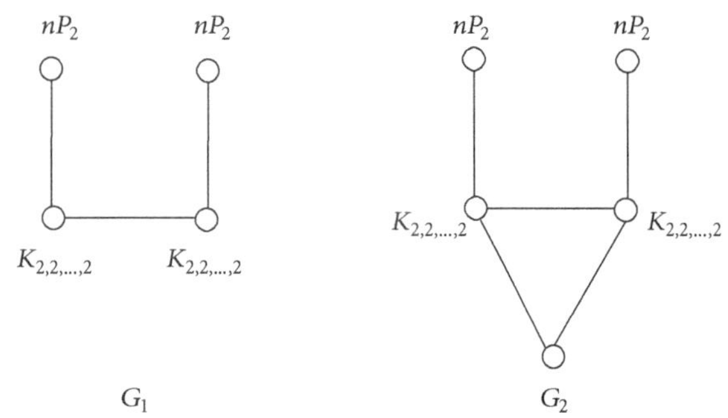

FIGURE 2

$K_{2,2,...,2}$. Since $\langle Z(G_1)\rangle$ does not have a universal vertex, by Theorem 17, G_1 is an odpu-graph. Let M be a minimal odpu-set of G_1. For any vertex $u \in Z(G_1)$, there is exactly one vertex $v \in Z(G_1)$ such that $d(u,v) = 2$. Hence, all the vertices of $Z(G_1)$ must be in M. Hence, $\text{od}(G_1) = 4n$.

Now, consider G_2. By the same argument above, all the vertices of $Z(G_2)$ must be in the minimal odpu-set M. Hence, the odpu-number of G_2 is $4n+2$, hence, the theorem. □

Theorem 20. *Any graph G with $r(G) \geq 2$ can be embedded into a self-complementary, odpu-graph H with both G and $\overline{G}$ as induced subgraphs of H.*

Proof. We construct the graph H as follows. First, consider a path P_4, and replace the end vertices of the path P_4 by copies of $\overline{G}$ and the interior vertices by copies of G. Whenever two vertices of P_4 were joined by an edge, the corresponding graphs are joined by all possible edges between them so that we get a complete bipartite graph between the vertices of the corresponding graphs. Then, the resulting graph H is self-complementary with G (and $\overline{G}$) as induced subgraph of H with $r(H) = 2$ and $d(H) = 3$. Also, all the vertices of G belong to the centre $Z(H)$, and the vertices of $\overline{G}$ belong to $H - Z(H)$. Let $M = Z(H)$.

Let $u \in Z(H)$. Since $r(G) \geq 2$, there exists a vertex v not adjacent to u in the same copy of G. But there exists a path (uwv) in H, where w is in G, for which u, v do not belong. Hence, $d_H(u,v) = 2$, and therefore, $2 \in f_M^o(u)$. Since $xy \in E(H)$ for $X \in V(G_1)$ and $y \in V(G_2)$, $1 \in f_M^o(u)$. Hence, $f_M^o(u) = \{1,2\}$ for all $u \in Z(H)$.

Let $u \in V(G) - Z(G)$. Then, u is adjacent to a vertex $w \in Z(H)$. Therefore, $1 \in f_M^o(u)$. Also by construction, there exists a vertex $v \in M$ such that $d(u,v) = 2$. Hence, $2 \in f_M^o(u)$. Hence, $f_M^o(u) = \{1,2\}$ for all $u \in V(H) - Z(H)$. Hence, $f_M^o(x) = \{1,2\}$ for all $x \in V(H)$. Hence, H is an odpu-graph. □

Remark 21. Suppose that the given graph is of radius 1. Then, the construction given in Theorem 20 gives that $\langle Z(H)\rangle$ has a universal vertex. Hence, by Theorem 17, the resulting graph is not an odpu-graph. So, if the given graph is of radius 1, then first embed the graph into a graph G of radius greater than or equal two, by adding a path in any one of the vertex or by any other methods in such a way that the given graph is an induced subgraph of G. Then, the graph G is of radius greater than or equal to 2, and hence, we can apply the above construction given in Theorem 20. Hence, we have the generalized form of Theorem 20, as follows.

Theorem 22. *Any graph G can be embedded into a self-complementary, odpu-graph H with G and $\overline{G}$ being induced subgraphs of H.*

The following result gives the characterization for odpu-cographs and proves that the odpu-number of an odpu-cograph is always even.

Theorem 23. *A cograph (P_4-free graph) G is an odpu-graph if and only if $|Z(G)| \geq 2$.*

Proof. Let G be a P_4-free odpu-graph. Then, clearly $|Z(G)| \geq 2$.

Conversely, let G be a cograph with $|Z(G)| \geq 2$. Since the diameter of a cograph is less than or equal to 2, the following are the only three possibilities. If (i) $r(G) = d(G) = 1$ or (ii) $r(G) = d(G) = 2$, then G is self-centered, and hence, it is an odpu-graph. If (iii) $r(G) = 1$ and $d(G) = 2$, then by assumption $|Z(G)| \geq 2$, there exist at least two universal vertices u and v in G. Therefore, $\{u, v\}$ forms an odpu-set of G, hence the theorem. □

Corollary 24. *A cograph G is an odpu-graph if and only if $r(G) = r(\langle Z(G)\rangle)$.*

Theorem 25. *The odpu-number of an odpu-cograph is always even.*

Proof. By Theorem 8, each cograph is uniquely represented as a cotree and conversely. Hence, we prove the theorem using cotree characterization of cographs. Consider the cotree T of the odpu-cograph G. Since the odpu-graphs are connected, the root which is labeled by (1) in the cotree has at least two children.

If the root (1) has a child which is a leaf in the cotree, then this leaf vertex u is adjacent to all vertices of the cograph G. Hence, u is a universal vertex in G, and hence, $r(G) = 1$. Since G is an odpu-cograph, $|Z(G)| \geq 2$, and hence, there exist two universal vertices in G. Hence, there exist at least two leaves attached to the root (1) of the cotree T. Hence, in this case $\text{od}(G) = 2$.

So assume that G does not have a universal vertex. Thus, there is no leaf attached to the root (1), and hence, $r(G) = 2$. Since G is connected and all children of the root (1) are labeled by (0), the root (1) has at least two children which are labeled by (0). Let the root (1) have k children, namely, $R_1, R_2, \ldots, R_k$, which are labeled by (0). Let M be the minimal odpu-set of G. Since $r(G) = 2$, $f_M^o(u) = \{1,2\}$ for all $u \in V(G)$. Also, for a cotree, each node R_i labeled by (0) at least has two leaves x and y descending from that node which are nonadjacent. Also, each leaf u descending from the node R_i is adjacent to all the leaves descending from R_j, $j \neq i$, hence, leaves x and y with $d(x,y) = 2$ only when x and y are descending from same R_i. So, there exist at least two vertices x and y descending from each R_i $(1 \le i \le k)$ and belonging to the minimal odpu-set M. Hence, $|M| \ge 2k$. □

Now, we prove that M has exactly $2k$ elements. Let M be the set of $2k$ leaves such that the exactly two leaves x and y are descending from the same node R_i, $(1 \le i \le k)$ with $xy \notin E(G)$. Let $x \in V(G)$. Then, there exists a $y \in M$ which is descending from the same node R_i, where the leaf x belongs, such that $xy \notin E(G)$. Hence, $d(x, y) = 2$, and hence, $2 \in f_M^o(x)$. Since x is adjacent to all vertices in M which are descending from the nodes R_j, for all $j \ne i$, $1 \in f_M^o(x)$. Hence, $f_M^o(x) = \{1, 2\}$ for all $x \in V(G)$. Hence, $\mathrm{od}(G) = 2k$.

Remark 26. Given any positive even integer $2n$, the complete n-partite graphs $K_{2,2,\dots,2}$ are odpu-cographs with odpu-number $2n$.

Next, we establish the characterization for odpu-chordal graphs.

Theorem 27. *A chordal graph G is an odpu-graph if and only if* $r(G) = r(\langle Z(G)\rangle)$.

Proof. Let G be a chordal graph with $r(G) = r(\langle Z(G)\rangle)$. Since $r(\langle Z(G)\rangle) \le 2$, there are two possibilities:

(i) $r(G) = r(\langle Z(G)\rangle) = 1$. Then, there are two universal vertices in G, and hence by Proposition 1, G is an odpu-graph,

(ii) $r(G) = r(\langle Z(G)\rangle) = 2$. By Proposition 10, $d_G(u, v) = d_{\langle Z(G)\rangle}(u, v)$ for all $u, v \in Z(G)$. Let $u \in Z(G)$. Since $r(\langle Z(G)\rangle) = 2$, there exists a vertex $v \in Z(G)$ such that $d(u, v) = 2$. Hence, $2 \in f_M^o(u)$ for all $u \in Z(G)$. Now, the connectedness of $\langle Z(G)\rangle$ ensures the existence of a vertex $w \in Z(G)$ such that $d(u, w) = 1$. Therefore, $f_M^o(u) = \{1, 2\}$ for all $u \in Z(G)$. Now, let $u \in V(G) - Z(G)$.

□

Claim 1 ($2 \in f_M^o(u)$). If there is no vertex $v \in Z(G)$ such that $d(u, v) = 2$, then u is adjacent to all the vertices of $Z(G)$. Since $r(\langle Z(G)\rangle) = 2$, there exist two vertices x and y in $Z(G)$ such that $d(x, y) = 2$. Now, consider $S_0(x, y)$. Since $u \notin Z(G)$, $u \notin S_0(x, y)$. But $G - S_0(x, y)$ contains an $x - y$ (xuy)-path, which is a contradiction to the fact that G is chordal, and hence, $S_0 \subseteq Z(G)$. Hence, there exists a vertex $v \in Z(G)$ such that $d(u, v) = 2$ implies $2 \in f_M^o(u)$.

Claim 2 ($1 \in f_M^o(u)$). Let u be nonadjacent to any of the vertices of $Z(G)$. Let $x, y \in Z(G)$ such that $d(x, y) = 2$. Since $r(G) = r(\langle Z(G)\rangle) = 2$, $d(x, u) = d(y, u) = 2$. Thus, there exist vertices w and z in $V(G) - Z(G)$ such that (xwu) and (yzu) are paths of length 2 in G. If $w = z$, then the path $(xwy) \in V(G) - S_0(x, y)$ leads to the contradiction. Otherwise the path $(xwuzy) \in V(G) - S_0(x, y)$, a contradiction that $S_0(x, y) \subseteq Z(G)$. Hence, there exists a vertex $v \in Z(G)$ such that $d(u, v) = 1$, and hence, $1 \in f_M^o(u)$ for all $u \in V(G) - Z(G)$. Therefore, $f_M^o(u) = \{1, 2\}$ for all $u \in V(G) - Z(G)$. Hence, G is an odpu-graph.

Conversely, assume that G is a chordal-odpu-graph. Since $r(G) \le 2$ and $\langle Z(G)\rangle$ are connected, $r(\langle Z(G)\rangle) \le 2$.

If $r(\langle Z(G)\rangle) = 1$, then $f_M^o(u) = \{1\}$ for all $u \in V(G)$. Thus, there exist at least two universal vertices in G. Hence, $r(G) = 1$.

If $r(\langle Z(G)\rangle) = 2$, $f_M^o(u) = \{1, 2\}$ for all $u \in V(G)$. Since $Z(G)$ is an odpu-set, the distance from any vertex of G to any vertex of $Z(G)$ is less than or equal to 2. Thus, $r(G) = 2$, and hence, $r(G) = r(\langle Z(G)\rangle)$.

Corollary 28. *A chordal graph G is an odpu-graph, and then* $\langle Z(G)\rangle$ *is self-centered.*

Proof. Let G be chordal, odpu-graphs. If $\langle Z(G)\rangle$ is not self-centered, then $r(\langle Z(G)\rangle) \ne d(\langle Z(G)\rangle)$. Hence, there exist vertices $u, v \in Z(G)$ such that $e_{\langle Z(G)\rangle}(u) \ne e_{\langle Z(G)\rangle}(v)$. Let $e_{\langle Z(G)\rangle}(u) = d(\langle Z(G)\rangle)$, $e_{\langle Z(G)\rangle}(v) = r(\langle Z(G)\rangle)$, and $M = Z(G)$. Since $d_{\langle Z(G)\rangle}(u, v) = d_G(u, v)$ for all $u, v \in Z(G)$, $f_M^o(u)$ has an element $d(\langle Z(G)\rangle)$, a contradiction. □

Remark 29. The converse of Corollary 28 need not be true. For example, P_4 is chordal with $\langle Z(P_4)\rangle = P_2$, which is self-centered, but P_4 is not an odpu-graph.

Since interval graphs, split graphs, block graphs, ptolemaic graphs, strongly chordal graphs, and maximal outerplanar graphs are subclasses of chordal graphs, (cf. [10]), the following corollary is immediate from Theorem 27 and Corollary 28.

Corollary 30. *The following classes of graphs G: interval graphs, split graphs, block graphs, ptolemaic graphs, strongly chordal graphs, or maximal outerplanar graphs are odpu-graphs if and only if* $r(G) = r(\langle Z(G)\rangle)$, *and hence,* $\langle Z(G)\rangle$ *is self-centered.*

The following theorem establishes the characterization of odpu-distance-hereditary graphs. Further, we show that the central subgraph $\langle Z(G)\rangle$ of distance-hereditary-odpu-graph G is either self-centered or disconnected.

Theorem 31. *Let G be a distance-hereditary graph with connected* $\langle Z(G)\rangle$. *Then, G is an odpu-graph if and only if* $r(G) = r(\langle Z(G)\rangle)$.

Proof. Let G be a distance-hereditary graph with $\langle Z(G)\rangle$ which is connected. Let G be an odpu-graph. If $r(G) \ne r(\langle Z(G)\rangle)$, then $r(\langle Z(G)\rangle) < r(G)$. Since $\langle Z(G)\rangle$ is connected, $d_{\langle Z(G)\rangle}(u, v) = d_G(u, v)$ for all $u, v \in Z(G)$. But since $r(\langle Z(G)\rangle) < r(G)$, $r(G) \notin f_M^o(x)$ for all $x \in Z(G)$, which is not possible. Hence, $r(G) = r(\langle Z(G)\rangle)$.

Conversely, let $r(G) = r(\langle Z(G)\rangle) = r$. Let $u \in Z(G)$. Then, there exists a vertex $v \in Z(G)$ such that $d(u, v) = r$. Since $\langle Z(G)\rangle$ is connected, $f_M^o(u) = \{1, 2, \dots, r\}$.

Now, let $u \in V(G) - Z(G)$. If $1 \notin f_M^o(u)$, then $N_1(u) \cap Z(G) = \phi$, where $N_t(u) = \{v \in V(G) : d(u, v) = t\}$. Let k be the largest integer such that $N_k(u) \cap Z(G) \ne \phi$, and let $w \in N_k(u) \cap Z(G)$. Then, since $d_G(u, w) = k$, $e_G(w) \ge k$. But since $u, N_1(u) \notin Z(G)$, by Proposition 15, $e_{Z(G)}(w) \le k - 2$.

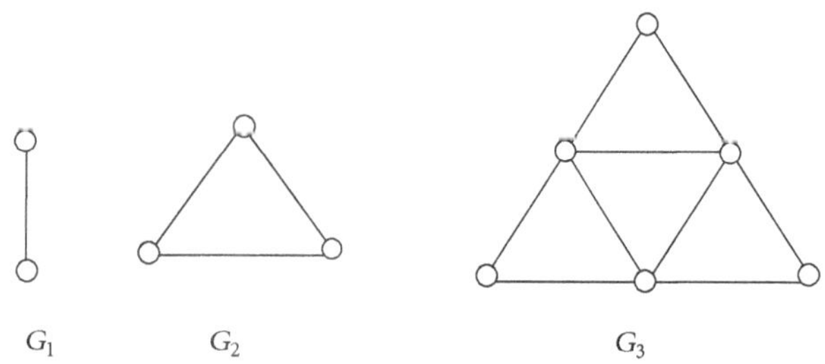

FIGURE 3

Thus, $r(G) \neq r(\langle Z(G)\rangle)$, a contradiction. Therefore $1 \in f_M^o(u)$ for all $u \in V(G) - Z(G)$.

Now, if $r \notin f_M^o(u)$, then $N_r(u) \cap Z(G) = \phi$. Since $u \notin Z(G)$, by Proposition 15, $e_{\langle Z(G)\rangle}(w) \leq r-1$ for every $w \in Z(G)$. Thus, $r(\langle Z(G)\rangle) < r(G)$, a contradiction. Hence, $r \in f_M^o(u)$.

Since $1, r \in f_M^o(u)$, $f_M^o(u) = \{1, 2, \ldots, r\}$ for all $v \in V(G) - Z(G)$, hence the theorem. □

Corollary 32. *For a distance-hereditary-odpu-graph G, either $\langle Z(G)\rangle$ is disconnected or it is self-centered.*

Proof. Let G be a distance-hereditary-odpu-graph. If $\langle Z(G)\rangle$ is connected, then we prove that $\langle Z(G)\rangle$ is self-centered. If not, let $r(\langle Z(G)\rangle) \neq d(\langle Z(G)\rangle)$. Let $u, v \in Z(G)$ such that $e_{\langle Z(G)\rangle}(u) = r(\langle Z(G)\rangle)$ and $e_{\langle Z(G)\rangle}(v) = d(\langle Z(G)\rangle)$. Since G is distance-hereditary, $d_{(\langle Z(G)\rangle)}(x, y) = d_G(x, y)$ for all $x, y \in Z(G)$. Thus, $d(\langle Z(G)\rangle) \in f_M^o(v)$ and $d(\langle Z(G)\rangle) \in f_M^o(u)$, which is a contradiction, hence the theorem. □

Remark 33. The converse of Corollary 32 need not be true. For example, P_4 is distance-hereditary and $\langle Z(P_4)\rangle = P_2$, which is self-centered, but P_4 is not an odpu-graph.

The following theorem gives a necessary condition for a maximal outerplanar graph to be an odpu-graph in terms of a specific structure of the central subgraph $\langle Z(G)\rangle$.

Theorem 34. *If G is a maximal outerplanar-odpu-graph, then its central subgraph $\langle Z(G)\rangle$ is isomorphic to one of the graphs in Figure 3.*

Proof. By Proposition 16, the central subgraph $\langle Z(G)\rangle$ of a maximal outerplanar graph G is isomorphic to one of the seven graphs given in Figure 1.

Since every maximal outerplanar graph is chordal, by Corollary 28, the central subgraph $\langle Z(G)\rangle$ is self-centered. Therefore, $\langle Z(G)\rangle$ is isomorphic to one of the three graphs given in Figure 3, hence the theorem. □

Theorem 35. *A maximal outerplanar graph G is an odpu-graph if and only if it is isomorphic to one of the graphs in Figure 4.*

Proof. Let G be a maximal outerplanar-odpu-graph. By Theorem 34, $\langle Z(G)\rangle$ is isomorphic to one of the graphs in Figure 3, and by Corollary 30, $r(G) = r(\langle Z(G)\rangle)$. Thus, there are two cases: $r(G) = r(\langle Z(G)\rangle) = 1$ and $r(G) = r(\langle Z(G)\rangle) = 2$. □

Case 1 ($r(G) = r(\langle Z(G)\rangle) = 1$). By Proposition 1, there exist at least two universal vertices in G, and hence, $\langle Z(G)\rangle$ is isomorphic to either G_1 or G_2 in Figure 3. Consider all graphs G with two universal vertices. Then, P_2 is the least of them. Let $P_2 \cong H_1$ be an edge uv. Now, add vertices one by one to get a new maximal outerplanar graph G in such a way that u and v are universal in G. That is, $G = uv + \{v_1, v_2, \ldots, v_k\}$, where "+" denotes the operator join (The join of two graphs $G_1 = (V_1, E_1)$ and $G_2 = (V_2, E_2)$ denoted by $G_1 + G_2$ has the vertex set as $V = V_1 \cup V_2$, and the edge set E contains all the edges of G_1 and G_2 together with all edges joining the vertices of V_1 with the vertices of V_2.) of two graphs. When $i = 1$, $G = uv + v_1 \cong C_3 = H_2$, which is a maximal outerplanar-odpu-graph. When $i = 2$, $G = uv + \{v_1, v_2\} \cong K_2 + \overline{K_2} = H_3$, which is also a maximal outerplanar-odpu-graph. When $i \geq 3$, $G = uv + \{v_1, v_2, \ldots, i\} \cong K_2 + \overline{K_i}$, $i \geq 3$ are not maximal outerplanar graphs. Thus, H_1, H_2, and H_3 are the only maximal outerplanar graphs with $r(G) = r(\langle Z(G)\rangle) = 1$.

Case 2 ($r(G) = r(\langle Z(G)\rangle) = 2$). By Theorem 34, the central subgraph is isomorphic to $G_3 \cong H_4$ only. Since H_4 is self-centered, it is an odpu-graph. Suppose that there exists a maximal outerplanar-odpu-graph H other than H_4 with $\langle Z(H)\rangle \cong G_3$. Then, there exists a vertex $x \notin V(H) - Z(H)$ such that x is adjacent to a vertex of $\langle Z(H)\rangle \cong G_3$; say the vertex a. Since H is maximal outerplanar, x cannot be adjacent to any of the vertices b, c, and a'. If there exists a vertex w such that (xwa') is a path in the graph H, then $S_0(a, a', H) \nsubseteq Z(H) = G_3$, a contradiction. Hence, $d(x, a') = 3$. Thus, $e_H(a') \geq 3$, and hence, $a' \notin Z(H)$, a contradiction. Hence, there does not exist a vertex x other than the vertices of G_3. Hence, there is no maximal outerplanar-odpu-graph H other than H_4 with $r(G) = r(\langle Z(G)\rangle) = 2$.

Theorem 36. *For a maximal outerplanar graph G, the odpu-number is either 2 or 5.*

Proof. By Theorem 25, the maximal outerplanar odpu-graphs are one of the graphs in Figure 4.

It is enough to prove that each of H_1, H_2, and H_3 has odpu-number 2. Next, the graph H_4 is self-centered with radius 2 and hence, $f_M^o(u) = \{1, 2\}$, for all $u \in V(H_4)$. Let M be a minimal odpu-set of H_4. Hence, $a' \in V(H_4)$ is the only vertex at a distance of two from the vertex a, and hence, a' must be in M. Similarly b' and c' are the only vertices of H_4 at a distance of two from b and c, respectively, and hence, $b', c' \in M$. Hence, $|M| \geq 3$. If $M = \{a', b', c'\}$, then $1 \notin f_M(u)$, for $u = a', b', c'$. Thus, without loss of generality, let $a \in M$. Then, $M = \{a, a', b', c'\}$. But $1 \notin f_M(a')$, and hence, at least one of the vertices b and c must be in M. So let $b \in M$. Then, $M = \{a, b, a', b', c'\}$. Hence, $f_M^o(u) = \{1, 2\}$, for all $u \in V(H_4)$. Thus, $\text{od}(H) = 5$. Hence, the odpu-number of maximal outerplanar graph is either 2 or 5. □

Theorem 37. *For every integer $n \geq 2$, there is an odpu-graph G of order $n(n+2)$ such that its central subgraph $\langle Z(G)\rangle$ is disconnected.*

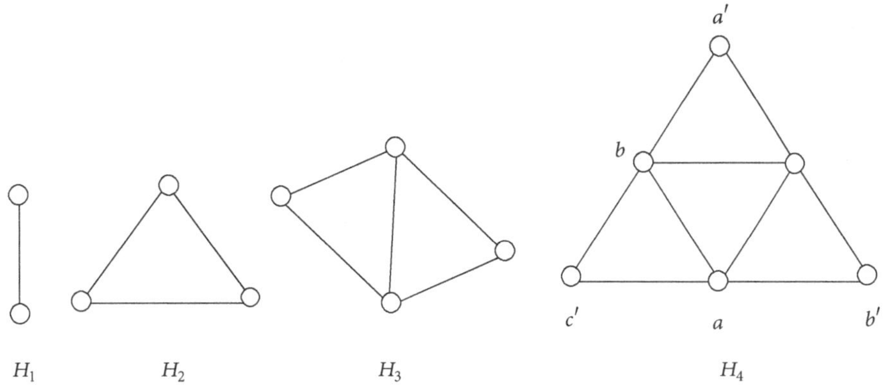

FIGURE 4

Proof. Consider two disjoint complete graphs $G_1 = K_n$ and $G_2 = K_n$ of order n. Now, add all edges between these two complete graphs, and subdivide each of the new edges of the bipartite subgraph between G_1 and G_2 by one (a degree 2 vertex) to get a graph G of order $n(n+2)$. Let the vertices of $G_1\{v_1, v_2, \ldots, v_n\}$, the vertices of $G_2\{u_1, u_2, \ldots, u_n\}$, and $w_{i,j}$ the vertex which subdivides the earlier edge v_iu_j of the bipartite graph. Then, all vertices $w_{i,j}$'s have eccentricity 3 in G, and the new graph has radius 2 and diameter 3. Also the central subgraph $\langle Z(G)\rangle$ is the disjoint union of the complete graphs G_1 and G_2. Hence, $\langle Z(G)\rangle$ is disconnected.

Now, we prove that G is an odpu-graph. Let $M = Z(G) = V(G_1) \cup V(G_2)$. For each $v_i \in V(G_1)$, there exists a $v_j \in V(G_1)$ such that $d(v_i, v_j) = 1$, and hence, $1 \in f_M^o(v_i)$. Now, $d(v_i, u_j) = 2$, for all $u_j \in V(G_2)$, and hence, $2 \in f_M^o(v_i)$. Hence, $f_M^o(v_i) = \{1, 2\}$ for all $v_i \in V(G_1)$. Similarly, $f_M^o(u_j) = \{1, 2\}$ for all $u_j \in V(G_2)$. Now, each vertex $w_{i,j}$ is adjacent to exactly v_i and u_j, and hence, $d(w_{i,j}, v_i) = 1$ and $d(w_{i,j}, u_j) = 1$. Hence, $1 \in f_M^o(w_{i,j})$ for all i, j. Since $d(w_{i,j}, v_k) = 2$, for all $i \neq k$ and $d(w_{i,j}, u_k) = 2$, for all $j \neq k$, $2 \in f_M^o(w_{i,j})$ for all i, j. Hence, $f_M^o(w_{i,j}) = \{1, 2\}$ for all i, j, and hence, G is an odpu-graph. □

3. Conclusion

The characterization of odpu-graphs leads to an interesting condition $r(G) = r(\langle Z(G)\rangle)$, for many important classes of graphs such as chordal graphs, interval graphs, split graphs, strongly chordal graphs, self-complementary graphs, P_4-free graphs, maximal outerplanar graphs, ptolemaic graphs, and distance-hereditary graphs. However, this characterization is not in general a characterization for all odpu-graphs. For example, by Theorem 37, there are classes of odpu-graphs with radius 2 and disconnected centre. That is, $r(\langle Z(G)\rangle) = \infty$. Thus, there are more classes of odpu-graphs which do not come under this characterization. We leave it for further scope of investigations.

References

[1] F. Harary, *Graph Theory*, Addison-Wesley, Reading, Mass, USA, 1969.

[2] B. K. Jose, "Open distance-pattern uniform graphs," *International Journal of Mathematical Combinatorics*, vol. 3, pp. 103–115, 2009.

[3] A. Farrugia, *Self-Complementary Graphs and Generalizations: A Comprehensive Reference Manual*, University of Malta, 1999.

[4] D. G. Corneil, H. Lerchs, and L. S. Burlingham, "Complement reducible graphs," *Discrete Applied Mathematics*, vol. 3, no. 3, pp. 163–174, 1981.

[5] D. G. Corneil, Y. Perl, and L. K. Stewart, "A linear recognition algorithm for cographs," *SIAM Journal on Computing*, vol. 14, no. 4, pp. 926–934, 1985.

[6] E. Dahlhaus, "Efficient parallel recognition algorithms of cographs and distance hereditary graphs," *Discrete Applied Mathematics*, vol. 57, no. 1, pp. 29–44, 1995.

[7] R. Lin and S. Olariu, "A fast parallel algorithm to recognize P4-sparse graphs," *Discrete Applied Mathematics*, vol. 81, no. 1–3, pp. 191–215, 1998.

[8] P. Das and S. B. Rao, "Center graphs of chordal graphs," in *Combinatorics and Applications (Kolkata, 1982)*, pp. 81–94, Indian Statistical Institute, Kolkata, India, 1984.

[9] E. Howorka, "A characterization of distance-hereditary graphs," *Quarterly Journal of Mathematics*, vol. 28, no. 4, pp. 417–420, 1977.

[10] F. Buckley and F. Harary, *Distance in Graphs*, Addison-Wesley Publishing Company Advanced Book Program, Redwood City, Calif, USA, 1990.

[11] E. Howorka, "A characterization of Ptolemaic graphs," *Journal of Graph Theory*, vol. 5, no. 3, pp. 323–331, 1981.

[12] A. Proskurowski, "Centers of maximal outerplanar graphs," *Journal of Graph Theory*, vol. 4, no. 1, pp. 75–79, 1980.

7

The m-Path Cover Polynomial of a Graph and a Model for General Coefficient Linear Recurrences

John P. McSorley and Philip Feinsilver

Department of Mathematics, Southern Illinois University, Carbondale, IL 62901-4408, USA

Correspondence should be addressed to John P. McSorley; mcsorley60@hotmail.com

Academic Editor: Jiang Zeng

An m-path cover $\Gamma = \{P_{\ell_1}, P_{\ell_2}, \ldots, P_{\ell_r}\}$ of a simple graph G is a set of vertex disjoint paths of G, each with $\ell_k \le m$ vertices, that span G. With every P_ℓ we associate a weight, $\omega(P_\ell)$, and define the weight of Γ to be $\omega(\Gamma) = \prod_{k=1}^{r} \omega(P_{\ell_k})$. The m-path cover polynomial of G is then defined as $\mathbb{P}_m(G) = \sum_\Gamma \omega(\Gamma)$, where the sum is taken over all m-path covers Γ of G. This polynomial is a specialization of the path-cover polynomial of Farrell. We consider the m-path cover polynomial of a weighted path $P(m-1, n)$ and find the $(m+1)$-term recurrence that it satisfies. The matrix form of this recurrence yields a formula equating the trace of the recurrence matrix with the m-path cover polynomial of a suitably weighted cycle $C(n)$. A directed graph, $T(m)$, the edge-weighted m-trellis, is introduced and so a third way to generate the solutions to the above $(m+1)$-term recurrence is presented. We also give a model for general-term linear recurrences and time-dependent Markov chains.

1. Introduction, m-Path Cover Polynomial, and Notation

Let G be a graph with no loops or multiple edges, with vertex set $V(G)$.

First we review some basic concepts to establish notation.

A *path* P_ℓ in G is a sequence of distinct vertices $P_\ell = [v_1, v_2, \ldots, v_\ell]$ where each pair (v_i, v_{i+1}) for $1 \le i \le \ell - 1$ is an edge. The *length* of a path is the number of vertices in it. Thus a path of length 1 is a vertex and a path of length 2 is an edge, and P_ℓ has length ℓ. Path P_ℓ *begins* at vertex v_1, its *first* vertex, and *ends* at vertex v_ℓ, its *last* vertex. The path $[v_1, v_2, \ldots, v_\ell]$ and its reverse $[v_\ell, v_{\ell-1}, \ldots, v_1]$ are considered to be the same path. The set of vertices in P_ℓ is $V(P_\ell) = \{v_1, v_2, \ldots, v_\ell\}$. Two paths P_ℓ and $P_{\ell'}$ in G are *disjoint* if $V(P_\ell) \cap V(P_{\ell'}) = \emptyset$. The *empty path* has 0 vertices. Finally, recall that a subgraph of G *spans* G if it has the same vertex set as G.

Now we introduce the central concept of this paper.

An *m-path* P_ℓ has $\ell \le m$; *that is,* it is a path of length at most m for some fixed m with $1 \le m \le |V(G)|$.

An *m-path cover* $\Gamma = \{P_{\ell_1}, P_{\ell_2}, \ldots, P_{\ell_r}\}$ of G is a set of pairwise disjoint m-paths of G that span G. Thus each ℓ_k satisfies $1 \le \ell_k \le m$, and every vertex of G lies in exactly one m-path; that is, $V(G) = \cup_{k=1}^{r} V(P_{\ell_k})$ is a partition of $V(G)$.

With every m-path P_ℓ we associate a *weight*, $\omega(P_\ell)$, and then the weight of Γ is $\omega(\Gamma) = \prod_{k=1}^{r} \omega(P_{\ell_k})$.

Definition 1. The *m-path cover polynomial* of G, $\mathbb{P}_m(G)$, is the sum of the weights of all m-path covers of G; that is,

$$\mathbb{P}_m(G) = \sum_{\Gamma} \omega(\Gamma), \tag{1}$$

where Γ is an m-path cover of G.

The path-cover polynomial (or path polynomial) of a graph G is a specialization of the F-cover polynomial of Farrell [1] where F is restricted to be a path; see Farrell [2]. Thus our m-path cover polynomial $\mathbb{P}_m(G)$ is a further specialization to paths of length $\ell \le m$. See also Chow [3] and D'Antona and Munarini [4].

It seems that this research is the first direct consideration of the m-path cover polynomial of a graph. See McSorley et al. [5] for specialization to the case $m = 2$, where all classical orthogonal polynomials are generated as 2-path cover polynomials of suitably weighted paths. For related work see the theory of weighted linear species developed in Joyal [6] and Bergeron et al. [7]. In particular, Munarini [8] uses the m-filtered linear partitions of a linearly ordered set to achieve some similar results; see especially our Sections 7 and 8.

In Section 2 we introduce a weighted path $P(m-1,n)$ and find the $(m+1)$-term recurrence that its m-path polynomial satisfies. In Section 3 the matrix form of this recurrence is presented and yields a trace formula that, in Section 4, gives the m-path cover polynomial of a suitably weighted cycle $C(n)$. Section 5 interprets our results in terms of a model for time-dependent Markov chains. In Section 6 a directed graph, $T(m)$, the edge-weighted m-trellis, is introduced and so a third way to generate the solutions to the above recurrence and trace is found. In Section 7 we model general constant coefficient linear recurrences, and we derive various relevant formulas with both algebraic and combinatorial proofs. Finally, in Section 8, we obtain a relevant new integer sequence and relate this sequence to known sequences in the literature.

Notation. We write $\mathbb{P}_m[v_1, v_2, \ldots, v_\ell]$, instead of $\mathbb{P}_m([v_1, v_2, \ldots, v_\ell])$, for the m-cover polynomial of the path $[v_1, v_2, \ldots, v_\ell]$; similarly we write $\omega[v_1, v_2, \ldots, v_\ell]$ instead of $\omega([v_1, v_2, \ldots, v_\ell])$, and so forth.

Vertices in $P(m-1,n)$ (Section 2) and in subpaths of $P(s,n)$ will be labelled u_i; vertices in $C(n)$ (Section 4) will be labelled v_i; and vertices in $T(m)$ (Section 6) will be labelled w_i.

For $1 \leq \ell \leq m$ we use indeterminate $x_{\ell,i}$ as the weight of a path of length ℓ in G. Throughout the paper $m \geq 1$ is fixed. In all the examples we set $m = 3$, and many examples have $n = 4$.

2. Weighted Path $P(m-1,\ n)$

For $m \geq 1$ and $n \geq 0$ the path $P(m-1,n)$ has $m-1+n$ vertices $\{u_1, u_2, \ldots, u_{m-1+n}\}$. The first $m-1$ vertices are weighted with weight 1 and the remaining n vertices are weighted, one by one, with the indeterminates from the set $\{x_{1,1}, x_{1,2}, \ldots, x_{1,n}\}$. Thus all vertices, that is, all paths of length $\ell = 1$, in $P(m-1,n)$ are weighted. For $2 \leq \ell \leq m$ a path of length ℓ in $P(m-1,n)$ is weighted with 0 if its last vertex has weight 1 and with $x_{\ell,i}$ if its last vertex has weight $x_{1,i}$. The path $P(0,0)$ is the empty path with no vertices.

Definition 2. For $n \geq 1$ let $f_{m,n}$ be the m-path cover polynomial of the weighted $P(m-1,n)$.

Starting conditions are $f_{m,n} = 1$ for $-(m-1) \leq n \leq 0$.

As mentioned in Section 1, throughout this paper the path $[u_a, u_{a+1}, \ldots, u_b]$ is a subpath of the weighted $P(m-1,n)$.

We now derive our main $(m+1)$-term recurrence.

Theorem 3. *For a fixed $m \geq 1$ and any $n \geq -(m-1)$,*

$$f_{m,n} = x_{1,n} f_{m,n-1} + x_{2,n} f_{m,n-2} + \cdots + x_{m,n} f_{m,n-m} = \sum_{\ell=1}^{m} x_{\ell,n} f_{m,n-\ell}. \qquad (2)$$

Proof. The last vertex u_{m-1+n} of $P(m-1,n)$ lies in every m-path cover of $P(m-1,n)$. Suppose, in such an m-path cover, it is present as the last vertex in an m-path of length ℓ. Then this m-path has weight $x_{\ell,n}$ and begins at $u_{m+n-\ell}$. The sum of the weights of all such m-path covers is therefore

$$x_{\ell,n} \mathbb{P}_m [u_1, u_2, \ldots, u_{m-1+n-\ell}] = x_{\ell,n} f_{m,n-\ell}, \qquad (3)$$

where $[u_1, u_2, \ldots, u_{m-1+n-\ell}]$ is a subpath of $P(m-1,n)$. Now summing over ℓ gives the result. The initial conditions $f_{m,n} = 1$ for $-(m-1) \leq n \leq 0$ ensure that this equation holds when $\ell \geq n$. □

Example 4. For $m = 3$ the weighted path $P(2,3)$ is

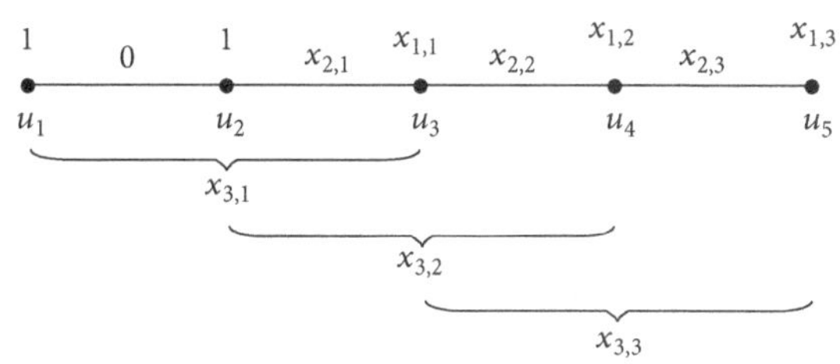

(4)

The weights of paths of lengths $\ell = 1$ and 2 (vertices and edges) are shown above the path. Vertex labels and weights of paths of length $\ell = 3$ are shown below the path. Let

$$\begin{aligned}
\ell = 1 : \omega[u_1] &= \omega[u_2] = 1, \\
\omega[u_3] &= x_{1,1}, \\
\omega[u_4] &= x_{1,2}, \\
\omega[u_5] &= x_{1,3}, \\
\ell = 2 : \omega[u_1, u_2] &= 0, \\
\omega[u_2, u_3] &= x_{2,1}, \\
\omega[u_3, u_4] &= x_{2,2}, \\
\omega[u_4, u_5] &= x_{2,3}, \\
\ell = 3 : \omega[u_1, u_2, u_3] &= x_{3,1}, \\
\omega[u_2, u_3, u_4] &= x_{3,2}, \\
\omega[u_3, u_4, u_5] &= x_{3,3}.
\end{aligned} \qquad (5)$$

All 3-path covers of $P(2,3)$ and their weights are shown below:

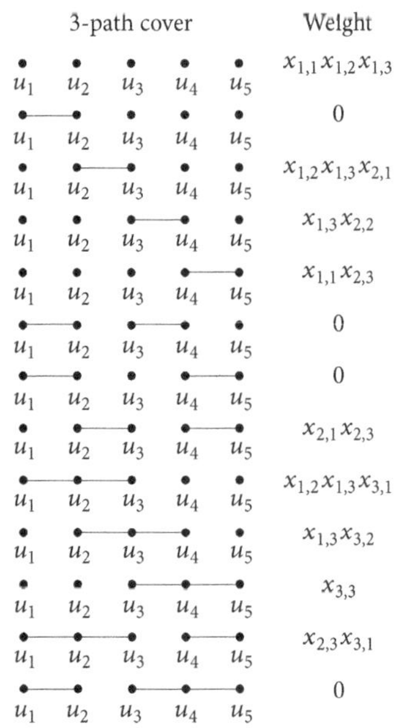

(6)

So $f_{3,3} = x_{1,1}x_{1,2}x_{1,3} + x_{1,2}x_{1,3}x_{2,1} + x_{1,3}x_{2,2} + x_{1,1}x_{2,3} + x_{2,1}x_{2,3} + x_{1,2}x_{1,3}x_{3,1} + x_{1,3}x_{3,2} + x_{2,3}x_{3,1} + x_{3,3}$.

Example 5. Theorem 3 with $m = 3$ gives the 4-term recurrence for a fixed $n \geq 1$,

$$f_{3,n} = x_{1,n}f_{3,n-1} + x_{2,n}f_{3,n-2} + x_{3,n}f_{3,n-3}. \quad (7)$$

Then the starting conditions $f_{3,-2} = f_{3,-1} = f_{3,0} = 1$ give

$$\begin{aligned}
f_{3,1} &= x_{1,1} + x_{2,1} + x_{3,1},\\
f_{3,2} &= x_{1,1}x_{1,2} + x_{1,2}x_{2,1} + x_{1,2}x_{3,1} + x_{2,2} + x_{3,2},\\
f_{3,3} &= x_{1,1}x_{1,2}x_{1,3} + x_{1,2}x_{1,3}x_{2,1} + x_{1,3}x_{2,2} + x_{1,1}x_{2,3}\\
&\quad + x_{2,1}x_{2,3} + x_{1,2}x_{1,3}x_{3,1} + x_{1,3}x_{3,2} + x_{2,3}x_{3,1} + x_{3,3},\\
f_{3,4} &= x_{1,1}x_{1,2}x_{1,3}x_{1,4} + x_{1,2}x_{1,3}x_{1,4}x_{2,1} + x_{1,2}x_{1,3}x_{1,4}x_{3,1}\\
&\quad + x_{1,1}x_{1,2}x_{2,4} + x_{1,1}x_{1,4}x_{2,3} + x_{1,2}x_{2,1}x_{2,4}\\
&\quad + x_{1,2}x_{2,4}x_{3,1} + x_{1,3}x_{1,4}x_{2,2} + x_{1,3}x_{1,4}x_{3,2}\\
&\quad + x_{1,4}x_{2,1}x_{2,3} + x_{1,4}x_{2,3}x_{3,1} + x_{1,1}x_{3,4}\\
&\quad + x_{1,4}x_{3,3} + x_{2,1}x_{3,4} + x_{2,2}x_{2,4}\\
&\quad + x_{2,4}x_{3,2} + x_{3,1}x_{3,4}\\
&\ \ \vdots
\end{aligned} \quad (8)$$

We check $f_{3,3}$ from Example 4.

Definition 6. For $0 \leq r \leq m-1$ we define $P(r,n)$ as above for $P(m-1,n)$, except that we have r vertices instead of $m-1$ vertices of weight 1 at the beginning of the path. Thus $P(r,n)$ has $r+n$ vertices and is formed from $P(m-1,n)$ by truncating from the right. All m-paths in $P(r,n)$ are weighted as in $P(m-1,n)$. We let $\mathcal{P}_m(r,n)$ be the m-path cover polynomial of the weighted $P(r,n)$. We note that $f_{m,n} = \mathcal{P}_m(m-1,n)$.

Example 7. For $m = 3$ and $n = 4$,

$$\begin{aligned}
\mathcal{P}_3(0,4) &= x_{1,1}x_{1,2}x_{1,3}x_{1,4} + x_{1,1}x_{1,2}x_{2,4}\\
&\quad + x_{1,1}x_{1,4}x_{2,3} + x_{1,3}x_{1,4}x_{2,2}\\
&\quad + x_{1,1}x_{3,4} + x_{1,4}x_{3,3} + x_{2,2}x_{2,4},\\
\mathcal{P}_3(1,4) &= x_{1,1}x_{1,2}x_{1,3}x_{1,4} + x_{1,2}x_{1,3}x_{1,4}x_{2,1} + x_{1,3}x_{1,4}x_{2,2}\\
&\quad + x_{1,1}x_{1,4}x_{2,3} + x_{1,1}x_{1,2}x_{2,4} + x_{1,4}x_{2,1}x_{2,3}\\
&\quad + x_{1,2}x_{2,1}x_{2,4} + x_{1,3}x_{1,4}x_{3,2}\\
&\quad + x_{1,1}x_{3,4} + x_{1,4}x_{3,3} + x_{2,1}x_{3,4} + x_{2,2}x_{2,4}\\
&\quad + x_{2,4}x_{3,2},\\
\mathcal{P}_3(2,4) &= f_{3,4};
\end{aligned} \quad (9)$$

see Example 5.

For a fixed r with $0 \leq r \leq m-1$ we define the starting conditions,

$$\mathcal{P}_m(r,n) = \begin{cases} 0, & \text{if } -(m-1) \leq n \leq -r-1, \\ 1, & \text{if } -r \leq n \leq 0. \end{cases} \quad (10)$$

We then have the following recurrence; the proof is similar to the proof of Theorem 3, and setting $r = m-1$ recovers Theorem 3.

Theorem 8. *For a fixed r with $0 \leq r \leq m-1$ and any $n \geq 1$,*

$$\mathcal{P}_m(r,n) = \sum_{\ell=1}^{m} x_{\ell,n}\mathcal{P}_m(r,n-\ell). \quad (11)$$

We now work with the fundamental solutions to recurrence (2).

For $1 \leq j \leq m$ let $f_{m,n}^{(j)}$ denote the jth fundamental solution to (2). Thus the $f_{m,n}^{(j)}$ obey the recurrence

$$f_{m,n}^{(j)} = \sum_{\ell=1}^{m} x_{\ell,n}f_{m,n-\ell}^{(j)}, \quad (12)$$

with starting conditions

$$f_{m,-(m-1)+k}^{(j)} = \begin{cases} 1, & \text{if } k = m-j, \\ 0, & \text{if } k \neq m-j, \end{cases} \quad (13)$$

where $0 \leq k \leq m-1$.

We have

$$f_{m,n} = \sum_{j=1}^{m} f_{m,n}^{(j)}. \quad (14)$$

Our next result expresses $f_{m,n}^{(j)}$ as the difference of two m-path cover polynomials. Consistent with (10) we set $\mathcal{P}_m(-1,n) = 0$ for every $n \geq -(m-1)$.

Lemma 9. *For $n \geq 1$ and $1 \leq j \leq m$,*

$$f_{m,n}^{(j)} = \mathscr{P}_m(j-1,n) - \mathscr{P}_m(j-2,n). \quad (15)$$

Proof. By induction on n, first consider $n = 1$. Now $f_{m,1-\ell}^{(j)} = 1$ when $\ell = j$ and $f_{m,1-\ell}^{(j)} = 0$ otherwise. Each $f_{m,n}^{(j)}$ satisfies (12), so $f_{m,1}^{(j)} = \sum_{\ell=1}^{m} x_{\ell,1} f_{m,1-\ell}^{(j)} = x_{j,1}$. Now consider the path $P(j-1,1)$ shown below:

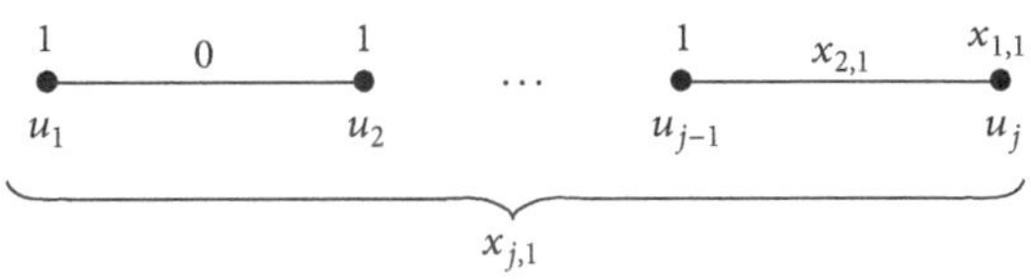

(16)

The first vertex u_1 lies in every m-path cover of $P(j-1,1)$ so, similar to the proof of Theorem 3, we have

$$\begin{aligned}\mathscr{P}_m(j-1,1) &= \omega[u_1]\mathscr{P}_m(j-2,1) \\ &\quad + \omega[u_1,u_2]\mathscr{P}_m(j-3,1) \\ &\quad + \cdots + \omega[u_1,u_2,\ldots,u_j] \\ &= 1\cdot\mathscr{P}_m(j-2,1) + 0\cdot\mathscr{P}_m(j-3,1) \\ &\quad + \cdots + x_{j,1}.\end{aligned} \quad (17)$$

Thus, from above, $f_{m,1}^{(j)} = x_{j,1} = \mathscr{P}_m(j-1,1) - \mathscr{P}_m(j-2,1)$; that is, (15) is true for $n = 1$.

Now we have

$$\begin{aligned} f_{m,n+1}^{(j)} &= \sum_{\ell=1}^{m} x_{\ell,n+1} f_{m,n+1-\ell}^{(j)} \\ &= \sum_{\ell=1}^{m} x_{\ell,n+1}\{\mathscr{P}_m(j-1,n+1-\ell) - \mathscr{P}_m(j-2,n+1-\ell)\} \\ &= \sum_{\ell=1}^{m} x_{\ell,n+1}\mathscr{P}_m(j-1,n+1-\ell) \\ &\quad - \sum_{\ell=1}^{m} x_{\ell,n+1}\mathscr{P}_m(j-2,n+1-\ell) \\ &= \mathscr{P}_m(j-1,n+1) - \mathscr{P}_m(j-2,n+1), \end{aligned} \quad (18)$$

using (12) again at the first line, the induction hypothesis at the second line and Theorem 8 at the last line. Hence the induction goes through and (15) is true for all $n \geq 1$. □

Example 10. Using (12) and the starting conditions following (12) for $m = 3$ and $n = 4$ the 3 fundamental solutions to recurrence (2) are

$$\begin{aligned} f_{3,4}^{(1)} &= x_{1,1}x_{1,2}x_{1,3}x_{1,4} + x_{1,1}x_{1,2}x_{2,4} + x_{1,1}x_{1,4}x_{2,3} \\ &\quad + x_{1,3}x_{1,4}x_{2,2} + x_{1,1}x_{3,4} + x_{1,4}x_{3,3} \\ &\quad + x_{2,2}x_{2,4}, \\ f_{3,4}^{(2)} &= x_{1,2}x_{1,3}x_{1,4}x_{2,1} + x_{1,3}x_{1,4}x_{3,2} + x_{1,4}x_{2,1}x_{2,3} \\ &\quad + x_{1,2}x_{2,1}x_{2,4} + x_{2,4}x_{3,2} + x_{2,1}x_{3,4}, \\ f_{3,4}^{(3)} &= x_{1,2}x_{1,3}x_{1,4}x_{3,1} + x_{1,4}x_{2,3}x_{3,1} + x_{1,2}x_{2,4}x_{3,1} \\ &\quad + x_{3,1}x_{3,4}. \end{aligned} \quad (19)$$

We check (14) using Example 5,

$$f_{3,4} = f_{3,4}^{(1)} + f_{3,4}^{(2)} + f_{3,4}^{(3)}. \quad (20)$$

We also check Lemma 9 using $\mathscr{P}_3(-1,4) = 0$ and Example 7,

$$\begin{aligned} f_{3,4}^{(1)} &= \mathscr{P}_3(0,4) - \mathscr{P}_3(-1,4) = \mathscr{P}_3(0,4), \\ f_{3,4}^{(2)} &= \mathscr{P}_3(1,4) - \mathscr{P}_3(0,4), \\ f_{3,4}^{(3)} &= \mathscr{P}_3(2,4) - \mathscr{P}_3(1,4). \end{aligned} \quad (21)$$

By iteration of such formulas, we have Corollary 11, where (ii) is a specialization of (i) with $r = 0$.

Corollary 11. *(i) For $1 \leq j \leq m$,*

$$\mathscr{P}_m(r,n) = \sum_{j=1}^{r+1} f_{m,n}^{(j)}, \quad (22)$$

(ii) the first fundamental solution to recurrence (2) is given by

$$f_{m,n}^{(1)} = \mathscr{P}_m(0,n). \quad (23)$$

Corollary 12 is a useful technical result.

Corollary 12. *For $n \geq 1$ and $1 \leq j \leq m$,*

$$f_{m,n+1-j}^{(j)} = \sum_{\ell=j}^{m} x_{\ell,\ell+1-j}\mathbb{P}_m[u_{m+\ell+1-j},\ldots,u_{m+n-j}]. \quad (24)$$

Proof. For $j = 1$ from Corollary 11(ii) we have $f_{m,n}^{(1)} = \mathscr{P}_m(0,n)$. Now in the weighted path $P(0,n)$ let vertex u_1 be covered by a path Q_ℓ of length ℓ where $1 \leq \ell \leq m$. Then Q_ℓ begins at vertex u_1 and ends at vertex u_ℓ, which has weight $x_{1,\ell}$; so $\omega(Q_\ell) = x_{\ell,\ell}$. Now in every m-path cover of $P(0,n)$ vertex u_1 must be covered by such a path Q_ℓ, so $f_{m,n}^{(1)} = \sum_{\ell=1}^{m} x_{\ell,\ell}\mathbb{P}_m[u_{\ell+1},\ldots,u_n]$, which is the above formula for $j = 1$.

For any $2 \leq j \leq m$ the path $[u_{m+1-j},\ldots,u_{m-1+n}]$ is a subpath of $P(m-1,n)$. In fact the weighted paths $P(j -$

$1,n)$ and $[u_{m+1-j},\ldots,u_{m-1+n}]$ (except for vertex labels) are identical, so $\mathscr{P}_m(j-1,n) = \mathbb{P}_m[u_{m+1-j},\ldots,u_{m-1+n}]$. From Lemma 9 we have

$$\begin{aligned} f^{(j)}_{m,n+1-j} &= \mathscr{P}_m(j-1,n+1-j) - \mathscr{P}_m(j-2,n+1-j) \\ &= \mathbb{P}_m\left[u_{m+1-j},\ldots,u_{m+n-j}\right] \\ &\quad - 1\cdot\mathbb{P}_m\left[u_{m+2-j},\ldots,u_{m+n-j}\right] \\ &= \text{sum of terms of } \mathbb{P}_m\left[u_{m+1-j},\ldots,u_{m+n-j}\right] \\ &\quad \text{in which vertex } u_{m+1-j} \text{ is covered} \\ &\quad \text{by a path whose weight is an indeterminate} \\ &\quad \text{as opposed to a path with weight 1.} \end{aligned} \tag{25}$$

So let vertex u_{m+1-j} be covered by a path Q_ℓ of length $\ell \geq 1$. Then Q_ℓ begins at vertex u_{m+1-j} and ends at vertex $u_{m+\ell-j}$, which has weight $x_{1,\ell+1-j}$. Hence $\omega(Q_\ell) = x_{\ell,\ell+1-j}$. Furthermore, because Q_ℓ ends at $u_{m+\ell-j}$ if $\ell < j$, then $m+\ell-j \leq m-1$; hence $w(Q_\ell) = 0$, a contradiction; so $\ell \geq j$.

Now, similar to the above, the sum of the terms of $\mathbb{P}_m[u_{m+1-j},\ldots,u_{m+n-j}]$ that contain $x_{\ell,\ell+1-j}$ is $x_{\ell,\ell+1-j}\mathbb{P}_m[u_{m+\ell+1-j},\ldots,u_{m+n-j}]$. Finally, summing over the lengths ℓ of all possible paths Q_ℓ, namely, summing over ℓ with $j \leq \ell \leq m$, gives the result. □

This completes study of the weighted path $P(m-1,n)$.

3. Matrix Formulation and Trace

We set up our $(m+1)$-term recurrence (2) in matrix form.

Let $X_{m,0} = I_m$ be the $m\times m$ identity matrix, and for $n \geq 1$ let $X_{m,n}$ be the $m\times m$ matrix

$$X_{m,n} = \begin{pmatrix} 0 & 1 & \cdot & \cdots & 0 \\ 0 & 0 & 1 & \cdots & 0 \\ \vdots & \vdots & \vdots & \vdots & \vdots \\ 0 & 0 & 0 & \cdots & 1 \\ x_{m,n} & x_{m-1,n} & x_{m-2,n} & \cdots & x_{1,n} \end{pmatrix}. \tag{26}$$

Let T denote transpose, and let $F_{m,n}$ be the vector $F_{m,n} = (f_{m,n-(m-1)},\ldots,f_{m,n})^T$. Then recurrence (2) can be written as

$$F_{m,n} = X_{m,n}F_{m,n-1}, \tag{27}$$

where $F_{m,0} = (f_{m,-(m-1)},\ldots,f_{m,0})^T = (1,\ldots,1)^T$. By iterating this equation we have $F_{m,n} = Y_{m,n}\,F_{m,0}$, where

$$Y_{m,n} = X_{m,n}X_{m,n-1}\cdots X_{m,0}$$
$$= \begin{pmatrix} f^{(m)}_{m,n-(m-1)} & \cdot\cdots\cdot & f^{(1)}_{m,n-(m-1)} \\ f^{(m)}_{m,n-(m-2)} & \cdot\cdots\cdot & f^{(1)}_{m,n-(m-2)} \\ \vdots & \vdots\ \vdots\ \vdots & \vdots \\ f^{(m)}_{m,n-1} & \cdot\cdots\cdot & f^{(1)}_{m,n-1} \\ f^{(m)}_{m,n} & \cdot\cdots\cdot & f^{(1)}_{m,n} \end{pmatrix}. \tag{28}$$

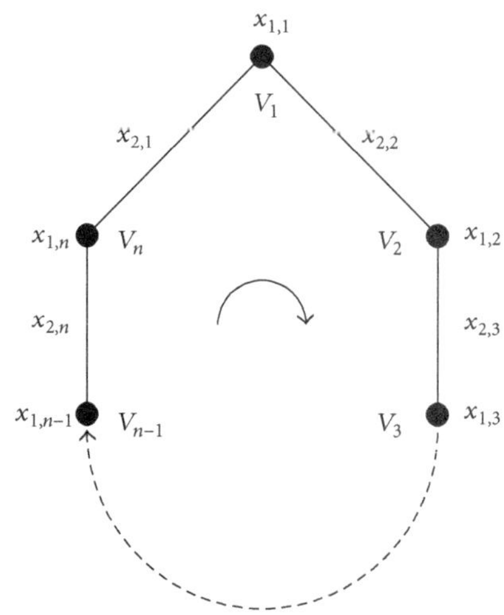

FIGURE 1: Weighted $C(n)$.

With tr denoting trace, we have the following.

Lemma 13. *For $n \geq 1$,*

$$\operatorname{tr}(Y_{m,n}) = \sum_{j=1}^{m} f^{(j)}_{m,n+1-j}. \tag{29}$$

We now apply these results to the weighted cycle $C(n)$.

4. Weighted Cycle $C(n)$ and Trace

We introduce the weighted cycle $C(n)$ for $n \geq 1$ shown in Figure 1. It has n vertices labelled $\{v_1,v_2,\ldots,v_n\}$ and n edges.

It is weighted as follows: for $1 \leq \ell \leq m$, let P_ℓ be a path of length ℓ that traverses $C(n)$ clockwise and ends at vertex v_i. We define $\omega(P_\ell) = x_{\ell,i}$.

Thus the weighted cycle $C(1)$ is an isolated vertex v_1 with weight $\omega(v_1) = x_{1,1}$; and the weighted cycle $C(2)$ has 2 vertices $\{v_1,v_2\}$ with $\omega(v_1) = x_{1,1}$ and $\omega(v_2) = x_{1,2}$ and 2 edges: edge (v_1,v_2) with $\omega(v_1,v_2) = x_{2,2}$ and edge (v_2,v_1) with $\omega(v_2,v_1) = x_{2,1}$.

In Figure 1 only the weights of paths of lengths $\ell = 1$ and 2 are shown.

Lemma 14. *For $1 \leq a \leq b \leq n$ the following m-path cover polynomials, the first which comes from $C(n)$ and the second from $P(m-1,n)$, are equal:*

$$\mathbb{P}_m\left[v_a,\ldots,v_b\right] = \mathbb{P}_m\left[u_{m-1+a},\ldots,u_{m-1+b}\right]. \tag{30}$$

Proof. Except for vertex labels, the weighted paths $[v_a,\ldots,v_b]$ in $C(n)$ and $[u_{m-1+a},\ldots,u_{m-1+b}]$ in $P(m-1,n)$ are identical. Hence the result is obtained. □

Definition 15. For $n \geq 1$ let $\mathscr{C}_m(n)$ be the m-path cover polynomial of the weighted $C(n)$.

In the following, when necessary, we reduce subscripts on u, v, and the second subscript on x, all modulo n. We write $u_{n+t} = u_t$, $v_{n+t} = v_t$, $x_{\ell,n+t} = x_{\ell,t}$, and so forth.

Theorem 16 is the main result of this section. Recall the matrix $Y_{m,n}$ from (28).

Theorem 16. *For* $n \geq 1$,

$$\mathscr{C}_m(n) = \operatorname{tr}(\mathrm{Y}_{\mathrm{m,n}}). \quad (31)$$

Proof. Consider the weighted $C(n)$. Vertex v_1 lies in every m-path cover of $C(n)$. Suppose, in such an m-path cover, it is covered by a path P_ℓ of length ℓ that begins at v_{n-p} and ends at $v_{n-p-1+\ell}$, for some $p \in \{-1, 0, 1, \dots, \ell-2\}$. Now $1 \leq \ell \leq m$; that is, $p+2 \leq \ell \leq m$. The sum of the weights of all such paths is then

$$\sum_{\ell=p+2}^{m} x_{\ell,n-p-1+\ell}\mathbb{P}_m\left[v_{n-p+\ell}, \dots, v_{n-p-1}\right]. \quad (32)$$

But $p \in \{-1, 0, 1, \dots, m-2\}$, so

$$\begin{aligned}
\mathscr{C}_m(n) &= \sum_{p=-1}^{m-2}\sum_{\ell=p+2}^{m} x_{\ell,n-p-1+\ell}\mathbb{P}_m\left[v_{n-p+\ell}, \dots, v_{n-p-1}\right] \\
&= \sum_{j=1}^{m}\sum_{\ell=j}^{m} x_{\ell,n+\ell+1-j}\mathbb{P}_m\left[v_{n+\ell+2-j}, \dots, v_{n+1-j}\right] \\
&= \sum_{j=1}^{m}\sum_{\ell=j}^{m} x_{\ell,\ell+1-j}\mathbb{P}_m\left[u_{m+\ell+1-j}, \dots, u_{m-j}\right] \\
&= \sum_{j=1}^{m} f^{(j)}_{m,n+1-j} \\
&= \operatorname{tr}(Y_{m,n}),
\end{aligned} \quad (33)$$

letting $j = p+2$ at the second line, and using subscript reduction modulo n and Lemma 14 at the third line, then Corollary 12 at the fourth line, and Lemma 13 at the last line. □

Example 17. For $m = 3$ and $n = 4$ consider the weighted $C(4)$ in Figure 2.

The 3-paths are weighted as follows:

$$\begin{aligned}
\ell = 1 &: \omega[v_1] = x_{1,1}, \\
&\omega[v_2] = x_{1,2}, \\
&\omega[v_3] = x_{1,3}, \\
&\omega[v_4] = x_{1,4}, \\
\ell = 2 &: \omega[v_1, v_2] = x_{2,2}, \\
&\omega[v_2, v_3] = x_{2,3}, \\
&\omega[v_3, v_4] = x_{2,4}, \\
&\omega[v_4, v_1] = x_{2,1}, \\
\ell = 3 &: \omega[v_1, v_2, v_3] = x_{3,3}, \\
&\omega[v_2, v_3, v_4] = x_{3,4}, \\
&\omega[v_3, v_4, v_1] = x_{3,1}, \\
&\omega[v_4, v_1, v_2] = x_{3,2}.
\end{aligned} \quad (34)$$

By considering all 3-path covers, the 3-path cover polynomial of the weighted $C(4)$ is

$$\begin{aligned}
\mathscr{C}_3(4) &= x_{1,1}x_{1,2}x_{1,3}x_{1,4} + x_{1,1}x_{1,2}x_{2,4} + x_{1,2}x_{1,3}x_{2,1} \\
&+ x_{1,3}x_{1,4}x_{2,2} + x_{1,1}x_{1,4}x_{2,3} + x_{1,3}x_{3,2} + x_{1,4}x_{3,3} \\
&+ x_{1,1}x_{3,4} + x_{1,2}x_{3,1} + x_{2,1}x_{2,3} + x_{2,2}x_{2,4}.
\end{aligned} \quad (35)$$

Similar to Example 10, the recurrence (12) and the starting conditions following (12) give

$$\begin{aligned}
f^{(1)}_{3,4} &= x_{1,1}x_{1,2}x_{1,3}x_{1,4} + x_{1,1}x_{1,2}x_{2,4} + x_{1,1}x_{1,4}x_{2,3} \\
&+ x_{1,3}x_{1,4}x_{2,2} + x_{1,1}x_{3,4} + x_{1,4}x_{3,3} + x_{2,2}x_{2,4}, \\
f^{(2)}_{3,3} &= x_{1,2}x_{1,3}x_{2,1} + x_{1,3}x_{3,2} + x_{2,1}x_{2,3}, \\
f^{(3)}_{3,2} &= x_{1,2}x_{3,1}.
\end{aligned} \quad (36)$$

Together with the following matrices

$$Y_{3,4} = X_{3,4}X_{3,3}X_{3,2}X_{3,1}X_{3,0}$$

$$= \begin{pmatrix} 0 & 1 & 0 \\ 0 & 0 & 1 \\ x_{3,4} & x_{2,4} & x_{1,4} \end{pmatrix} \begin{pmatrix} 0 & 1 & 0 \\ 0 & 0 & 1 \\ x_{3,3} & x_{2,3} & x_{1,3} \end{pmatrix} \begin{pmatrix} 0 & 1 & 0 \\ 0 & 0 & 1 \\ x_{3,2} & x_{2,2} & x_{1,2} \end{pmatrix} \begin{pmatrix} 0 & 1 & 0 \\ 0 & 0 & 1 \\ x_{3,1} & x_{2,1} & x_{1,1} \end{pmatrix}$$

$$= \begin{pmatrix} x_{1,2}x_{3,1} & x_{1,2}x_{2,1} + x_{3,2} & x_{1,1}x_{1,2} + x_{2,2} \\ x_{1,2}x_{1,3}x_{3,1} + x_{2,3}x_{3,1} & \begin{matrix} x_{1,3}x_{3,2} + x_{2,1}x_{2,3} \\ +x_{1,2}x_{1,3}x_{2,1} \end{matrix} & \begin{matrix} x_{1,1}x_{2,3} + x_{1,3}x_{2,2} \\ +x_{1,1}x_{1,2}x_{1,3} + x_{3,3} \end{matrix} \\ \begin{matrix} x_{1,2}x_{2,4}x_{3,1} + x_{1,2}x_{1,3}x_{1,4}x_{3,1} \\ +x_{1,4}x_{2,3}x_{3,1} + x_{3,1}x_{3,4} \end{matrix} & \begin{matrix} x_{1,2}x_{2,4}x_{2,1} + x_{1,2}x_{1,3}x_{1,4}x_{2,1} + x_{1,3}x_{1,4}x_{3,2} \\ +x_{1,4}x_{2,1}x_{2,3} + x_{2,1}x_{3,4} + x_{2,4}x_{3,2} \end{matrix} & \begin{matrix} x_{1,1}x_{3,4} + x_{1,1}x_{1,2}x_{2,4} + x_{1,1}x_{1,4}x_{2,3} \\ +x_{1,1}x_{1,2}x_{1,3}x_{1,4} + x_{1,3}x_{1,4}x_{2,2} \\ +x_{1,4}x_{3,3} + x_{2,2}x_{2,4} \end{matrix} \end{pmatrix}, \tag{37}$$

we may check the results from Lemma 13 and Theorem 16,

$$\mathscr{C}_3(4) = \operatorname{tr}(Y_{3,4}) = \sum_{j=1}^{3} f_{3,5-j}^{(j)} = f_{3,4}^{(1)} + f_{3,3}^{(2)} + f_{3,2}^{(3)}. \tag{38}$$

5. Markov Chain Interpretation

In this section we consider an interesting special case, where in the matrix formulation of the recurrence we have stochastic matrices. A matrix of Form (26) can be considered a transition matrix for a Markov chain with m states under the conditions

$$\sum_j x_{j,n} = 1, \quad x_{j,n} \geq 0, \ \forall j. \tag{39}$$

Because the probabilities $x_{j,n}$ vary with n, these are the transition matrices for a nonhomogeneous Markov chain. Note also that, as transition matrices are multiplied from left to right, the process is effectively time reversed. In fact,

$$P[\text{jump at time } \nu \text{ from state } m \text{ to state } j] = x_{m-j+1,n-\nu+1}. \tag{40}$$

This process is often referred to as a *ladder process*. From any state j, with $j < m$, the process jumps with certainty to $j+1$, then to $j+2$, and so forth, up the ladder, till it reaches state m. At that point it jumps randomly back down the ladder to one of the intermediates states j, $1 \leq j < m$, and the procedure repeats. Because all of the matrices are stochastic, the row sums of matrices such as $Y_{m,n}$, see (28), will all equal 1. Recall from Section 3 that

$$\begin{pmatrix} f_{m,n-m+1} \\ \vdots \\ f_{m,n} \end{pmatrix} = Y_{m,n} \begin{pmatrix} 1 \\ \vdots \\ 1 \end{pmatrix}. \tag{41}$$

Thus, we have the following.

Proposition 18. *In the stochastic case, all of the path polynomials $f_{m,n}$ evaluate to 1.*

5.1. Homogeneous Case. In the case of constant coefficients (see (26)), sending $x_{\ell,i} \rightarrow x_\ell$, forall i, we drop the dependence on n and write

$$X_m = \begin{pmatrix} 0 & 1 & \cdot & \cdots & 0 \\ 0 & 0 & 1 & \cdots & 0 \\ \vdots & \vdots & \vdots & \vdots & \vdots \\ 0 & 0 & 0 & \cdots & 1 \\ x_m & x_{m-1} & x_{m-2} & \cdots & x_1 \end{pmatrix}, \tag{42}$$

with $\sum x_j = 1$. Now

$$Y_{m,n} = (X_m)^n \tag{43}$$

is the n-step transition matrix. It is easy to see that a row vector (on the left) fixed by X_m is

$$(x_m, x_m + x_{m-1}, \ldots, x_m + x_{m-1} + \cdots + x_2, 1). \tag{44}$$

Furthermore, under the assumption $x_j > 0$, forall j, it is immediate that the chain is irreducible and aperiodic, hence ergodic. That is,

$$\lim_{n\to\infty} Y_{m,n} = \Omega \tag{45}$$

exists and has equal rows, and each row proportional to the left-invariant vector indicated above normalized to row sum 1.

Example 19. Take the uniform case $x_j = 1/m$, $1 \leq j \leq m$. Then we have the fixed vector $(1, 2, 3, \ldots, m)$ and the limits

$$\lim_{n\to\infty} f_{m,n}^{(j)} = \frac{2(m-j+1)}{m(m+1)}. \tag{46}$$

Thus, for large n, if we randomly choose an m-path cover of $P(m-1, n)$ then the probability that it belongs to the j-fundamental solution is $2(m-j+1)/m(m+1)$. In particular, the first fundamental solution satisfies

$$\lim_{n\to\infty} f_{m,n}^{(1)} = \frac{2}{m+1}. \tag{47}$$

So the m-path cover polynomial model provides a combinatorial model for nonhomogeneous Markov chains. A closely related model, the trellis, is discussed in detail below.

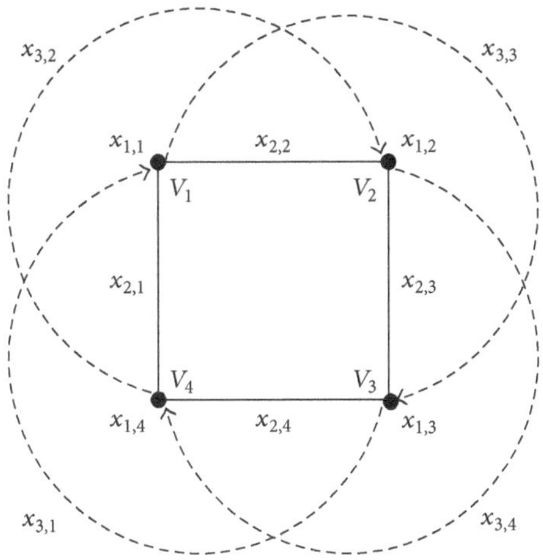

FIGURE 2: Weighted $C(4)$.

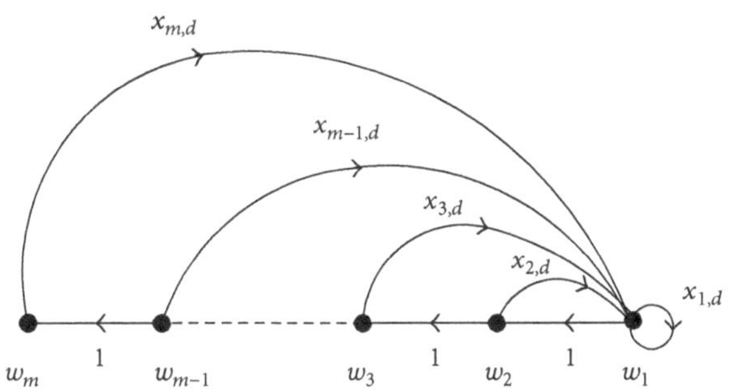

FIGURE 3: Edge-weighted m-trellis $T(m)$.

6. Edge-Weighted m-Trellis $T(m)$

In this section we deal with the *edge-weighted m-trellis*, $T(m)$, shown in Figure 3, and give another method of generating $f_{m,n}^{(j)}$ and $\mathscr{C}_m(n)$.

The vertices of $T(m)$ are labelled $\{w_1, w_2, \ldots, w_m\}$. All edges in $T(m)$ are *directed*, with arrows as shown. All circuits in $T(m)$ are directed and are traversed in the direction of the arrows. We use S to denote a directed circuit in $T(m)$, which we simply call a circuit. A circuit is *based* at vertex w_j if it begins and ends at vertex w_j. A circuit may pass through the same vertex more than once. The length of a circuit S is the number of edges in it.

The weights on the edges of $T(m)$ are taken from $\{1, x_{1,d}, \ldots, x_{m,d}\}$ where $d \geq 1$, as shown. The *weight* of circuit S, $w(S)$, is the product of the weights of all the edges in S. If the edge with weight $x_{j,d}$ is traversed as the kth edge in S, then $x_{j,k}$ is a factor in $w(S)$; thus the meaning of $x_{j,d}$ here is different from that in Sections 2 and 4. We allow empty circuits with length 0.

Definition 20. Let $\mathscr{T}_m(w_j, 0) = 1$ and, for $s \geq 1$, let $\mathscr{T}_m(w_j, s)$ be the sum of the weights of all circuits in $T(m)$ that are based at vertex w_j with length s.

Notation. We use standard multiset notation: $1^k = \underbrace{1 \cdot 1 \cdots 1 \cdot 1}_{k}$, and 1^0 means no occurrences of 1.

Theorem 21. *For $s \geq 0$,*

$$\mathscr{T}_m(w_1, s) = \mathscr{P}_m(0, s). \quad (48)$$

Proof. By strong induction on s. Now $\mathscr{T}_m(w_1, 0) = \mathscr{P}_m(0, 0) = 1$; hence (48) is true for $s = 0$. We now assume that $\mathscr{T}_m(w_1, s') = \mathscr{P}_m(0, s')$ for all $0 \leq s' \leq s$. Consider any term in $\mathscr{T}_m(w_1, s+1)$ is the weight of some circuit S in $T(m)$ based at vertex w_1 with length $s+1$. Clearly S ends with a k-cycle based at vertex w_1, for some k with $1 \leq k \leq m$. Thus the last edge of S is (w_k, w_1), with weight $x_{k,s+1}$, and the previous $k-1$ edges are $(w_k, w_{k-1}), (w_{k-1}, w_{k-2}), \ldots, (w_2, w_1)$, each of weight 1. Hence $\omega(S) = \mathscr{T}_m(w_1, s+1-k)1^{k-1}x_{k,s+1}$. Thus

$$\begin{aligned} \mathscr{T}_m(w_1, s+1) &= \sum_{k=1}^{m} x_{k,s+1} \mathscr{T}_m(w_1, s+1-k) \\ &= \sum_{k=1}^{m} x_{k,s+1} \mathscr{P}_m(0, s+1-k) \\ &= \mathscr{P}_m(0, s+1), \end{aligned} \quad (49)$$

using the strong induction hypothesis and then Theorem 8. So the induction goes through and (48) is true for all $s \geq 0$. □

Let $\mathscr{T}_m^{+c}(w_1, s)$ be the expression obtained when every indeterminate $x_{a,b}$ in $\mathscr{T}_m(w_1, s)$ is replaced by $x_{a,b+c}$; similarly for other expressions.

Recall that $[u_m, \ldots, u_{m-1+s}]$ is a subpath of $P(m-1, n)$ for $s \geq 0$; for $s = 0$ the path $[u_m, u_{m-1}]$ is the empty path $P(0, 0)$, and $\mathscr{P}_m(0, 0) = 1$.

Corollary 22. *For $s \geq 0$ and $0 \leq c \leq n - s$,*

$$\mathscr{T}_m^{+c}(w_1, s) = \mathbb{P}_m[u_{m+c}, \ldots, u_{m-1+s+c}]. \quad (50)$$

Proof. For $s = 0$ we have $\mathscr{T}_m^{+c}(w_1, 0) = \mathbb{P}_m[u_{m+c}, u_{m-1+c}] = 1$. For $s \geq 1$ then $[u_m, \ldots, u_{m-1+s}]$ is a subpath of $P(m-1, n)$ so, for every $n \geq s$, we have $\mathscr{P}_m(0, s) = \mathbb{P}_m[u_m, \ldots, u_{m-1+s}]$. Now, from Theorem 21, $\mathscr{T}_m(w_1, s) = \mathscr{P}_m(0, s)$, so $\mathscr{T}_m^{+c}(w_1, s) = \mathscr{P}_m^{+c}(0, s) = \mathbb{P}_m[u_{m+c}, \ldots, u_{m-1+s+c}]$, as required. □

We now connect $\mathscr{T}_m(w_j, n)$ and the fundamental solutions of the $(m+1)$-term recurrence (2).

Theorem 23. *For $n \geq 0$,*

$$\mathscr{T}_m(w_j, n) = f_{m,n+1-j}^{(j)}. \quad (51)$$

Proof. Consider a circuit S in $T(m)$ based at vertex w_j with n edges. Then, for some $0 \leq k \leq m - j$, the first k edges in this circuit are $(w_j, w_{j+1}), (w_{j+1}, w_{j+2}), \ldots, (w_{j+k-1}, w_{j+k})$, followed by edge (w_{j+k}, w_1) ending at vertex w_1. These edges contribute $1^k x_{j+k,k+1}$ to $w(S)$. Now, starting at vertex w_1, the last $j-1$ edges traversed in S are

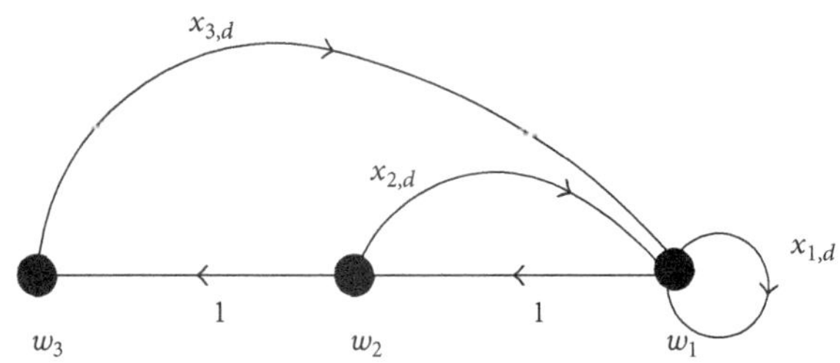

FIGURE 4: Edge-weighted 3-trellis $T(3)$.

$(w_1, w_2), (w_2, w_3), \ldots, (w_{j-1}, w_j)$, contributing 1^{j-1} to $w(S)$. Hence $\omega(S) = x_{j+k,k+1}\mathcal{T}_m^{+(k+1)}(w_1, n-j-k)$.

Thus

$$\begin{aligned}\mathcal{T}_m\left(w_j, n\right) &= \sum_{k=0}^{m-j} x_{j+k,k+1}\mathcal{T}_m^{+(k+1)}\left(w_1, n-j-k\right)\\ &= \sum_{\ell=j}^{m} x_{\ell,\ell+1-j}\mathcal{T}_m^{+(\ell+1-j)}\left(w_1, n-\ell\right)\\ &= \sum_{\ell=j}^{m} x_{\ell,\ell+1-j}\mathbb{P}_m\left[u_{m+\ell+1-j}, \ldots, u_{m+n-j}\right]\\ &= f_{m,n+1-j}^{(j)},\end{aligned} \tag{52}$$

putting $\ell = j + k$ at the second line, then using Corollary 22 with $c = \ell + 1 - j$ and $s = n - \ell$ at the third line, and finally using Corollary 12 at the last line. □

Example 24. Consider $T(3)$, the edge-weighted 3-trellis; see Figure 4.

(a) $\mathcal{T}_3(w_2, 5)$ = sum of weights of circuits of $T(3)$ based at w_2 with length 5 = $x_{2,1}x_{1,2}x_{1,3}x_{1,4}1 + x_{2,1}x_{1,2}1x_{2,4}1 + x_{2,1}1 \cdot 1\, x_{3,4}1 + x_{2,1}1x_{2,3}x_{1,4}1 + 1\, x_{3,2}x_{1,3}x_{1,4}1 + 1\, x_{3,2}1x_{2,4}1 = f_{3,4}^{(2)}$, as in Example 10.

(b) $\mathcal{T}_3(w_3, 6)$ = sum of weights of circuits based at w_3 with length 6.

We observe that the first edge in such a circuit is edge (w_3, w_1) of weight $x_{3,1}$; hence $x_{3,1}$ is a factor of every term in $\mathcal{T}_3(w_3, 6) = f_{3,4}^{(3)}$, consistent with Example 10 again.

Finally, we bring the results from Lemma 13 and Theorems 16 and 23 together in Theorem 25.

Theorem 25. *For* $1 \le n \le m$,

$$\mathcal{C}_m(n) = \operatorname{tr}\left(\mathrm{Y}_{\mathrm{m,n}}\right) = \sum_{\mathrm{j}=1}^{\mathrm{m}} \mathcal{T}_{\mathrm{m}}\left(\mathrm{w_j}, \mathrm{n}\right). \tag{53}$$

Example 26. Again, from $T(3)$, we have $\mathcal{C}_3(4) = \sum_{j=1}^{3}\mathcal{T}_3(w_j, 4)$:

$$\begin{aligned}\mathcal{T}_3\left(w_1, 4\right) &= x_{1,1}x_{1,2}x_{1,3}x_{1,4} + x_{1,1}x_{1,2}1x_{2,4}\\ &\quad + x_{1,1}1x_{2,3}x_{1,4} + x_{1,1}1 \cdot 1x_{3,4}\\ &\quad + 1x_{2,2}x_{1,3}x_{1,4} + 1x_{2,2}1x_{2,4}\\ &\quad + 1 \cdot 1x_{3,3}x_{1,4} = f_{3,4}^{(1)},\end{aligned} \tag{54}$$

$$\begin{aligned}\mathcal{T}_3\left(w_2, 4\right) &= x_{2,1}x_{1,2}x_{1,3}1 + x_{2,1}1\, x_{2,3}1 + 1x_{3,2}x_{1,3}1\\ &= f_{3,3}^{(2)},\\ \mathcal{T}_3\left(w_3, 4\right) &= x_{3,1}x_{1,2}1 \cdot 1 = f_{3,2}^{(3)},\end{aligned}$$

which are consistent with the above definitions and results and with Example 17.

7. Homogeneous Case, $x_{\ell,i} \rightarrow x_\ell$

In this section, we consider the case of constant coefficients, that is, where the indeterminates $x_{\ell,i}$ are independent of i.

Notation. We use $*$ to modify a path or expression or matrix in which weights or indeterminates $x_{\ell,i}$ are replaced with x_ℓ.

First we review some known properties of m-path polynomials using standard techniques. Then we show how our model recovers these results combinatorially.

7.1. Constant Coefficient Recurrences. This subsection mainly establishes notation and recalls basic results of interest.

Consider the recurrence

$$y_n = \sum_{i=1}^{m} x_i y_{n-i}. \tag{55}$$

We begin with the first fundamental solution. The following is standard and readily derived via geometric series and multinomial expansion.

Proposition 27. *One has the generating function and formula*

$$\begin{aligned}\sum_{n\ge0} h_n\, t^n &= \frac{1}{1 - \sum_{i=1}^{m} x_i\, t^i}\\ &= \sum_{n\ge0}\sum_{\sum \ell s_\ell = n}\binom{s_1 + s_2 + \cdots + s_m}{s_1, s_2, \ldots, s_m} x_1^{s_1}x_2^{s_2}\cdots x_m^{s_m}\, t^n\end{aligned} \tag{56}$$

giving the (first) fundamental solution, h_n, to the recurrence, that is, with initial values $h_i = 0$, $-(m-1) \le i < 0$, $h_0 = 1$.

The matrix X_m takes the form, confor Section 5.1,

$$X_m = \begin{pmatrix} 0 & 1 & 0 & \cdots & 0\\ 0 & 0 & 1 & \cdots & 0\\ \vdots & \vdots & \vdots & \ddots & \vdots\\ 0 & 0 & 0 & \ddots & 1\\ x_m & x_{m-1} & x_{m-2} & \cdots & x_1 \end{pmatrix} \tag{57}$$

so that $\det(I - tX_m) = 1 - \sum_{i=1}^{m} x_i t^i$. Define the $(r+1)$st fundamental solution to recurrence (55) to be the one with initial conditions

$$y_i = 0, \quad \text{for } -(m-1) \le i \le 0,\ i \ne -r$$
$$y_{-r} = 1, \tag{58}$$

and denote this fundamental solution by $h_n^{(r+1)}$, with $h_n = h_n^{(1)}$. Then the entries in the bottom row of $(X_m)^n$ are exactly the values

$$\left((X_m)^n\right)_{(m,j)} = h_n^{(m-j+1)}. \tag{59}$$

In general,

$$\left((X_m)^n\right)_{(i,j)} = h_{n-m+i}^{(m-j+1)}. \tag{60}$$

The fundamental solutions for $r > 0$ can be expressed in terms of the first fundamental solution as follows.

Proposition 28. *The $(r+1)$st fundamental solution to the recurrence (55) is given by*

$$h_n^{(r+1)} = h_{n+r} - \sum_{k=0}^{r-1} h_{n+k} x_{r-k}, \tag{61}$$

where h_n denotes the first fundamental solution.

Proof. We will illustrate for $r \le 2$ that shows how the general case works. We have

$$h_n^{(1)} = h_n,$$
$$h_n^{(2)} = h_{n+1} - x_1 h_n, \tag{62}$$
$$h_n^{(3)} = h_{n+2} - x_1 h_{n+1} - x_2 h_n.$$

For $r = 1$, we obtain 0 for nonpositive n, except for $n = -1$, as required. Similarly, for $r = 2$, for nonpositive n we obtain 1 precisely for $n = -2$; otherwise we get 0. Note that the subtractions are necessary to cancel off terms when $0 \ge n > -r$. Since the coefficients are independent of n, these are indeed solutions to the recurrence. Thus the result is obtained. □

Now for the trace, we have the following.

Proposition 29. *The trace of $(X_m)^n$ is given by*

$$\operatorname{tr}(X_m)^n = \sum_{j=1}^{m} j h_{n-j} x_j. \tag{63}$$

Proof. From (60), we have, using Proposition 28,

$$\begin{aligned}
\operatorname{tr}(X_m)^n &= \sum_{i=1}^{m} h_{n-m+i}^{(m-i+1)} = \sum_{i=0}^{m-1} h_{n-i}^{(i+1)} \\
&= \sum_{i=0}^{m-1} \left[h_n - \sum_{k=0}^{i-1} h_{n-i+k} x_{i-k} \right] \\
&= \sum_{i=0}^{m-1} \left[h_n - \sum_{j=1}^{i} h_{n-j} x_j \right] \\
&= m h_n - \sum_{i=0}^{m-1} \sum_{j=1}^{i} h_{n-j} x_j \\
&\quad \text{(next, interchanging the order of summation)} \\
&= m h_n - \sum_{j=1}^{m-1} \sum_{i=j}^{m-1} h_{n-j} x_j \\
&= m h_n - \sum_{j=1}^{m-1} (m-j) h_{n-j} x_j \\
&= m \left[h_n - \sum_{j=1}^{m-1} h_{n-j} x_j \right] + \sum_{j=1}^{m-1} j h_{n-j} x_j \\
&= \sum_{j=1}^{m} j h_{n-j} x_j \quad \left(\text{by the recurrence for } \{h_n\}\right).
\end{aligned} \tag{64}$$

□

Remark 30. These are a variation on Newton's Identities relating power sum symmetric functions and elementary symmetric functions. Here, the homogeneous symmetric functions, h_n, play a role as well.

7.2. Combinatorial Proofs. We now show how these formulas may be derived combinatorially by our model with the specialization $x_{\ell,i} \to x_\ell$. The weighted path $P^*(2,3)$ looks like

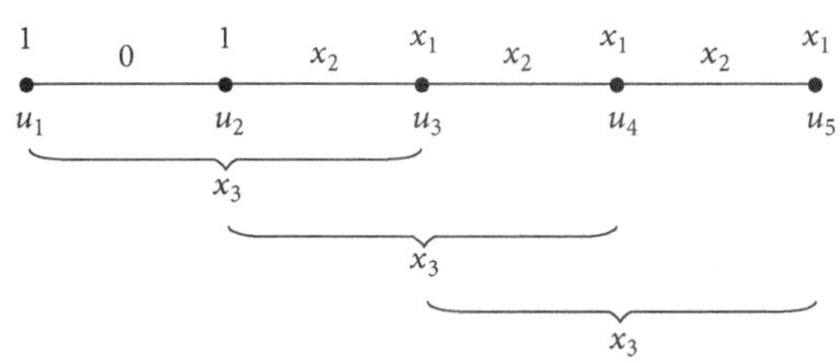

(65)

Notation. Consistent with the above, we use h or $\mathscr{H}$ to represent expressions in which we have replaced $x_{\ell,i}$ with x_ℓ. Thus $\mathscr{H}_m(r,n) = \mathscr{P}_m^*(r,n)$, for $0 \le r \le m-1$; see Definition 6 of weighted path $P(r,n)$.

7.2.1. First Fundamental Solution. Proposition 27 is readily seen from the weighting of path $P^*(m-1,n)$. For the first fundamental solution, there are no vertices with weight 1, and no edges weighted 0. The first vertex has weight x_1, and so on. In an m-path cover the exponent s_ℓ is the number of paths of length ℓ, for each $1 \le \ell \le m$, and the multinomial coefficient counts the number of m-path covers obtained from any fixed set of m-paths. So this model gives a visual interpretation to the analytic formula.

7.2.2. Higher Fundamental Solutions. Start with the following.

Lemma 31. *For a fixed r with $1 \le r \le m-1$ and any $n \ge 1$,*

$$\mathscr{H}_m(r,n) - \mathscr{H}_m(r-1,n) = \sum_{\ell=r+1}^{m} x_\ell \mathscr{H}_m(0, n+r-\ell). \quad (66)$$

Proof. For $r \ge 1$, consider the weighted path $P^*(r,n)$. The first vertex u_1 must lie in every m-path cover of this path, say on a path Q_ℓ of length ℓ for $1 \le \ell \le m$, starting at u_1. If $\ell = 1$ then $\omega(Q_1) = \omega(u_1) = 1$, and the sum of all such m-path covers is thus $1 \cdot \mathscr{H}_m(r-1,n)$. If $2 \le \ell \le r$ then Q_ℓ finishes at vertex u_ℓ where $\omega(u_\ell) = 1$, so $\omega(Q_\ell) = 0$. And if $r+1 \le \ell \le m$ then Q_ℓ finishes at vertex u_ℓ where $\omega(u_\ell) = x_1$ and so $\omega(Q_\ell) = x_\ell$, and the sum of all such m-path covers is $x_\ell \mathscr{H}_m(0, n+r-\ell)$. Hence $\mathscr{H}_m(r,n) = \mathscr{H}_m(r-1,n) + \sum_{\ell=r+1}^{m} x_\ell \mathscr{H}_m(0,n+r-\ell)$, and so the result is obtained. □

Now for a combinatorial proof of Proposition 28.

Theorem 32. *For the fundamental solutions to the recurrence for the homogeneous path polynomials, one has*

$$h_n^{(r+1)} = h_{n+r} - \sum_{\ell=1}^{r} x_\ell h_{n+r-\ell}. \quad (67)$$

Proof. By our definitions and Corollary 11(ii) we have $h_n = f_{m,n}^{(1)*} = \mathscr{P}_m^*(0,n) = \mathscr{H}_m(0,n)$. And, from Lemmas 9 and 31, we have

$$h_n^{(r+1)} = \mathscr{H}_m(r,n) - \mathscr{H}_m(r-1,n) = \sum_{\ell=r+1}^{m} x_\ell h_{n+r-\ell}. \quad (68)$$

Now

$$\begin{aligned} h_{n+r} &= \mathscr{H}_m(0,n+r) \\ &= \sum_{\ell=1}^{m} x_\ell h_{n+r-\ell} \\ &= \sum_{\ell=1}^{r} x_\ell h_{n+r-\ell} + \sum_{\ell=r+1}^{m} x_\ell h_{n+r-\ell} \\ &= \sum_{\ell=1}^{r} x_\ell h_{n+r-\ell} + h_n^{(r+1)}, \end{aligned} \quad (69)$$

where, at the second line, we note that in every m-path cover of the weighted path $P^*(0,n+r)$ vertex u_{n+r} must lie on a path Q_ℓ of length ℓ and weight x_ℓ where $1 \le \ell \le m$, and at the last line we use (68). This gives the result. □

7.2.3. Trace Formula. We now give a combinatorial derivation of the trace formula in Proposition 29.

First let $\mathscr{T}_m(n)$ be the sum of the weights of all circuits of length n in $T^*(m)$ and the m-trellis with edge-weights $x_{\ell,i}$ replaced by x_ℓ; that is, $\mathscr{T}_m(n) = \sum_{j=1}^{m} \mathscr{T}_m^*(w_j, n)$; see Section 6.

Theorem 33. *For any $n \ge 1$,*

$$\operatorname{tr}(X_m)^n = \sum_{j=1}^{m} j x_j h_{n-j}. \quad (70)$$

Proof. We recall that the indeterminates in any term of $\mathscr{T}_m(n)$ are initially ordered according to the edges traversed in the corresponding circuit; see Example 26. Let $\mathscr{X} = x_j x_{\ell_1} x_{\ell_2} \cdots x_{\ell_r}$ be a typical ordered term in $\mathscr{T}_m(n)$ with all 1's removed and with first indeterminate x_j. We first show that term $\mathscr{X}$ occurs j times in $\mathscr{T}_m(n)$.

When there are two successive indeterminates x_ℓ and $x_{\ell'}$ in $\mathscr{X}$, then, in the corresponding circuit, the edges traversed are first (w_ℓ, w_1) of weight x_ℓ, followed by the $\ell'-1$ edges $(w_1,w_2), (w_2,w_3), \ldots, (w_{\ell'-1}, w_{\ell'})$ each of weight 1, and then finishing with the edge $(w_{\ell'}, w_1)$ of weight $x_{\ell'}$. Hence pair $x_\ell x_{\ell'}$ becomes $x_\ell 1^{\ell'-1} x_{\ell'}$ when the indeterminates are considered as weights on edges in a circuit in $T^*(m)$.

Now, because the first indeterminate in $\mathscr{X}$ is x_j, any circuit corresponding to $\mathscr{X}$ must be based at vertex $w_{j'}$ for some $j' \in \{1,2,\ldots,j\}$. Hence $\mathscr{X}$ will appear in $\mathscr{T}_m(n)$ as

$$1^{j-j'} x_j 1^{\ell_1-1} x_{\ell_1} 1^{\ell_2-1} x_{\ell_2} 1^{\ell_3-1} \cdots 1^{\ell_r-1} x_{\ell_r} 1^{j'-1}, \quad (71)$$

for each $j' \in \{1,2,\ldots,j\}$ in $\mathscr{T}_m(n)$. There are j such j', so there are j occurrences of term $\mathscr{X}$ in $\mathscr{T}_m(n)$.

Now consider an occurrence of $\mathscr{X}$ in which $j' = j$, namely,

$$x_j 1^{\ell_1-1} x_{\ell_1} 1^{\ell_2-1} x_{\ell_2} 1^{\ell_3-1} \cdots 1^{\ell_r-1} x_{\ell_r} 1^{j-1}. \quad (72)$$

Let

$$\frac{\mathscr{X}}{x_j 1^{j-1}} = 1^{\ell_1-1} x_{\ell_1} 1^{\ell_2-1} x_{\ell_2} 1^{\ell_3-1} \cdots 1^{\ell_r-1} x_{\ell_r} = \mathscr{Z}. \quad (73)$$

Then the sequence of edges traversed in $T^*(m)$ corresponding to $\mathscr{Z}$ begins at w_1 and ends at w_1, and so is a circuit based at w_1, with length $n-1-(j-1) = n-j$. Thus $\mathscr{Z} \in \mathscr{T}_m^*(w_1, n-j)$. Conversely given any $\mathscr{Z} \in \mathscr{T}_m^*(w_1, n-j)$ then $x_j \mathscr{Z} 1^{j-1}$ is an occurrence of term $\mathscr{X}$ starting with 1^0 and ending with 1^{j-1}. Thus $(\sum_{j'=j} \mathscr{X})/x_j = \mathscr{T}_m^*(w_1, n-j)$ and $\sum_{j'=j} \mathscr{X} = x_j \mathscr{T}_m^*(w_1, n-j)$.

Now we can partition the weighted circuits of $T^*(m)$ of length n by their first indeterminate x_j (ignoring the edges of weight 1 preceding this first indeterminate). That is, we can partition the terms of $\mathscr{T}_m(n)$ by their first indeterminate x_j. So, using the above arguments, we have

$$\mathscr{T}_m(n) = \sum_{j=1}^{m} j x_j \mathscr{T}_m^*(w_1, n-j). \quad (74)$$

Furthermore, $\mathcal{T}^*_m(w_1, n-j) = \mathcal{P}^*_m(0, n-j) = f^{(1)*}_{m,n-j} = h_{n-j}$; the first equality is Theorem 21, the second is Corollary 11(ii), and the third is by definition of h_n. So finally,

$$\operatorname{tr}(X^n_m) = \mathcal{T}_m(n) = \sum_{j=1}^{m} jx_j\mathcal{T}^*_m(w_1, n-j) = \sum_{j=1}^{m} jx_j h_{n-j}. \tag{75}$$

□

Example 34. See Examples 17 and 26. Here $m = 3$ and $n = 4$:

$$\begin{aligned}\operatorname{tr}\left(X^4_3\right) &= \mathcal{T}_3(4)\\ &= x_1^4 + 4x_1^2x_2 + 4x_1x_3 + 2x_2^2\\ &= x_1\left(x_1^3 + 2x_1x_2 + x_3\right) + 2x_2\left(x_1^2 + x_2\right) + 3x_3\left(x_1\right)\\ &= x_1\left(f^{(1)*}_{3,3}\right) + 2x_2\left(f^{(1)*}_{3,2}\right) + 3x_3\left(f^{(1)*}_{3,1}\right)\\ &= x_1h_3 + 2x_2h_2 + 3x_3h_1,\end{aligned} \tag{76}$$

where, at line 2, we have rearranged the terms according to their first indeterminate x_j, using Example 26, and combined like terms.

Remark 35. From Theorem 25 and our definitions of matrices $Y_{m,n}$ and X_m from Sections 3 and 5.1, respectively, we have the following equalities:

$$\begin{aligned}\mathcal{C}^*_m(n) &= \operatorname{tr}\left(Y^*_{m,n}\right) = \sum_{j=1}^{m}\mathcal{T}^*_m\left(w_j, n\right),\\ \operatorname{tr}\left(Y^*_{m,n}\right) &= \operatorname{tr}\left(X^n_m\right).\end{aligned} \tag{77}$$

Thus, from Theorem 33,

$$\sum_{j=1}^{m}\mathcal{T}^*_m\left(w_j, n\right) = \sum_{j=1}^{m} jx_j\mathcal{T}^*_m\left(w_1, n-j\right). \tag{78}$$

8. Sequences, $x_{\ell,i} \rightarrow 1$

In Section 7 we specialized by replacing weights $x_{\ell,i}$ with x_ℓ. In this section we specialize further by replacing all weights $x_{\ell,i}$ with 1. We denote this operation by #. We then use these # matrices to count m-path covers of the path and cycle.

Recall matrix $X_{m,n}$ from (26), we define matrix Z_m:

$$Z_m = X^\#_{m,n} = \begin{pmatrix} 0 & 1 & \cdot & \cdots & 0\\ 0 & 0 & 1 & \cdots & 0\\ \vdots & \vdots & \vdots & \vdots & \vdots\\ 0 & 0 & 0 & \cdots & 1\\ 1 & 1 & 1 & \cdots & 1\end{pmatrix}. \tag{79}$$

Similarly, let $c_m(n) = \mathcal{C}^\#_m(n)$ be the expression $\mathcal{C}_m(n)$ evaluated when all $x_{\ell,i} = 1$. So $Y^\#_{m,n} = Z^n_m$ and $c_m(n) = \operatorname{tr}(Y^\#_{m,n}) = \operatorname{tr}(Z^n_m)$. Thus $c_m(n)$ counts the number of m-path covers of the weighted $C(n)$ or of an arbitrary n-cycle. (cf., Corollary 11.1, Section 8, Farrell [2].)

Theorem 36. *For $1 \le n \le m$, one has $c_m(n) = 2^n - 1$.*

Proof. Let $[n] = \{1, 2, \ldots, n\}$ and let $C[n]$ denote the cycle whose vertices are the elements of $[n]$ arranged clockwise in a circle. Now $n \le m$ so any path cover of $C[n]$ will be an m-path cover. We show that the number of path covers of $C[n]$ is $2^n - 1$.

Given a subset $\{i_1, i_2, \ldots, i_k\}$ of $[n]$ with $\{i_1 < i_2 < \cdots < i_k\}$ we define a path cover $[i_1, i_1+1, \ldots, i_2-1], [i_2, i_2+1, \ldots, i_3-1], \ldots, [i_k, i_k+1, \ldots, i_1-1]$, of $C[n]$. Conversely, given a path cover $[i_1, i_1+1, \ldots, i_2-1], [i_2, i_2+1, \ldots, i_3-1], \ldots, [i_k, i_k+1, \ldots, i_1-1]$ of $C[n]$ we take the first vertex from each path to form a subset $\{i_1, i_2, \ldots, i_k\}$ of $[n]$ and then rearrange its elements to form a subset of $[n]$ with increasing elements. These two operations illustrate a bijection from the set of non-empty subsets of $[n]$ to the set of m-path covers of $C[n]$. Hence $c_m(n) = 2^n - 1$. □

From recurrence (2), Lemma 13, and Theorems 16 and 36, for $n \ge m+1$ we see that $c_m(n)$ obeys the m-anacci recurrence,

$$\begin{aligned}c_m(n) &= c_m(n-1) + c_m(n-2) + \cdots + c_m(n-m)\\ &= \sum_{\ell=1}^{m} c_m(n-\ell),\end{aligned} \tag{80}$$

with starting conditions $c_m(n) = 2^n - 1$ for $1 \le n \le m$.

In the square array (see Table 1) $c_m(n)$ is the (n, m) entry, for $n, m \ge 1$. Column m is determined by the above m-anacci recurrence. We observe that the (m, m) main diagonal entry is $c_m(m) = 2^m - 1$.

Consider the triangle, in bold, where $c_m(n)$ is the (n, m) entry for all $n \ge 1$ and $1 \le m \le n$; it counts the number of m-path covers of a cycle with n vertices. We have entered the sequence obtained from reading this triangle row-by-row to the Online Encyclopedia of Integer Sequences [9]; it is sequence A185722.

Each of the 10 columns of the square array (see Table 1) appears as a sequence in [9]; for example, the second column ($m = 2$) gives sequence A000204 and the third column ($m = 3$) gives A001644. Thus we have a new combinatorial interpretation for each of these sequences and a connection between them.

A closely related sequence is A126198 (replace "k" by "m" in its description). Let $T(n, m)$ be the (n, m) entry of the triangle corresponding to A126198, then $T(n, m)$ counts the number of compositions of integer n into parts of size $\le m$. Now consider n vertices arranged in a path. A composition of n into parts of size $\le m$ corresponds naturally to an m-path cover of this path with n vertices by identifying a part of size ℓ in the composition with a path of length ℓ in the corresponding m-path cover. This correspondence can also be reversed. Thus in our terminology, $T(n, m)$ is the number of m-path covers of a path with n vertices, and, from Corollary 11(ii) and our operation #, we have $T(n, m) = f^{(1)\#}_{m,n} = \mathcal{P}^\#_m(0, n)$. The (m, m) main diagonal entry in this triangle is $T(m, m) = 2^{m-1}$ (as is well known, there are

Table 1

$n\backslash m$	1	2	3	4	5	6	7	8	9	10	...
1	1	1	1	1	1	1	1	1	1	1	...
2	1	3	3	3	3	3	3	3	3	3	...
3	1	4	7	7	7	7	7	7	7	7	...
4	1	7	11	15	15	15	15	15	15	15	...
5	1	11	21	26	31	31	31	31	31	31	...
6	1	18	39	51	57	63	63	63	63	63	...
7	1	29	71	99	113	120	127	127	127	127	...
8	1	47	131	191	223	239	247	255	255	255	...
9	1	76	241	367	439	475	493	502	511	511	...
10	1	123	443	708	863	943	983	1003	1013	1023	...
⋮	⋮	⋮	⋮	⋮	⋮	⋮	⋮	⋮	⋮	⋮	

2^{m-1} compositions of m) and column m of this triangle is determined by the m-anacci recurrence,

$$T(n,m) = T(n-1,m) + T(n-2,m) + \cdots + T(n-m,m) = \sum_{\ell=1}^{m} T(n-\ell,m), \quad (81)$$

for $n \geq m+1$, with starting conditions $T(n,m) = 2^{n-1}$ for $1 \leq n \leq m$.

The (n,m) entry in our triangle, $c_m(n)$, counts the number of m-path covers of a cycle with n vertices. We have starting conditions $c_m(n) = 2^n - 1$ as opposed to $T(n,m) = 2^{n-1}$ above, for $1 \leq n \leq m$.

Furthermore, from above and the definition of matrix $Y_{m,n}$ from (28), we have $T(n,m) = f_{m,n}^{(1)\#}$ = the (m,m) entry of matrix $Y_{m,n}^{\#} = Z_m^n$. Thus both

$$c_m(n) = \operatorname{tr}(Z_m^n), \qquad T(n,m) = (Z_m^n)_{(m,m)}, \quad (82)$$

can be obtained from matrix Z_m^n. This gives a new derivation of $T(n,m)$, and so of sequence A126198.

Example 37. $m = 3$ and $n = 4$. We have

$$Z_3 = \begin{pmatrix} 0 & 1 & 0 \\ 0 & 0 & 1 \\ 1 & 1 & 1 \end{pmatrix}, \qquad Z_3^4 = \begin{pmatrix} 1 & 2 & 2 \\ 2 & 3 & 4 \\ 4 & 6 & 7 \end{pmatrix}, \quad (83)$$

gives

$$c_3(4) = \operatorname{tr}(Z_3^4) = 11, \qquad T(4,3) = (Z_3^4)_{(3,3)} = 7; \quad (84)$$

see Examples 17, 26, and 7.

Conflict of Interests

The authors declare that there is no conflict of interests regarding the publication of the paper.

Acknowledgment

The authors thank the referee for bringing to their attention various relevant papers.

References

[1] E. J. Farrell, "On a general class of graph polynomials," *Journal of Combinatorial Theory B*, vol. 26, no. 1, pp. 111–122, 1979.

[2] E. J. Farrell, "Path decompositions of chains and circuits," *International Journal of Mathematics and Mathematical Sciences*, vol. 6, no. 3, pp. 521–533, 1983.

[3] T. Y. Chow, "The path-cycle symmetric function of a digraph," *Advances in Mathematics*, vol. 118, no. 1, pp. 71–98, 1996.

[4] O. M. D'Antona and E. Munarini, "The cycle-path indicator polynomial of a digraph," *Advances in Applied Mathematics*, vol. 25, no. 1, pp. 41–56, 2000.

[5] J. P. McSorley, P. Feinsilver, and R. Schott, "Generating orthogonal polynomials and their derivatives using vertex|matching-partitions of graphs," *Ars Combinatoria*, vol. 87, pp. 75–95, 2008.

[6] A. Joyal, "Une théorie combinatoire des séries formelles," *Advances in Mathematics*, vol. 42, no. 1, pp. 1–82, 1981.

[7] F. Bergeron, G. Labelle, and P. Leroux, *Combinatorial Species and Tree-Like Structures*, vol. 67 of *Encyclopedia of Mathematics and its Applications*, Cambridge University Press, Cambridge, Mass, USA, 1998.

[8] E. Munarini, "A combinatorial interpretation of the generalized Fibonacci numbers," *Advances in Applied Mathematics*, vol. 19, no. 3, pp. 306–318, 1997.

[9] Online Encyclopedia of Integer Sequences, published electronically, http://oeis.org.

8

Embedding Structures Associated with Riordan Arrays and Moment Matrices

Paul Barry

School of Science, Waterford Institute of Technology, Waterford, Ireland

Correspondence should be addressed to Paul Barry; pbarry@wit.ie

Academic Editor: Toufik Mansour

Every ordinary Riordan array contains two naturally embedded Riordan arrays. We explore this phenomenon, and we compare it to the situation for certain moment matrices of families of orthogonal polynomials.

1. Introduction

Riordan arrays [1] have been used mainly to prove combinatorial identities [2, 3]. Recently, their links to orthogonal polynomials have been investigated [4, 5], while there is a growing literature surrounding their structural properties [6–10]. In this paper we investigate an embedding structure, common to all ordinary Riordan arrays. We also look at this embedding structure in the context of moment matrices of families of orthogonal polynomials. In addition to some knowledge of Riordan arrays, we assume that the reader has a basic familiarity with the theory of orthogonal polynomials on the real line [11–13], production matrices [14, 15], and continued fractions [16]. We will meet a number of integer sequences and integer triangles in this paper. The On-Line Encyclopedia may be consulted for many of them [17, 18]. In this paper we will understand by an ordinary Riordan array an integer number triangle whose (n, k)-th element $T_{n,k}$ is defined by a pair of power series $g(x)$ and $f(x)$ over the integers with $g(x) = 1 + g_1x + g_2x^2 + \cdots$, $f(x) = x + f_2x^2 + f_3x^3 + \cdots$, in the following manner:

$$T_{n,k} = [x^n]\, g(x)\, f(x)^k. \tag{1}$$

The group law for Riordan arrays is given by

$$(g, f) \cdot (h, l) = (g(h \circ f), l \circ f). \tag{2}$$

The identity for this law is $I = (1, x)$ and the inverse of (g, f) is $(g, f)^{-1} = (1/(g \circ \bar{f}), \bar{f})$, where $\bar{f}$ is the compositional inverse of f. For a power series $f(x) = \sum_{n=0}^{\infty} a_n x^n$ with $f(0) = 0$, we define the reversion or compositional inverse of f to be the power series $\bar{f}(x)$ such that $f(\bar{f}(x)) = x$. We sometimes write this as $\bar{f} = \text{Rev} f$.

If a matrix A is the inverse of the coefficient array of a family of orthogonal polynomials, then we will call it a moment matrix, and we will single out the first column as the moment sequence.

2. The Canonical Embedding

Let (g, f) be an ordinary Riordan array R, with general term

$$T_{n,k} = [x^n]\, g f^k. \tag{3}$$

Then we observe that there are two naturally associated Riordan arrays "embedded" in the array R as follows.

Beginning at the first column of R, we take every second column, "raising" the columns appropriately to obtain a lower-triangular matrix A. The matrix A is then the Riordan array

$$A = \left(g, \frac{f^2}{x}\right) \tag{4}$$

with general term $A_{n,k}$ given by

$$\begin{aligned} A_{n,k} &= [x^n]\, g\left(\frac{f^2}{x}\right)^k \\ &= [x^n]\, g x^{-k} f^{2k} \\ &= \left[x^{n+k}\right] g f^{2k} \\ &= T_{n+k,2k}. \end{aligned} \tag{5}$$

Similarly, starting at the second column of R, taking every second column and "raising" all columns appropriately to obtain a lower-triangular matrix, we obtain a matrix B. This matrix B is then a Riordan array, given by

$$B = \left(g\frac{f}{x}, \frac{f^2}{x}\right). \tag{6}$$

We have

$$B = \left(\frac{f}{x}, x\right) \cdot A. \tag{7}$$

The general term $B_{n,k}$ of B is given by

$$\begin{aligned} B_{n,k} &= [x^n]\, g\frac{f}{x}\left(\frac{f^2}{x}\right)^k \\ &= [x^n]\, g x^{-k-1} f^{2k+1} \\ &= \left[x^{n+k+1}\right] g f^{2k+1} \\ &= T_{n+k+1,2k+1}. \end{aligned} \tag{8}$$

Example 1. We take the example of the binomial matrix

$$R = \left(\frac{1}{1-x}, \frac{x}{1-x}\right). \tag{9}$$

We then have

$$\begin{aligned} A &= \left(\frac{1}{1-x}, \frac{x}{(1-x)^2}\right) \\ &\text{with general term } \binom{n+k}{2k}, \\ B &= \left(\frac{1}{(1-x)^2}, \frac{x}{(1-x)^2}\right) \\ &\text{with general term } \binom{n+k+1}{2k+1}. \end{aligned} \tag{10}$$

The following decomposition makes this clear:

$$\begin{pmatrix} 1 & 0 & 0 & 0 & 0 & 0 & \cdots \\ 1 & 1 & 0 & 0 & 0 & 0 & \cdots \\ 1 & 2 & 1 & 0 & 0 & 0 & \cdots \\ 1 & 3 & 3 & 1 & 0 & 0 & \cdots \\ 1 & 4 & 6 & 4 & 1 & 0 & \cdots \\ 1 & 5 & 10 & 10 & 5 & 1 & \cdots \\ \vdots & \vdots & \vdots & \vdots & \vdots & \vdots & \ddots \end{pmatrix}. \tag{11}$$

The matrices A and B are the coefficient arrays of the Morgan-Voyce polynomials $b_n(x)$ and $B_n(x)$, respectively.

Example 2. We take the Riordan array

$$R = (c(x), xc(x)), \tag{12}$$

where

$$c(x) = \frac{1-\sqrt{1-4x}}{2x}. \tag{13}$$

Then we find that

$$A = \left(c(x), xc(x)^2\right), \qquad B = \left(c(x)^2, xc(x)^2\right). \tag{14}$$

The matrix R begins with

$$\begin{pmatrix} 1 & 0 & 0 & 0 & 0 & 0 & \cdots \\ 1 & 1 & 0 & 0 & 0 & 0 & \cdots \\ 2 & 2 & 1 & 0 & 0 & 0 & \cdots \\ 5 & 5 & 3 & 1 & 0 & 0 & \cdots \\ 14 & 14 & 9 & 4 & 1 & 0 & \cdots \\ 42 & 42 & 28 & 14 & 5 & 1 & \cdots \\ \vdots & \vdots & \vdots & \vdots & \vdots & \vdots & \ddots \end{pmatrix}. \tag{15}$$

We note that the matrix $A = (c(x), xc(x)^2)$ is the moment array for the family of orthogonal polynomials with coefficient array given by

$$A^{-1} = \left(c(x), xc(x)^2\right)^{-1} = \left(\frac{1}{1+x}, \frac{x}{(1+x)^2}\right). \tag{16}$$

Denoting this family by $P_n(x)$, we have

$$P_n(x) = (x-2)\,P_{n-1}(x) - P_{n-2}(x), \tag{17}$$

with $P_0(x) = 1$ and $P_1(x) = x - 1$. Similarly the matrix $B = (c(x)^2, xc(x)^2)$ is the moment array for the family of orthogonal polynomials with coefficient array given by

$$B^{-1} = \left(\frac{1}{(1+x)^2}, \frac{x}{(1+x)^2}\right). \tag{18}$$

Denoting this family by $Q_n(x)$, we have

$$Q_n(x) = (x-2)\,Q_{n-1}(x) - Q_{n-2}(x), \tag{19}$$

with $Q_0(x) = 1$ and $Q_1(x) = x - 2$.

The inverse matrix R^{-1} is given by

$$\begin{pmatrix} 1 & 0 & 0 & 0 & 0 & 0 & \cdots \\ -1 & 1 & 0 & 0 & 0 & 0 & \cdots \\ 0 & -2 & 1 & 0 & 0 & 0 & \cdots \\ 0 & 1 & -3 & 1 & 0 & 0 & \cdots \\ 0 & 0 & 3 & -4 & 1 & 0 & \cdots \\ 0 & 0 & -1 & 6 & -5 & 1 & \cdots \\ \vdots & \vdots & \vdots & \vdots & \vdots & \vdots & \ddots \end{pmatrix}, \tag{20}$$

which is the Riordan array $(1-x, x(1-x))$. In it we see the elements of A^{-1} and B^{-1} in staggered fashion.

3. A Counter Example

It is natural to ask the question: is a matrix that contains two embedded Riordan arrays as above itself a Riordan array? The following example shows that this is not a sufficient condition on an array to be Riordan.

Example 3. We will construct an invertible integer lower-triangular matrix which has two embedded Riordan arrays in the fashion above, but which is not itself a Riordan array. We start with the essentially two-period sequence $(a_n)_{n\geq 0}$

$$1, 2, 3, 2, 3, 2, 3, \ldots. \tag{21}$$

We form the matrix

$$\begin{pmatrix} 1 & 0 & 0 & 0 & 0 & 0 & \cdots \\ -2 & 1 & 0 & 0 & 0 & 0 & \cdots \\ 0 & -3 & 1 & 0 & 0 & 0 & \cdots \\ 0 & 0 & -2 & 1 & 0 & 0 & \cdots \\ 0 & 0 & 0 & -3 & 1 & 0 & \cdots \\ 0 & 0 & 0 & 0 & -2 & 1 & \cdots \\ \vdots & \vdots & \vdots & \vdots & \vdots & \vdots & \ddots \end{pmatrix}. \tag{22}$$

The inverse of this matrix begins with

$$\begin{pmatrix} 1 & 0 & 0 & 0 & 0 & 0 & \cdots \\ 2 & 1 & 0 & 0 & 0 & 0 & \cdots \\ 6 & 3 & 1 & 0 & 0 & 0 & \cdots \\ 12 & 6 & 2 & 1 & 0 & 0 & \cdots \\ 36 & 18 & 6 & 3 & 1 & 0 & \cdots \\ 72 & 36 & 12 & 6 & 2 & 1 & \cdots \\ \vdots & \vdots & \vdots & \vdots & \vdots & \vdots & \ddots \end{pmatrix}, \tag{23}$$

where we note an alternating pattern of constant columns (with generating functions $(1+2x)/(1-6x^2)$ and $(1+3x)/(1-6x^2)$, resp.). Removing the first row of this matrix provides us with a production matrix, which is not of the form that produces a Riordan array (after the first column, subsequent columns would be shifted versions of the second column [14, 15]). Thus the resulting matrix will not be a Riordan array. This resulting matrix begins with

$$\begin{pmatrix} 1 & 0 & 0 & 0 & 0 & 0 & \cdots \\ 2 & 1 & 0 & 0 & 0 & 0 & \cdots \\ 10 & 5 & 1 & 0 & 0 & 0 & \cdots \\ 62 & 31 & 7 & 1 & 0 & 0 & \cdots \\ 430 & 215 & 51 & 10 & 1 & 0 & \cdots \\ 3194 & 1597 & 389 & 87 & 12 & 1 & \cdots \\ \vdots & \vdots & \vdots & \vdots & \vdots & \vdots & \ddots \end{pmatrix}. \tag{24}$$

We now observe that for this matrix, we have

$$\begin{aligned} A &= \left(\frac{1}{1+2x}, \frac{x}{1+5x+6x^2}\right)^{-1}, \\ B &= \left(\frac{1}{1+5x+6x^2}, \frac{x}{1+5x+6x^2}\right)^{-1}. \end{aligned} \tag{25}$$

We notice that the sequence

$$1, 2, 10, 62, 430, 3194, \ldots \tag{26}$$

has generating function given by the continued fraction

$$\cfrac{1}{1-\cfrac{2x}{1-\cfrac{3x}{1-\cfrac{2x}{1-\cdots}}}}, \tag{27}$$

and secondly that

$$1+5x+6x^2 = 1+(2+3)x+2.3x^2 = (1+2x)(1+3x). \tag{28}$$

This construction is easily generalized.

4. Embedding a Riordan Array

Another natural question to ask is: if we are given a Riordan array A, is it possible to embed it as above into a Riordan array R? For this, we let

$$A = (u, v), \tag{29}$$

and seek to determine

$$R = (g, f) \tag{30}$$

such that A embeds into R. For this, we need

$$u = g, \qquad v = \frac{f^2}{x}. \tag{31}$$

Thus we require that

$$f = \sqrt{xv} = x\sqrt{\frac{v}{x}}. \tag{32}$$

Since we are working in the context of integer valued Riordan arrays, we require that $\sqrt{v/x}$ generates an integer sequence. We can state our result as follows.

Proposition 4. *The Riordan array*

$$A = (u, v) \tag{33}$$

can be embedded in the Riordan array

$$R = \left(g, x\sqrt{\frac{v}{x}}\right) \tag{34}$$

on condition that $\sqrt{v/x}$ is the generating function of an integer sequence.

Example 5. The Riordan array

$$A = \left(\frac{1}{\sqrt{1-4x}}, \frac{x}{1-4x}\right) \tag{35}$$

can be embedded in the Riordan array

$$R = \left(\frac{1}{\sqrt{1-4x}}, \frac{x}{\sqrt{1-4x}} \right). \quad (36)$$

For this example, the matrix A begins with

$$\begin{pmatrix} 1 & 0 & 0 & 0 & 0 & 0 & \cdots \\ 2 & 1 & 0 & 0 & 0 & 0 & \cdots \\ 6 & 6 & 1 & 0 & 0 & 0 & \cdots \\ 20 & 30 & 10 & 1 & 0 & 0 & \cdots \\ 70 & 140 & 70 & 14 & 1 & 0 & \cdots \\ 252 & 630 & 420 & 126 & 18 & 1 & \cdots \\ \vdots & \vdots & \vdots & \vdots & \vdots & \vdots & \ddots \end{pmatrix}, \quad (37)$$

while R begins with

$$\begin{pmatrix} 1 & 0 & 0 & 0 & 0 & 0 & \cdots \\ 2 & 1 & 0 & 0 & 0 & 0 & \cdots \\ 6 & 4 & 1 & 0 & 0 & 0 & \cdots \\ 20 & 16 & 6 & 1 & 0 & 0 & \cdots \\ 70 & 64 & 30 & 8 & 1 & 0 & \cdots \\ 252 & 256 & 140 & 48 & 10 & 1 & \cdots \\ \vdots & \vdots & \vdots & \vdots & \vdots & \vdots & \ddots \end{pmatrix}. \quad (38)$$

5. A Cascading Decomposition

We note that we can "cascade" this embedding process, in the sense that, given a Riordan array R, with embedded Riordan arrays A and B, we can consider decomposing A and B in their turns and then continue this process. For instance, we can decompose

$$A = \left(g, \frac{f^2}{x} \right) \quad (39)$$

into the two matrices

$$A_A = \left(g, \frac{f^4}{x^3} \right), \qquad B_A = \left(g\frac{f^2}{x^2}, \frac{f^4}{x^3} \right). \quad (40)$$

In their turn A_A and B_A can be decomposed and so on.

6. Embedding and Orthogonal Polynomials

The phenomenon of embedding as described above is not confined to Riordan arrays, as the continued fraction example above shows. To further amplify this point, we give another example involving a continued fraction. Although we take a particular case, the general case can be inferred easily from it. Thus we take the particular case of the continued fraction

$$\cfrac{1}{1-\cfrac{2x}{1-\cfrac{3x}{1-\cfrac{5x}{1-\cfrac{2x}{1-\cfrac{3x}{1-\cfrac{5x}{1-\cdots}}}}}}}. \quad (41)$$

This continued fraction is equal to

$$\cfrac{1}{1-2x-\cfrac{6x^2}{1-8x-\cfrac{10x^2}{1-5x-\cfrac{15x^2}{1-7x-\cfrac{6x^2}{1-8x-\cfrac{10x^2}{1-5x-\cdots}}}}}}. \quad (42)$$

By the theory of orthogonal polynomials, the power series expressed by both continued fractions is the generating function for the moment sequence of the family of orthogonal polynomials whose moment matrix (the inverse of the coefficient array of the orthogonal polynomials) has production matrix given by

$$\begin{pmatrix} 2 & 1 & 0 & 0 & 0 & 0 & \cdots \\ 6 & 8 & 1 & 0 & 0 & 0 & \cdots \\ 0 & 10 & 5 & 1 & 0 & 0 & \cdots \\ 0 & 0 & 15 & 7 & 1 & 0 & \cdots \\ 0 & 0 & 0 & 6 & 8 & 1 & \cdots \\ 0 & 0 & 0 & 0 & 10 & 5 & \cdots \\ \vdots & \vdots & \vdots & \vdots & \vdots & \vdots & \ddots \end{pmatrix}. \quad (43)$$

This production matrix generates the moment matrix A that begins with

$$\begin{pmatrix} 1 & 0 & 0 & 0 & 0 & 0 & \cdots \\ 2 & 1 & 0 & 0 & 0 & 0 & \cdots \\ 10 & 10 & 1 & 0 & 0 & 0 & \cdots \\ 80 & 100 & 15 & 1 & 0 & 0 & \cdots \\ 760 & 1030 & 190 & 22 & 1 & 0 & \cdots \\ 7700 & 10900 & 2310 & 350 & 30 & 1 & \cdots \\ \vdots & \vdots & \vdots & \vdots & \vdots & \vdots & \ddots \end{pmatrix}. \quad (44)$$

In order to produce an embedding for this matrix, we proceed as follows. We form the matrix

$$\begin{pmatrix} 1 & 0 & 0 & 0 & 0 & 0 & \cdots \\ -2 & 1 & 0 & 0 & 0 & 0 & \cdots \\ 0 & -3 & 1 & 0 & 0 & 0 & \cdots \\ 0 & 0 & -5 & 1 & 0 & 0 & \cdots \\ 0 & 0 & 0 & -2 & 1 & 0 & \cdots \\ 0 & 0 & 0 & 0 & -3 & 1 & \cdots \\ \vdots & \vdots & \vdots & \vdots & \vdots & \vdots & \ddots \end{pmatrix}. \quad (45)$$

We invert this matrix, remove the first row of the resulting matrix, and use this new matrix as a production matrix. The generated matrix R then begins with

$$\begin{pmatrix} 1 & 0 & 0 & 0 & 0 & 0 & \cdots \\ 2 & 1 & 0 & 0 & 0 & 0 & \cdots \\ 10 & 5 & 1 & 0 & 0 & 0 & \cdots \\ 80 & 40 & 10 & 1 & 0 & 0 & \cdots \\ 760 & 380 & 100 & 12 & 1 & 0 & \cdots \\ 7700 & 3850 & 1030 & 130 & 15 & 1 & \cdots \\ \vdots & \vdots & \vdots & \vdots & \vdots & \vdots & \ddots \end{pmatrix}. \quad (46)$$

The moment matrix A is evidently embedded in the matrix R. We can show that the corresponding matrix B is the moment matrix for the family of orthogonal polynomials whose moments have generating function given by

$$\cfrac{1}{1-5x-\cfrac{15x^2}{1-7x-\cfrac{6x^2}{1-8x-\cfrac{10x^2}{1-5x-\cdots}}}}. \quad (47)$$

The matrix B begins with

$$\begin{pmatrix} 1 & 0 & 0 & 0 & 0 & 0 & \cdots \\ 5 & 1 & 0 & 0 & 0 & 0 & \cdots \\ 40 & 12 & 1 & 0 & 0 & 0 & \cdots \\ 380 & 130 & 20 & 1 & 0 & 0 & \cdots \\ 3850 & 1410 & 300 & 25 & 1 & 0 & \cdots \\ 40400 & 15520 & 4060 & 440 & 32 & 1 & \cdots \\ \vdots & \vdots & \vdots & \vdots & \vdots & \vdots & \ddots \end{pmatrix}, \quad (48)$$

and it has production matrix

$$\begin{pmatrix} 5 & 1 & 0 & 0 & 0 & 0 & \cdots \\ 15 & 7 & 1 & 0 & 0 & 0 & \cdots \\ 0 & 6 & 8 & 1 & 0 & 0 & \cdots \\ 0 & 0 & 10 & 5 & 1 & 0 & \cdots \\ 0 & 0 & 0 & 15 & 7 & 1 & \cdots \\ 0 & 0 & 0 & 0 & 6 & 8 & \cdots \\ \vdots & \vdots & \vdots & \vdots & \vdots & \vdots & \ddots \end{pmatrix}. \quad (49)$$

The matrix R^{-1} can now be characterized as the coefficient array of a family of polynomials $R_n(x)$ defined as follows. We let $P_n(x)$ be the family of orthogonal polynomials with coefficient array A^{-1}, and we let $Q_n(x)$ be the family of orthogonal polynomials with coefficient B^{-1}. Then we have

$$R_n(x) = \begin{cases} Q_{n/2}(x)\, x^{n/2}, & \text{if } n \text{ is even;} \\ P_{\lceil n/2 \rceil}(x)\, x^{\lfloor n/2 \rfloor}, & \text{otherwise.} \end{cases} \quad (50)$$

In the general case of a moment sequence generated by the continued fraction

$$\cfrac{1}{1-\cfrac{\alpha x}{1-\cfrac{\beta x}{1-\cfrac{\gamma x}{1-\cfrac{\alpha x}{1-\cdots}}}}}, \quad (51)$$

the matrix A will be generated by the production matrix

$$\begin{pmatrix} \alpha & 1 & 0 & 0 & 0 & 0 & \cdots \\ \alpha\beta & \beta+\gamma & 1 & 0 & 0 & 0 & \cdots \\ 0 & \alpha\gamma & \alpha+\beta & 1 & 0 & 0 & \cdots \\ 0 & 0 & \beta\gamma & \alpha+\gamma & 1 & 0 & \cdots \\ 0 & 0 & 0 & \alpha\beta & \beta+\gamma & 1 & \cdots \\ 0 & 0 & 0 & 0 & \alpha\gamma & \alpha+\beta & \cdots \\ \vdots & \vdots & \vdots & \vdots & \vdots & \vdots & \ddots \end{pmatrix}, \quad (52)$$

while the matrix B is generated by the production matrix

$$\begin{pmatrix} \alpha+\beta & 1 & 0 & 0 & 0 & 0 & \cdots \\ \beta\gamma & \alpha+\gamma & 1 & 0 & 0 & 0 & \cdots \\ 0 & \alpha\beta & \beta+\gamma & 1 & 0 & 0 & \cdots \\ 0 & 0 & \alpha\gamma & \alpha+\beta & 1 & 0 & \cdots \\ 0 & 0 & 0 & \beta\gamma & \alpha+\gamma & 1 & \cdots \\ 0 & 0 & 0 & 0 & \alpha\beta & \beta+\gamma & \cdots \\ \vdots & \vdots & \vdots & \vdots & \vdots & \vdots & \ddots \end{pmatrix}. \quad (53)$$

7. Embedding, Orthogonal Polynomials, and Riordan Arrays

In this section, we consider the case of two related families of orthogonal polynomials $P_n(x)$ and $Q_n(x)$ defined by

$$P_n(x) = (x-7)\, P_{n-1}(x) - 12P_{n-2}, \quad (54)$$

with $P_0(x) = 1$, $P_1(x) = x - 3$, and

$$Q_n(x) = (x-7)\, Q_{n-1}(x) - 12Q_{n-2}, \quad (55)$$

with $Q_0(x) = 1$, $Q_1(x) = x - 7$. We note that

$$(1+3x)(1+4x) = 1 + 7x + 12x^2. \quad (56)$$

The coefficient array of the polynomials $P_n(x)$ is then given by the Riordan array

$$A^{-1} = \left(\frac{1}{1+3x}, \frac{x}{1+7x+12x^2}\right), \quad (57)$$

while that of $Q_n(x)$ is given by

$$B^{-1} = \left(\frac{1}{1+7x+12x^2}, \frac{x}{1+7x+12x^2}\right). \quad (58)$$

The moment matrix A begins with

$$\begin{pmatrix} 1 & 0 & 0 & 0 & 0 & 0 & \cdots \\ 3 & 1 & 0 & 0 & 0 & 0 & \cdots \\ 21 & 10 & 1 & 0 & 0 & 0 & \cdots \\ 183 & 103 & 17 & 1 & 0 & 0 & \cdots \\ 1785 & 1108 & 234 & 24 & 1 & 0 & \cdots \\ 18651 & 12349 & 3034 & 414 & 31 & 1 & \cdots \\ \vdots & \vdots & \vdots & \vdots & \vdots & \vdots & \ddots \end{pmatrix}, \quad (59)$$

while the moment matrix B starts with

$$\begin{pmatrix} 1 & 0 & 0 & 0 & 0 & 0 & \cdots \\ 7 & 1 & 0 & 0 & 0 & 0 & \cdots \\ 61 & 14 & 1 & 0 & 0 & 0 & \cdots \\ 595 & 171 & 21 & 1 & 0 & 0 & \cdots \\ 6217 & 2044 & 330 & 28 & 1 & 0 & \cdots \\ 68047 & 24485 & 4690 & 538 & 35 & 1 & \cdots \\ \vdots & \vdots & \vdots & \vdots & \vdots & \vdots & \ddots \end{pmatrix}. \quad (60)$$

Letting

$$R_n(x) = \begin{cases} Q_{n/2}(x)\, x^{n/2}, & \text{if } n \text{ is even;} \\ P_{(n+1)/2}(x)\, x^{\lfloor n/2 \rfloor}, & \text{otherwise,} \end{cases} \quad (61)$$

we find that the inverse R of the coefficient array of the family $R_n(x)$ is given by

$$\begin{pmatrix} 1 & 0 & 0 & 0 & 0 & 0 & \cdots \\ 3 & 1 & 0 & 0 & 0 & 0 & \cdots \\ 21 & 7 & 1 & 0 & 0 & 0 & \cdots \\ 183 & 61 & 10 & 1 & 0 & 0 & \cdots \\ 1785 & 595 & 103 & 14 & 1 & 0 & \cdots \\ 18651 & 6217 & 1108 & 171 & 17 & 1 & \cdots \\ \vdots & \vdots & \vdots & \vdots & \vdots & \vdots & \ddots \end{pmatrix}. \quad (62)$$

Thus the two Riordan arrays A and B, which are the moment arrays of the two families of orthogonal polynomials $P_n(x)$ and $Q_n(x)$, respectively, embed into the generalized moment array R for the family of polynomials $R_n(x)$. Now the production matrix of R begins with

$$\begin{pmatrix} 3 & 1 & 0 & 0 & 0 & 0 & \cdots \\ 12 & 4 & 1 & 0 & 0 & 0 & \cdots \\ 36 & 12 & 3 & 1 & 0 & 0 & \cdots \\ 144 & 48 & 12 & 4 & 1 & 0 & \cdots \\ 432 & 144 & 36 & 12 & 3 & 1 & \cdots \\ 1728 & 576 & 144 & 48 & 12 & 4 & \cdots \\ \vdots & \vdots & \vdots & \vdots & \vdots & \vdots & \ddots \end{pmatrix}, \quad (63)$$

where the columns have generating functions $(1 + 4x)/(1 - 12x^2)$, $(1 + 3)/(1 - 12x^2)$, respectively. We now observe that this production matrix is obtained by removing the first row of the inverse of the matrix

$$\begin{pmatrix} 1 & 0 & 0 & 0 & 0 & 0 & \cdots \\ -3 & 1 & 0 & 0 & 0 & 0 & \cdots \\ 0 & -4 & 1 & 0 & 0 & 0 & \cdots \\ 0 & 0 & -3 & 1 & 0 & 0 & \cdots \\ 0 & 0 & 0 & -4 & 1 & 0 & \cdots \\ 0 & 0 & 0 & 0 & -3 & 1 & \cdots \\ \vdots & \vdots & \vdots & \vdots & \vdots & \vdots & \ddots \end{pmatrix}. \quad (64)$$

We note finally that the sequence

$$1, 3, 21, 183, 1785, 18651, 204141, \ldots \quad (65)$$

has generating function given by

$$g(x) = \cfrac{1}{1 - \cfrac{3x}{1 - \cfrac{4x}{1 - \cfrac{3x}{1 - \cdots}}}}. \quad (66)$$

We have in fact that

$$g(x) = \frac{1}{x}\mathrm{Rev}\frac{x(1-4x)}{1-x}. \quad (67)$$

On the other hand, if we let

$$P_n(x) = (x-7)P_{n-1}(x) - 12P_{n-2}(x), \quad (68)$$

but this time take $P_0(x) = 1$ and $P_1(x) = x - 4$, then the matrix A becomes

$$A = \left(\frac{1}{1+4x}, \frac{x}{1+7x+12x^2}\right)^{-1}. \quad (69)$$

The matrix A then begins with

$$\begin{pmatrix} 1 & 0 & 0 & 0 & 0 & 0 & \cdots \\ 4 & 1 & 0 & 0 & 0 & 0 & \cdots \\ 28 & 11 & 1 & 0 & 0 & 0 & \cdots \\ 244 & 117 & 18 & 1 & 0 & 0 & \cdots \\ 2380 & 1279 & 255 & 25 & 1 & 0 & \cdots \\ 24868 & 14393 & 3364 & 442 & 32 & 1 & \cdots \\ \vdots & \vdots & \vdots & \vdots & \vdots & \vdots & \ddots \end{pmatrix}, \quad (70)$$

where the moment sequence

$$1, 4, 28, 244, 2380, \ldots \quad (71)$$

has generating function

$$\cfrac{1}{1 - \cfrac{4x}{1 - \cfrac{3x}{1 - \cfrac{4x}{1 - \cdots}}}}, \quad (72)$$

or equivalently

$$\cfrac{1}{1 - 4x - \cfrac{12x^2}{1 - 7x - \cfrac{12x^2}{1 - 7x - \cdots}}}. \quad (73)$$

In this case, the matrix R begins with

$$\begin{pmatrix} 1 & 0 & 0 & 0 & 0 & 0 & \cdots \\ 4 & 1 & 0 & 0 & 0 & 0 & \cdots \\ 28 & 7 & 1 & 0 & 0 & 0 & \cdots \\ 244 & 61 & 11 & 1 & 0 & 0 & \cdots \\ 2380 & 595 & 117 & 14 & 1 & 0 & \cdots \\ 24868 & 6217 & 1279 & 171 & 18 & 1 & \cdots \\ \vdots & \vdots & \vdots & \vdots & \vdots & \vdots & \ddots \end{pmatrix}. \quad (74)$$

This matrix is then associated with the matrix

$$\begin{pmatrix} 1 & 0 & 0 & 0 & 0 & 0 & \cdots \\ -4 & 1 & 0 & 0 & 0 & 0 & \cdots \\ 0 & -3 & 1 & 0 & 0 & 0 & \cdots \\ 0 & 0 & -4 & 1 & 0 & 0 & \cdots \\ 0 & 0 & 0 & -3 & 1 & 0 & \cdots \\ 0 & 0 & 0 & 0 & -4 & 1 & \cdots \\ \vdots & \vdots & \vdots & \vdots & \vdots & \vdots & \ddots \end{pmatrix}. \quad (75)$$

Conflict of Interests

The author declares that there is no conflict of interests regarding the publication of this paper.

References

[1] L. W. Shapiro, S. Getu, W.-J. Woan, and L. C. Woodson, "The Riordan group," *Discrete Applied Mathematics*, vol. 34, no. 1–3, pp. 229–239, 1991.

[2] C. Corsani, D. Merlini, and R. Sprugnoli, "Left-inversion of combinatorial sums," *Discrete Mathematics*, vol. 180, no. 1–3, pp. 107–122, 1998.

[3] R. Sprugnoli, "Riordan arrays and combinatorial sums," *Discrete Mathematics*, vol. 132, no. 1–3, pp. 267–290, 1994.

[4] P. Barry and A. Hennessy, "Meixner-type results for Riordan arrays and associated integer sequences," *Journal of Integer Sequences*, vol. 13, no. 9, article 10.9.4, 2010.

[5] P. Barry, "Riordan arrays, orthogonal polynomials as moments, and Hankel transforms," *Journal of Integer Sequences*, vol. 14, no. 2, article 11.2.2, 2011.

[6] G.-S. Cheon, H. Kim, and L. W. Shapiro, "Riordan group involutions," *Linear Algebra and Its Applications*, vol. 428, no. 4, pp. 941–952, 2008.

[7] D. E. Davenport, L. W. Shapiro, and L. C. Woodson, "The double Riordan group," *The Electronic Journal of Combinatorics*, vol. 18, no. 2, Paper 33, 2012.

[8] T.-X. He and R. Sprugnoli, "Sequence characterization of Riordan arrays," *Discrete Mathematics*, vol. 309, no. 12, pp. 3962–3974, 2009.

[9] S.-T. Jin, "A characterization of the Riordan Bell subgroup by C-sequences," *The Korean Journal of Mathematics*, vol. 17, no. 2, pp. 147–154, 2009.

[10] D. Merlini, D. G. Rogers, R. Sprugnoli, and M. C. Verri, "On some alternative characterizations of Riordan arrays," *Canadian Journal of Mathematics*, vol. 49, no. 2, pp. 301–320, 1997.

[11] T. S. Chihara, *An Introduction to Orthogonal Polynomials*, Gordon and Breach, New York, NY, USA, 1978.

[12] W. Gautschi, *Orthogonal Polynomials: Computation and Approximation*, Clarendon Press, Oxford, UK, 2004.

[13] G. Szegö, *Orthogonal Polynomials*, American Mathematical Society, Providence, RI, USA, 4th edition, 1975.

[14] E. Deutsch, L. Ferrari, and S. Rinaldi, "Production matrices," *Advances in Applied Mathematics*, vol. 34, no. 1, pp. 101–122, 2005.

[15] E. Deutsch, L. Ferrari, and S. Rinaldi, "Production matrices and Riordan arrays," *Annals of Combinatorics*, vol. 13, no. 1, pp. 65–85, 2009.

[16] H. S. Wall, *Analytic Theory of Continued Fractions*, AMS Chelsea, New York, NY, USA, 1967.

[17] N. J. A. Sloane, "The On-Line Encyclopedia of Integer Sequences," 2011, http://oeis.org.

[18] N. J. A. Sloane, "The on-line Encyclopedia of Integer Sequences," *Notices of the American Mathematical Society*, vol. 50, no. 8, pp. 912–915, 2003.

Bounds on the Size of the Minimum Dominating Sets of Some Cylindrical Grid Graphs

Mrinal Nandi,[1] Subrata Parui,[2] and Avishek Adhikari[3]

[1] *Department of Statistics, West Bengal State University, Barasat, North 24 Parganas 700126, India*
[2] *IMBIC, AH 317, Salt Lake City, Calcutta 700091, India*
[3] *Department of Pure Mathematics, University of Calcutta, 35 Ballygunge Circular Road, Calcutta 700019, India*

Correspondence should be addressed to Avishek Adhikari; avishek.adh@gmail.com

Academic Editor: Liying Kang

Let $\gamma(P_m \square C_n)$ denote the domination number of the cylindrical grid graph formed by the Cartesian product of the graphs P_m, the path of length m, $m \geq 2$, and the graph C_n, the cycle of length n, $n \geq 3$. In this paper we propose methods to find the domination numbers of graphs of the form $P_m \square C_n$ with $n \geq 3$ and $m = 5$ and propose tight bounds on domination numbers of the graphs $P_6 \square C_n$, $n \geq 3$. Moreover, we provide rough bounds on domination numbers of the graphs $P_m \square C_n$, $n \geq 3$ and $m \geq 7$. We also point out how domination numbers and minimum dominating sets are useful for wireless sensor networks.

1. Introduction

The problem of domination is one of the most widely studied topics in graph theory: the 1998 book by Haynes et al. [1] contains a bibliography with over 1200 papers on the subject. The domination problem was studied from the 1950s onwards, but the rate of research on domination significantly increased in the mid-1970s.

The signed domination number of a graph G was defined in [2] and has been studied by several authors including [2, 3]. Independent dominating sets were introduced into the theory of games by Morgenstern in [4]. For an extensive survey of domination problems and comprehensive bibliography the readers are referred to [5]. The study of domination numbers of products of graphs was initiated by Vizing [6]. He conjectured that the domination number of the Cartesian product of two graphs is always greater than or equal to the product of the domination numbers of the two factors, a conjecture which is still unproven. In [7], a link is shown between the existence of tilings in Manhattan metric with $\{1\}$, bowls and minimum total dominating sets of Cartesian products of paths and cycles. Domination numbers of Cartesian products were intensively investigated in [7–9].

The graphs considered here are finite, nonempty, connected, undirected without loops and without multiple edges. Besides these, any undefined terms in this paper may be found in Harary [10].

Let G be a simple graph whose vertex set and edge set are $V(G)$ and $E(G)$, respectively. The set $D \subseteq V(G)$ of a simple graph G is called a dominating set if every vertex $v \in V(G) \setminus D$ is adjacent to some vertex $u \in D$. The domination number of the graph G is the cardinality of a smallest dominating set of the graph G; it is usually denoted by $\gamma(G)$ and dominating set with smallest cardinality is called a minimum dominating set of the graph G.

For any two graphs G and H, the Cartesian product $G \square H$ is the graph with vertex set $V(G) \times V(H)$ and with edge set $E(G \times H)$ such that $(u_1, v_1)(u_2, v_2) \in E(G \times H)$, whenever $v_1 = v_2$ and $u_1 u_2 \in E(G)$, or $u_1 = u_2$ and $v_1 v_2 \in E(H)$ [11].

In this paper we follow the following notations and terminologies. The numbers $0, 1, 2, \ldots, n-1$ always denote the vertices of a path P_n or a cycle C_n. Let $\gamma(P_m \square P_n)$ and $\gamma(P_m \square C_n)$

denote the domination numbers of Cartesian product graphs $P_m \Box P_n$ and $P_m \Box C_n$, respectively. Let $(G)_v = G \times \{v\}$, where $v \in V(H)$ and $(H)_u = \{u\} \times H$, where $u \in V(G)$. $(G)_v$ and $(H)_u$ are called the layers of G and H, respectively. Moreover, layer of a dominating set means $D \cap (P_m)_i$ for $i \in V(C_n)$. Next we shall define the term modified concatenation of two dominating sets of $P_m \Box C_{n_1}$ and $P_m \Box C_{n_2}$. If D_1 and D_2 are two dominating sets of $P_m \Box C_{n_1}$ and $P_m \Box C_{n_2}$, respectively, then the modified concatenation of D_1 and D_2 is a subset D of $P_m \Box C_{n_1+n_2}$ such that $D \cap (P_m)_i = D_1 \cap (P_m)_i, i = 0, 1, \ldots, n_1 - 1$, and $D \cap (P_m)_{n_1+i} = D_2 \cap (P_m)_i, i = 0, 1, \ldots, n_2 - 1$; that is, the ith (P_m)-layer of D is coming from the ith (P_m)-layer of D_1 if $0 \le i \le n_1 - 1$ and from the ith (P_m)-layer of D_2 if $n_1 \le i \le n_1 + n_2 - 1$. The illustration of modified concatenation is shown in Figure 1.

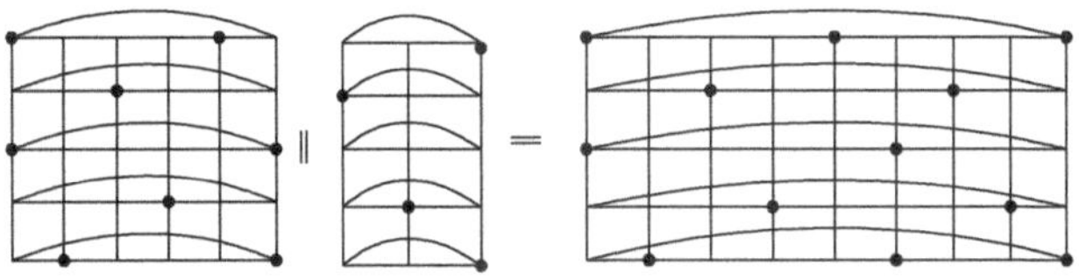

FIGURE 1: Modified concatenation of dominating sets of $P_5 \Box C_6$ and $P_5 \Box C_3$ to get dominating set for $P_5 \Box C_9$.

One of the most challenging problems concerning the domination number of Cartesian products of graphs is the proof of the Vizing Conjecture, namely, $\gamma(G \Box H) \ge \gamma(G)\gamma(H)$ [6]. Despite numerous results showing its validity in some special cases, till date the conjecture remains an open problem. Partial works have been made towards finding the domination numbers of some particular Cartesian products. This problem also seems to be difficult one and the authors of [12] proved that even for subgraphs of $P_m \Box P_n$, this problem is NP-complete. In [13], Jacobson and Kinch established the following results:

For all $n \ge 1$,

(i) $\gamma(P_2 \Box P_n) = \lfloor (n+2)/2 \rfloor$.

(ii) $\gamma(P_3 \Box P_n) = \lfloor (3n+4)/4 \rfloor$.

(iii)

$$\gamma(P_4 \Box P_n) = \begin{cases} n+1, & \text{for } n = 1, 2, 3, 5, 6, 9; \\ n, & \text{otherwise.} \end{cases} \quad (1)$$

In [14], Chang and Clark established the following results:

(i)

$$\gamma(P_5 \Box P_n) = \begin{cases} \left\lfloor \dfrac{6n+6}{5} \right\rfloor, & \text{for } n = 2, 3, 7; \\ \left\lfloor \dfrac{6n+8}{5} \right\rfloor, & \text{otherwise.} \end{cases} \quad (2)$$

(ii)

$$\gamma(P_6 \Box P_n) = \begin{cases} \left\lfloor \dfrac{10n+10}{7} \right\rfloor, & \text{for } n \ge 6 \text{ and } n \equiv 1 \pmod{7}; \\ \left\lfloor \dfrac{10n+12}{7} \right\rfloor, & \text{otherwise if } n \ge 4. \end{cases} \quad (3)$$

In [11], the authors established the following results regarding the Cartesian product of two cycles.

(i) For $n \ge 4$, $\gamma(C_3 \Box C_n) = n - \lfloor n/4 \rfloor$.

(ii) For $n \ge 4$, $\gamma(C_4 \Box C_n) = n$.

(iii) For $n \ge 5$,

$$\gamma(C_5 \Box C_n) = \begin{cases} n, & n = 5k; \\ n+1, & n \in \{5k+1, 5k+2, 5k+4\}. \end{cases} \quad (4)$$

Furthermore, $\gamma(C_5 \Box C_{5k+3}) \le 5(k+1)$.

More works may be found in [8, 9, 15, 16].

In the paper [17], Nandi et al. dealt with the domination number of some special types of graphs, known as cylindrical grid graphs $P_m \Box C_n, m \ge 2, n \ge 3$ as shown in Figure 2. An alternative way of looking at the same cylindrical grid graph is also shown in Figure 2, where the leftmost column in all figures denotes the layer $(P_m)_0$. In that paper the authors found the domination numbers as well as minimum dominating sets of the graphs $P_m \Box C_n$, for $m = 2, 3, 4$ and for all $n \ge 3$ and provided bounds on $\gamma(P_m \Box C_n)$ for $m = 5$ and for all $n \ge 3$. They pose an open problem for finding the domination numbers of the $P_m \Box C_n$, for $m \ge 5$.

In the current paper, we deal with the above-mentioned open problem as posed in [17] and towards solving the problem, we get some partial results. We find the domination numbers as well as minimum dominating sets of the graphs $P_m \Box C_n$, for $m = 5$ and for all $n \ge 3$. We also give tight bounds on $\gamma(P_m \Box C_n)$ for $m = 6$ and for all $n \ge 3$. Moreover, we provide rough bounds on domination numbers of the graphs $P_m \Box C_n$, $n \ge 3$ and $m \ge 7$. We also point out how domination numbers and minimum dominating sets are useful for wireless sensor networks. As a brief summary, we state the following results that we prove in the subsequent sections.

For all $n \ge 3$,

(i) $\gamma(P_5 \Box C_3) = 4$ and for $n \ge 4$, $\gamma(P_5 \Box C_n) = n + \lfloor n/5 \rfloor + p_n$, where

$$p_n = \begin{cases} 0, & \text{if } n \equiv 0 \pmod{10}; \\ 2, & \text{if } n \equiv 3, 9 \pmod{10}; \\ 1, & \text{otherwise.} \end{cases} \quad (5)$$

(ii) $\gamma(P_6 \Box C_3) = 5$, $\gamma(P_6 \Box C_4) = 6$, $\gamma(P_6 \Box C_5) = 8$, $\gamma(P_6 \Box C_6) = 9$, $\gamma(P_6 \Box C_7) = 11$, and for $n > 7$,

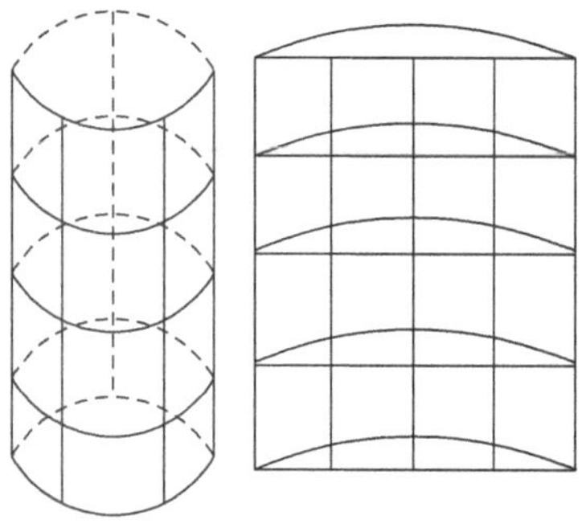

FIGURE 2: Two different looks of the cylindrical grid graph $P_5 \square C_5$.

$$\frac{4n}{3} \le \gamma(P_6 \square C_n) \le \begin{cases} \left\lceil \frac{10n}{7} \right\rceil, & \text{for } n \equiv 0, 4, 5, 8, 10, 11, 12 \pmod{14}; \\ \left\lceil \frac{10n}{7} \right\rceil + 1, & \text{otherwise.} \end{cases} \quad (6)$$

(iii) If $m = 5p + k$ and $n = 5q + l$, where p, q are natural numbers and $k,\ l = 0, 1, 2, 3, 4$, then

$$\left\lceil \frac{mn}{5} \right\rceil = 5pq + kq + pl + \left\lceil \frac{kl}{5} \right\rceil \le \gamma(P_m \square C_n) \le \begin{cases} 5pq + (k+2)q + (l+1)p, & \text{if } l \ne 0 \\ 5pq + (k+2)q, & \text{if } l = 0. \end{cases} \quad (7)$$

2. Finding Minimum Dominating Sets of $P_5 \square C_n$, for All $n \ge 3$

In this section we find the domination numbers as well as minimum dominating sets of particular cylindrical grid graphs of the form $P_m \square C_n$, for all $n \ge 3$ and for $m = 5$. To prove the main results we state the following lemmas and theorems that are proved in [17]. Throughout the paper we use the arithmetic operations of the indices over modulo n.

Lemma 1 (see [17]). *Let $m \ge 2$. Then there exists a minimum dominating set D of $P_m \square C_n$ such that for every $i \in V(C_n)$, $|(P_m)_i \cap D| \le m - 1$.*

Lemma 2 (see [17]). *There cannot be two consecutive P_m-layers having empty intersection with a minimum dominating set of $P_m \square C_n$, for $m \ge 3$ and $n \ge 4$.*

Lemma 3 (see [17]). *For every dominating set D of $P_m \square C_n$, the following m inequalities hold:*

$$x_{i-1} + 3x_i + x_{i+1} \ge n, \qquad \forall i = 0, \ldots, m-1, \quad (8)$$

where $x_i = |(C_n)_i \cap D|$ for $i = 0, 1, \ldots, m-1$ and $x_{-1} = x_m = 0$. Moreover, if $x_{i-1} + 3x_i + x_{i+1} = n$, then there does not exist any pair of vertices from $((C_n)_{i-1} \cup (C_n)_i \cup (C_n)_{i+1}) \cap D$ such that they dominate a common vertex of $(C_n)_i$. Finally, $\sum_{i=0}^{m-1} x_i = |D| \ge \gamma(P_m \square C_n)$.

Remark 4. The similar result holds for $x_i = |(P_m)_i \cap D|$.

Lemma 5 (see [17]). *For $n \ge 3$, there exists a minimum dominating set D of $P_5 \square C_n$ such that for every $i \in V(C_n)$, $|(P_5)_i \cap D| \le 3$.*

Lemma 6 (see [17]). *For $n \ge 5$, there exists a minimum dominating set D of $P_5 \square C_n$ such that for every $i \in V(C_n)$, either (a) $|(P_5)_i \cap D| \le 2$ or (b) $|(P_5)_i \cap D| = 3$ with $(P_5)_{i-1} \cap D = \phi$ and $(P_5)_{i+1} \cap D = \phi$ for $i \in V'$ and $|(P_5)_i \cap D| \le 2$ for all $i \notin V'$, for some $V' \subseteq V(C_n)$.*

Lemma 7 (see [17]). *For $n \ge 5$, there cannot be a dominating set D with five consecutive P_5-layers having exactly one vertex in common with D.*

Lemma 8 (see [17]). *Let D be a minimum dominating set with the property as stated in Lemma 6. Again let $(P_5)_i$ and $(P_5)_j$ be two layers having two vertices in common with D and $|(P_5)_{i+1} \cap D| \ne 2$, $|(P_5)_{i+2} \cap D| \ne 2, \ldots, |(P_5)_{j-1} \cap D| \ne 2$. Then either (a) $|(P_5)_{i+1} \cap D| = 0$, $|(P_5)_{i+2} \cap D| = 3$, $|(P_5)_{i+3} \cap D| = 0$, $|(P_5)_{i+4} \cap D| = 3, \ldots, |(P_5)_{j-1} \cap D| = 0$, or (b) $j - i \le 5$ and $|(P_5)_l \cap D| = 1$ for all $l = (i+1), (i+2), \ldots, (j-1)$.*

Theorem 9 (see [17]). *For $n \ge 6$, $\gamma(P_5 \square C_n) \ge \lceil 6n/5 \rceil$.*

Note 1. Let us call the collection of P_5-layers $\{(P_5)_k : k = i+1, i+2, \ldots, j-1\}$ as a block, where i and j are as in Lemma 8.

Then $|D| = (6n/5) + ((x_3 + 2(x_3 - x_2'') + (4x_2' - x_1) + 4x_2^0)/5)$, where x_1, x_3 denotes the number of P_5-layers having 1 and 3 vertices, respectively, in common with D, x_2' denotes the number of blocks in which every P_5-layer has exactly one vertex in common with D, x_2'' denotes the number of blocks in which every P_5-layer has either 0 or 3 vertices in common with D, and x_2^0 denotes the number of blocks, where $j = i + 1$, that is, when the block contains no P_5-layer.

The above note will be useful to prove Theorem 13.

Theorem 10 (see [17]). *Consider the following:*

(a) $\gamma(P_5 \square C_3) = 4$, (b) $\gamma(P_5 \square C_4) = 5$, and (c) $\gamma(P_5 \square C_5) = 7$.

Using the above-mentioned lemmas and theorems, we are going to prove the following main results.

Theorem 11. *For $n \ge 6$, $\gamma(P_5 \square C_n) \le n + \lfloor n/5 \rfloor + p_n$, where*

$$p_n = \begin{cases} 0, & \text{if } n \equiv 0 \pmod{10}; \\ 2, & \text{if } n \equiv 3, 9 \pmod{10}; \\ 1, & \text{otherwise.} \end{cases} \quad (9)$$

Proof. Consider the dominating sets of $P_5 \square C_4$ and $P_5 \square C_5$ as in Theorem 10 and the dominating sets of $P_5 \square C_n$ for $n = 6, 7, \ldots, 13$ as shown in Figure 3.

Now using modified concatenation of the dominating set for $n = 10$ repeatedly with suitably choosing one of these ten dominating sets, we get dominating sets beyond $n = 13$. For example, to find a dominating set for $C_5 \square P_{17}$, we use modified concatenation of the dominating set $C_5 \square P_{10}$ with

$C_5 \square P_7$. Again to find a dominating set for $C_5 \square P_{27}$, we use modified concatenation of the dominating set $C_5 \square P_{10}$ with $C_5 \square P_{17}$. □

Corollary 12. *For $n \geq 4$,*

$$\gamma(P_5 \square C_n) = \begin{cases} n + \left\lfloor \frac{n}{5} \right\rfloor, & \text{if } n \equiv 0 \,(\text{mod } 10); \\ n + \left\lfloor \frac{n}{5} \right\rfloor + 1, & \text{if } n \equiv 1,2,4,6,7,8 \,(\text{mod } 10). \end{cases} \quad (10)$$

Proof. Using Theorems 9, 10, and 11 we get the desired result. □

Theorem 13. *For $n \geq 5$,*

(1) $\gamma(P_5 \square C_n) = n + \lfloor n/5 \rfloor + 1$, *if* $n \equiv 5(\text{mod } 10)$,

(2) $\gamma(P_5 \square C_n) = n + \lfloor n/5 \rfloor + 2$, *if* $n \equiv 3(\text{mod } 10)$,

(3) $\gamma(P_5 \square C_n) = n + \lfloor n/5 \rfloor + 2$, *if* $n \equiv 9(\text{mod } 10)$.

Proof. To prove the theorem we first recall Note 1. Here for any minimum dominating set D with the property of $P_5 \square C_n$, as stated in Lemma 8, we get $|D| = (6n/5) + ((x_3 + 2(x_3 - x_2'') + (4x_2' - x_1) + 4x_2^0)/5)$, where the symbols are as in Note 1.

(a) Let $n \equiv 5(\text{mod}10)$; that is, $n = 5(2p+1)$, $p = 0, 1, 2, \ldots$.

We claim that for any minimum dominating set D of $P_5 \square C_n$ with the property as stated in Lemma 8, $x_3 + 2(x_3 - x_2'') + (4x_2' - x_1) + 4x_2^0 > 0$. Otherwise, there will exist a minimum dominating set D of $P_5 \square C_n$ with the same property such that $x_3 + 2(x_3 - x_2'') + (4x_2' - x_1) + 4x_2^0 = 0$. Then $x_3 = x_2'' = x_2^0 = 0$ and $4x_2' = x_1$ (since $x_3 \geq x_2''$, $4x_2' \geq x_1$, $x_2^0 \geq 0$).

Hence D has the following properties:

(1) each P_5-layers has exactly 1 or 2 vertices in common with D,

(2) each block has exactly 4 P_5-layers; that is, without loss of generality we can write

$$|(P_5)_i \cap D| = \begin{cases} 2, & \text{if } i \equiv 0 \,(\text{mod } 5); \\ 1, & \text{otherwise.} \end{cases} \quad (11)$$

Since $|(P_5)_{5k} \cap D| = 2$ and $|(P_5)_{5k+1} \cap D| = |(P_5)_{5k+2} \cap D| = |(P_5)_{5k+3} \cap D| = |(P_5)_{5k+4} \cap D| = 1$ and $|(P_5)_{5k+5} \cap D| = 2$, the number of vertices dominated by each of the vertices of $(P_5)_{5k+2} \cap D$ and $(P_5)_{5k+3} \cap D$ will be 4 (excluding themselves) and no vertex will be dominated by both of them simultaneously. Therefore $\bigcup_{j=0}^{5}[(P_5)_{5k+j} \cap D] = \{(0,5k), (0,5k+4), (1,5k+2), (2,5k), (2,5k+5), (3,5k+3), (4,5k+1), (4,5k+5)\}$ or $\{(0,5k+1), (0,5k+5), (1,5k+3), (2,5k), (2,5k+5), (3,5k+2), (4,5k), (4,5k+4)\}$ are the only two possibilities (as shown in Figure 4) of $\bigcup_{j=0}^{5}[(P_5)_{5k+j} \cap D]$.

Now, without loss of generality let $(0,0), (2,0) \in D$. Then $(0,10p), (2,10p) \in D$.

Therefore $(1,10p+2), (3,10p+3) \in D$ and $(0,10p+2) \notin D$; $(l,10p+3) \notin D$ for all $l = 0,1,2,4$. Consider $(0,10p+4) \in D$ and $(l,10p+4) \notin D$ for all $l = 1,2,3,4$.

Hence $(4,10p+4)$ cannot be dominated by the vertices of D, contradicting that D is a dominating set.

Therefore $x_3 + 2(x_3 - x_2'') + (4x_2' - x_1) + 4x_2^0 > 0$.

Hence for any minimum dominating set D, $|D| > 6n/5 = 6(2p+1)$; that is, $|D| \geq 6(2p+1) + 1$. Therefore $\gamma(P_5 \square C_n) \geq 6(2p+1)+1 = n + \lfloor n/5 \rfloor + 1$. The rest of the proof of (a) follows from Theorem 11.

We note the following observation.

Observation 1. If $|(P_5)_j \cap D| = |(P_5)_{j+5} \cap D| = 2$, $|(P_5)_{j+1} \cap D| = |(P_5)_{j+2} \cap D| = |(P_5)_{j+3} \cap D| = |(P_5)_{j+4} \cap D| = 1$, then (i) $(0,j)$ and $(2,j) \in D$ imply $(2,j+5), (4,j+5) \in D$ and (ii) $(2,j)$ and $(4,j) \in D$ imply $(0,j+5), (2,j+5) \in D$.

(b) Let $n \equiv 3(\text{mod } 10)$; that is, $n = 10p + 3$.

We claim that $x_3 + 2(x_3 - x_2'') + (4x_2' - x_1) + 4x_2^0 > 2$ for any minimum dominating set D of $P_5 \square C_n$ with the property as stated in Lemma 8.

If possible let there exist a minimum dominating set D with the property as stated in Lemma 8 and $x_3 + 2(x_3 - x_2'') + (4x_2' - x_1) + 4x_2^0 = 0$. Then $x_2'' = x_3 = 0$ and $x_1 = 4x_2'$, $x_2^0 = 0$ which will be impossible since $n \not\equiv 0(\text{mod } 5)$.

If possible let there exist a minimum dominating set D with the property as stated in Lemma 8 and $x_3 + 2(x_3 - x_2'') + (4x_2' - x_1) + 4x_2^0 = 1$. Then $x_2'' = x_3$ and $x_2^0 = 0$. Therefore $x_2'' = x_3 = 1$, $x_1 = 4x_2'$, and $x_2^0 = 0$ or $x_2'' = x_3 = x_2^0 = 0$ and $x_1 = 4x_2' - 1$.

Now if $x_2'' = x_3 = 1$ and $x_1 = 4x_2'$ then let x_0 denote the number of P_5-layers having 0 vertices in common with D and $x_0 = x_2'' + x_3 = 2$.

By Lemma 8, $x_2 = x^0 + x_2' + x_2''$ is the number of P_5-layers having 2 vertices in common with D and $n = x_0 + x_1 + x_2 + x_3 = x_0 + x_1 + x_2^0 + x_2' + x_2'' + x_3 = 2 + 4x_2' + x_2' + 1 + 1 = 5x_2' + 4 \not\equiv 3(\text{mod } 5)$ which is a contradiction.

Again if $x_2'' = x_3 = 0$ and $x_1 = 4x_2' - 1$ then $n = x_0 + x_1 + x_2^0 + x_2' + x_2'' + x_3 = 4x_2' - 1 + x_2' \not\equiv 3(\text{mod } 5)$, which is also a contradiction.

If possible let there exist a minimum dominating set D with the property as stated in Lemma 8 and $x_3 + 2(x_3 - x_2'') + (4x_2' - x_1) + 4x_2^0 = 2 \ldots (A)$. Then $x_2'' = x_3$ (otherwise, $x_3 \neq x_2''$ implies $x_3 - x_2'' \geq 1$ and $x_3 \geq 1$, contradicting (A)) and $x_2^0 = 0$.

Now there are the following three cases.

Case 1. Consider $x_3 = x_2'' = x_2^0 = 0$ and $x_1 = 4x_2' - 2$.

Since $x_3 = x_2'' = x_2^0 = 0$ each block contains only those P_5-layers which have exactly one vertex in common with D.

Again since $x_1 = 4x_2' - 2$, there are two sub cases.

Subcase 1.1. When only one block contains two P_5-layers and other block contains four P_5-layers then without loss of generality let $|(P_5)_j \cap D| = 2$ for $j = 0, 5, 10, \ldots, 10p$ and $|(P_5)_j \cap D| = 1$, otherwise. Now $(0,0), (2,0) \in D$ or $(2,0), (4,0) \in D$. If possible let $(0,0), (2,0) \in D$. Then $(0,10p), (2,10p) \in D$ and $(j,10p) \notin D$ for $j = 1,3,4$. But

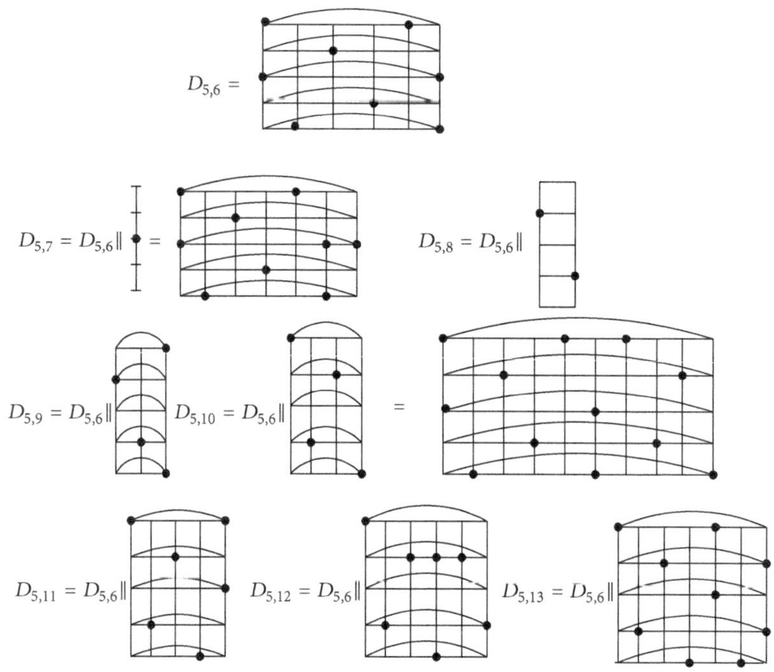

FIGURE 3: Figures showing a particular dominating set for $P_5\square C_i, i = 6, 7, \ldots, 13$, with $|D_{5,6}| = 8$, $|D_{5,7}| = 9$, $|D_{5,8}| = 10$, $|D_{5,9}| = 12$, $|D_{5,10}| = 12$, $|D_{5,11}| = 14$, $|D_{5,12}| = 15$, and $|D_{5,13}| = 17$.

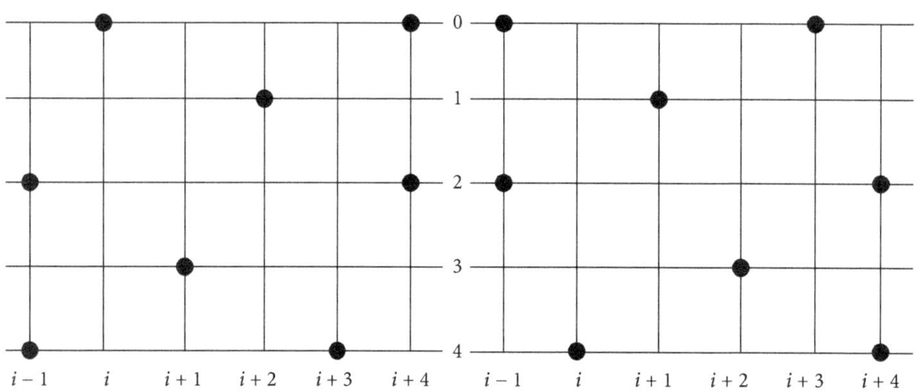

FIGURE 4: The only two possible configurations of the vertices of a minimum dominating set of the graph $P_5\square C_n$ in four consecutive layers P_i, P_{i+1}, P_{i+2}, and P_{i+3}.

this is not possible since $|(P_5)_{10p+1} \cap D| = |(P_5)_{10p+2} \cap D| = 1$ and $(0, 10p + 3), (2, 10p + 3) \in D$ and $(j, 10p + 3) \notin D$ for $j = 1, 3, 4$. Note that $(P_5)_{10p+3} = (P_5)_0$.

Similar contradiction will be arrived for $(2, 0), (4, 0) \in D$.

Subcase 1.2. When only two blocks contain three P_5-layers and other blocks contain four P_5-layers then if

(i) the blocks containing three P_5-layers, each occurs consecutively then without loss of generality let $|(P_5)_j \cap D| = 2$ for $j = 0, 5, 10, \ldots, 10p-5, 10p-1$ and $|(P_5)_j \cap D| = 1$, otherwise. Now if $(0, 0), (2, 0) \in D$ then $(2, 10p - 5), (4, 10p - 5) \in D \Rightarrow (1, 10p + 1), (3, 10p - 3) \in D \Rightarrow (0, 10p - 1), (4, 10p - 1) \in D$ which is a contradiction.Similar contradiction will be arrived for $(2, 0), (4, 0) \in D$,

(ii) the blocks containing three P_5-layers do not occur consecutively then among them one of the blocks consist of $j + 1, j + 2, j + 3$ P_5-layer for some j and either (a) $(0, j), (2, j) \in D$ and $(2, j+4), (4, j+4) \in D$ or (b) $(2, j), (4, j) \in D$ and $(0, j + 4), (2, j + 4) \in D$. In each case, we arrive at a contradiction. Hence we conclude that Case 1 cannot occur.

Case 2. Consider $x_3 = x_2'' = 1, x_2^0 = 0$, and $x_1 = 4x_2' - 1$.

In this case only two blocks contain three P_5-layers and other blocks contain four P_5-layers. Among these two blocks one contains only those P_5-layers which have exactly one vertex in common with D and other block contains three layers whose number of vertices are 0, 3, 0 consecutively.

Now we note the following observations.

Observation 2. If $|(P_5)_j \cap D| = |(P_5)_{j+4} \cap D| = 2$, $|(P_5)_{j+2} \cap D| = 3$, and $|(P_5)_{j+1} \cap D| = |(P_5)_{j+3} \cap D| = 0$ then $(0, j)$, $(2, j) \in D$ implies $(0, j+4), (2, j+4) \in D$ and $(2, j), (4, j) \in D$ implies $(2, j+4), (4, j+4) \in D$.

Observation 3. If $|(P_5)_j \cap D| = |(P_5)_{j+4} \cap D| = 2$, $|(P_5)_{j+1} \cap D| = |(P_5)_{j+2} \cap D| = |(P_5)_{j+3} \cap D| = 1$ then $(0, j)$, $(2, j) \in D$ implies $(0, j+4)$, $(2, j+4) \in D$ or $(0, j+4)$, $(3, j+4) \in D$ and $(2, j)$, $(4, j) \in D$ implies $(2, j+4)$, $(4, j+4) \in D$ or $(1, j+4)$, $(4, j+4) \in D$.

Without loss of generality we now assume that $|(P_5)_0 \cap D| = |(P_5)_4 \cap D| = 2$, $|(P_5)_1 \cap D| = |(P_5)_2 \cap D| = |(P_5)_3 \cap D| = 1$. Therefore by Observation 2 and Observation 1 in Case (a) we have either $(0,0), (2,0) \in D$ and $(2,4), (4,4) \in D$ or $(2,0), (4,0) \in D$ and $(0,4), (2,4) \in D$ which is again a contradiction by Observation 3.

Case 3. Consider $x_3 = x_2'' = 2, x_2^0 = 0$, and $x_1 = 4x_2'$.

In this case only two blocks contain three P_5-layers and other blocks contain four P_5-layers and these two blocks contain three layers whose number of vertices is 0, 3, 0 consecutively. Therefore by the Observation 2, as in Case 2, we arrive at a contradiction.

Then $|D| = (6n/5) + ((x_3 + 2(x_3 - x_2'') + (4x_2' - x_1) + 4x_2^0)/5) > ((60p + 18)/5) + (2/5) = 12p + 4$.

Therefore $|D| \geq 12p + 5 = 10p + 3 + 2p + 2 = n + \lfloor n/5 \rfloor + 2$. This completes the proof of (*b*).

(c) Let $n \equiv 9 (\text{mod } 10)$; that is, $n = 10p + 9$.

We claim that for any minimum dominating set D of $P_5 \square C_n$ with the property as stated in Lemma 8, $x_3 + 2(x_3 - x_2'') + (4x_2' - x_1) + 4x_2^0 > 1$.

If possible let $x_3 + 2(x_3 - x_2'') + (4x_2' - x_1) + 4x_2^0 = 0$. Then as in (b), $n \not\equiv 0 (\text{mod } 5)$. As a result, we have the same contradiction.

If possible let $x_3 + 2(x_3 - x_2'') + (4x_2' - x_1) + 4x_2^0 = 1$. Then $x_3 = x_2''$ and therefore the following two cases arise.

Case 1. Consider $x_3 = x_2'' = x_2^0 = 0$ and $x_1 = 4x_2' - 1$.

Case 2. Consider $x_3 = x_2'' = 1$, $x_1 = 4x_2'$, and $x_2^0 = 0$.

In these two cases contradiction can be shown similarly as in (b), where $n \equiv 3 (\text{mod } 10)$.

Hence $|D| = (6n/5) + ((x_3 + 2(x_3 - x_2'') + (4x_2' - x_1) + 4x_2^0)/5) > ((60p + 54)/5) + (1/5) = 12p + 11$.

Therefore $|D| \geq 12p + 12 = n + \lfloor n/5 \rfloor + 2$. This completes the proof of (c). □

Remark 14. For $n \geq 3$ and $n \neq 7$,

$$\gamma (P_5 \square C_n) = \begin{cases} \gamma (P_5 \square P_n) - 1, & \text{when } n \equiv 0, 2, 4, 7, 8 \ (\text{mod } 10); \\ \gamma (P_5 \square P_n), & \text{otherwise,} \end{cases} \quad (12)$$

and $\gamma(P_5 \square C_7) = \gamma(P_5 \square P_7)$. Moreover, for $n \geq 3$, $\gamma(P_5 \square C_n) > \gamma(P_5) \cdot \gamma(C_n) = 2 \cdot \lceil n/3 \rceil$.

3. Bounds on Domination Numbers of $P_6 \square C_n$ for All $n \geq 3$

In this section we give upper and lower bounds of the domination numbers of $P_6 \square C_n$. Towards this direction, we first prove the following lemmas.

Lemma 15. *Let $n \geq 4$. Then there exists a minimum dominating set D of $P_6 \square C_n$ with $|(P_6)_i \cap D| \leq 4$ $\forall i$.*

Proof. Let D be a minimum dominating set of $P_6 \square C_n$ with the property $|(P_6)_i \cap D| \leq 5$ $\forall i$. Such a D exists by Lemma 1. Suppose further that $|(P_6)_j \cap D| = 5$ for k many P_6 layers.

Assume that $|(P_6)_i \cap D| = 5$ for some i, then $|(P_6)_{i+1} \cap D| \leq 2$; otherwise $D' = D \setminus [(P_6)_i \cup (P_6)_{i+1}] \cup \{(0, i), (1, i+2), (2, i-1), (3, i+1), (4, i-1), (5, i), (5, i+2)\}$ is a dominating set with $|D'| < |D|$, which is a contradiction.

Similar contradiction shows that $|(P_6)_{i-1} \cap D| \leq 2$. Now construct $D'' = D \setminus (P_6)_i \cup \{(0, i), (2, i-1), (2, i+1), (4, i), (5, i)\}$. Then $|D''| \leq |D|$ and hence D'' is a minimum dominating set with $k - 1$ many P_6 layers having 5 vertices in common with D''. Repeating this construction we get the desired minimum dominating set and hence the lemma follows. □

Lemma 16. *For any minimum dominating set D, $|(P_6)_i \cap D| = 1 = |(P_6)_{i+1} \cap D|$ $\forall i$ implies $|(P_6)_{i+2} \cap D| \geq 2$ and $|(P_6)_{i-1} \cap D| \geq 2$.*

Proof. The proof of the lemma follows from Remark 4. □

Lemma 17. *For $n \geq 4$, there exists a minimum dominating set D of $P_6 \square C_n$ with the property (i) if $|(P_6)_i \cap D| = 1$, $|(P_6)_{i+1} \cap D| = 1$, $|(P_6)_{i+2} \cap D| = 2$ for some i, then $|(P_6)_{i+3} \cap D| \geq 1$;*

(ii) if $|(P_6)_i \cap D| = 2$, $|(P_6)_{i+1} \cap D| = 1$, $|(P_6)_{i+2} \cap D| = 1$ for some i, then $|(P_6)_{i-1} \cap D| \geq 1$ and (iii) $|(P_6)_i \cap D| \leq 4 \forall i$.

Proof. Let D be a minimum dominating set of $P_6 \square C_n$ with the property $|(P_m)_i \cap D| \leq 4$ $\forall i$. Such a D exists by Lemma 15. Let $x_i = |(P_6)_i \cap D|, i = 0, 1, \ldots, n-1$.

When $n = 4$, then $x_i = 1, x_{i+1} = 1$, and $x_{i+2} = 2$ imply that $x_{i+3} \geq 1$ since by Remark 4, $x_{i-1} + 3x_i + x_{i+1} \geq 6$. Note that in this case, $(P_6)_{i-1} = (P_6)_{i+3}$ and hence $x_{i-1} = x_{i+3}$.

Let $n \geq 5$. If $x_i = 1, x_{i+1} = 1, x_{i+2} = 2$, and $x_{i+3} = 0$ then $x_{i+4} = 4$. Also, $x_{i+5} \leq 3$; otherwise $D' = D \setminus [(P_6)_{i+4} \cup (P_6)_{i+5}] \cup \{(0, i+4), (1, i+6), (2, i+3), (2, i+5), (4, i+4), (4, i+6), (5, i+4)\}$ will be a dominating set with $|D'| < |D|$, a contradiction. Clearly, $(0, i+1), (5, i+1) \notin D$. Also $(1, i+1), (4, i+1) \notin D$, because if $(1, i+1) \in D$, then $(0, i+2) \in D$ or $(1, i+2) \in D$, each of which will be a contradiction, since $(3, i+1)$, $(4, i+1)$ and $(5, i+1)$ cannot be dominated by one vertex from $(P_6)_i$ and one vertex from $(P_6)_{i+2}$. Similar contradiction will be arrived for $(4, i+1) \in D$. Let $(2, i+1) \in D$. Then $(0, i+2), (4, i+2) \in D$. Hence we have $(1, i+4), (2, i+4), (3, i+4), (5, i+4) \in D$. Therefore we can construct $D' = [D \setminus \{(2, i+4), (3, i+4)\}] \cup \{(2, i+3), (3, i+5)\}$. In a similar way we can construct D' when $(3, i+2) \in D$. Repeating this construction for all i for which $x_i = x_{i+1} = 1, x_{i+2} = 2, x_{i+3} = 0$, and $x_{i-1} = 0, x_i = 2, x_{i+1} = 1, x_{i+2} = 1$, we get a minimum dominating set with the desired property. □

Lemma 18. *Consider a minimum dominating set D of $P_6\Box C_n$ with the property as stated in Lemma 17. Let $x_i = |(P_6)_i \cap D|, i = 0, 1, \ldots, n-1$.*

Then $(x_i/2) + x_{i+1} + x_{i+2} + ((x_{i+3})/2) \geq 4$.

Proof. Case 1. Consider $x_{i+1} = 0$; this implies $x_{i+2} \geq 2$ (since $x_i \leq 4$).

Subcase 1.1. Consider $x_{i+2} = 2$; this implies $x_i = 4$.

Subcase 1.2. Consider $x_{i+2} = 3$; this implies $x_i \geq 3$.

Subcase 1.3. Consider $x_{i+2} = 4$.

For every subcase we get the desired result.

Case 2. Consider $x_{i+1} = 1$; this implies $x_{i+2} \geq 1$.

Subcase 2.1. Consider $x_{i+2} = 1$; this implies $x_i \geq 2$, $x_{i+3} \geq 2$.

Subcase 2.2. Consider $x_{i+2} = 2$; this implies $x_i \geq 1$ which implies $x_i + x_{i+3} \geq 2$ (since if $x_i = 1$ then by Lemma 17 $x_{i+3} \geq 1$).

Subcase 2.3. Consider $x_{i+2} \geq 3$.

For every subcase we get the desired result.

Case 3. Consider $x_{i+1} = 2$.

Subcase 3.1. Consider $x_{i+2} = 0$; this implies $x_{i+3} \geq 4$.

Subcase 3.2. Consider $x_{i+2} = 1$; this implies $x_{i+3} \geq 1$ which implies $x_i + x_{i+3} \geq 2$ (by Lemma 17).

Subcase 3.3. Consider $x_{i+2} \geq 2$.

For every subcase we get the desired result.

Case 4. Consider $x_{i+1} = 3$.

Subcase 4.1. Consider $x_{i+2} = 0$; this implies $x_{i+3} \geq 3$ as in Subcase 1.2.

Subcase 4.2. Consider $x_{i+2} \geq 1$.

For every subcase we get the desired result.

Case 5. Consider $x_{i+1} = 4$.

In this case also $(x_i/2) + x_{i+1} + x_{i+2} + (x_{i+3}/2) \geq 4$. □

Based on Lemma 17, we prove the following theorem which provides a lower bound on $\gamma(P_6\Box C_n)$, for $n \geq 3$.

Theorem 19. *For $n \geq 3$, $\gamma(P_6\Box C_n) \geq 4n/3$.*

Proof. Case 1. When $n = 3$, for any dominating set D, $|D| \geq 18/5 > 3$ (since the total number of vertices is $6 \times 3 = 18$ and one vertex of D can dominate at most 5 vertices including itself). Therefore $|D| \geq 4 = ((4 \times 3)/3)$. Hence the theorem is true for $n = 3$.

Case 2. For $n \geq 4$ consider a minimum dominating set D of $P_6\Box C_n$ with the property as stated in Lemma 17. Then $(x_i/2) + x_{i+1} + x_{i+2} + (x_{i+3}/2) \geq 4\ \forall i = 0, 1, \ldots, n-1$. Therefore $\sum_{i=0}^{n-1}\{(x_i/2) + x_{i+1} + x_{i+2} + (x_{i+3}/2)\} \geq 4n$ that implies $3\sum_{i=0}^{n-1} x_i \geq 4n$ that implies $\sum_{i=0}^{n-1} x_i \geq 4n/3$; that is, $|D| \geq 4n/3$, hence the theorem. □

Remark 20. For simplification, henceforth, in all the figures, we are only considering the grid structure avoiding the circular arks to represent the circular grid as shown in Figure 5.

The following theorem provides domination numbers and dominating sets for some particular cyclic grid graphs which will help in providing the upper bounds for $\gamma(P_6\Box C_n)$, for $n > 7$ as addressed in Theorem 23.

Theorem 21. *Consider the following:*

(a) $\gamma(P_6\Box C_3) = 5$, (b) $\gamma(P_6\Box C_4) = 6$, (c) $\gamma(P_6\Box C_5) = 8$, (d) $\gamma(P_6\Box C_6) = 9$, and (e) $\gamma(P_6\Box C_7) = 11$.

Proof. (a) A dominating set for $P_6\Box C_3$ is $D_{6,3} = \{(0,1), (2,0), (2,2), (4,1), (5,1)\}$, as shown in Figure 6. As $|D_{6,3}| = 5$, $\gamma(P_6\Box C_3) \leq 5$. We want to show $\gamma(P_6\Box C_3) = 5$.

If possible let there exist a dominating set D, of $P_6\Box C_3$ with $|D| \leq 4$.

Let $x_i = |(C_3)_i \cap D|$, $i = 0, 1, 2, 3, 4, 5$.

Therefore,

(i) $x_0 + x_1 + \cdots + x_5 \leq 4$,

(ii) $3x_0 + x_1 \geq 3$,

(iii) $x_0 + 3x_1 + x_2 \geq 3$,

(iv) $x_1 + 3x_2 + x_3 \geq 3$,

(v) $x_2 + 3x_3 + x_4 \geq 3$,

(vi) $x_3 + 3x_4 + x_5 \geq 3$,

(vii) $x_4 + 3x_5 \geq 3$.

We claim that $x_0 \geq 1$; otherwise, if possible let $x_0 = 0$; then $x_1 \geq 3$.

Therefore, $x_2 + x_3 + x_4 + x_5 \leq 1$.

This implies $x_4 + x_5 \leq 1$ and hence $x_5 \geq 1$ (from (vii)).

Therefore, $x_5 = 1$ and then $x_1 + x_5 \geq 4$ implies $x_0 = x_2 = x_3 = x_4 = 0$, contradicting (v), hence the claim.

Similarly, we can show that $x_5 \geq 1$.

Therefore, $x_1 + x_2 + x_3 + x_4 \leq 2 \Rightarrow x_3 \geq 1, x_2 \geq 1$ (using (iv) and (v)).

Hence, $x_0 = x_2 = x_3 = x_5 = 1$ which implies $x_1 = x_4 = 0$, contradicting (iii).

Therefore, $\gamma(P_6\Box C_3) = 5$.

(b) A dominating set for $P_6\Box C_4$ is $D_{6,4} = \{(0,3), (1,1), (2,3), (3,0), (4,2), (5,0)\}$, as shown in Figure 6. As $|D_{6,4}| = 6$, $\gamma(P_6\Box C_4) \leq 6$. We want to show $\gamma(P_6\Box C_4) = 6$.

If possible let there exist a dominating set D of $P_6\Box C_4$ with $|D| \leq 5$.

Let $x_i = |(C_4)_i \cap D|$, $i = 0, 1, 2, 3, 4, 5$.

Therefore,

(i) $x_0 + x_1 + \cdots + x_5 \leq 5$,

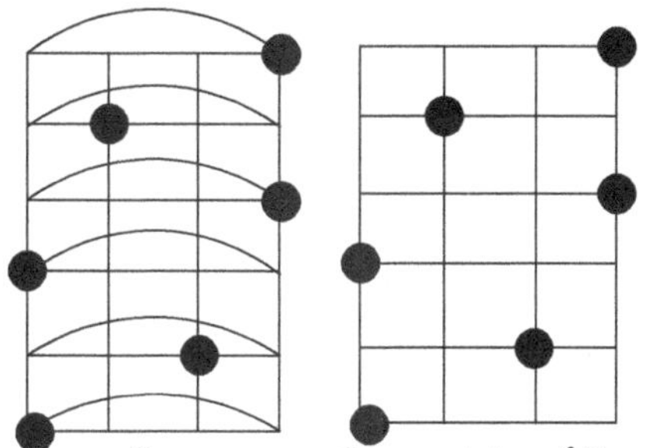

FIGURE 5: Dominating set for $P_6 \square C_4$ and its representation in grid form without circular arks.

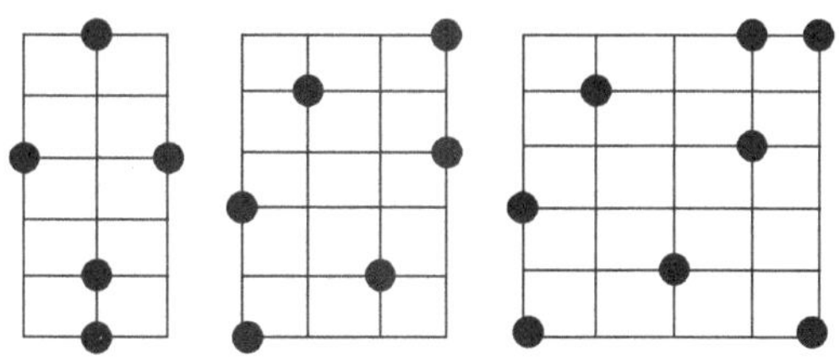

FIGURE 6: Dominating sets for $P_6 \square C_3$, $P_6 \square C_4$, and $P_6 \square C_5$.

(ii) $3x_0 + x_1 \geq 4$,

(iii) $x_0 + 3x_1 + x_2 \geq 4$,

(iv) $x_1 + 3x_2 + x_3 \geq 4$,

(v) $x_2 + 3x_3 + x_4 \geq 4$,

(vi) $x_3 + 3x_4 + x_5 \geq 4$,

(vii) $x_4 + 3x_5 \geq 4$.

We claim that $x_0 \geq 1$; otherwise, if possible let $x_0 = 0$. Then $x_1 \geq 4$, which implies $x_2+x_3+x_4+x_5 \leq 1$, contradicting (vii), hence the claim.

Therefore, $x_0 \geq 1$. Similarly, we can show that $x_5 \geq 1$. Next we claim that $x_1 \geq 1$. If possible let $x_1 = 0$. Therefore, $x_0 + x_2 \geq 4 \Longrightarrow x_3 + x_4 + x_5 \leq 1$, contradicting (vii), hence the claim. Therefore $x_1 \geq 1$. Similarly, $x_5, x_3, x_4 \geq 1$, contradicting (i).

Therefore, $\gamma(P_6 \square C_4) = 6$.

(c) A dominating set for $P_6 \square C_5$ is $D_{6,5} = \{(0,3), (0,4), (1,1), (2,3), (3,0), (4,2), (5,0), (5,4)\}$ as shown in Figure 6.

As $|D_{6,5}| = 8$, $\gamma(P_6 \square C_5) \leq 8$. We want to show $\gamma(P_6 \square C_5) = 8$.

If possible let there exist a dominating set D of $P_6 \square C_5$ with $|D| \leq 7$.

Let $x_i = |(C_4)_i \cap D|$, $i = 0, 1, 2, 3, 4, 5$.

Therefore,

(i) $x_0 + x_1 + \cdots + x_5 \leq 7$,

(ii) $3x_0 + x_1 \geq 5$,

(iii) $x_0 + 3x_1 + x_2 \geq 5$,

(iv) $x_1 + 3x_2 + x_3 \geq 5$,

(v) $x_2 + 3x_3 + x_4 \geq 5$,

(vi) $x_3 + 3x_4 + x_5 \geq 5$,

(vii) $x_4 + 3x_5 \geq 5$.

We claim that $x_0 \geq 1$; otherwise, if possible let $x_0 = 0$. Then $x_1 \geq 5$, which implies $x_2 + x_3 + x_4 + x_5 \leq 2$.

From (vii) we get, $x_5 = 2$ and $x_4 = 0$; hence, $x_2 = x_3 = 0$, a contradiction.

Similarly, we can show that $x_5 \geq 1$. Now we claim that $x_1 \geq 1$; otherwise, $x_0 + x_2 \geq 5$, which implies $x_3 + x_4 + x_5 \leq 2$. Therefore, from (vii), $x_5 = 2$ and $x_4 = 0 = x_3$, contradicting (vi). Similarly, $x_4, x_2, x_3 \geq 1$. Next we claim that $x_0 \geq 2$; otherwise, $x_0 = 1$ and $x_1 \geq 2$. Therefore, $x_0 = x_2 = x_3 = x_4 = x_5 = 1$ and $x_1 = 2$, contradicting (vii). Similarly, $x_5 \geq 2$, contradicting (i).

(d) A dominating set for $P_6 \square C_6$ is $D_{6,6} = \{(0,0), (0,4), (1,2), (2,5), (3,1), (3,3), (4,0), (5,2), (5,4)\}$ as shown in Figure 7. As $|D_{6,6}| = 9$, $\gamma(P_6 \square C_6) \leq 9$. We want to show $\gamma(P_6 \square C_6) = 9$.

If possible let there exist a dominating set D of $P_6 \square C_6$ with $|D| \leq 8$.

Let $x_i = |(C_4)_i \cap D|$, $i = 0, 1, 2, 3, 4, 5$.

Therefore,

(i) $x_0 + x_1 + \cdots + x_5 \leq 8$,

(ii) $3x_0 + x_1 \geq 6$,

(iii) $x_0 + 3x_1 + x_2 \geq 6$,

(iv) $x_1 + 3x_2 + x_3 \geq 6$,

(v) $x_2 + 3x_3 + x_4 \geq 6$,

(vi) $x_3 + 3x_4 + x_5 \geq 6$,

(vii) $x_4 + 3x_5 \geq 6$.

We claim that $x_0 \geq 1$; otherwise, if possible let $x_0 = 0$. Then $x_1 \geq 6$, which implies $x_4 + x_5 \leq 2$ and hence $x_5 = 2$ (by (vii)).

Hence, $x_0 = x_2 = x_3 = x_4 = 0$, a contradiction.

Therefore, $x_0 \geq 1$. Similarly we can show that $x_5 \geq 1$.

Now we claim that $x_1 \geq 1$; otherwise, $x_0 + x_2 \geq 6$, which implies that $x_4 + x_5 \leq 2$. Therefore, $x_5 = 2$ (by (vii)) and hence $x_1 = x_3 = x_4 = 0$, contradicting (vi). Hence, $x_1 \geq 1$. Similarly $x_2, x_3, x_4 \geq 1$.

Next we claim that $x_0 \geq 2$; otherwise, let $x_0 = 1$. Then $x_1 \geq 3$.

Therefore, $x_0 = x_2 = x_3 = x_4 = x_5 = 1$ and hence $x_1 = 3$ which contradicts (vii).

Hence, $x_0 \geq 2$. Similarly, $x_5 \geq 2$. Therefore, $x_0 = x_5 = 2$ and $x_1 = x_2 = x_3 = x_4 = 1$, which contradicts (iv). Hence, $\gamma(P_6 \square C_6) = 9$.

(e) A dominating set for $P_6 \square C_7$ is $D_{6,7} = \{(0,0), (0,4), (1,2), (2,5), (2,6), (3,1), (3,3), (4,0), (4,6), (5,2), (5,4)\}$ as shown in Figure 7. As $|D_{6,7}| = 11$, $\gamma(P_6 \square C_7) \leq 11$. We want to show $\gamma(P_6 \square C_7) = 11$.

If possible let there exist a dominating set D of $P_6 \square C_7$ with $|D| \leq 10$.

Let $x_i = |(C_4)_i \cap D|$, $i = 0, 1, 2, 3, 4, 5$.

Therefore,

(i) $x_0 + x_1 + \cdots + x_5 \leq 10$,

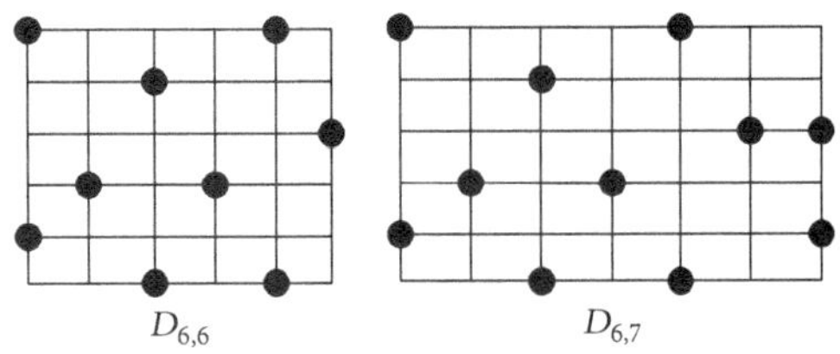

FIGURE 7: Dominating sets for $P_6\square C_6$ and $P_6\square C_7$.

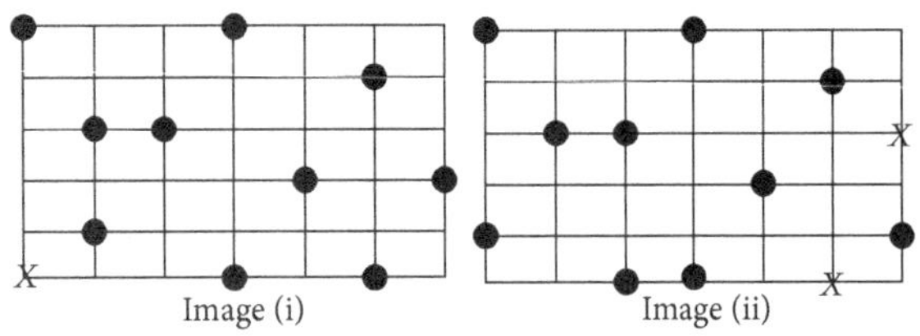

FIGURE 8: Figures show nonpossibility of dominating sets in some cases.

(ii) $3x_0 + x_1 \geq 7$,

(iii) $x_0 + 3x_1 + x_2 \geq 7$,

(iv) $x_1 + 3x_2 + x_3 \geq 7$,

(v) $x_2 + 3x_3 + x_4 \geq 7$,

(vi) $x_3 + 3x_4 + x_5 \geq 7$,

(vii) $x_4 + 3x_5 \geq 7$.

By similar arguments, as in (d) we can prove, $x_0, x_5 \geq 2$ and $x_1, x_2, x_3, x_4 \geq 1$.

We claim that $x_1 \leq 2$. If possible let $x_1 \geq 3$. Then $x_0 = x_5 = 2$, $x_1 = 3$, and $x_2 = x_3 = x_4 = 1$, contradicting (v), which implies that $x_1 \leq 2$. Similarly, $x_2, x_3, x_4 \leq 2$.

Next we claim that $x_0 \leq 2$; otherwise, let $x_0 \geq 3$. Now clearly $x_0 \leq 3$ and hence $x_0 = 3$.

Hence, we have the following cases: $(x_1, x_2, x_3, x_4, x_5) = (2, 1, 1, 1, 2)$ or $(1, 2, 1, 1, 2)$ or $(1, 1, 2, 1, 2)$ or $(1, 1, 1, 2, 2)$ or $(1, 1, 1, 1, 3)$ and each of the cases lead to contradictions.

Therefore, $x_0 = 2$. Similarly, we can show that $x_5 = 2$.

Hence the possibilities are $(x_1, x_2, x_3, x_4) = (1, 1, 2, 2)$ or $(2, 2, 1, 1)$ or $(2, 1, 1, 2)$ or $(1, 2, 2, 1)$ or $(2, 1, 2, 1)$ or $(1, 2, 1, 2)$.

It is easy to show that first three cases lead to contradictions.

Hence the possibilities become $(x_0, x_1, x_2, x_3, x_4, x_5) = (2, 1, 2, 2, 1, 2)$ or $(2, 2, 1, 2, 1, 2)$ or $(2, 1, 2, 1, 2, 2)$.

If possible let there exist a dominating set D with $(x_0, x_1, x_2, x_3, x_4, x_5) = (2, 1, 2, 2, 1, 2)$.

Without loss of generality, let $(0, 0) \in D$.

Therefore, $(0, 1), (0, 2), (0, 5), (0, 6) \notin D$.

Without loss of generality, let $(0, 3) \in D$. Therefore $(1, 5) \in D$. This implies $(2, 1), (2, 2) \in D$.

Hence, $(3, 4), (3, 6) \in D$; then, $(4, 0), (4, 3), (4, 4), (4, 5), (4, 6) \notin D$.

Now if $(4, 1) \in D$, then, $(5, 3), (5, 5), \in D$. But then $(5, 0)$ is not dominated by any vertex of D as shown in Figure 8.

And if $(4, 2) \in D$, then, $(5, 0), (5, 5), \in D$. But then $(5, 3)$ is not dominated by any vertex of D.

Hence there does not exist any dominating set D with $(x_0, x_1, x_2, x_3, x_4, x_5) = (2, 1, 2, 2, 1, 2)$.

By similar manner we can show there does not exist any dominating set with $(x_0, x_1, x_2, x_3, x_4, x_5) = (2, 1, 2, 1, 2, 2)$, as shown in Figure 8.

Again there does not exist any dominating set D with $(x_0, x_1, x_2, x_3, x_4, x_5) = (2, 2, 1, 2, 1, 2)$ as this case is similar as the above case. Hence, $\gamma(P_6\square C_7) = 11$. □

The following theorem provides bounds for dominating sets that will help in proving the main Theorem 23.

Theorem 22. *Consider the following:*

(i) $\gamma(P_6\square C_8) \leq 12$, *(ii)* $\gamma(P_6\square C_9) \leq 14$, *(iii)* $\gamma(P_6\square C_{10}) \leq 15$, *(iv)* $\gamma(P_6\square C_{11}) \leq 16$, *(v)* $\gamma(P_6\square C_{12}) \leq 18$, *(vi)* $\gamma(P_6\square C_{13}) \leq 20$, *(vii)* $\gamma(P_6\square C_{14}) \leq 20$, *(viii)* $\gamma(P_6\square C_{15}) \leq 23$, *(ix)* $\gamma(P_6\square C_{16}) \leq 24$, *(x)* $\gamma(P_6\square C_{17}) \leq 26$, *(xi)* $\gamma(P_6\square C_{18}) \leq 26$, *(xii)* $\gamma(P_6\square C_{19}) \leq 28$, *(xiii)* $\gamma(P_6\square C_{20}) \leq 30$, *(xiv)* $\gamma(P_6\square C_{21}) \leq 31$, *(xv)* $\gamma(P_6\square C_{22}) \leq 31$, *(xvi)* $\gamma(P_6\square C_{23}) \leq 34$, *(xvii)* $\gamma(P_6\square C_{24}) \leq 35$, *(xviii)* $\gamma(P_6\square C_{25}) \leq 36$, *(xix)* $\gamma(P_6\square C_{26}) \leq 38$, *and (xx)* $\gamma(P_6\square C_{27}) \leq 40$.

Proof. Dominating sets for each of the cases are shown in Figures 12, 13, 14, 15, 16, 17, and 18. □

Theorem 23. *For $n > 7$,*

$$\gamma(P_6\square C_n) \leq \begin{cases} \left\lceil \dfrac{10n}{7} \right\rceil, & \text{for } n \equiv 0, 4, 5, 8, 10, 11, 12 \pmod{14}; \\ \left\lceil \dfrac{10n}{7} \right\rceil + 1, & \text{otherwise.} \end{cases} \quad (13)$$

Proof. For $3 \leq n \leq 27$, the above inequalities have already been proved in Theorems 21 and 22. For $n \geq 28$ one can easily find a dominating set for $\gamma(P_6\square C_n)$ using repeated modified concatenation between the dominating set for $\gamma(P_6\square C_{14})$ with dominating set for $\gamma(P_6\square C_{n-14})$. □

The following theorem provides a lower bound for $\gamma(P_6\square C_n)$, for $n > 7$.

Theorem 24. *For $n > 7$, if there exists a minimum dominating set D with $|(P_6)_i \cap D| = 1$ or 2 for all i, then $\gamma(P_6\square C_n) \geq \lceil 10n/7 \rceil$ resulting $\gamma(P_6\square C_n) = \lceil 10n/7 \rceil + 0$ or 1, accordingly.*

Proof. It is enough to show that, for a dominating set D with $|(P_6)_i \cap D| = 1$ or 2 for all i, it holds that $|D| \geq \lceil 10n/7 \rceil$.

We say the configuration of the type $k_1 k_2 k_3 \ldots$ occurs in a dominating set D, where $k_1, k_2, k_3, \ldots$ are 1 or 2, if there are consecutive columns in the graph with $k_1, k_2, k_3 \ldots$ many common vertex (or vertices) with D.

We have the following observations which can be verified considering all possible cases (we omit the verification because it has several cases and subcases and require similar types of arguments).

(1) There cannot be three consecutive columns with only one vertex common with D; that is, the configuration

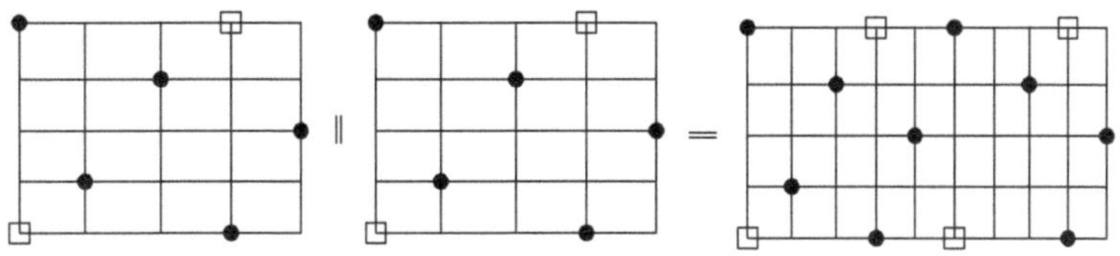

FIGURE 9: Modified concatenation of subsets of dominating sets columnwise, where □ denotes the vertex not captured by the subset.

of the type 111 cannot occur in D. This is clear from Remark 4.

(2) If the configuration of the type 2112112 occurs, then just before and after these seven columns the configuration of the type 22 should occur; that is, the configuration of the type 22211211222 should occur where the previous configuration occurs inside this.

(3) If there are two configurations of the type 2112 which occurs and between these two configurations only 212 type configuration occurs then the configuration of the type 212 occurs at least twice.

(4) The configuration of the type 211221122112 cannot occur.

Now let there be $2x + y$ many configurations of the type 2112 occur where $2x$ many configurations occur in x many pair and y many occur without pairing. Also let z many configurations of the type 22 occur and t many configuration of the type 212 occur.

Then, $z \geq x$ and $t \geq 2(x + y - z)$. Now the number of columns which have one vertex in common with D is $2(2x + y) + t$ and the number of columns which have two vertices in common with D is $2x + y + z + t$.

Therefore, $n = 2(2x + y) + t + (2x + y + z + t)$ and $|D| = 2(2x + y) + t + 2(2x + y + z + t)$.

Hence, $7|D| - 10n = 4(2x + y + z + t) - 3(2(2x + y) + t) = -4x - 2y + t + 4z = 2(z - x) + t - 2(x + y - z) \geq 0$. So $|D| \geq 10n/7$. □

Remark 25. For $n \geq 3$, $\gamma(P_6 \square C_n) > \gamma(P_6) \cdot \gamma(C_n) = 2 \cdot \lceil n/3 \rceil$.

4. Bounds for General Case

In this section we give rough bounds of $\gamma(P_m \square C_n)$ for any $m, n \geq 5$. If $m = 5p + k$ and $n = 5q + l$, where p, q are natural numbers and $k, l = 0, 1, 2, 3, 4$, a lower bound for $\gamma(P_m \square C_n)$ is $\lceil mn/5 \rceil = 5pq + kq + pl + \lceil kl/5 \rceil$, as a vertex can dominate at most 5 vertices. An upper bound of $\gamma(P_m \square C_n)$ is given by $5pq + (k+2)q + (l+1)p$ if $l \neq 0$ and $5pq + (k+2)q$ if $l = 0$. A dominating set with above cardinality can be constructed as follows.

Step 1. Consider the subset $D = \{(0,0), (1,2), (2,4), (3,1), (4,3)\}$ of $P_5 \square C_n$ as shown in Figure 9. Modified concatenation columnwise and concatenation rowwise are shown in Figures 9 and 10, respectively.

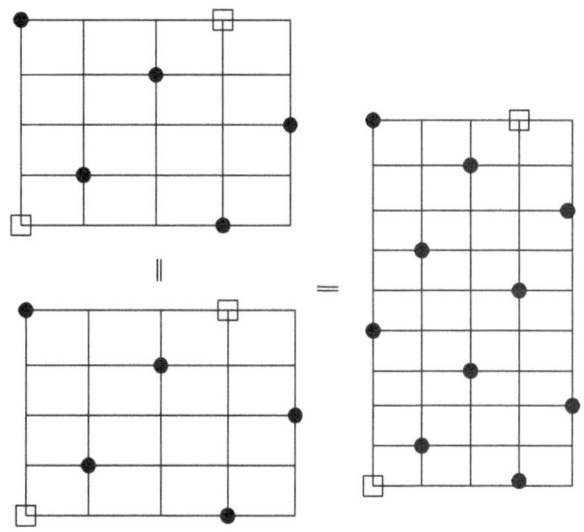

FIGURE 10: Concatenation of subsets of dominating sets row wise.

Step 2. Construct a subset of $P_{5p} \square C_{5q}$ using modified concatenation columnwise q many times and using concatenation rowwise p many times the set D as shown in Figure 11.

Step 3. Add extra q suitably chosen vertices in the first row and extra q suitably chosen vertices in the last row to construct dominating set for $P_{5p} \square C_{5q}$. Hence a dominating set of $P_{5p} \square C_{5q}$ is $\{(5i, 5j), (5i + 1, 5j + 2), (5i + 2, 5j + 4), (5i + 3, 5j + 1), (5i + 4, 5j + 3) : i = 0, 1, \ldots, p - 1, j = 0, 1, \ldots, q - 1\} \cup \{(0, 5j + 4) : j = 0, 1, \ldots, q - 1\} \cup \{(5p - 1, 5j) : j = 0, 1, \ldots, q - 1\}$.

Step 4. One can construct a dominating set for $P_{5p+k} \square C_{5q+l}$, $(k, l) \neq (0, 0)$, using concatenation and modified concatenation between the above dominating set for $P_{5p} \square C_{5q}$ and a suitably chosen subset D' of vertices for extra k rows and l columns. The cardinality of D' will be $kq + (l + 1)p$ if $l \neq 0$ and kq if $l = 0$.

Remark 26. We have already seen that if $m = 5p + k$ and $n = 5q + l$, where p, q are natural numbers and k, $l = 0, 1, 2, 3, 4$, then

$$\left\lceil \frac{mn}{5} \right\rceil = 5pq + kq + pl + \left\lceil \frac{kl}{5} \right\rceil \leq \gamma(P_m \square C_n) \leq \begin{cases} 5pq + (k+2)q + (l+1)p, & \text{if } l \neq 0 \\ 5pq + (k+2)q, & \text{if } l = 0. \end{cases} \quad (14)$$

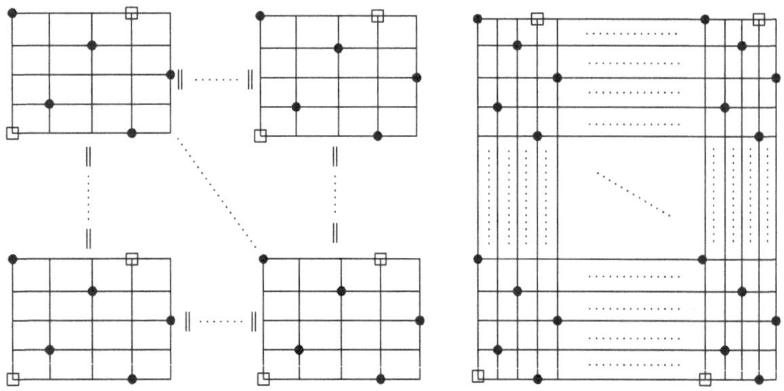

FIGURE 11: Modified concatenation and concatenation of subsets of dominating sets both column- and row wise.

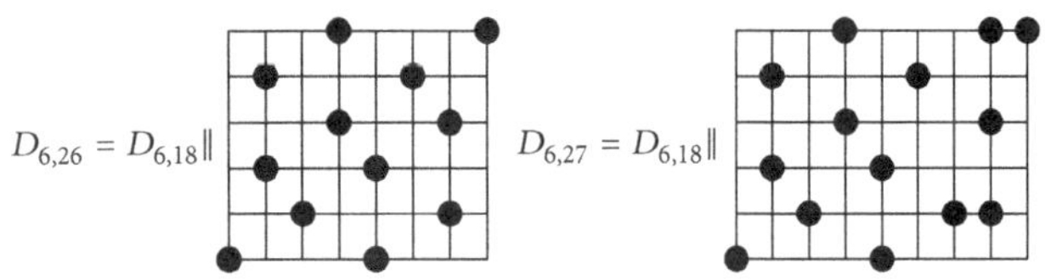

FIGURE 12: Dominating sets for $P_6\square C_{26}$ and $P_6\square C_{27}$.

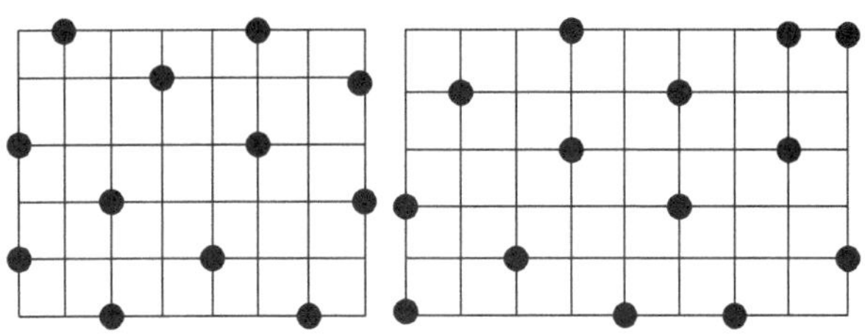

FIGURE 13: Dominating sets for $P_6\square C_8$ and $P_6\square C_9$.

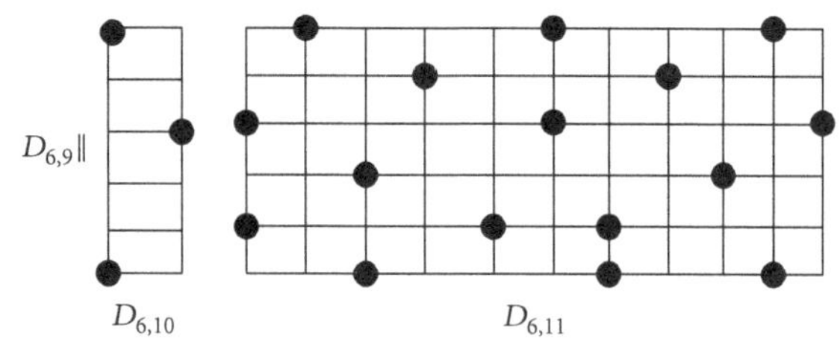

FIGURE 14: Dominating sets for $P_6\square C_{10}$ and $P_6\square C_{11}$.

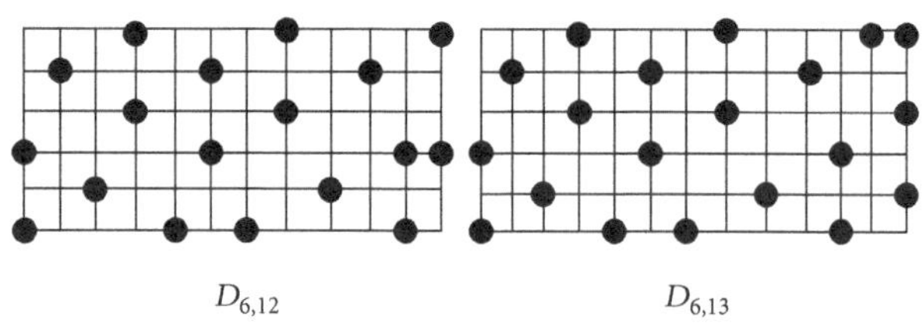

FIGURE 15: Dominating sets for $P_6\square C_{12}$ and $P_6\square C_{13}$.

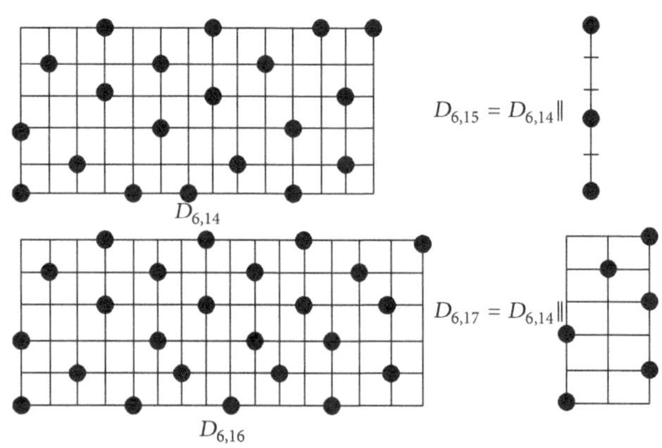

FIGURE 16: Dominating sets for $P_6\square C_{14}$, $P_6\square C_{15}$, $P_6\square C_{16}$, and $P_6\square C_{17}$.

Thus we have

$$1 + \frac{l}{5q} \leq \lim_{p\to\infty} \frac{\gamma(P_m C_n)}{5pq} \leq 1 + \frac{l+1}{5q}, \quad \text{for } l \neq 0,$$
$$\lim_{p\to\infty} \frac{\gamma(P_m\square C_n)}{5pq} = 1, \quad \text{for } l = 0. \tag{15}$$

Therefore, $1 + (k/5p) \leq \lim_{q\to\infty}(\gamma(P_m\square C_n)/5pq) \leq 1 + ((k+2)/5p)$, for $l = 0$ and $\lim_{p,q\to\infty}(\gamma(P_m\square C_n)/5pq) = 1$.

5. Application of Domination Number in Wireless Sensor Networks

Wireless sensor networks (WSN) provide a new communication and networking paradigms and myriad new applications. The wireless sensors have small size, low battery capacity, nonrenewable power supply, small processing power, limited buffer capacity, and low-power radio. They may measure distance, direction, speed, humidity, wind speed, soil makeup, temperature, chemicals, light and various other parameters.

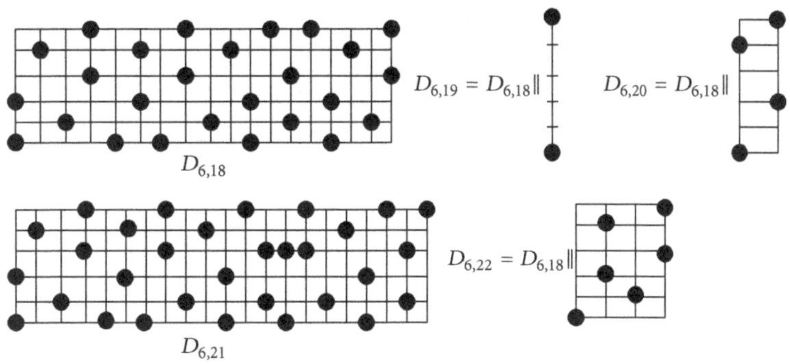

FIGURE 17: Dominating sets for $P_6\square C_{18}$, $P_6\square C_{19}$, $P_6\square C_{20}$, $P_6\square C_{21}$, and $P_6\square C_{22}$.

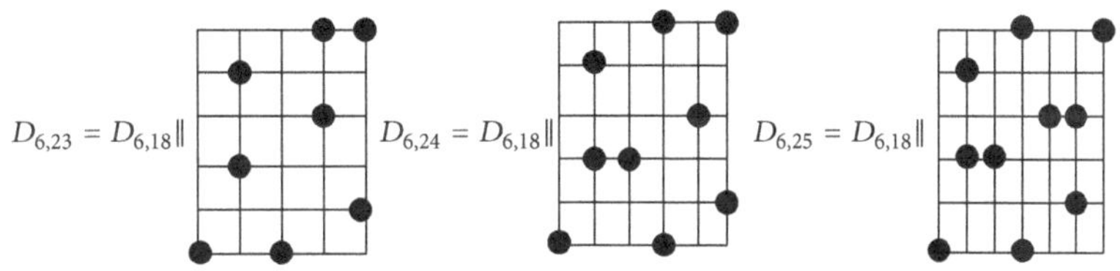

FIGURE 18: Dominating sets for $P_6\square C_{23}$, $P_6\square C_{24}$, and $P_6\square C_{25}$.

Recent advancements in wireless communications and electronics have enabled the development of low-cost WSN. A WSN usually consists of a large number of small sensor nodes, which are equipped with one or more sensors, some processing circuit and a wireless transceiver. One of the unique features of a WSN is random deployment in inaccessible terrains and cooperative effort that offers unprecedented opportunities for a broad spectrum of civilian and military applications; such as industrial automation, military tactical surveillance and national security [18]. Sensor Networks are useful in detecting topological events such as forest fires [19]. Sensor networks aim at monitoring their surroundings for event detection and object tracking [20]. Because of this surveillance goal, *coverage* is the functional basis of any sensor network. In order to fulfill its designated surveillance tasks, a sensor network must fully cover the Region of Interest (ROI) without leaving any *internal sensing hole* [21]. The ROI may be a rectangular grid, which may be divided into several squares.

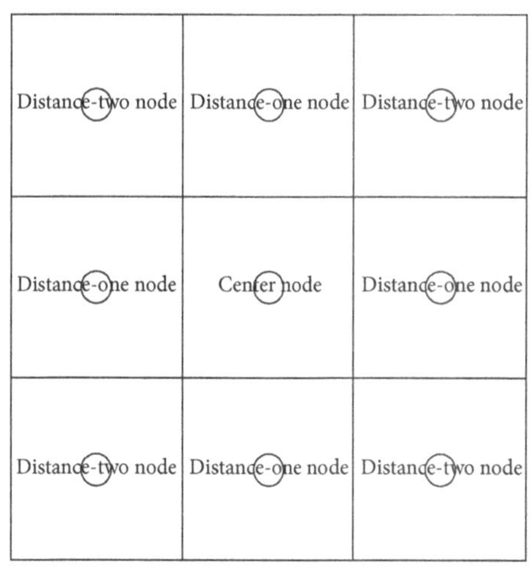

FIGURE 19: Nodes placed in cells of the ROI.

In general, a sensor is placed at the center of a square, known as center node as shown in Figure 19. This sensor can detect events or tracking objects at the center node along with the four centers of the four adjacent squares which have a common edge with the center square. These four centers are known as Distance-one nodes as shown in Figure 19. The sensor placed at the center node cannot detect events or tracking objects placed at the center of the other squares, for example, Distance-two node as shown in Figure 19. Our objective is to place minimum number of sensors at the center of some selected squares in such a way that they can detect the events or tracking the objects at the center of all the squares. Then the minimum number of sensor required is the same as the domination number of the corresponding rectangular grid and a minimum dominating set will suggest which squares we have to choose.

6. Conclusion

In this paper we find the domination numbers of the graphs $P_m\square C_n$, $m = 5$, $n \geq 3$. We also provide bounds on $\gamma(P_6\square C_n)$, $n \geq 3$. Minimum dominating sets corresponding to the above-mentioned graphs are also constructed. Moreover, we provide rough bounds on domination numbers of the graphs $\gamma(P_m\square C_n)$, $m \geq 7$ and $n \geq 3$, and in future, we would like to provide sharper bounds. We also point out how domination numbers and minimum dominating sets may be useful to wireless sensor networks.

Conflict of Interests

The authors declare that there is no conflict of interests regarding the publication of this paper.

References

[1] T. Haynes, S. Hedetniemi, and P. Slater, *Fundamentals of Domination in Graphs*, Marcel Dekker, New York, NY, USA, 1998.

[2] J. E. Dunbar, S. T. Hedetniemi, M. A. Henning, and P. J. Slater, "Signed domination in graphs," in *Graph Theory, Combinatorics and Applications*, pp. 311–322, John Wiley & Sons, 1995.

[3] O. Favaron, "Signed domination in regular graphs," *Discrete Mathematics*, vol. 158, no. 1–3, pp. 287–293, 1996.

[4] O. Morgenstern, "The colaboration between Oskar Morgenstern and John von Neumann on the theory of games," *Journal of Economic Literature*, vol. 14, pp. 805–816, 1976.

[5] S. T. Hedetniemi and R. C. Laskar, "Bibliography on domination in graphs and some basic definitions of domination parameters," *Discrete Mathematics*, vol. 86, no. 1–3, pp. 257–277, 1990.

[6] V. G. Vizing, "The cartesian product of graphs," *Vychisl Sistemy*, vol. 9, pp. 30–43, 1963.

[7] S. Gravier, "Total domination number of grid graphs," *Discrete Applied Mathematics*, vol. 121, no. 1–3, pp. 119–128, 2002.

[8] R. J. Faudree and R. H. Schelp, "The domination number for the product of graphs," *Congressus Numerantium*, vol. 79, pp. 29–33, 1990.

[9] M. El-Zahar and C. M. Pareek, "Dominationnumber of products of graphs," *Ars Combinatoria*, vol. 31, pp. 223–227, 1991.

[10] F. Harary, *Graph Theory*, Addison-Weslay, Reading, Mass, USA, 1969.

[11] S. Klavžar and N. Seifter, "Dominating cartesian products of cycles," *Discrete Applied Mathematics*, vol. 59, no. 2, pp. 129–136, 1995.

[12] B. N. Clark, C. J. Colbourn, and D. S. Johnson, "Unit disk graphs," *Discrete Mathematics*, vol. 86, no. 1–3, pp. 165–177, 1990.

[13] M. S. Jacobson and L. F. Kinch, "On the domination of the products of graphs I," *Ars Combinatoria*, vol. 18, pp. 33–44, 1983.

[14] T. Y. Chang and W. E. Clark, "The domination numbers of the $5 \times n$ and $6 \times n$ grid graphs," *Journal of Graph Theory*, vol. 17, pp. 81–107, 1993.

[15] M. H. El-Zahar, S. M. Khamis, and K. M. Nazzal, "On the domination number of the cartesian product of the cycle of length n and any graph," *Discrete Applied Mathematics*, vol. 155, no. 4, pp. 515–522, 2007.

[16] B. L. Hartnell and D. F. Rall, "On Vizing's conjecture," *Congressus Numerantium*, vol. 82, pp. 87–96, 1991.

[17] M. Nandi, S. Parui, and A. Adhikari, "Domination number of cylindrical grids," *Applied Mathematics and Computation*, vol. 217, no. 10, pp. 4879–4889, 2011.

[18] G. J. Pottie and W. J. Kaiser, "Wireless integrated network sensors," *Communications of the ACM*, vol. 43, no. 5, pp. 51–58, 2000.

[19] C. Farah, F. Schwaner, A. Abedi, and M. Worboys, "Distributed homology algorithm to detect topological events via wireless sensor networks," *IET Wireless Sensor Systems*, vol. 1, no. 3, pp. 151–160, 2011.

[20] F. Martincic and L. Schwiebert, "Introduction to wireless sensor networking," in *Handbook of Sensor Networks: Algorithms and Architectures*, I. Stojmenovic, Ed., chapter 1, John Wiley & Sons, 2005.

[21] X. Bai, S. Kumar, D. Xuan, Z. Yun, and T. H. Lai, "Deploying wireless sensors to achieve both coverage and connectivity," in *Proceedings of the 7th ACM International Symposium on Mobile Ad Hoc Networking and Computing (MOBIHOC '06)*, pp. 131–142, May 2006.

10

The Terminal Hosoya Polynomial of Some Families of Composite Graphs

Emeric Deutsch[1] and Juan Alberto Rodríguez-Velázquez[2]

[1] *Polytechnic Institute of New York University, Brooklyn, NY 11201, USA*
[2] *Departament d'Enginyeria Informàtica i Matemàtiques, Universitat Rovira i Virgili, Avenida Països Catalans 26, 43007 Tarragona, Spain*

Correspondence should be addressed to Juan Alberto Rodríguez-Velázquez; juanalberto.rodriguez@urv.cat

Academic Editor: Jiang Zeng

Let G be a connected graph and let $\Omega(G)$ be the set of pendent vertices of G. The *terminal Hosoya polynomial* of G is defined as $\mathrm{TH}(G,t) := \sum_{x,y\in\Omega(G):x\neq y} t^{d_G(x,y)}$, where $d_G(x,y)$ denotes the distance between the pendent vertices x and y. In this note paper we obtain closed formulae for the terminal Hosoya polynomial of rooted product graphs and corona product graphs.

1. Introduction

Throughout this paper $G = (V, E)$ denotes a connected, simple, and finite graph with vertex set $V = V(G)$ and edge set $E = E(G)$. The *distance* $d_G(u,v)$ between two vertices u and v is the minimum of the lengths of paths between u and v.

The *Wiener index* $W(G)$ of a graph G with vertex set $\{v_1, v_2, \ldots, v_n\}$, defined as the sum of distances between all pairs of vertices of G,

$$W(G) := \frac{1}{2}\sum_{i=1,j=1}^{n} d_G(v_i, v_j), \tag{1}$$

is the first mathematical invariant reflecting the topological structure of a molecular graph.

This topological index has been extensively studied; for instance, a comprehensive survey on the direct calculation, applications, and the relation of the Wiener index of trees with other parameters of graphs can be found in [1]. Moreover, a list of 120 references of the main works on the Wiener index of graphs can be found in the referred survey.

The Hosoya polynomial of a graph was introduced in Hosoya's seminal paper [2] in 1988 and received a lot of attention afterwards. The polynomial was later independently introduced and considered by Sagan et al. [3] under the name Wiener polynomial of a graph. Both names are still used for the polynomial but the term Hosoya polynomial is nowadays used by the majority of researchers. The main advantage of the Hosoya polynomial is that it contains a wealth of information about distance based graph invariants. For instance, knowing the Hosoya polynomial of a graph, it is straightforward to determine the Wiener index of a graph as the first derivative of the polynomial at the point $t = 1$. Cash [4] noticed that the hyper-Wiener index can be obtained from the Hosoya polynomial in a similar simple manner. A comprehensive survey on the main properties of the Hosoya polynomial, which includes a wide list of bibliographic references, can be found in Gutman et al. [5]. Also, Estrada et al. [6] studied several chemical applications of the Hosoya polynomial.

The *Hosoya polynomial* of a connected graph G is defined as

$$H(G,t) := \sum_{x,y\in V(G):x\neq y} t^{d_G(x,y)}. \tag{2}$$

As we pointed out above, the Wiener index of a graph G is determined as the first derivative of the polynomial $H(G,t)$ at $t = 1$.

The Hosoya polynomial has been obtained for trees, composite graphs, benzenoid graphs, tori, zig-zag open-ended nanotubes, certain graph decorations, armchair open-ended nanotubes, zigzag polyhex nanotorus, $TUC_4C_8(S)$ nanotubes, pentachains, polyphenyl chains, the circumcoronene series, Fibonacci and Lucas cubes, Hanoi graphs, and so forth. See the references in [7].

The *degree* of a vertex v in G is the number of edges incident to it and is denoted by $\deg_G(v)$. If $\deg_G(v) = 1$, then v is called a *pendent* vertex of G. The set of pendent vertices of G will be denoted by $\Omega(G)$.

Gutman et al. proposed the concept of *terminal Wiener index*, which is defined as the sum of distances between all pairs of pendent vertices of G [8]. For detailed information on the terminal Wiener index the readers may refer to the recent papers [8–11] and the references cited therein.

By analogy to the definition of Hosoya polynomial, the *terminal Hosoya polynomial* was defined by Narayankar et al. [12] as

$$\mathrm{TH}(G,t) := \sum_{x,y\in\Omega(G):x\neq y} t^{d_G(x,y)}, \tag{3}$$

and recently Ramane et al. studied $\mathrm{TH}(G,t)$ for the case of line graphs [13]. As we can expect, the terminal Wiener index, denoted by $\mathrm{TW}(G)$, is obtained from $\mathrm{TH}(G,t)$ as

$$\mathrm{TW}(G) = \left.\frac{\partial(\mathrm{TH}(G,t))}{\partial t}\right|_{t=1}. \tag{4}$$

Among the most known products of graphs, namely, Cartesian product, lexicographic product, strong product, rooted product, and corona product, the two latest ones are the only cases in which we can obtain connected graphs having pendent vertices. So, in this paper we obtain formulae for the terminal Hosoya polynomial of both types of graphs, rooted product graphs and corona product graphs.

The remaining definitions will be given the first time that the concept is found in the text.

2. The Terminal Hosoya Polynomial of Rooted Product Graphs

A *rooted graph* is a graph in which one vertex is labeled in a special way so as to distinguish it from the other vertices. The special vertex is called the *root* of the graph. Let G be a labeled graph on n vertices. Let $\mathcal{F}$ be a sequence of n rooted graphs $G'_1, G'_2, \dots, G'_n$. The *rooted product graph* $G(\mathcal{F})$ is the graph obtained by identifying the root of G'_i with the ith vertex of G [14]. In this paper we consider a particular case of rooted product graphs where $\mathcal{F}$ consists of n isomorphic rooted graphs [15]. More formally, assuming that $V(G) = \{u_1, \dots, u_n\}$ and that the root vertex of G' is v, we define the *rooted product graph* $G\circ_v G'$ as the graph whose vertex set is $V(G\circ_v G') = V(G)\times V(G')$ and whose edge set is

$$E(G\circ_v G') = \bigcup_{i=1}^{n}\{(u_i,b)(u_i,y) : by \in E(G')\} \cup \{(u_i,v)(u_j,v) : u_iu_j \in E(G)\}. \tag{5}$$

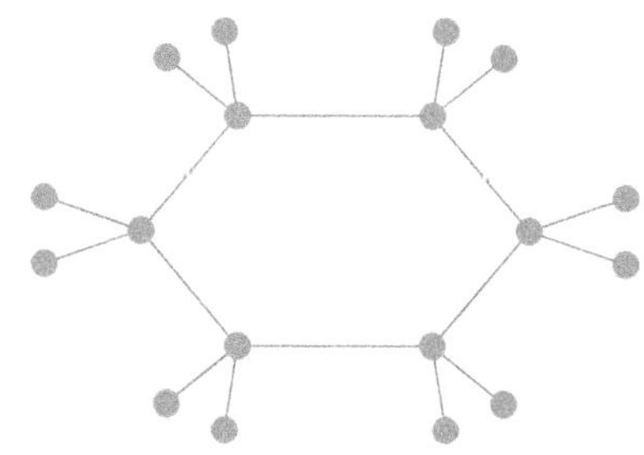

FIGURE 1: The graph $C_6\circ_v P_3$, where C_6 is the cycle graph of order 6, P_3 is a path of order 3, and v is the central vertex of P_3. This chemical graph that corresponds to the Cyclohexane (C6H12) is a naturally occurring chemical that is also produced synthetically and used as a solvent in numerous industries.

As an example of a chemical compound whose graph is obtained as a rooted product graph we consider the Cycloalkanes with a single ring, whose chemical formula is C_nH_{2n}, and whose molecular graph can be expressed as $C_n\circ_v P_3$, where C_n is the cycle graph of order n, P_3 is a path of order 3, and v is the central vertex of P_3 (see Figure 1).

To begin with, we need some additional notation. We define the *terminal Hosoya polynomial with respect to u* as

$$\mathrm{TH}_u(G,t) := \sum_{w\in\Omega(G)-\{u\}} t^{d_G(u,w)}. \tag{6}$$

Theorem 1. *Let G be a connected graph of order $n \geq 2$ and let G' be a connected nontrivial graph. Then for any $v \in V(G') - \Omega(G')$,*

$$TH(G\circ_v G',t) = n\cdot TH(G',t) + H(G,t)\left(TH_v(G',t)\right)^2. \tag{7}$$

Proof. Let $V(G) = \{u_1, u_2, \dots, u_n\}$ and let G'_i be the ith-"copy" of G' in $G\circ_v G'$; that is, G'_i is the subgraph of $G\circ_v G'$ generated by the set $\{u_i\}\times V(G')$, which is isomorphic to G'. Then,

$$\Omega(G\circ_v G') = \bigcup_{i=1}^{n}\Omega(G'_i). \tag{8}$$

So, if $\Omega(G') = \emptyset$, then $\Omega(G\circ_v G') = \emptyset$ and, as a consequence, $\mathrm{TH}_v(G',t) = \mathrm{TH}(G',t) = \mathrm{TH}(G\circ_v G',t) = 0$. So, we assume that $\Omega(G') \neq \emptyset$.

Notice that for every pair x, y of pendent vertices of G' and $u_i \in V(G)$ we have

$$d_{G\circ_v G'}((u_i,x),(u_i,y)) = d_{G'_i}((u_i,x),(u_i,y)) = d_{G'}(x,y). \tag{9}$$

Also, for $u_i, u_j \in V(G)$, $i \neq j$, we have

$$d_{G\circ_v G'}((u_i,x),(u_j,y)) = d_{G'}(x,v) + d_G(u_i,u_j) + d_{G'}(v,y). \tag{10}$$

Hence, $\mathrm{TH}(G\circ_v G',t)$ splits as the sum of two polynomials $P(t) + Q(t)$, where $P(t)$ is obtained from pairs of pendent vertices in the same G_i and $Q(t)$ is obtained from pairs $(u_i, x), (u_j, y)$, where $i \neq j$. On the one hand,

$$P(t) = \sum_{i=1}^{n} \left(\sum_{x,y \in \Omega(G'): x \neq y} t^{d_{G'}(x,y)} \right) = n \cdot \mathrm{TH}(G', t), \quad (11)$$

and, on the other hand,

$$\begin{aligned} Q(t) &= \sum_{u_i,u_j \in V(G): i \neq j} t^{d_G(u_i,u_j)} \left(\sum_{x \in \Omega(G')} t^{d_{G'}(x,v)} \right) \\ &\quad \times \left(\sum_{y \in \Omega(G')} t^{d_{G'}(y,v)} \right) \\ &= \left(\mathrm{TH}_v(G', t) \right)^2 \sum_{u_i,u_j \in V(G): i \neq j} t^{d_G(u_i,u_j)} \\ &= \left(\mathrm{TH}_v(G', t) \right)^2 H(G, t). \end{aligned} \quad (12)$$

Therefore, the result follows. □

Given a vertex $v \in V(G')$, we define

$$d_T(v) := \sum_{w \in \Omega(G') - \{v\}} d_{G'}(v, w). \quad (13)$$

Notice that we can see $d_T(v)$ as the *terminal Wiener index with respect to* v; that is, $d_T(v)$ is the first derivative of the polynomial $\mathrm{TH}_v(G', t)$ at $t = 1$. As a direct consequence of Theorem 1 we obtain the following result on the terminal Wiener index.

Corollary 2. *Let G' be a connected nontrivial graph. Then for any $v \in V(G') - \Omega(G')$ and any connected graph G of order $n \geq 2$,*

$$\begin{aligned} TW(G\circ_v G', t) &= n \cdot TW(G') + W(G) \left| \Omega(G') \right|^2 \\ &\quad + n(n-1) \left| \Omega(G') \right| d_T(v). \end{aligned} \quad (14)$$

Now we consider the case of rooted product graphs $G\circ_v G'$ where the root v is a pendent vertex of G'.

Theorem 3. *Let G be a connected graph of order $n \geq 2$ and let G' be a connected nontrivial graph. Then for any $v \in \Omega(G')$,*

$$\begin{aligned} TH(G\circ_v G', t) &= n \left(TH(G', t) - TH_v(G', t) \right) \\ &\quad + H(G, t) \left(TH_v(G', t) \right)^2. \end{aligned} \quad (15)$$

Proof. We proceed as in the proof of Theorem 1, with the only difference that in this case $P(t)$ is obtained as follows:

$$\begin{aligned} P(t) &= \sum_{i=1}^{n} \left(\sum_{x,y \in \Omega(G') - \{v\}: x \neq y} t^{d_{G'}(x,y)} \right) \\ &= n \left(\sum_{x,y \in \Omega(G'): x \neq y} t^{d_{G'}(x,y)} - \sum_{x \in \Omega(G') - \{v\}} t^{d_{G'}(x,v)} \right) \\ &= n \left(\mathrm{TH}(G', t) - \mathrm{TH}_v(G', t) \right). \end{aligned} \quad (16)$$

□

Corollary 4. *Let G' be a connected nontrivial graph. Then for any $v \in \Omega(G')$ and any connected graph G of order $n \geq 2$, one has*

$$\begin{aligned} TW(G\circ_v G', t) &= n \left(TW(G') - d_T(v) \right) \\ &\quad + W(G) \left(\left| \Omega(G') \right| - 1 \right)^2 \\ &\quad + n(n-1) \left(\left| \Omega(G') \right| - 1 \right) d_T(v). \end{aligned} \quad (17)$$

3. The Terminal Hosoya Polynomial of Corona Graphs

Let G and G' be two graphs of order n and n', respectively. The corona product $G \odot G'$ is defined as the graph obtained from G and G' by taking one copy of G and n copies of G' and joining by an edge each vertex from the ith-copy of G' with the ith-vertex of G [16]. We will denote by $V = \{v_1, v_2, \ldots, v_n\}$ the set of vertices of G and by $G'_i = (V_i, E_i)$ the copy of G' such that v_i is adjacent to v for every $v \in V_i$. Thus, the vertex set of $G \odot G'$ is $V \cup (\cup_{i=1}^{n'} V_i)$ and, consequently, $G \odot G'$ has order $n(n'+1)$ and, if G is connected, then $G \odot G'$ is also connected and it has diameter $D(G \odot G') = D(G) + 2$.

Notice that the corona graph $K_1 \odot G'$ is isomorphic to the join graph $K_1 + G'$. Now, if G is a tree and $G' = N_r$ is an empty graph (An empty graph on r nodes consists of r isolated nodes with no edges. Such graphs are sometimes also called edgeless graphs or null graphs), then the corona graph $G \odot N_r$ is a tree and the set of leaves (pendent vertices) of $G \odot N_r$ is $\cup_{i=1}^{n} V_i$. In particular, if $G = P$ is a path, then $P \odot N_r$ is a caterpillar graph and the path P, that is formed by the nonleaves of $P \odot N_r$, is known as the backbone of the caterpillar.

The formula for the Hosoya polynomial of corona graphs was previously obtained in [17]. In this section we obtain the formula for the terminal Hosoya polynomial of corona graphs $G \odot G'$. Of course, $G \odot G'$ has pendent vertices if and only if G' has isolated vertices.

Corollary 5. *For any connected graph G of order $n \geq 2$ and any graph G' having r isolated vertices,*

$$TH(G \odot G', t) = n \binom{r}{2} t^2 + r^2 t^2 H(G, t). \quad (18)$$

Proof. First of all, notice that $G \odot G' = G\circ_v(\langle v \rangle + G')$, where $\langle v \rangle$ is the trivial graph composed by the vertex v. Now, since $\mathrm{TH}(\langle v \rangle + G', t) = \binom{r}{2} t^2$ and $\mathrm{TH}_v(\langle v \rangle + G', t) = rt$, the result immediately follows from Theorem 1. □

Finally, Corollary 5 (and also Corollary 2) leads to the following formula for the terminal Wiener index of $G \odot G'$.

Remark 6. For any connected graph G of order $n \geq 2$ and any graph G' having r isolated vertices,

$$\mathrm{TW}(G \odot G') = nr(nr-1) + r^2 W(G). \quad (19)$$

Conflict of Interests

The authors declare that there is no conflict of interests regarding the publication of this paper.

References

[1] A. A. Dobrynin, R. Entringer, and I. Gutman, "Wiener index of trees: theory and applications," *Acta Applicandae Mathematicae. An International Survey Journal on Applying Mathematics and Mathematical Applications*, vol. 66, no. 3, pp. 211–249, 2001.

[2] H. Hosoya, "On some counting polynomials in chemistry," *Discrete Applied Mathematics*, vol. 19, no. 1-3, pp. 239–257, 1988.

[3] B. E. Sagan, Y. Yeh, and P. Zhang, "The Wiener polynomial of a graph," *International Journal of Quantum Chemistry*, vol. 60, no. 5, pp. 959–969, 1996.

[4] G. G. Cash, "Relationship between the Hosoya polynomial and the hyper-Wiener index," *Applied Mathematics Letters*, vol. 15, no. 7, pp. 893–895, 2002.

[5] I. Gutman, Y. Zhang, M. Dehmer, and A. Ilic, "Altenburg, Wiener, and Hosoya polynomials," in *Distance in Molecular Graphs-Theory*, I. Gutman and B. Furtula, Eds., pp. 49–70, University of Kragujevac, Kragujevac, Serbia, 2012.

[6] E. Estrada, O. Ivanciuc, I. Gutman, A. Gutierrez, and L. Rodríguez, "Extended Wiener indices. A new set of descriptors for quantitative structure-property studies," *New Journal of Chemistry*, vol. 22, no. 8, pp. 819–822, 1998.

[7] E. Deutsch and S. Klavžar, "Computing the Hosoya polynomial of graphs from primary subgraphs," *MATCH Communications in Mathematical and in Computer Chemistry*, vol. 70, no. 2, pp. 627–644, 2013.

[8] I. Gutman, B. Furtula, and M. Petrović, "Terminal Wiener index," *Journal of Mathematical Chemistry*, vol. 46, no. 2, pp. 522–531, 2009.

[9] L. A. Székely, H. Wang, and T. Wu, "The sum of the distances between the leaves of a tree and the "semi-regular" property," *Discrete Mathematics*, vol. 311, no. 13, pp. 1197–1203, 2011.

[10] X. Deng and J. Zhang, "Equiseparability on terminal Wiener index," *Applied Mathematics Letters*, vol. 25, no. 3, pp. 580–585, 2012.

[11] Y.-H. Chen and X.-D. Zhang, "On Wiener and terminal Wiener indices of trees," *MATCH Communications in Mathematical and in Computer Chemistry*, vol. 70, no. 2, pp. 591–602, 2013.

[12] K. P. Narayankar, S. S. Shirkol, H. S. Ramane, and S. B. Lokesh, "Terminal Hosoya polynomial of thorn graphs," in *International Conference on Discrete Mathematics (ICDM '13)*, Karnatak University, Dharwad, India, 2013.

[13] H. S. Ramane, A. B. Ganagi, K. P. Narayankar, and S. S. Shirkol, "Terminal Hosoya polynomial of line graphs," *Journal of Discrete Mathematics*, vol. 2013, Article ID 857908, 3 pages, 2013.

[14] C. D. Godsil and B. D. McKay, "A new graph product and its spectrum," *Bulletin of the Australian Mathematical Society*, vol. 18, no. 1, pp. 21–28, 1978.

[15] A. J. Schwenk, "Computing the characteristic polynomial of a graph," in *Graphs and Combinatorics*, R. Bari and F. Harary, Eds., pp. 153–172, Springer, Berlin, Germany, 1974.

[16] R. Frucht and F. Harary, "On the corona of two graphs," *Aequationes Mathematicae*, vol. 4, pp. 322–325, 1970.

[17] D. Stevanović, "Hosoya polynomial of composite graphs," *Discrete Mathematics*, vol. 235, no. 1-3, pp. 237–244, 2001.

11

Decomposition Formulas for Triple q-Hypergeometric Functions

Thomas Ernst

Department of Mathematics, Uppsala University, P.O. Box 480, SE-751 06 Uppsala, Sweden

Correspondence should be addressed to Thomas Ernst; thomas@math.uu.se

Academic Editor: R. Yuster

In the spirit of Hasanov, Srivastava, and Turaev (2006), we introduce new inverse operators together with a more general operator and find a summation formula for the last one. Based on these operators and the earlier known q-analogues of the Burchnall-Chaundy operators, we find 15 symbolic operator formulas. Then, 10 expansions for the q-analogues of Srivastava's three triple hypergeometric functions in terms of ${}_4\phi_3$ q-hypergeometric and q-Kampé de Fériet functions are derived. These expansions readily reduce to 10 new expansions for the three triple Srivastava hypergeometric functions in terms of ${}_4F_3$ hypergeometric and Kampé de Fériet functions.

1. Introduction

The concept of decomposition formulas for multiple hypergeometric functions is well known from the articles by Hasanov et al. [1] and Bin-Saad [2]. This paper follows the first one by using q-analogues of 15 symbolic operator formulas.

The paper is organized as follows. In Section 2, we give the basic definitions and introduce a new inverse q-analogue of the Burchnall-Chaundy operators for an arbitrary number of variables, together with an explicit summation formula for one of these; the proof uses a q-Lauricella function summation formula. In Section 3, we express the three triple Srivastava q-hypergeometric functions in terms of these operators and q-hypergeometric and q-Appell functions.

Finally, we find 10 expansions for these triple q-hypergeometric functions. The proofs show a certain symmetry, which is explained in detail in the last section.

2. Definitions

We start by defining the umbral notation [3], a mixture of Heine 1846 and [4].

Definition 1. The power function is defined by $q^a \equiv e^{a\log(q)}$. We always assume that $0 < q < 1$. Let $\delta > 0$ be an arbitrary small number. We will use the following branch of the logarithm: $-\pi + \delta < \mathrm{Im}(\log q) \leq \pi + \delta$. This defines a simply connected space in the complex plane.

The variables $a, b, c, \ldots \in \mathbb{C}$ denote certain parameters. The variables i, j, k, l, m, n will denote natural numbers except for certain cases where it will be clear from the context that i will denote the imaginary unit. The q-shifted factorial is given by

$$\langle a; q\rangle_n \equiv \prod_{m=0}^{n-1} \left(1 - q^{a+m}\right); \ (a; q)_n \equiv \prod_{m=0}^{n-1} \left(1 - aq^m\right). \quad (1)$$

Since products of q-shifted factorials occur so often, to simplify them, we will frequently use the more compact notation:

$$\langle a_1, \ldots, a_m; q\rangle_n \equiv \prod_{j=1}^{m} \langle a_j; q\rangle_n. \quad (2)$$

The operator

$$\sim: \frac{\mathbb{C}}{\mathbb{Z}} \longmapsto \frac{\mathbb{C}}{\mathbb{Z}} \quad (3)$$

is defined by

$$a \longmapsto a + \frac{\pi i}{l \log q}. \quad (4)$$

By (4), it follows that

$$\widetilde{\langle a; q\rangle}_n = \prod_{m=0}^{n-1} \left(1 + q^{a+m}\right). \quad (5)$$

Assume that $(m, l) = 1$; that is, m and l relatively prime. The operator

$$\widetilde{\frac{m}{l}} : \frac{\mathbb{C}}{\mathbb{Z}} \longmapsto \frac{\mathbb{C}}{\mathbb{Z}} \tag{6}$$

is defined by

$$a \longmapsto a + \frac{2\pi i m}{l \log q}. \tag{7}$$

Furthermore,

$$\begin{aligned} (a;q)_\infty &= \prod_{m=0}^{\infty} (1 - aq^m), \quad 0 < |q| < 1, \\ (a;q)_\alpha &\equiv \frac{(a;q)_\infty}{(aq^\alpha;q)_\infty}, \quad a \neq q^{-m-\alpha}, \; m = 0, 1, \ldots. \end{aligned} \tag{8}$$

The q-gamma function is given by

$$\Gamma_q(z) \equiv \frac{\langle 1;q\rangle_\infty}{\langle z;q\rangle_\infty} (1-q)^{1-z}, \quad 0 < |q| < 1. \tag{9}$$

To save space, the following notation for quotients of Γ_q functions will often be used:

$$\Gamma_q \left[\begin{matrix} a_1, \ldots, a_p \\ b_1, \ldots, b_r \end{matrix} \right] \equiv \frac{\Gamma_q(a_1) \cdots \Gamma_q(a_p)}{\Gamma_q(b_1) \cdots \Gamma_q(b_r)}. \tag{10}$$

On the ring of polynomials $\mathbb{C}[x]$, we define the functions $\epsilon_i : \mathbb{C}[x] \mapsto \mathbb{C}[x]$ by

$$[\epsilon_i] f(x_i) \equiv f(qx_i). \tag{11}$$

The following notation is often used when we have long exponents:

$$QE(x) \equiv q^x. \tag{12}$$

The following notation will sometimes be used:

$$\widehat{a} \equiv a \vee \widetilde{a} \vee \widetilde{{}^{m/n} a}. \tag{13}$$

Definition 2. Generalizing Heine's series, we will define a q-hypergeometric series by

$$\begin{aligned} &{}_{p+p'}\phi_{r+r'} \left[\begin{matrix} \widehat{a}_1, \ldots, \widehat{a}_p \\ \widehat{b}_1, \ldots, \widehat{b}_1 \end{matrix} \middle| q; z \middle\| \begin{matrix} \prod_i f_i(k) \\ \prod_i g_i(k) \end{matrix} \right] \\ &\equiv \sum_{k=0}^{\infty} \frac{\langle \widehat{a}_1;q\rangle_k \cdots \langle \widehat{a}_p;q\rangle_k}{\langle 1, \widehat{b}_1;q\rangle_k \cdots \langle \widehat{b}_r;q\rangle_k} \left[(-1)^k q^{\binom{k}{2}} \right]^{1+r+r'-p-p'} \\ &\quad \times z^k \frac{\prod_i f_i(k)}{\prod_i g_i(k)}. \end{aligned} \tag{14}$$

We assume that the $f_i(k)$ and $g_j(k)$ contain p' and r' factors of the form $\langle \widehat{a(k)};q\rangle_k$ or $(s(k);q)_k$, respectively. The notation $\sum_{\vec{m}}$ denotes a multiple summation with the indices $m_1, \ldots, m_n$ running over all nonnegative integer values. In this connection, we put $m \equiv \sum_{j=1}^{n} m_j$.

Definition 3. The vectors

$$(a), (b), (g_i), (h_i), (a'), (b'), (g_i'), (h_i') \tag{15}$$

have dimensions

$$A, B, G, H, A', B', G', H'. \tag{16}$$

Let

$$1 + B + B' + H + H' - A - A' - G - G' \geq 0. \tag{17}$$

Then, the generalized q-Kampé de Fériet function is defined by

$$\begin{aligned} &\Phi^{A+A':G+G'}_{B+B':H+H'} \left[\begin{matrix} (\widehat{a}) : (\widehat{g}_1); \ldots; (\widehat{g}_n) \\ (\widehat{b}) : (\widehat{h}_1); \ldots; (\widehat{h}_n) \end{matrix} \middle| \vec{q}; \vec{x} \middle\| \begin{matrix} (a') : (g_1'); \ldots; (g_n') \\ (b') : (h_1'); \ldots; (h_n') \end{matrix} \right] \\ &\equiv \sum_{\vec{m}} \left(\langle (\widehat{a}); q_0\rangle_m (a')(q_0, m) \prod_{j=1}^{n} \langle (\widehat{g}_j); q_j\rangle_{m_j} \right. \\ &\qquad \left. \times \left((g_j')(q_j, m_j) x_j^{m_j} \right) \right) \\ &\quad \times \left(\langle (\widehat{b}); q_0\rangle_m (b')(q_0, m) \right. \\ &\qquad \left. \times \prod_{j=1}^{n} \left(\langle (\widehat{h}_j); q_j\rangle_{m_j} (h_j')(q_j, m_j) \langle 1; q_j\rangle_{m_j} \right) \right)^{-1} \\ &\quad \times (-1)^{\sum_{j=1}^{n} m_j (1+H+H'-G-G'+B+B'-A-A')} \\ &\quad \times QE\left((B + B' - A - A') \binom{m}{2}, q_0 \right) \\ &\quad \times \prod_{j=1}^{n} QE\left((1 + H + H' - G - G') \binom{m_j}{2}, q_j \right). \end{aligned} \tag{18}$$

It is assumed that there are no zero factors in the denominator. We assume that

$$(a')(q_0, m), (g_j')(q_j, m_j), (b')(q_0, m), (h_j')(q_j, m_j) \tag{19}$$

contain factors of the form $\langle \widehat{a}(k);q\rangle_k$, $(s;q)_k$, $(s(k);q)_k$, or $QE(f(\vec{m}))$. In the rest of the paper, we write $\langle a;q\rangle_m$ instead of $\langle \widehat{a};q\rangle_m$.

Definition 4. The four q-Appell functions are given by

$$\Phi_1\left(a;b,b';c \mid q;x_1,x_2\right) \equiv \sum_{m_1,m_2=0}^{\infty} \frac{\langle a;q\rangle_{m_1+m_2} \langle b;q\rangle_{m_1} \langle b';q\rangle_{m_2}}{\langle 1;q\rangle_{m_1} \langle 1;q\rangle_{m_2} \langle c;q\rangle_{m_1+m_2}} x_1^{m_1} x_2^{m_2} \quad \max\left(|x_1|,|x_2|\right)<1,$$

$$\Phi_2\left(a;b,b';c,c' \mid q;x_1,x_2\right) \equiv \sum_{m_1,m_2=0}^{\infty} \frac{\langle a;q\rangle_{m_1+m_2} \langle b;q\rangle_{m_1} \langle b';q\rangle_{m_2}}{\langle 1;q\rangle_{m_1} \langle 1;q\rangle_{m_2} \langle c;q\rangle_{m_1} \langle c';q\rangle_{m_2}} x_1^{m_1} x_2^{m_2}, \quad |x_1|+|x_2|<1,$$

$$\Phi_3\left(a,a';b,b';c \mid q;x_1,x_2\right) \equiv \sum_{m_1,m_2=0}^{\infty} \frac{\langle a;q\rangle_{m_1} \langle a';q\rangle_{m_2} \langle b;q\rangle_{m_1} \langle b';q\rangle_{m_2}}{\langle 1;q\rangle_{m_1} \langle 1;q\rangle_{m_2} \langle c;q\rangle_{m_1+m_2}} x_1^{m_1} x_2^{m_2}, \quad \max\left(|x_1|,|x_2|\right)<1,$$

$$\Phi_4\left(a;b;c,c' \mid q;x_1,x_2\right) \equiv \sum_{m_1,m_2=0}^{\infty} \frac{\langle a;q\rangle_{m_1+m_2} \langle b;q\rangle_{m_1+m_2}}{\langle 1;q\rangle_{m_1} \langle 1;q\rangle_{m_2} \langle c;q\rangle_{m_1} \langle c';q\rangle_{m_2}} x_1^{m_1} x_2^{m_2}, \quad \left|\sqrt{x_1}\right|+\left|\sqrt{x_2}\right|<1. \tag{20}$$

The q-analogues of the Srivastava triple hypergeometric functions are

$$H_A\left(a,b_1,b_2;c_1,c_2 \mid q;x_1,x_2,x_3\right) \equiv \sum_{m,n,p=0}^{\infty} \frac{\langle a;q\rangle_{m+p} \langle b_1;q\rangle_{m+n} \langle b_2;q\rangle_{n+p}}{\langle 1,c_1;q\rangle_m \langle 1;q\rangle_n \langle 1;q\rangle_p \langle c_2;q\rangle_{n+p}} x_1^m x_2^n x_3^p,$$

$$H_B\left(a,b_1,b_2;c_1,c_2,c_3 \mid q;x_1,x_2,x_3\right) \equiv \sum_{m,n,p=0}^{\infty} \frac{\langle a;q\rangle_{m+p} \langle b_1;q\rangle_{m+n} \langle b_2;q\rangle_{n+p}}{\langle 1,c_1;q\rangle_m \langle 1,c_2;q\rangle_n \langle 1,c_3;q\rangle_p} x_1^m x_2^n x_3^p, \tag{21}$$

$$H_C\left(a,b_1,b_2;c \mid q;x_1,x_2,x_3\right) \equiv \sum_{m,n,p=0}^{\infty} \frac{\langle a;q\rangle_{m+p} \langle b_1;q\rangle_{m+n} \langle b_2;q\rangle_{n+p}}{\langle 1;q\rangle_m \langle 1;q\rangle_n \langle 1;q\rangle_p \langle c;q\rangle_{m+n+p}} x_1^m x_2^n x_3^p.$$

Definition 5. In the spirit of Burchnall and Chaundy [5], the following inverse pair of symbolic operators was introduced in [3]; compare Jackson [6]. Let

$$\theta_{q,j} \equiv x_j D_{q,x_j}. \tag{22}$$

Then,

$$\nabla_{x_1,x_2;q}(h) \equiv \Gamma_q \begin{bmatrix} h, h+\theta_{q,1}+\theta_{q,2} \\ h+\theta_{q,1}, h+\theta_{q,2} \end{bmatrix}, \quad \Delta_{x_1,x_2;q}(h) \equiv \Gamma_q \begin{bmatrix} h+\theta_{q,1}, h+\theta_{q,2} \\ h+\theta_{q,1}+\theta_{q,2}, h \end{bmatrix}. \tag{23}$$

We limit our considerations to the case $h \neq -n$.

Then, one has

$$\nabla_{x_1,x_2;q}(h) \langle h;q\rangle_m \langle h;q\rangle_n x_1^m x_2^n = \langle h;q\rangle_{m+n} x_1^m x_2^n, \quad \nabla_{x_1,x_2;q}(h) = \sum_{r=0}^{\infty} \frac{\langle -\theta_{q,1}, -\theta_{q,2};q\rangle_r}{\langle 1,h;q\rangle_r} q^{rh} \epsilon_1^r \epsilon_2^r. \tag{24}$$

Similarly, we obtain

$$\Delta_{x_1,x_2;q}(h) = \sum_{r=0}^{\infty} \frac{\langle -\theta_{q,1}, -\theta_{q,2};q\rangle_r}{\langle 1, 1-h-\theta_{q,1}-\theta_{q,2};q\rangle_r} q^r,$$

$$\Delta_{x_1,x_2;q}(h) = \sum_{r=0}^{\infty} \frac{(-1)^r \langle -\theta_{q,1}, -\theta_{q,2};q\rangle_r \langle h;q\rangle_{2r}}{\langle 1, h+r-1, h+\theta_{q,1}, h+\theta_{q,2};q\rangle_r} q^{\binom{r}{2}+rh} \epsilon_1^r \epsilon_2^r,$$

$$\nabla_{x_1,x_2;q}(h) \Delta_{x_1,x_2;q}(k) = \sum_{r=0}^{\infty} \frac{\langle -\theta_{q,1}, -\theta_{q,2}, k-h;q\rangle_r \langle k;q\rangle_{2r}}{\langle 1,h,k+r-1,k+\theta_{q,1},k+\theta_{q,2};q\rangle_r} q^{rh} \epsilon_1^r \epsilon_2^r. \tag{25}$$

Definition 6. Introduce the inverse pair of operators

$$\widetilde{\Delta}_{x_1;x_2x_3;q}(h) \frac{1}{\langle h;q\rangle_m \langle h;q\rangle_{n+p}} x_1^m x_2^n x_3^p \equiv \frac{1}{\langle h;q\rangle_{m+n+p}} x_1^m x_2^n x_3^p, \tag{26}$$

$$\widetilde{\nabla}_{x_1;x_2x_3;q}(h) \frac{1}{\langle h;q\rangle_{m+n+p}} x_1^m x_2^n x_3^p \equiv \frac{1}{\langle h;q\rangle_m \langle h;q\rangle_{n+p}} x_1^m x_2^n x_3^p. \tag{27}$$

A general operator of the type (27) is given by

$$\widetilde{\nabla}_{x_1;x_2\cdots x_r;q}(h) \equiv \Gamma_q \begin{bmatrix} h, h+\theta_{q,1}+\cdots+\theta_{q,r} \\ h+\theta_{q,1}, h+\theta_{q,2}+\cdots+\theta_{q,r} \end{bmatrix}. \tag{28}$$

We have the following summation formula for the operator $\widetilde{\nabla}_{x_1;x_2\cdots x_r;q}(h)$:

$$\widetilde{\nabla}_{x_1;x_2\cdots x_r;q}(h) = \sum_{m_2,\ldots,m_r=0}^{\infty} \frac{\langle -\theta_{q,1};q\rangle_m \langle -\theta_{q,2};q\rangle_{m_2} \cdots \langle -\theta_{q,r};q\rangle_{m_r}}{\langle h;q\rangle_m \langle 1;q\rangle_{m_2} \cdots \langle 1;q\rangle_{m_r}} q^{hm} \epsilon^m \epsilon_2^{-m_3} \left(\epsilon_2\epsilon_3\right)^{-m_4} \cdots \left(\epsilon\epsilon_1^{-1}\epsilon_r^{-1}\right)^{-m_r}, \tag{29}$$

where

$$m \equiv m_2 + \cdots + m_r, \qquad \epsilon \equiv \epsilon_1 \cdots \epsilon_r. \tag{30}$$

This follows from the summation formula

$$\begin{aligned}\Phi_D^{(n)} &= \left(b, -k_1, \ldots, -k_n; c \mid q; q^{c-b+K}, q^{c-b+K-k_1}, \ldots, q^{c-b+k_n}\right) \\ &= \frac{\langle c-b; q\rangle_K}{\langle c; q\rangle_K}, \qquad K \equiv \sum_{i=1}^{n} k_i,\end{aligned} \tag{31}$$

where the fourth q-Lauricella function is given by

$$\begin{aligned}\Phi_D^{(n)} &(a, b_1, \ldots, b_n; c \mid q; x_1, \ldots, x_n) \\ &\equiv \sum_{\vec{m}} \frac{\langle a; q\rangle_m \prod_{j=1}^{n} \langle b_j; q\rangle_{m_j} x_j^{m_j}}{\langle c; q\rangle_m \prod_{j=1}^{n} \langle 1; q\rangle_{m_j}}, \\ &\max\left(|x|, \ldots, |x_n|\right) < 1.\end{aligned} \tag{32}$$

3. Computations

The proofs of the following 15 formulas are straightforward from the definitions.

Theorem 7 (q-analogues of [1, pp. 959-960]). *Consider*

$$\begin{aligned}H_A &(a, b_1, b_2; c_1, c_2 \mid q; x_1, x_2, x_3) \\ &= \nabla_{x_1,x_3;q}(a) \nabla_{x_1,x_2;q}(b_1) {}_2\phi_1(a, b_1; c_1 \mid q; x_1) \\ &\times \Phi_1(b_2; b_1, a; c_2 \mid q; x_2, x_3),\end{aligned} \tag{33}$$

$$\begin{aligned}H_B &(a, b_1, b_2; c_1, c, c \mid q; x_1, x_2, x_3) \\ &= \nabla_{x_1,x_3;q}(a) \nabla_{x_1,x_2;q}(b_1) \nabla_{x_2,x_3;q}(c) \\ &\times {}_2\phi_1(a, b_1; c_1 \mid q; x_1) \\ &\times \Phi_1(b_2; b_1, a; c \mid q; x_2, x_3),\end{aligned} \tag{34}$$

$$\begin{aligned}H_C &(a, b_1, b_2; c \mid q; x_1, x_2, x_3) \\ &= \nabla_{x_1,x_3;q}(a) \nabla_{x_1,x_2;q}(b_1) \widetilde{\nabla}_{x_1;x_2x_3;q}(c) \\ &\times {}_2\phi_1(a, b_1; c \mid q; x_1) \\ &\times \Phi_1(b_2; b_1, a; c \mid q; x_2, x_3),\end{aligned} \tag{35}$$

$$\begin{aligned}H_A &(a, b_1, b_2; c_1, c_2 \mid q; x_1, x_2, x_3) \\ &= \nabla_{x_1,x_3;q}(a) \nabla_{x_1,x_2;q}(b_1) \Delta_{x_2,x_3;q}(c_2) \\ &\times {}_2\phi_1(a, b_1; c_1 \mid q; x_1) \Phi_2(b_2; b_1, a; c_2, c_2 \mid q; x_2, x_3).\end{aligned} \tag{36}$$

$$\begin{aligned}H_B &(a, b_1, b_2; c_1, c_2, c_3 \mid q; x_1, x_2, x_3) \\ &= \nabla_{x_1,x_3;q}(a) \nabla_{x_1,x_2;q}(b_1) {}_2\phi_1(a, b_1; c_1 \mid q; x_1) \\ &\times \Phi_2(b_2; b_1, a; c_2, c_3 \mid q; x_2, x_3),\end{aligned} \tag{37}$$

$$\begin{aligned}H_C &(a, b_1, b_2; c \mid q; x_1, x_2, x_3) \\ &= \nabla_{x_1,x_3;q}(a) \nabla_{x_1,x_2;q}(b_1) \Delta_{x_2,x_3;q}(c) \widetilde{\Delta}_{x_1;x_2x_3;q}(c) \\ &\times {}_2\phi_1(a, b_1; c \mid q; x_1) \Phi_2(b_2; b_1, a; c, c \mid q; x_2, x_3),\end{aligned} \tag{38}$$

$$\begin{aligned}H_A &(a, b_1, b_2; c_1, c_2 \mid q; x_1, x_2, x_3) \\ &= \nabla_{x_1,x_3;q}(a) \nabla_{x_1,x_2;q}(b_1) \nabla_{x_2,x_3;q}(b_2) \\ &\times {}_2\phi_1(a, b_1; c_1 \mid q; x_1) \Phi_3(b_1, b_2; b_2, a; c_2 \mid q; x_2, x_3),\end{aligned} \tag{39}$$

$$\begin{aligned}H_B &(a, b_1, b_2; c_1, c, c \mid q; x_1, x_2, x_3) \\ &= \nabla_{x_1,x_3;q}(a) \nabla_{x_1,x_2;q}(b_1) \nabla_{x_2,x_3;q}(b_2) \nabla_{x_2,x_3;q}(c) \\ &\times {}_2\phi_1(a, b_1; c_1 \mid q; x_1) \Phi_3(b_2, a; b_1, b_2; c \mid q; x_2, x_3),\end{aligned} \tag{40}$$

$$\begin{aligned}H_C &(a, b_1, b_2; c \mid q; x_1, x_2, x_3) \\ &= \nabla_{x_1,x_3;q}(a) \nabla_{x_1,x_2;q}(b_1) \nabla_{x_2,x_3;q}(b_2) \widetilde{\Delta}_{x_1;x_2x_3;q}(c) \\ &\times {}_2\phi_1(a, b_1; c \mid q; x_1) \Phi_3(b_2, b_2; b_1, a; c \mid q; x_2, x_3),\end{aligned} \tag{41}$$

$$\begin{aligned}H_A &(a, a, b_2; c_1, c_2 \mid q; x_1, x_2, x_3) \\ &= \nabla_{x_1,x_2;q}(a) \nabla_{x_1,x_3;q}(a) \Delta_{x_2,x_3;q}(a) \Delta_{x_2,x_3;q}(c_2) \\ &\times {}_2\phi_1(a, a; c_1 \mid q; x_1) \Phi_4(a; b_2; c_2, c_2 q; x_2, x_3),\end{aligned} \tag{42}$$

$$\begin{aligned}H_B &(a, a, b_2; c_1, c_2, c_3 \mid q; x_1, x_2, x_3) \\ &= \nabla_{x_1,x_3;q}(a) \nabla_{x_1,x_2;q}(a) \Delta_{x_2,x_3;q}(a) \\ &\times {}_2\phi_1(a, a; c_1 \mid q; x_1) \Phi_4(a; b_2; c_2, c_3 \mid q; x_2, x_3),\end{aligned} \tag{43}$$

$$\begin{aligned}H_C &(a, a, b_2; c \mid q; x_1, x_2, x_3) \\ &= \nabla_{x_1,x_3;q}(a) \nabla_{x_1,x_2;q}(a) \Delta_{x_2,x_3;q}(a) \Delta_{x_2,x_3;q}(c) \widetilde{\Delta}_{x_1;x_2x_3;q}(c) \\ &\times {}_2\phi_1(a, a; c \mid q; x_1) \Phi_4(a; b_2; c, c \mid q; x_2, x_3),\end{aligned} \tag{44}$$

$$\begin{aligned}
&H_A(a,b_1,b_2;c_1,c_2 \mid q;x_1,x_2,x_3)\\
&= \nabla_{x_1,x_3;q}(a)\nabla_{x_1,x_2;q}(b_1)\nabla_{x_2x_3;q}(b_2)\Delta_{x_2,x_3;q}(c_2)\\
&\quad\times {}_2\phi_1(a,b_1;c_1 \mid q;x_1)\,{}_2\phi_1(b_1,b_2;c_2 \mid q;x_2)\\
&\quad\times {}_2\phi_1(a,b_2;c_2 \mid q;x_3),
\end{aligned} \tag{45}$$

$$\begin{aligned}
&H_B(a,b_1,b_2;c_1,c_2,c_3 \mid q;x_1,x_2,x_3)\\
&= \nabla_{x_1,x_3;q}(a)\nabla_{x_1,x_2;q}(b_1)\nabla_{x_2,x_3;q}(b_2)\\
&\quad\times {}_2\phi_1(a,b_1;c_1 \mid q;x_1)\,{}_2\phi_1(b_1,b_2;c_2 \mid q;x_2)\\
&\quad\times {}_2\phi_1(a,b_2;c_3 \mid q;x_3),
\end{aligned} \tag{46}$$

$$\begin{aligned}
&H_C(a,b_1,b_2;c \mid q;x_1,x_2,x_3)\\
&= \nabla_{x_1,x_3;q}(a)\nabla_{x_1,x_2;q}(b_1)\nabla_{x_2,x_3;q}(b_2)\Delta_{x_2,x_3;q}(c)\\
&\quad\times \widetilde{\Delta}_{x_1;x_2x_3;q}(c)\,{}_2\phi_1(a,b_1;c \mid q;x_1)\,{}_2\phi_1(b_1,b_2;c \mid q;x_2)\\
&\quad\times {}_2\phi_1(a,b_2;c \mid q;x_3).
\end{aligned} \tag{47}$$

Remark 8. We note that there are $3\times 5 = 15$ equations and H_A, H_B, and H_C always follow after each other five times. The operator $\widetilde{\nabla}_{x_1;x_2x_3;q}$ always follows after H_C. The slightly more involved proofs of formulas (42) to (44) are somehow connected. The formulas are symmetrically ordered (in triples to the far right) according to the scheme Φ_1, Φ_2, Φ_3, Φ_4, ${}_2\phi_1$. We have corrected a slight error in formula [1, p. 959 (3.6)].

We will now present 10 q-analogues of the improved versions of [1, pp. 960–962]. The proofs are quite similar; in each case, we obtain a 2-, 3-, or 4-sum (denoted by the letters i, j, k, l), followed by a three-sum mnp, which denotes a q-hypergeometric function times a q-Appell function. The most difficult part in each proof is the variable transformation which takes place at the step denoted by. If we denote the variables at this stage by i, j, k, l, r, m, n, p (the old summation indices) and i', j', k', l', r', m', n', p' (the new summation indices), we have to find a symmetric variable transformation with nonzero determinant. The two first transformations take the forms:

$$\begin{aligned}
&m = i' + j' + m', \qquad n = j' + n', \qquad p = j' + p',\\
&\qquad\qquad i = i',\ j = j',\\
&m = i' + j' + m', \qquad n = j' + k' + n', \qquad p = i' + k' + p',\\
&\qquad\qquad i = i',\ j = j',\ k = k'.
\end{aligned} \tag{48}$$

Throughout, we use the following abbreviations:

$$X^M \equiv x_1^m x_2^n x_3^p, \qquad C_2 \equiv c_2 + k, \tag{49}$$

$$\begin{aligned}
&H_A(a,b_1,b_2;c_1,c_2 \mid q;x_1,x_2,x_3)\\
&= \sum_{ij} \frac{\langle 1,a,b_1,b_2;q\rangle_{i+j}}{\langle 1.1;q\rangle_i \langle c_1,c_2;q\rangle_{i+j}\langle 1,1;q\rangle_j}\\
&\quad\times \frac{x_1^{i+j}x_2^j x_3^i q^{i(a+i-1)+j(b_1+j-1)}}{\langle 1,b_1;q\rangle_{i+k}\langle 1;q\rangle_{i-k}\langle 1;q\rangle_n\langle 1;q\rangle_p}\\
&\quad\times {}_4\phi_3\left[\begin{matrix} a+i+j, b_1+i+j, 1+i+j, 1\\ 1+i, 1+j, c_1+i+j\end{matrix} \mid q; x_1\right]\\
&\quad\times \Phi_1(b_2+i+j; b_1+j, a+i; c_2+i+j \mid q; x_2, x_3).
\end{aligned} \tag{50}$$

Proof. Use formula (33). The computation goes as follows:

$$\begin{aligned}
&H_A(a,b_1,b_2;c_1,c_2 \mid q;x_1,x_2,x_3)\\
&= \sum_{ijmnp} \frac{\langle -m,-p;q\rangle_i q^{i(a+m+p)}\langle -m,-n;q\rangle_j q^{j(b_1+m+n)}}{\langle 1,a;q\rangle_i\langle 1,b_1;q\rangle_j}\\
&\quad\times X^M \frac{\langle a,b_1;q\rangle_m\langle b_2;q\rangle_{n+p}\langle b_1;q\rangle_n\langle a;q\rangle_p}{\langle 1,c_1;q\rangle_m\langle c_2;q\rangle_{n+p}\langle 1;q\rangle_n\langle 1;q\rangle_p}\\
&= \sum_{ijmnp} \frac{q^{i(a+i-1)+j(b_1+j-1)}}{\langle 1;q\rangle_{m-i}\langle 1,a;q\rangle_i\langle 1;q\rangle_{m-j}\langle 1;q\rangle_{p-i}\langle 1,b_1;q\rangle_j}\\
&\quad\times X^M \frac{\langle 1,a,b_1;q\rangle_m\langle b_2;q\rangle_{n+p}\langle b_1;q\rangle_n\langle a;q\rangle_p}{\langle 1;q\rangle_{n-j}\langle c_1;q\rangle_m\langle c_2;q\rangle_{n+p}}\\
&\overset{\star}{=} \sum_{ijmnp} \Big(\langle 1,a,b_1;q\rangle_{m+i+j}\langle a;q\rangle_{i+p}\langle b_1;q\rangle_{j+n}\\
&\qquad\times \langle b_2;q\rangle_{i+j+n+p} x_1^{m+i+j} x_2^{j+n} x_3^{i+p}\Big)\\
&\quad\times \left(\langle 1;q\rangle_{m+i}\langle 1;q\rangle_p\langle 1,a;q\rangle_i\langle 1,b_1;q\rangle_j\langle 1;q\rangle_{m+j}\right)^{-1}\\
&\quad\times \frac{q^{i(a+i-1)+j(b_1+j-1)}}{\langle 1;q\rangle_n\langle c_2;q\rangle_{i+j+n+p}\langle c_1;q\rangle_{m+i+j}} = \text{RHS}.
\end{aligned} \tag{51}$$

□

The limit case $x_3 = 0$ in (50) leads to (75). Consider

$$\begin{aligned}
&H_B(a,b_1,b_2;c_1,c,c \mid q;x_1,x_2,x_3)\\
&= \sum_{ijk}\langle 1,a,b_1;q\rangle_{i+j}\langle b_2;q\rangle_{i+j+2k}\mathrm{QE}(j(b_1+j-1))\\
&\quad\times \frac{q^{i(a+i-1)+k(c+k-1)}x_1^{i+j}x_2^{j+k}x_3^{i+k}\langle 1,a;q\rangle_{i+k}\langle 1,b_1;q\rangle_{j+k}}{\langle 1,1,1,b_1;q\rangle_j\langle 1,1,1,a;q\rangle_i\langle 1,1,1,c;q\rangle_k\langle c_1;q\rangle_{i+j}\langle c;q\rangle_{i+j+2k}}\\
&\quad\times {}_4\phi_3\left[\begin{matrix} 1, a+i+j, b_1+i+j, 1+i+j\\ 1+j, c_1+i+j, 1+i\end{matrix} \mid q; x_1\right]\\
&\quad\times \Phi^{1:3}_{1:2}\left[\begin{matrix} b_2+i+j+2k : 1, b_1+j+k, 1+j+k; 1, a+i+k, 1+i+k\\ c+i+j+2k : 1+k, 1+j; 1+k, 1+i\end{matrix} \mid q; x_2, x_3\right]
\end{aligned} \tag{52}$$

Proof. Use formula (34). The very symmetric computation goes as follows:

$$\begin{aligned}
&H_B(a,b_1,b_2;c_1,c,c \mid q;x_1,x_2,x_3)\\
&= \sum_{ijkmnp} \frac{\langle -m,-p;q\rangle_i q^{i(a+m+p)} \langle -m,-n;q\rangle_j q^{j(b_1+m+n)} \langle -n,-p;q\rangle_k q^{k(c+n+p)}}{\langle 1,a;q\rangle_i \langle 1,b_1;q\rangle_j \langle 1,c;q\rangle_k}\\
&\quad \times X^M \frac{\langle a,b_1;q\rangle_m \langle b_2;q\rangle_{n+p} \langle b_1;q\rangle_n \langle a;q\rangle_p}{\langle 1,c_1;q\rangle_m \langle c;q\rangle_{n+p} \langle 1;q\rangle_n \langle 1;q\rangle_p}\\
&= \sum_{ijkmnp} \frac{q^{i(a+i-1)+j(b_1+j-1)+k(c+k-1)}}{\langle 1;q\rangle_{m-i} \langle 1,a;q\rangle_i \langle 1;q\rangle_{m-j} \langle 1;q\rangle_{p-i} \langle 1;q\rangle_{p-k} \langle 1,b_1;q\rangle_j \langle 1,c;q\rangle_k}\\
&\quad \times X^M \frac{\langle 1,a,b_1;q\rangle_m \langle b_2;q\rangle_{n+p} \langle 1,b_1;q\rangle_n \langle 1,a;q\rangle_p}{\langle 1;q\rangle_{n-j} \langle 1;q\rangle_{n-k} \langle c_1;q\rangle_m \langle c;q\rangle_{n+p}}\\
&\stackrel{\star}{=} \sum_{ijkmnp} \frac{\langle 1,a,b_1;q\rangle_{i+j+m} \langle 1,a;q\rangle_{i+k+p} \langle 1,b_1;q\rangle_{j+k+n} \langle b_2;q\rangle_{i+j+2k+n+p} x_1^{i+j+m}}{\langle 1;q\rangle_{m+j} \langle 1;q\rangle_{m+i} \langle 1;q\rangle_{k+p} \langle 1;q\rangle_{i+p} \langle 1,a;q\rangle_i \langle 1,b_1;q\rangle_j \langle 1,c;q\rangle_k \langle 1;q\rangle_{j+n}}\\
&\quad \times \frac{x_2^{j+n+k} x_3^{i+k+p} q^{i(a+i-1)+j(b_1+j-1)+k(c+k-1)}}{\langle 1;q\rangle_{k+n} \langle c;q\rangle_{i+j+2k+n+p} \langle c_1;q\rangle_{i+j+m}} = \text{RHS}.
\end{aligned} \tag{53}$$

□

The limit case $x_1 = 0$ in (52) leads to (76). Consider

$$\begin{aligned}
&H_A(a,b_1,b_2;c_1,c_2 \mid q;x_1,x_2,x_3)\\
&= \sum_{ijk} \frac{q^{i(a+i-1)+j(b_1+j-1)} \langle 1,a,b_1;q\rangle_{i+j} x_1^{i+j} x_2^{k+j} x_3^{i+k} q^{(3/2)(k^2-k)+kc_2}}{\langle C_2;q\rangle_{j+k} \langle C_2;q\rangle_{i+k}}\\
&\quad \times \frac{\langle 1,b_1;q\rangle_{k+j} \langle b_2;q\rangle_{i+j+2k} (-1)^k \langle c_2;q\rangle_{2k} \langle 1,a;q\rangle_{i+k} \langle 1+i+k, a+i+k;q\rangle_p}{\langle 1,1,1,a;q\rangle_i \langle 1,1,1,b_1;q\rangle_j \langle c_1;q\rangle_{i+j} \langle 1,1,1,C_2-1,c_2,c_2;q\rangle_k}\\
&\quad \times {}_4\phi_3\left[\begin{matrix} 1,1+i+j,a+i+j,b_1+i+j \\ 1+i,1+j,c_1+i+j \end{matrix} \mid q;x_1\right]\\
&\quad \times \Phi^{1:4}_{1:3}\left[\begin{matrix} b_2+i+j+2k : 1,1+k+j,b_1+k+j,\infty;1,1+i+k,a+i+k,\infty \\ \infty : c_2+j+2k,1+k,1+j;c_2+i+2k,1+i,1+k \end{matrix} \mid q;x_2,x_3\right].
\end{aligned} \tag{54}$$

Proof. Use formula (36). The computation goes as follows:

$$
\begin{aligned}
&H_A(a,b_1,b_2;c_1,c_2 \mid q;x_1,x_2,x_3) \\
&= \sum_{ijkmnp} \frac{\langle -m,-p;q\rangle_i q^{i(a+m+p)} \langle -m,-n;q\rangle_j q^{j(b_1+m+n)} (-1)^k \langle -n,-p;q\rangle_k \langle c_2;q\rangle_{2k}}{\langle 1,a;q\rangle_i \langle 1,b_1;q\rangle_j \langle 1,C_2-1,c_2+n,c_2+p;q\rangle_k} \\
&\quad \times X^M \frac{\langle a,b_1;q\rangle_m \langle b_2;q\rangle_{n+p} \langle b_1;q\rangle_n \langle a;q\rangle_p}{\langle 1,c_1;q\rangle_m \langle 1,c_2;q\rangle_n \langle 1,c_2;q\rangle_p} q^{\binom{k}{2}+k(c_2+n+p)} \\
&= \sum_{ijkmnp} \frac{\langle 1,a,b_1;q\rangle_m q^{i(a+i-1)+j(b_1+j-1)} \langle 1,a;q\rangle_p (-1)^k \langle c_2;q\rangle_{2k}}{\langle 1;q\rangle_{m-i} \langle 1;q\rangle_{m-j} \langle 1,a;q\rangle_i \langle 1,b_1;q\rangle_j \langle 1;q\rangle_{p-i}} \\
&\quad \times \frac{X^M \langle b_2;q\rangle_{n+p} \langle 1,b_1;q\rangle_n q^{(3/2)(k^2-k)+kc_2}}{\langle 1;q\rangle_{n-j} \langle 1;q\rangle_{n-k} \langle c_1;q\rangle_m \langle C_2;q\rangle_n \langle C_2;q\rangle_p \langle 1;q\rangle_{p-k} \langle 1,C_2-1,c_2,c_2;q\rangle_k} \\
&\stackrel{\star}{=} \sum_{ijkmnp} \frac{q^{i(a+i-1)+j(b_1+j-1)} \langle 1,a,b_1;q\rangle_{m+i+j} x_1^{m+i+j} x_2^{n+k+j} x_3^{p+i+k} q^{(3/2)(k^2-k)+kc_2}}{\langle 1,C_2-1,c_2,c_2;q\rangle_k \langle C_2;q\rangle_{n+j+k} \langle C_2;q\rangle_{p+i+k} \langle 1,a;q\rangle_i \langle 1,b_1;q\rangle_j} \\
&\quad \times \frac{(-1)^k \langle c_2;q\rangle_{2k} \langle b_2;q\rangle_{n+p+i+j+2k} \langle 1,a;q\rangle_{p+i+k} \langle 1,b_1;q\rangle_{n+k+j}}{\langle 1;q\rangle_{m+j} \langle 1;q\rangle_{m+i} \langle 1;q\rangle_{p+k} \langle 1;q\rangle_{p+i} \langle 1;q\rangle_{n+k} \langle 1;q\rangle_{n+j} \langle c_1;q\rangle_{m+i+j}} = \text{RHS},
\end{aligned}
\tag{55}
$$

$$
\begin{aligned}
&H_B(a,b_1,b_2;c_1,c_2,c_3 \mid q;x_1,x_2,x_3) \\
&= \sum_{ij} \frac{\langle 1,a,b_1,b_2;q\rangle_{i+j}}{\langle 1,1,c_3;q\rangle_i \langle c_1;q\rangle_{i+j} \langle 1,1,c_2;q\rangle_j} \\
&\quad \times x_1^{i+j} x_2^j x_3^i q^{i(a+i-1)+j(b_1+j-1)} {}_4\phi_3 \left[\begin{matrix} a+i+j, b_1+i+j, 1+i+j, 1 \\ 1+i, 1+j, c_1+i+j \end{matrix} \mid q;x_1 \right] \\
&\quad \times \Phi^{1:2}_{1:1} \left[\begin{matrix} b_2+i+j : b_1+j, \infty; a+i, \infty \\ \infty : c_2+j; c_3+i \end{matrix} \mid q;x_2,x_3 \right].
\end{aligned}
\tag{56}
$$

□

Proof. Use formula (37). The computation goes as follows:

$$
\begin{aligned}
&H_B(a,b_1,b_2;c_1,c_2,c_3 \mid q;x_1,x_2,x_3) \\
&= \sum_{ijmnp} \frac{\langle -m,-p;q\rangle_i q^{i(a+m+p)} \langle -m,-n;q\rangle_j q^{j(b_1+m+n)}}{\langle 1,a;q\rangle_i \langle 1,b_1;q\rangle_j} X^M \frac{\langle a,b_1;q\rangle_m \langle b_2;q\rangle_{n+p} \langle b_1;q\rangle_n \langle a;q\rangle_p}{\langle 1,c_1;q\rangle_m \langle 1,c_2;q\rangle_n \langle 1,c_3;q\rangle_p} \\
&= \sum_{ijmnp} \frac{\langle 1,a,b_1;q\rangle_m \langle b_2;q\rangle_{n+p} \langle b_1;q\rangle_n \langle a;q\rangle_p q^{i(a+i-1)+j(b_1+j-1)} X^M}{\langle 1;q\rangle_{m-i} \langle 1,a;q\rangle_i \langle 1;q\rangle_{m-j} \langle 1;q\rangle_{p-i} \langle 1,b_1;q\rangle_j \langle 1;q\rangle_{n-j} \langle c_1;q\rangle_m \langle c_2;q\rangle_n \langle c_3;q\rangle_p} \\
&\stackrel{\star}{=} \sum_{ijmnp} \frac{\langle 1,a,b_1;q\rangle_{m+i+j} \langle a;q\rangle_{i+p} \langle b_1;q\rangle_{j+n} \langle b_2;q\rangle_{i+j+n+p} x_1^{m+i+j} x_2^{j+n} x_3^{i+p}}{\langle 1;q\rangle_{m+i} \langle 1;q\rangle_{m+j} \langle 1;q\rangle_p \langle 1,a;q\rangle_i \langle 1,b_1;q\rangle_j \langle 1;q\rangle_n} \frac{q^{i(a+i-1)+j(b_1+j-1)}}{\langle c_1;q\rangle_{m+i+j} \langle c_2;q\rangle_{n+j} \langle c_3;q\rangle_{p+i}} = \text{RHS}.
\end{aligned}
\tag{57}
$$

□

The limit case $x_3 = 0$ in (56) leads to (75). Consider

$$
\begin{aligned}
&H_A(a,b_1,b_2;c_1,c_2 \mid q;x_1,x_2,x_3)\\
&=\sum_{ijk}\frac{\langle 1,a,b_1;q\rangle_{i+j}}{\langle c_1;q\rangle_{i+j}}x_1^{i+j}x_2^{j+k}x_3^{i+k}\\
&\times\frac{q^{i(a+i-1)+k(b_2+k-1)+j(b_1+j-1)}\langle 1,a,b_2;q\rangle_{i+k}\langle 1,b_1,b_2;q\rangle_{j+k}}{\langle 1,1,1,b_1;q\rangle_j\langle 1,1,1,a;q\rangle_i\langle 1,1,1,b_2;q\rangle_k\langle c_2;q\rangle_{i+j+2k}}\\
&\times {}_4\phi_3\left[\begin{matrix}1,a+i+j,b_1+i+j,1+i+j\\ 1+j,c_1+i+j,1+i\end{matrix}\mid q;x_1\right]\\
&\times\Phi^{1:4}_{1:3}\left[\begin{matrix}\infty:1,b_1+j+k,1+j+k,b_2+j+k;1,a+i+k,1+i+k,b_2+i+k\\ c_2+i+j+2k:\infty,1+k,1+j;\infty,1+i,1+k\end{matrix}\mid q;x_2,x_3\right].
\end{aligned}
\tag{58}
$$

Proof. Use formula (39). The computation goes as follows:

$$
\begin{aligned}
&H_A(a,b_1,b_2;c_1,c_2 \mid q;x_1,x_2,x_3)\\
&=\sum_{ijkmnp}\frac{\langle -m,-p;q\rangle_i q^{i(a+m+p)}\langle -m,-n;q\rangle_j q^{j(b_1+m+n)}\langle -n,-p;q\rangle_k q^{k(b_2+n+p)}}{\langle 1,a;q\rangle_i\langle 1,b_1;q\rangle_j\langle 1,b_2;q\rangle_k}\\
&\qquad\times X^M\frac{\langle a,b_1;q\rangle_m\langle b_1,b_2;q\rangle_n\langle a,b_2;q\rangle_p}{\langle 1,c_1;q\rangle_m\langle c_2;q\rangle_{n+p}\langle 1;q\rangle_n\langle 1;q\rangle_p}\\
&=\sum_{ijkmnp}\frac{q^{i(a+i-1)+j(b_1+j-1)+k(b_2+k-1)}}{\langle 1;q\rangle_{m-i}\langle 1,a;q\rangle_i\langle 1;q\rangle_{m-j}\langle 1;q\rangle_{p-i}\langle 1;q\rangle_{p-k}\langle 1,b_1;q\rangle_j\langle 1,b_2;q\rangle_k}\\
&\qquad\times X^M\frac{\langle 1,a,b_1;q\rangle_m\langle 1,b_1,b_2;q\rangle_n\langle 1,a,b_2;q\rangle_p}{\langle 1;q\rangle_{n-j}\langle 1;q\rangle_{n-k}\langle c_1;q\rangle_m\langle c_2;q\rangle_{n+p}}\\
&\overset{\star}{=}\sum_{ijkmnp}\frac{\langle 1,a,b_1;q\rangle_{i+j+m}\langle 1,a,b_2;q\rangle_{i+k+p}\langle 1,b_1,b_2;q\rangle_{j+k+n}x_1^{i+j+m}}{\langle 1;q\rangle_{m+j}\langle 1;q\rangle_{m+i}\langle 1;q\rangle_{k+p}\langle 1;q\rangle_{i+p}\langle 1,a;q\rangle_i\langle 1,b_1;q\rangle_j\langle 1,b_2;q\rangle_k}\\
&\qquad\times\frac{x_2^{j+n+k}x_3^{i+k+p}q^{i(a+i-1)+j(b_1+j-1)+k(b_2+k-1)}}{\langle 1;q\rangle_{j+n}\langle 1;q\rangle_{k+n}\langle c_2;q\rangle_{i+j+2k+n+p}\langle c_1;q\rangle_{i+j+m}}=\text{RHS}.
\end{aligned}
\tag{59}
$$

□

Put

$$
\begin{aligned}
\alpha &\equiv (1,1+j+k+l,1+j+k+l,b_1+j+k+l,b_2+j+k+l);\\
&\quad(1,1+i+k+l,1+i+k+l,a+i+k+l,b_2+i+k+l)\\
\beta &\equiv (1+j+k,1+k+l,1+j+l,\infty);\\
&\quad(1+k+l,1+i+l,1+i+k,\infty).
\end{aligned}
\tag{60}
$$

Then,

$$\begin{aligned} &H_B\left(a,b_1,b_2;c_1,c,c \mid q;x_1,x_2,x_3\right) \\ &= \sum_{ijkl} \frac{\langle 1,a,b_1;q\rangle_{i+j} q^{j(b_1+j-1)+l(c+l-1)}}{\langle 1,1;q\rangle_{k+l}\langle 1;q\rangle_{i+l}\langle 1;q\rangle_{j+l}\langle 1;q\rangle_{j+k}\langle 1;q\rangle_{i+k}\langle 1,c;q\rangle_l} \\ &\times \frac{q^{i(a+i-1)+k(b_2+k-1)} x_1^{i+j} x_2^{j+k+l} x_3^{i+k+l}\langle 1,1,a,b_2;q\rangle_{i+k+l}\langle 1,1,b_1,b_2;q\rangle_{j+k+l}}{\langle 1,1,b_1;q\rangle_j\langle 1,1,a;q\rangle_i\langle 1,b_2;q\rangle_k\langle c_1;q\rangle_{i+j}\langle c;q\rangle_{i+j+2k+2l}} \\ &\times {}_4\phi_3\left[\begin{matrix} 1,a+i+j,b_1+i+j,1+i+j \\ 1+j,c_1+i+j,1+i \end{matrix} \mid q;x_1\right] \\ &\times \Phi^{1:5}_{1:4}\left[\begin{matrix} \infty:\alpha \\ c+i+j+2k+2l:\beta \end{matrix} \mid q;x_2,x_3\right]. \end{aligned} \tag{61}$$

Proof. Use formula (40). The computation goes as follows:

$$\begin{aligned} &H_B\left(a,b_1,b_2;c_1,c,c \mid q;x_1,x_2,x_3\right) \\ &= \sum_{ijklmnp} \frac{\langle -m,-p;q\rangle_i q^{i(a+m+p)}\langle -m,-n;q\rangle_j q^{j(b_1+m+n)}\langle -n,-p;q\rangle_k}{\langle 1,a;q\rangle_i\langle 1,b_1;q\rangle_j\langle 1,b_2;q\rangle_k\langle 1,c;q\rangle_l} \\ &\times X^M \frac{\langle a,b_1;q\rangle_m\langle b_1,b_2;q\rangle_n\langle a,b_2;q\rangle_p\langle -n,-p;q\rangle_l q^{k(b_2+n+p)+l(c+n+p)}}{\langle 1,c_1;q\rangle_m\langle c;q\rangle_{n+p}\langle 1;q\rangle_n\langle 1;q\rangle_p} \\ &= \sum_{ijklmnp} \frac{\langle 1,a,b_1;q\rangle_m q^{i(a+i-1)+j(b_1+j-1)+k(b_2+k-1)+l(c+l-1)}}{\langle 1;q\rangle_{m-i}\langle 1,a;q\rangle_i\langle 1;q\rangle_{m-j}\langle 1;q\rangle_{p-i}\langle 1;q\rangle_{p-k}\langle 1,b_1;q\rangle_j\langle 1,b_2;q\rangle_k} \\ &\times X^M \frac{\langle 1,1,b_1,b_2;q\rangle_n\langle 1,1,a,b_2;q\rangle_p}{\langle 1;q\rangle_{n-j}\langle 1;q\rangle_{n-k}\langle 1;q\rangle_{n-l}\langle 1;q\rangle_{p-l}\langle c_1;q\rangle_m\langle c;q\rangle_{n+p}\langle 1,c;q\rangle_l} \\ &\stackrel{\star}{=} \sum_{ijklmnp} \frac{\langle 1,a,b_1;q\rangle_{i+j+m}\langle 1,1,a,b_2;q\rangle_{i+k+l+p}\langle 1,1,b_1,b_2;q\rangle_{j+k+l+n}}{\langle 1;q\rangle_{m+j}\langle 1;q\rangle_{m+i}\langle 1;q\rangle_{k+l+p}\langle 1;q\rangle_{i+l+p}\langle 1;q\rangle_{i+k+p}\langle 1,a;q\rangle_i\langle 1,b_1;q\rangle_j\langle 1,b_2;q\rangle_k} \\ &\times \frac{x_1^{i+j+m} x_2^{j+k+l+n} x_3^{i+k+l+p} q^{i(a+i-1)+j(b_1+j-1)+k(b_2+k-1)+l(c+l-1)}}{\langle 1;q\rangle_{j+k+n}\langle 1;q\rangle_{k+l+n}\langle c_1;q\rangle_{i+j+m}\langle c;q\rangle_{i+j+2k+2l+n+p}\langle 1,c;q\rangle_l\langle 1;q\rangle_{j+l+n}} = \text{RHS}. \end{aligned} \tag{62}$$

□

Put

$$\begin{aligned} \alpha &\equiv (1,1+j+k+l,1+j+k+l,a+j+k+l,\infty,\infty); \\ &\quad (1,1+i+k+l,1+i+k+l,a+i+k+l,\infty,\infty), \\ \beta &\equiv (1+l+k,1+j+l,1+j+k,C_2 \\ &\quad +j+k+l,a+j+k+2l); \\ &\quad (1+k+l,1+i+k,1+i+l,C_2 \\ &\quad +i+k+l,a+i+k+2l). \end{aligned} \tag{63}$$

Then,

$$H_A(a,a,b_2;c_1,c_2 \mid q;x_1,x_2,x_3) = \sum_{ijkl} \frac{q^{i(a+i-1)+j(a+j-1)+(3/2)(k^2-k)+kc_2+(3/2)(l^2-l)+la}\langle 1,a,a;q\rangle_{i+j} x_1^{i+j} x_2^{k+j+l} x_3^{i+k+l}}{\langle C_2;q\rangle_{j+k+l}\langle C_2;q\rangle_{i+k+l}\langle 1,1,a;q\rangle_i\langle 1,1,a;q\rangle_j\langle 1;q\rangle_{j+l}\langle 1;q\rangle_{j+k}}$$
$$\times \frac{\langle a,b_2;q\rangle_{i+j+2k+2l}(-1)^{k+l}\langle c_2;q\rangle_{2k}\langle a;q\rangle_{2l}}{\langle 1,1;q\rangle_{k+l}\langle 1;q\rangle_{i+k}\langle 1;q\rangle_{i+l}\langle c_1;q\rangle_{i+j}\langle 1,C_2-1,c_2,c_2;q\rangle_k\langle 1,a+l-1,a,a;q\rangle_l}$$
$$\times \frac{\langle 1,1,a;q\rangle_{j+k+l}\langle 1,1,a;q\rangle_{i+k+l}}{\langle a+l;q\rangle_{j+k+l}\langle a+l;q\rangle_{i+k+l}} {}_4\phi_3\left[\begin{matrix}1,1+i+j,a+i+j,a+i+j\\ 1+i,1+j,c_1+i+j\end{matrix} \mid q;x_1\right]$$
$$\times \Phi^{2:6}_{2:5}\left[\begin{matrix}a+i+j+2k+2l,b_2+i+j+2k+2l:\alpha\\ \infty,\infty:\beta\end{matrix} \mid q;x_2,x_3\right]. \tag{64}$$

Proof. Use formula (42). The computation goes as follows:

$$H_A(a,a,b_2;c_1,c_2 \mid q;x_1,x_2,x_3) = \sum_{ijklmnp} \frac{X^M\langle -m,-p;q\rangle_i\langle -m,-n;q\rangle_j q^{j(a+m+n)}(-1)^{k+l}\langle -n,-p;q\rangle_k\langle -n,-p;q\rangle_l}{\langle 1,a;q\rangle_j\langle 1,C_2-1,c_2+n,c_2+p;q\rangle_k\langle 1,a+l-1,a+n,a+p;q\rangle_l}$$
$$\times \frac{\langle a,a;q\rangle_m\langle a,b_2;q\rangle_{n+p}\langle a;q\rangle_{2l}\langle c_2;q\rangle_{2k} q^{\binom{k}{2}+i(a+m+p)+k(c_2+n+p)+\binom{l}{2}+l(a+n+p)}}{\langle 1,a;q\rangle_i\langle 1,c_1;q\rangle_m\langle 1,c_2;q\rangle_n\langle 1,c_2;q\rangle_p}$$
$$= \sum_{ijklmnp} \frac{\langle 1,a,a;q\rangle_m q^{i(a+i-1)+j(a+j-1)}\langle 1,1,a;q\rangle_n\langle 1,1,a;q\rangle_p(-1)^{k+l}\langle c_2;q\rangle_{2k}\langle a;q\rangle_{2l}}{\langle 1;q\rangle_{m-i}\langle 1;q\rangle_{m-j}\langle 1,a;q\rangle_i\langle 1,a;q\rangle_j\langle 1;q\rangle_{p-i}\langle 1,a+l-1,a,a;q\rangle_l}$$
$$\times \frac{X^M\langle a,b_2;q\rangle_{n+p} q^{(3/2)(k^2-k)+kc_2+(3/2)(l^2-l)+la}}{\langle 1;q\rangle_{n-j}\langle 1;q\rangle_{n-k}\langle 1;q\rangle_{n-l}\langle c_1;q\rangle_m\langle a+l,C_2;q\rangle_n\langle a+l,C_2;q\rangle_p}$$
$$\times \frac{1}{\langle 1;q\rangle_{p-k}\langle 1;q\rangle_{p-l}\langle 1,C_2-1,c_2,c_2;q\rangle_k}$$
$$\stackrel{\star}{=} \sum_{ijklmnp} \frac{x_1^{m+i+j}x_2^{n+j+k+l}x_3^{p+i+k+l} q^{i(a+i-1)+j(a+j-1)+(3/2)(k^2-k)+kc_2+(3/2)(l^2-l)+la}}{\langle 1;q\rangle_{p+k+l}\langle 1;q\rangle_{p+i+k}\langle 1,a;q\rangle_i\langle 1,a;q\rangle_j}$$
$$\times \frac{\langle 1,a,a;q\rangle_{m+i+j}\langle a,b_2;q\rangle_{n+p+i+j+2k+2l}\langle a;q\rangle_{2l}}{\langle 1;q\rangle_{p+i+l}\langle 1;q\rangle_{n+l+k}\langle 1;q\rangle_{m+j}\langle 1;q\rangle_{m+i}\langle 1;q\rangle_{n+j+l}\langle c_1;q\rangle_{m+i+j}\langle a+l,C_2;q\rangle_{n+j+k+l}}$$
$$\times \frac{\langle 1,1,a;q\rangle_{n+j+k+l}\langle 1,1,a;q\rangle_{p+i+k+l}(-1)^{k+l}\langle c_2;q\rangle_{2k}}{\langle 1;q\rangle_{n+j+k}\langle 1,C_2-1,c_2,c_2;q\rangle_k\langle 1,a+l-1,a,a;q\rangle_l\langle a+l,C_2;q\rangle_{p+i+k+l}} = \text{RHS}. \tag{65}$$

□

Put

$$\alpha \equiv (1,1+k+j,a+k+j,\infty,\infty);$$
$$(1,1+i+k,a+i+k,\infty,\infty),$$

$$\beta \equiv (a+j+2k, c_2+j+k, 1+k, 1+j); \\ (a+i+2k, c_3+i+k, 1+i, 1+k). \tag{66}$$

Then,

$$\begin{aligned} &H_B(a,a,b_2;c_1,c_2,c_3 \mid q; x_1,x_2,x_3) \\ &= \sum_{ijk} \frac{q^{i(a+i-1)+j(a+j-1)}\langle 1,a,a;q\rangle_{i+j} x_1^{i+j} x_2^{k+j} x_3^{i+k} q^{(3/2)(k^2-k)+ka}}{\langle a+k;q\rangle_{j+k}\langle a+k;q\rangle_{i+k}\langle c_2;q\rangle_{k+j}\langle c_3;q\rangle_{i+k}} \\ &\times \frac{\langle 1,a;q\rangle_{k+j}\langle a,b_2;q\rangle_{i+j+2k}(-1)^k\langle a;q\rangle_{2k}\langle 1,a;q\rangle_{i+k}\langle 1+i+k,a+i+k;q\rangle_p}{\langle 1,1,1,a;q\rangle_i\langle 1,1,1,a;q\rangle_j\langle c_1;q\rangle_{i+j}\langle 1,1,a+k-1,a,a;q\rangle_k} \\ &\times {}_4\phi_3\left[\begin{matrix} 1,1+i+j,a+i+j,a+i+j \\ 1+i,1+j,c_1+i+j \end{matrix} \mid q; x_1\right] \\ &\times \Phi^{2:5}_{2:4}\left[\begin{matrix} a+i+j+2k, b_2+i+j+2k : \alpha \\ \infty,\infty : \beta \end{matrix} \mid q; x_2, x_3\right]. \end{aligned} \tag{67}$$

Proof. Use formula (43). The computation goes as follows:

$$\begin{aligned} &H_B(a,a,b_2;c_1,c_2,c_3 \mid q; x_1,x_2,x_3) \\ &= \sum_{ijkmnp} \frac{\langle -m,-p;q\rangle_i q^{i(a+m+p)}\langle -m,-n;q\rangle_j q^{j(a+m+n)}(-1)^k\langle -n,-p;q\rangle_k\langle a;q\rangle_{2k}}{\langle 1,a;q\rangle_i\langle 1,a;q\rangle_j\langle 1,a+k-1,a+n,a+p;q\rangle_k} \\ &\times X^M \frac{\langle a,a;q\rangle_m\langle a,b_2;q\rangle_{n+p}}{\langle 1,c_1;q\rangle_m\langle 1,c_2;q\rangle_n\langle 1,c_3;q\rangle_p} q^{\binom{k}{2}+k(a+n+p)} \\ &= \sum_{ijkmnp} \frac{\langle 1,a,a;q\rangle_m q^{i(a+i-1)+j(a+j-1)}\langle 1,a;q\rangle_p(-1)^k\langle a;q\rangle_{2k}}{\langle c_2;q\rangle_n\langle c_3;q\rangle_p\langle 1;q\rangle_{m-i}\langle 1;q\rangle_{m-j}\langle 1,a;q\rangle_i\langle 1,a;q\rangle_j\langle 1;q\rangle_{p-i}} \\ &\times \frac{X^M\langle a,b_2;q\rangle_{n+p}\langle a,1;q\rangle_n q^{(3/2)(k^2-k)+ka}}{\langle 1;q\rangle_{n-j}\langle 1;q\rangle_{n-k}\langle c_1;q\rangle_m\langle a+k;q\rangle_n\langle a+k;q\rangle_p\langle 1;q\rangle_{p-k}\langle 1,a+k-1,a,a;q\rangle_k} \\ &\overset{\star}{=} \sum_{ijkmnp} \frac{q^{i(a+i-1)+j(a+j-1)}\langle 1,a,a;q\rangle_{m+i+j} x_1^{m+i+j} x_2^{n+k+j} x_3^{p+i+k} q^{(3/2)(k^2-k)+ka}}{\langle a+k,c_2;q\rangle_{n+j+k}\langle a+k,c_3;q\rangle_{p+i+k}\langle 1,a;q\rangle_i\langle 1,a;q\rangle_j} \\ &\times \frac{\langle a,b_2;q\rangle_{n+p+i+j+2k}(-1)^k\langle a;q\rangle_{2k}}{\langle 1,a+k-1,a,a;q\rangle_k} \\ &\times \frac{\langle 1,a;q\rangle_{n+k+j}\langle 1,a;q\rangle_{p+k+i}}{\langle 1;q\rangle_{m+j}\langle 1;q\rangle_{m+i}\langle 1;q\rangle_{p+k}\langle 1;q\rangle_{p+i}\langle 1;q\rangle_{n+k}\langle 1;q\rangle_{n+j}\langle c_1;q\rangle_{m+i+j}} = \text{RHS}. \end{aligned} \tag{68}$$

□

Put

$$\alpha \equiv (1, 1+k+j+l, 1+k+j+l, b_1 \\ +k+j+l, b_2+k+j+l); \\ (1, 1+i+k+l, 1+i+k+l, a \\ +i+k+l, b_2+i+k+l)$$

$$\beta \equiv (c_2 + j + 2k + l, 1 + j + k, 1 + k + l, 1 + j + l); \\ (c_2 + i + 2k + l, 1 + i + l, 1 + k + l, 1 + i + k). \tag{69}$$

Then,

$$\begin{aligned} &H_A(a, b_1, b_2; c_1, c_2 \mid q; x_1, x_2, x_3) \\ &= \sum_{ijkl} \frac{q^{i(a+i-1)+j(b_1+j-1)+l(b_2+l-1)} \langle 1, a, b_1; q\rangle_{i+j} x_1^{i+j} x_2^{k+j+l} x_3^{i+k+l} q^{(3/2)(k^2-k)+kc_2}}{\langle c_2 + k; q\rangle_{j+k+l} \langle c_2 + k; q\rangle_{i+k+l} \langle 1, a; q\rangle_i} \\ &\times \frac{\langle 1, 1, b_1, b_2; q\rangle_{k+j+l} \langle 1, 1, a, b_2; q\rangle_{i+k+l} (-1)^k \langle c_2; q\rangle_{2k} \langle 1, a; q\rangle_{i+k}}{\langle 1, b_2; q\rangle_l \langle 1, 1, b_1; q\rangle_j \langle c_1; q\rangle_{i+j} \langle 1, c_2 + k - 1, c_2, c_2; q\rangle_k} \\ &\times \frac{1}{\langle 1, 1, a; q\rangle_i \langle 1; q\rangle_{j+k} \langle 1; q\rangle_{i+k} \langle 1; q\rangle_{i+l} \langle 1, 1; q\rangle_{k+l} \langle 1; q\rangle_{j+l}} \\ &\times {}_4\phi_3 \left[\begin{matrix} 1, 1 + i + j, a + i + j, b_1 + i + j \\ 1 + i, 1 + j, c_1 + i + j \end{matrix} \mid q; x_1 \right] \Phi^{0:5}_{0:4} \left[\begin{matrix} - : \alpha \\ - : \beta \end{matrix} \mid q; x_2, x_3 \right]. \end{aligned} \tag{70}$$

Proof. Use formula (45). The computation goes as follows:

$$\begin{aligned} &H_A(a, b_1, b_2; c_1, c_2 \mid q; x_1, x_2, x_3) \\ &= \sum_{ijklmnp} \frac{\langle -m, -p; q\rangle_i q^{i(a+m+p)} \langle -m, -n; q\rangle_j q^{j(b_1+m+n)} (-1)^k \langle -n, -p; q\rangle_k \langle c_2; q\rangle_{2k}}{\langle 1, a; q\rangle_i \langle 1, b_1; q\rangle_j \langle 1, c_2 + k - 1, c_2 + n, c_2 + p; q\rangle_k} \\ &\quad \times X^M \frac{\langle a, b_1; q\rangle_m \langle b_1, b_2; q\rangle_n \langle a, b_2; q\rangle_p}{\langle 1, c_1; q\rangle_m \langle 1, c_2; q\rangle_n \langle 1, c_2; q\rangle_p} q^{\binom{k}{2}+k(c_2+n+p)} \frac{\langle -n, -p; q\rangle_l q^{l(b_2+n+p)}}{\langle 1, b_2; q\rangle_l} \\ &= \sum_{ijklmnp} \frac{\langle 1, a, b_1; q\rangle_m q^{i(a+i-1)+j(b_1+j-1)+l(b_2+l-1)} \langle 1, 1, a, b_2; q\rangle_p (-1)^k \langle c_2; q\rangle_{2k}}{\langle 1; q\rangle_{m-i} \langle 1; q\rangle_{m-j} \langle 1; q\rangle_{n-j} \langle 1; q\rangle_{n-k} \langle 1, a; q\rangle_i \langle 1, b_1; q\rangle_j \langle 1; q\rangle_{p-i} \langle 1, b_2; q\rangle_l} \\ &\quad \times \frac{X^M \langle 1, 1, b_1, b_2; q\rangle_n q^{(3/2)(k^2-k)+kc_2}}{\langle 1; q\rangle_{n-l} \langle 1; q\rangle_{p-l} \langle c_1; q\rangle_m \langle c_2 + k; q\rangle_n \langle c_2 + k; q\rangle_p \langle 1; q\rangle_{p-k} \langle 1, c_2 + k - 1, c_2, c_2; q\rangle_k} \\ &\stackrel{\star}{=} \sum_{ijklmnp} \frac{q^{i(a+i-1)+j(b_1+j-1)} \langle 1, a, b_1; q\rangle_{m+i+j} x_1^{m+i+j} x_2^{n+j+k+l} x_3^{p+i+k+l} q^{(3/2)(k^2-k)+kc_2}}{\langle c_2 + k; q\rangle_{n+j+k+l} \langle c_2 + k; q\rangle_{p+i+k+l} \langle 1, a; q\rangle_i \langle 1, b_1; q\rangle_j} \\ &\quad \times \frac{\langle 1, 1, b_1, b_2; q\rangle_{n+k+j+l}}{\langle 1; q\rangle_{n+j+k} \langle 1; q\rangle_{p+i+k} \langle 1, b_2; q\rangle_l \langle 1; q\rangle_{m+j} \langle 1; q\rangle_{m+i} \langle 1; q\rangle_{p+i+l} \langle 1; q\rangle_{n+k+l} \langle c_1; q\rangle_{m+i+j}} \\ &\quad \times \frac{\langle 1, 1, a, b_2; q\rangle_{p+i+k+l} (-1)^k \langle c_2; q\rangle_{2k} q^{l(b_2+l-1)}}{\langle 1; q\rangle_{n+j+l} \langle 1; q\rangle_{p+k+l} \langle 1, c_2 + k - 1, c_2, c_2; q\rangle_k} = \text{RHS}. \end{aligned} \tag{71}$$

□

Put

$$\begin{aligned}\alpha &\equiv (1, b_1 + j + k, 1 + j + k, b_2 + j + k); (1, a + i + k, 1 + i + k, b_2 + i + k)\\ \beta &\equiv (1 + k, 1 + j, c_2 + j + k); (1 + k, 1 + i, c_3 + i + k).\end{aligned} \tag{72}$$

Then,

$$\begin{aligned}&H_B(a, b_1, b_2; c_1, c_2, c_3 \mid q; x_1, x_2, x_3)\\ &= \sum_{ijk} \langle 1, a, b_1; q\rangle_{i+j} \mathrm{QE}(j(b_1 + j - 1)) \frac{q^{i(a+i-1)+k(b_2+k-1)} x_1^{i+j} x_2^{j+k} x_3^{i+k} q\langle 1, a, b_2; q\rangle_{i+k} \langle 1, b_1, b_2; q\rangle_{j+k}}{\langle 1, 1, 1, b_1; q\rangle_j \langle 1, 1, 1, a; q\rangle_i \langle 1, 1, 1, b_2; q\rangle_k \langle c_1; q\rangle_{i+j} \langle c_2; q\rangle_{j+k} \langle c_3; q\rangle_{i+k}}\\ &\quad \times {}_4\phi_3 \left[\begin{matrix} 1, a + i + j, b_1 + i + j, 1 + i + j \\ 1 + j, c_1 + i + j, 1 + i \end{matrix} \mid q; x_1 \right] \Phi^{0:4}_{0:3} \left[\begin{matrix} - : \alpha \\ - : \beta \end{matrix} \mid q; x_2, x_3 \right].\end{aligned} \tag{73}$$

Proof. Use formula (46). The computation goes as follows:

$$\begin{aligned}&H_B(a, b_1, b_2; c_1, c_2, c_3 \mid q; x_1, x_2, x_3)\\ &= \sum_{ijkmnp} \frac{\langle -m, -p; q\rangle_i q^{i(a+m+p)} \langle -m, -n; q\rangle_j q^{j(b_1+m+n)} \langle -n, -p; q\rangle_k q^{k(b_2+n+p)}}{\langle 1, a; q\rangle_i \langle 1, b_1; q\rangle_j \langle 1, b_2; q\rangle_k} X^M \frac{\langle a, b_1; q\rangle_m \langle b_1, b_2; q\rangle_n \langle a, b_2; q\rangle_p}{\langle 1, c_1; q\rangle_m \langle 1, c_2; q\rangle_n \langle 1, c_3; q\rangle_p}\\ &= \sum_{ijkmnp} \frac{q^{i(a+i-1)+j(b_1+j-1)+k(b_2+k-1)}}{\langle 1; q\rangle_{m-i} \langle 1, a; q\rangle_i \langle 1; q\rangle_{m-j} \langle 1; q\rangle_{p-i} \langle 1; q\rangle_{p-k} \langle 1, b_1; q\rangle_j \langle 1, b_2; q\rangle_k} X^M \frac{\langle 1, a, b_1; q\rangle_m \langle 1, b_1, b_2; q\rangle_n \langle 1, a, b_2; q\rangle_p}{\langle 1; q\rangle_{n-j} \langle 1; q\rangle_{n-k} \langle c_1; q\rangle_m \langle c_2; q\rangle_n \langle c_3; q\rangle_p}\\ &\stackrel{\star}{=} \sum_{ijkmnp} \frac{\langle 1, a, b_1; q\rangle_{i+j+m} \langle 1, a, b_2; q\rangle_{i+k+p} \langle 1, b_1, b_2; q\rangle_{j+k+n} x_1^{i+j+m}}{\langle 1; q\rangle_{m+j} \langle 1; q\rangle_{m+i} \langle 1; q\rangle_{k+p} \langle 1; q\rangle_{i+p} \langle 1, a; q\rangle_i \langle 1, b_1; q\rangle_j \langle 1, b_2; q\rangle_k \langle 1; q\rangle_{j+n}}\\ &\quad \times \frac{x_2^{j+n+k} x_3^{i+k+p} q^{i(a+i-1)+j(b_1+j-1)+k(b_2+k-1)}}{\langle 1; q\rangle_{k+n} \langle c_1; q\rangle_{i+j+m} \langle c_2; q\rangle_{n+j+k} \langle c_3; q\rangle_{p+i+k}} = \mathrm{RHS}.\end{aligned} \tag{74}$$

□

The following two expansion formulas from [3] are special cases of our new results.

Theorem 9 (a q-analogue of Burchnall, Chaundy [5, (26)]). *Consider*

$$\begin{aligned}&\Phi_2(a; b, b'; c, c' \mid q; x_1, x_2)\\ &= \sum_{r=0}^{\infty} \frac{\langle a, b, b'; q\rangle_r}{\langle 1, c, c'; q\rangle_r} x_1^r x_2^r q^{r(a+r-1)}\\ &\quad \times {}_2\phi_1(a + r, b + r; c + r \mid q; x_1)\\ &\quad \times {}_2\phi_1(a + r, b' + r; c' + r \mid q; x_2).\end{aligned} \tag{75}$$

Theorem 10 (a q-analogue of [5, (35)]). *Consider*

$$\begin{aligned}&\Phi_2(a; b, b'; c, c \mid q; x_1, x_2)\\ &= \sum_{r=0}^{\infty} \frac{\langle b, b'; q\rangle_r \langle a; q\rangle_{2r}}{\langle 1, c; q\rangle_r \langle c; q\rangle_{2r}} q^{rc+r(r-1)} x_1^r x_2^r\\ &\quad \times \Phi_1(a + 2r; b + r, b' + r; c + 2r \mid q; x_1, x_2).\end{aligned} \tag{76}$$

4. Discussion

To obtain the 10 new expansions for the three triple Srivastava hypergeometric functions in terms of ${}_4F_3$ hypergeometric and Kampé de Fériet functions mentioned in the abstract, let all ∞ disappear and adapt the indices for Φ accordingly. All of this is explained in detail in the book [3].

We have not used the expansion formulas for H_C, since all these involve the operator $\widetilde{\Delta}_{x_1;x_2x_3;q}$, which has no simple summation formula. Expansion formulas for the first q-Lauricella function by the $\widetilde{\nabla}_{x_1;x_2x_3;q}$ operator will be given in the forthcoming paper.

We notice that the three triple q-functions have limits q-Appell functions and not q-Horn functions. The q-Horn functions will appear in another paper.

Hasanov, Srivastava, and Turaev have published several other articles of this type and Bin-Saad [2] treated the four-variable case. The interested reader might consult papers by Exton on this last theme. Neither of these authors, however, used the q-operators from this paper.

The convergence regions for the q-Appell functions Φ_2 and Φ_4 given earlier are for the case $q = 1$. For general q, the convergence regions are much larger as has been explained at the ICNAAM Conference 2010; see also [7].

5. Conclusion

The expansion formulas often have a symmetric form. If a number of ∞ appear up or down in the first place of the Φ-function, an equal number of ∞ always appear opposite in the two second places of the Φ-function. The sum of the number of indices of q-shifted factorials in numerator and denominator is always the same for every step in the proofs (except formula (70)). The simplest expansion is formula (50) and the most complicated one is (64). If a factor (-1) appears, it comes from the operator $\Delta_{x_1,x_2;q}$. The number of variables in the exponent in this case is equal to the number of $\Delta_{x_1,x_2;q}$-operators.

Conflict of Interests

The author declares that there is no conflict of interests regarding the publication of this paper.

References

[1] A. Hasanov, H. M. Srivastava, and M. Turaev, "Decomposition formulas for some triple hypergeometric functions," *Journal of Mathematical Analysis and Applications*, vol. 324, no. 2, pp. 955–969, 2006.

[2] M. G. Bin-Saad, "Symbolic operational images and decomposition formulas for hypergeometric functions," *Journal of Mathematical Analysis and Applications*, vol. 376, no. 2, pp. 451–468, 2011.

[3] T. Ernst, *A Comprehensive Treatment of q-Calculus*, Birkhäuser, 2012.

[4] G. Gasper and M. Rahman, *Basic Hypergeometric Series*, Cambridge University Press, Cambridge, UK, 1990.

[5] J. L. Burchnall and T. W. Chaundy, "Expansions of appell's double hypergeometric functions," *Quarterly Journal of Mathematics*, vol. 11, no. 1, pp. 249–270, 1940.

[6] F. H. Jackson, "On basic double hypergeometric functions," *Quarterly Journal of Mathematics*, vol. 13, no. 1, pp. 69–82, 1942.

[7] T. Ernst, "Convergence aspects for q-Appell functions I," *The Journal of the Indian Mathematical Society*, vol. 81, no. 1-2, pp. 67–77, 2014.

12

On the Cardinality of the T_0-Topologies on a Finite Set

Messaoud Kolli

Department of Mathematics, Faculty of Science, King Khalid University, Abha, Saudi Arabia

Correspondence should be addressed to Messaoud Kolli; kmessaud@kku.edu.sa

Academic Editor: Laszlo A. Szekely

Let $T_0(n,k)$ be the number of all labeled T_0-topologies having k open sets that we can define on n points, and let $t_0(n,k)$ be the number of those which are nonhomeomorphic. In this paper, we compute these numbers for $k \geq 5 \cdot 2^{n-4}$ and arbitrary $n \geq 4$. The numbers $t_{n0}(n,k)$ of all unlabeled and non-T_0-topologies with k open sets are also given for $k \geq 2^{n-2}$.

1. Introduction

Let E be a finite set with n elements. The total number $T(n)$ of all labelled topologies one can define on E is a long standing open question. No reasonable explicit or recursive counting formula for $T(n)$ is known. Moreover, the number $T_0(n)$ of all labelled T_0-topologies definable on E is also an open problem. Recall that a T_0-topology is a topology satisfying the separate axiom: for all $x \neq y$, there is an open set containing one but not the other.

A classical and well known result is that each topology on E corresponds to a quasi-order on E (reflexive and transitive relation) and each T_0-topology corresponds to a partial order (reflexive, transitive, and antisymmetric relation). This bijective correspondence is due, independently, to Alexandroff [1] and Birkhoff [2]. On the other hand, several authors (Comtet [3], Evans et al. [4], Renteln [5] and others) have observed the following connexion between the $T(n)$ and the $T_0(n)$: $T(n) = \sum_{l=1}^{n} S_{n,l} T_0(l)$, where the $S_{n,l}$ are the Stirling numbers of the second kind. This formula was later refined by Erné [6] to $T^p(n) = \sum_{l=1}^{n} S_{n,l} T_0^p(l)$, where p is any topological property which is invariant under lattice isomorphisms. Numerically, the values $T(n)$ have been found for $n \leq 6$ by several authors (see, e.g., [3]), for $n \leq 7$ by Evans et al. [4], for $n \leq 9$ by Erné [6], for $n \leq 11$ by Das [7], and for $n \leq 14$ by Erné and Stege [8]. Later, the calculations were pursued by Brinkmann and McKay [9, 10] until $n = 18$.

Another approach is the study of the topologies by their number of open sets. For this, let $T(n,k)$ be the number of all labeled topologies with k open sets that we can define on E and let $T_0(n,k)$ be the number of those which are T_0. As for $T(n)$, there is no explicit formula giving the $T(n,k)$ or the $T_0(n,k)$, while some values are known. When k is large, Sharp [11] and Stephen [12] showed that there is no topology with $3 \cdot 2^{n-2} < k < 2^n$ open sets. In 1971, Stanley [13] enumerated the $T(n,k)$ for $k \geq 7 \cdot 2^{n-4}$ and arbitrary n, and in 2007, the author [14] extended the range to $k \geq 3 \cdot 2^{n-3}$. For small values of k, Erné and Stege [8, 15] determined the $T(n,k)$ and $T_0(n,k)$ for all k and $n \leq 12$. Independently, Benoumhani [16] computed by a direct method the $T(n,k)$ for $k \leq 12$ and arbitrary n. Later, the range was extended in [17] until $k \leq 17$.

For the unlabelled case, let $t(n,k)$ be the number of all unlabeled topologies having k open sets, let $t_0(n,k)$ be the number of those which are T_0, and let $t_{n0}(n,k)$ be the number of those which are non-T_0. Unfortunately, all these numbers are missing until now. However, some values are obtained. For example, Stanley [13] computed the $t_0(n,k)$ and the $t_{n0}(n,k)$ for $k \geq 7 \cdot 2^{n-4}$ and $k \geq 3 \cdot 2^{n-3}$, respectively. In the opposite sense, the numbers $t_0(n, n+k)$ are given for $k \leq 12$ by Erné and Stege [15], see also the paper [17] for $k \leq 8$.

Our present paper is a continuation of the previous effort in the field. Here we give the values $T_0(n,k)$ and $t_0(n,k)$ until $k \geq 5 \cdot 2^{n-4}$. For the $t_{n0}(n,k)$, the calculations are given until $k \geq 2^{n-2}$. Our approach is based on the notion of level partitions (Definition 2). Let $T_0^m(n,k)$ denote the number of all labeled T_0-topologies with k open sets and m levels; then, $T_0(n,k) = \sum_{m=1}^{n} T_0^m(n,k)$. In Section 2, we prove that if $k > 3 \cdot 2^{n-4}$; then, $m \leq 4$. The summation is then reduced to $T_0(n,k) = \sum_{m=1}^{4} T_0^m(n,k)$ and this reduction permits us to enumerate, in Section 3, all the possibilities and find the desired values. The unlabeled case is studied in Section 4. In Lemma 13, we prove that any two T_0-topologies are homeomorphic if and only if their corresponding levels are also homeomorphic. This property enables us to calculate the desired numbers.

2. Basic Results

In this section, we prove that if τ is a T_0-topology with $k > 3\cdot 2^{n-4}$ open sets, then τ has at most 4 levels.

Definition 1. Let τ be a topology on E. The nonempty open set A is called a minimal open set if, for all O in τ, we have

$$A \cap O = A \quad \text{or} \quad \emptyset. \tag{1}$$

In the lattice terminology, a minimal open set is an atom. Thereby, for any topology τ, we define inductively the decreasing sequence of the subsets E^j and the topologies τ^j on E^j as follows: for $j = 1$, we set $E^1 = E$, $\tau^1 = \tau$, and we denote by $A_1^1, \dots, A_{n_1}^1$ the minimal opens sets of τ^1. We put $A^1 = \bigcup_{i=1}^{n_1} A_i^1$, $E^2 = E^1 - A^1$, $\tau^2 = \{O - A^1,\ O \in \tau^1\}$, and we denote by $A_1^2, \dots, A_{n_2}^2$ the minimal open sets of τ^2. Again, $A^2 = \bigcup_{i=1}^{n_2} A_i^2$, $E^3 = E^2 - A^2$, $\tau^3 = \{O - A^2,\ O \in \tau^2\}$, and so on. This process should stop after at most n iterations. Let $U^j = \bigcup_{i=1}^{j} A^i$, and let F^j be the increasing family defined as follows:

$$F^0 = \{\emptyset\}, \qquad F^j = \left\{O \in \tau,\ O \subset U^j\right\}, \quad j \geq 1. \tag{2}$$

We can see that for all j, $U^j \in \tau$, $F^j = \tau|_{U^j}$, and then F^j is a topology on U^j. Now, we recall the notion of level.

Definition 2. Let τ be a topology on the finite set E, and let

$$L^0 = \{\emptyset\}, \qquad L^j = F^j - F^{j-1}, \quad j \geq 1. \tag{3}$$

For all $j \geq 0$, L^j is called the jth level of τ.

Lemma 3. *There exists a finite index $m \geq 1$, such that $\tau^{m+1} = \{\emptyset\}$, and τ^m is a Boolean algebra.*

The smallest integer m such that $\tau^{m+1} = \{\emptyset\}$ is the number of level of τ. The number m is at most n, and it is reached if and only if τ is the maximal chain. On the other hand, we observe that $\{L^j, j = 0, \dots, m\}$ is a partition of τ, $\{A_1^j, \dots, A_{n_j}^j\}$ is a partition of A^j, and $\{A_1^1, \dots, A_{n_1}^1, A_1^2, \dots, A_{n_2}^2, \dots, A_1^m, \dots, A_{n_m}^m\}$ and $\{A^1, A^2, \dots, A^m\}$ are both partitions of E. Let us now give an example.

Example 4. Take $E = \{x, y, z, t, u\}$ and

$$\begin{aligned}\tau = \{&\emptyset, \{x\}, \{y\}, \{x, y\}, \{x, z\}, \{x, y, z\}, \{x, y, t\},\\ &\{x, y, z, t\}, \{x, y, z, u\}, E\}.\end{aligned} \tag{4}$$

We have $\tau^2 = \{\emptyset, \{z\}, \{t\}, \{z, t\}, \{z, u\}, E^2\}$, $\tau^3 = \{\emptyset, E^3\}$, and $\tau^4 = \{\emptyset\}$. Then τ has three levels with $L^1 = \{\{x\}, \{y\}, \{x, y\}\}$, $L^2 = \{\{x, z\}, \{x, y, z\}, \{x, y, t\}, \{x, y, z, t\}\}$, and $L^3 = \{\{x, y, z, u\}, E\}$. We see that $A^1 = \{x, y\}$, with $A_1^1 = \{x\}$ and $A_2^1 = \{y\}$, $A^2 = \{z, t\}$, with $A_1^2 = \{z\}$ and $A_2^2 = \{t\}$, and finally $A^3 = \{u\}$, with $A_1^3 = \{u\}$. The partially ordered set corresponding to τ is as follows:

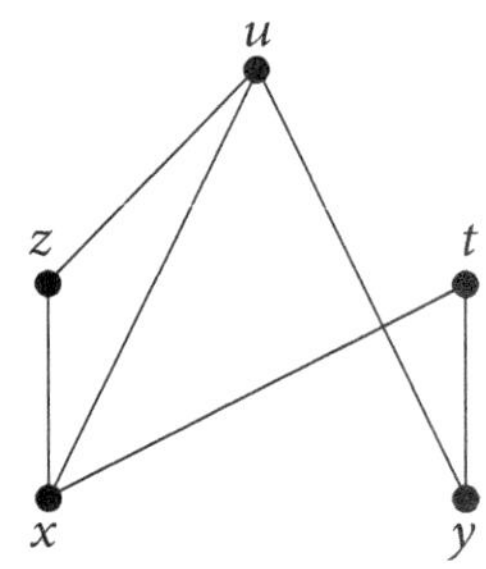

(5)

Since the elements of a minimal open set cannot be distinguishable, we have the following.

Lemma 5. *If τ is T_0, then every minimal open set of τ is a singleton. Moreover, for all O in τ, the topology $\tau' = \{A - O, A \in \tau\}$ is also T_0.*

A consequence of Lemma 5 is the following.

Theorem 6. *Let τ be a topology with m levels. Then τ is T_0 if and only if τ^j is T_0 for all $j = 1, \dots, m$.*

It follows again that a topology is T_0 if and only if all its blocks

$$\left\{A_1^1, \dots, A_{n_1}^1, A_1^2, \dots, A_{n_2}^2, \dots, A_1^m, \dots, A_{n_m}^m\right\} \tag{6}$$

are singletons and then $\sum_{j=1}^{m} n_j = n$. In Example 4, this equality is satisfied and of course τ is T_0.

Lemma 7. *Let τ be a T_0-topology with m levels. Then, τ has at least*

$$2^{n_1} + \cdots + 2^{n_m} - m + 1 \tag{7}$$

open sets and at most

$$2^{n_1-1} + 2^{n_1+n_2-2} + \cdots + 2^{n_1+\cdots+n_{m-1}-(m-1)} + 2^{n_1+\cdots+n_m-m+1} \tag{8}$$

open sets. In addition, these two bounds are optimal.

Proof. The jth level L^j is generated by $\{B_1, \dots, B_{n_j}\}$, where $B_l = A_l^j \cup O_l$ and O_l is arbitrary from the previous level L^{j-1}. So, L^j contains at least $2^{n_j} - 1$ open sets, and then τ has at least $2^{n_1} + \cdots + 2^{n_m} - m + 1$ open sets. On the other hand, if we take $B_l = A_l^j \cup U^{j-1}$, then L^j contains exactly $2^{n_j} - 1$ open sets and the lower bound is optimal. Let us prove the upper bound. We see that each B_l has the form $A_l^j \bigcup (\bigcup_{i=1}^{j-1} O_i)$, where each O_i is a union of a certain number of the minimal open sets $A^i_1, \ldots, A^i_{n_i}$. So each B_l generates at most $2^{n_1-1} \cdot 2^{n_2-1} \cdots 2^{n_{j-1}-1}$ open sets, and then L^j contains at most $(2^{n_j}-1)2^{n_1+\cdots+n_{j-1}-(j-1)}$ open sets. All these numbers added give the desired upper bound; in addition, if we take $B_l = A^j_l \bigcup (\bigcup_{i=1}^{j-1} A^i_1)$, then L^j contains exactly $2^{n_1+\cdots+n_j-(j-1)} - 2^{n_1+\cdots+n_{j-1}-(j-1)}$ open sets and the upper bound is reached. □

Lemma 8. *Let τ be a T_0-topology with m levels. Then τ has at most $(m+1)2^{n-m}$ open sets, and this bound is optimal.*

Proof. Since τ is T_0, then $\sum_{j=1}^{m} n_j = n$, and then, for all $j = 1, \ldots, m$,

$$2^{n_1+\cdots+n_j-j} \leq 2^{n-m}. \qquad (9)$$

By injecting all these inequalities in the upper bound appearing in Lemma 7, we obtain the desired bound. The optimality is also a consequence of the optimality in Lemma 7. □

It is trivial to see that two T_0-topologies having different number of levels are also different, so we have the following.

Theorem 9. *For all integers $n \geq 1$ and $k \in [2, 2^n]$, we have*

$$T_0(n,k) = \sum_{m=1}^{n} T_0^m(n,k). \qquad (10)$$

This last formula remains true in the non-T_0 case. For large values of k it reduces to the following.

Theorem 10. *Let $n \geq 4$. If $k > 3 \cdot 2^{n-4}$, then*

$$T_0(n,k) = \sum_{m=1}^{4} T_0^m(n,k). \qquad (11)$$

Proof. As $3 \cdot 2^{n-4} < k \leq (m+1)2^{n-m}$, then m should be less or equal to 4. □

This reduction enables us to enumerate easily all the possibilities and then to find the numbers $T_0(n,k)$. This is the content of the next section.

3. Computing $T_0(n,k)$ until $k \geq 5 \cdot 2^{n-4}$

As mentioned before, the values $T_0(n,k)$ are known for $k \geq 3 \cdot 2^{n-3}$. So, we have to compute them only for k in the interval $[5 \cdot 2^{n-4}, 3 \cdot 2^{n-3})$. Equivalently, we have to compute the numbers $T_0^m(n,k)$, for $m = 2,3,4$. Details are given only in the case $m = 2$, the others are similar. Thereby, let τ be a T_0-topology with two levels and let $A^1 = \{x_1, \ldots, x_{n-i}\}$, $A^2 = \{y_1, \ldots, y_i\}$ be the associated partition. The level L^1 is generated by $\{x_1\}, \ldots, \{x_{n-i}\}$, and then it contains $2^{n-i} - 1$ open sets. The level L^2 is generated by $\{U_{y_1}, \ldots, U_{y_i}\}$, where $U_{y_i} = O_i \cup \{y_i\}$ and O_i is arbitrary from L^1. The question is then how to choose the O_i in order to get a topology τ such that $|\tau| \in [5 \cdot 2^{n-4}, 3 \cdot 2^{n-3})$? For this, we consider all the cases $i = 1, \ldots, n-1$ one by one and we use the notation $(n)_i = n(n-1)\cdots(n-i+1)$ if $1 \leq i \leq n$, and 0 otherwise.

Case 1 ($(n_1,n_2) = (n-1,1)$). In this case, τ has at least $2^{n-1}+1$ open sets which is outside the interval $[5 \cdot 2^{n-4}, 3 \cdot 2^{n-3})$; so there is no solution in this case.

Case 2 ($(n_1,n_2) = (n-2,2)$). In this case, we prove that there are only 8 possibilities. For a given partition $A^1 = \{x_1, \ldots, x_{n-2}\}$, $A^2 = \{y_1, y_2\}$ of E, we try to reconstruct all the T_0-topologies τ having this partition and such that $|\tau|$ is in the desired interval. For this, we have to remark first that:

if one of the $y_i, i = 1,2$ is adjoined to the singleton $\{x_j\}$, then $|\tau| > 2^{n-2} + 2^{n-3} = 3 \cdot 2^{n-3}$, which is outside the desired interval;

if the $y_i, i = 1,2$ are both adjoined to more than three elements, then $|\tau| \leq 2^{n-2} + 3 \cdot 2^{n-6}$, which is also outside the desired interval.

So, at least one of the $y_i, i = 1,2$ should be adjoined to a set with two or three elements. Without loss of generality, suppose that y_1 is adjoined to $\{x_1, x_2\}$. Here we enumerate 3 possibilities. The first one is when y_2 is adjoined to $\{x_1, x_2, x_q, \ldots, x_{q+l+1}\}$, $l = 0, \ldots, n-6$. The corresponding poset is

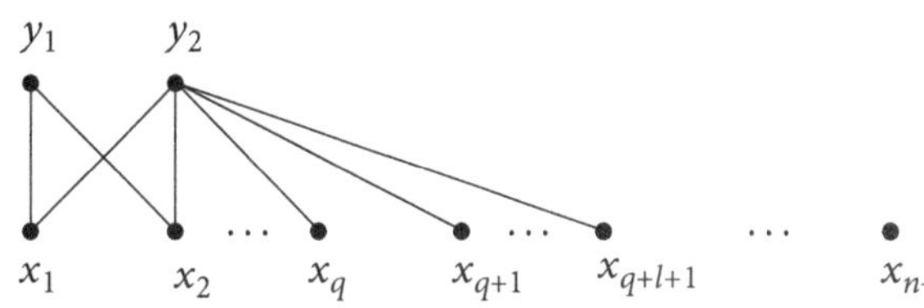

(12)

Obviously, the contribution of the open set $U_{y_1} = \{x_1, x_2, y_1\}$ to the topology τ is 2^{n-4} open sets, that of $U_{y_2} = \{x_1, x_2, x_q, \ldots, x_{q+l+1}, y_2\}$ is 2^{n-6-l} open sets, and that of their union is also 2^{n-6-l} open sets. So, τ has $k = 5 \cdot 2^{n-4} + 2^{n-5-l}$ open sets and the number of such topologies is $2\binom{n}{n-2}\binom{n-2}{2}\binom{n-4}{l+2} = (n)_{l+6}/2(l+2)!$

The second possibility is when y_2 is adjoined to $\{x_1, x_q, \ldots, x_{q+l+1}\}$, $l = 0, \ldots, n-6$, as follows:

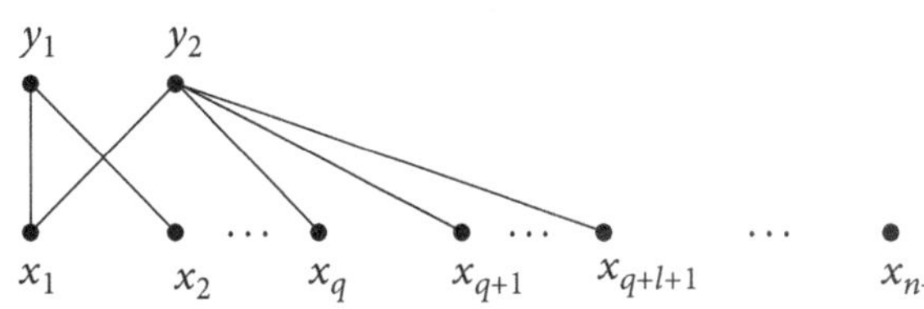

(13)

Here the contribution of U_{y_1} is 2^{n-4}, that of U_{y_2} is 2^{n-5-l}, and that of their union is 2^{n-6-l}. So, the topology τ has $k = 5 \cdot 2^{n-4} + 3 \cdot 2^{n-6-l}$ open sets and the number of such topologies is $4\binom{n}{n-2}\binom{n-2}{2}\binom{n-4}{l+2} = (n)_{l+6}/(l+2)!$

The third possibility is when y_2 is adjoined to $\{x_q, \ldots, x_{q+l+2}\}$, $l = 0, \ldots, n-7$, as follows:

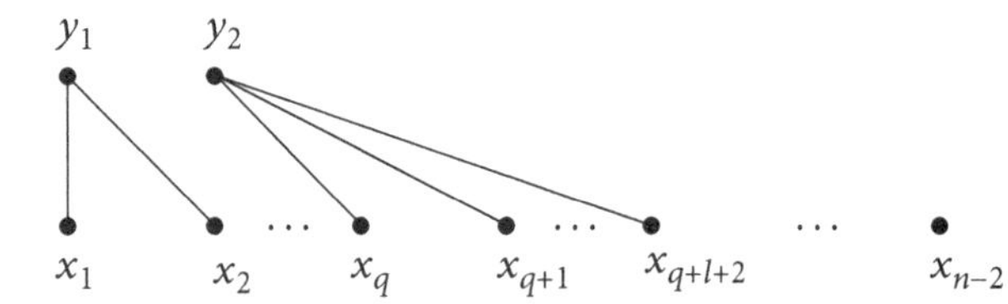

(14)

The contribution of U_{y_1} is 2^{n-4}, that of U_{y_2} is 2^{n-5-l}, and that of their union is 2^{n-l-7}. So, the topology τ has $k = 5 \cdot 2^{n-4} + 5 \cdot 2^{n-7-l}$ open sets and the number of such topologies is $2\binom{n}{n-2}\binom{n-2}{2}\binom{n-4}{l+3} = (n)_{l+7}/2(l+3)!$

Suppose now that y_1 is adjoined to three elements $\{x_1, x_2, x_3\}$. Remark that if y_2 is adjoined to more than four elements, then $|\tau| \le 2^{n-2} + 2^{n-5} + 2^{n-6}$, which is excluded. So, y_2 should be adjoined to three or four elements. When y_2 is adjoined to three elements, we find four possibilities. The first one is when y_2 is adjoined to $\{x_1, x_2, x_3\}$. The corresponding poset is

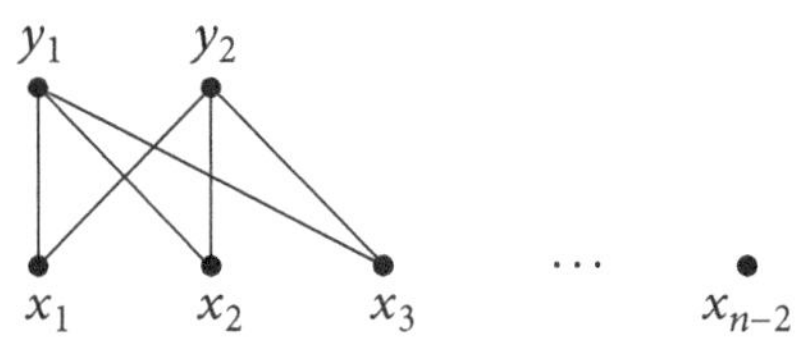

(15)

Here the contribution of U_{y_1} is 2^{n-5}, that of U_{y_2} is 2^{n-5}, and that of their union is also 2^{n-5}. So, τ has $k = 5 \cdot 2^{n-4} + 2^{n-5}$ open sets. The number of such topologies is $\binom{n}{n-2}\binom{n-2}{3} = (n)_5/12$.

The second possibility is when y_2 is adjoined to $\{x_1, x_2, x_4\}$ as follows:

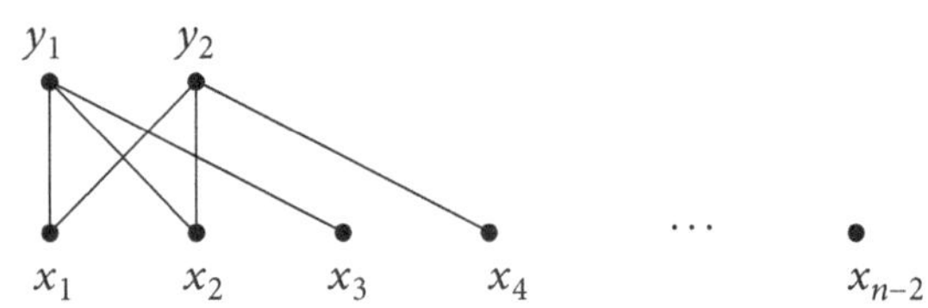

(16)

Here the contribution of U_{y_1} is 2^{n-5}, that of U_{y_2} is 2^{n-5}, and that of their union is 2^{n-6}. So, τ has $k = 5 \cdot 2^{n-4} + 2^{n-6}$ open sets. The number of such topologies is $12\binom{n}{n-2}\binom{n-2}{4} = (n)_6/4$.

The third possibility is when y_2 is adjoined to $\{x_1, x_4, x_5\}$ as follows:

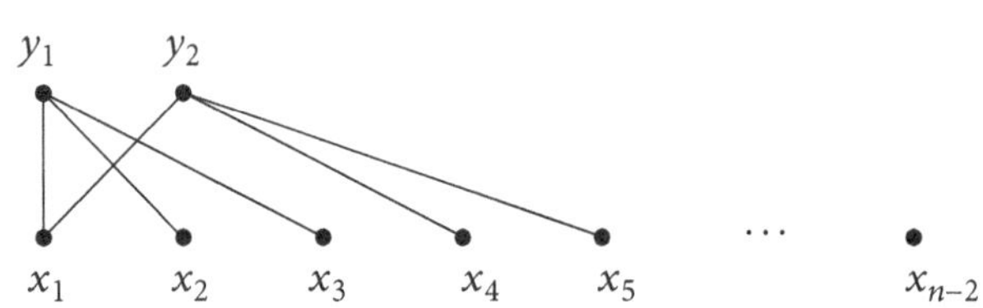

(17)

Here the contribution of U_{y_1} is 2^{n-5}, that of U_{y_2} is 2^{n-5}, and that of their union is 2^{n-7}. So, τ has $k = 5 \cdot 2^{n-4} + 2^{n-7}$ open sets. The number of such topologies is $30\binom{n}{n-2}\binom{n-2}{5} = (n)_7/8$.

The fourth possibility is when y_2 is adjoined to $\{x_4, x_5, x_6\}$ as follows:

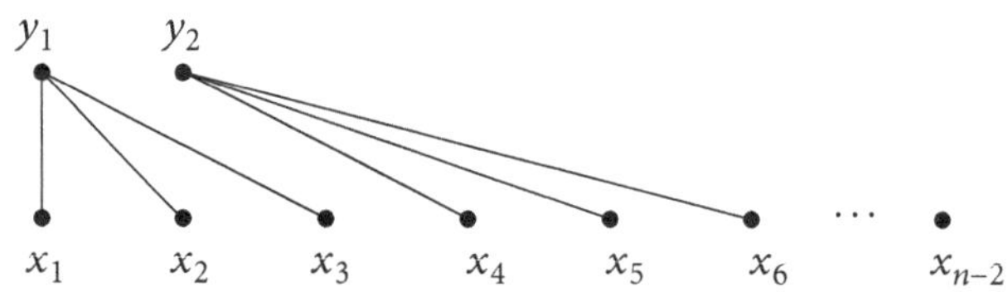

(18)

Here the contribution of U_{y_1} is 2^{n-5}, that of U_{y_2} is 2^{n-5}, and that of their union is 2^{n-8}. So, τ has $k = 5 \cdot 2^{n-4} + 2^{n-8}$ open sets. The number of such topologies is $20\binom{n}{n-2}\binom{n-2}{6} = (n)_8/72$.

When y_2 is adjoined to four elements $\{x_{j_1}, x_{j_2}, x_{j_3}, x_{j_4}\}$, we remark that if

$$\left|\{x_1, x_2, x_3\} \bigcap \{x_{j_1}, x_{j_2}, x_{j_3}, x_{j_4}\}\right| \le 2, \tag{19}$$

then $|\tau| \le 2^{n-2} + 2^{n-5} + 2^{n-6} + 2^{n-7}$, which is outside the desired interval; so the unique possibility is when y_2 is adjoined to $\{x_1, x_2, x_3, x_4\}$ as follows:

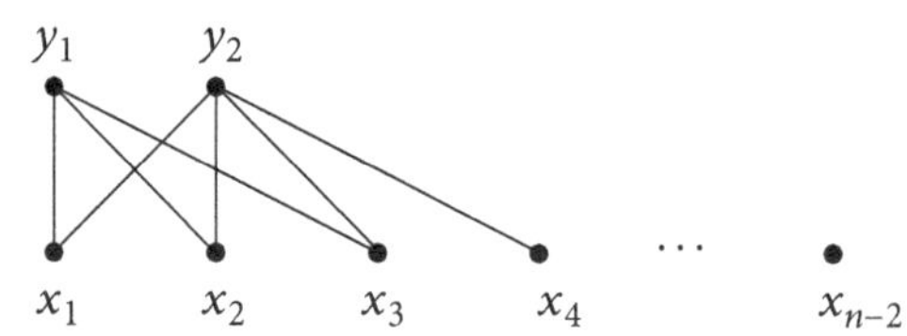

(20)

Here the contribution of U_{y_1} is 2^{n-5}, that of U_{y_2} is 2^{n-6}, and that of their union is also 2^{n-6}. So, τ has $k = 5 \cdot 2^{n-4}$ open sets and the number of such topologies is $8\binom{n}{n-2}\binom{n-2}{4} = (n)_6/6$.

Then we have proved that there are only 8 possibilities. We enumerate them in Table 1.

We use the same reasoning for all the rest of the cases $(n-i, i)$, $i = 3, \ldots, n-1$. Note that we obtain 13 possibilities for the case $(n-3, 3)$, also 13 for $(n-4, 4)$, 7 for $(n-5, 5)$, 6 for $(n-6, 6)$, and 5 for all the cases $(n-i, i)$, $i = 7, \ldots, n-1$. Since the machinery is the same as above, we have omitted the tedious details for the reader's convenience. The total number of T_0-topologies with two levels and $5 \cdot 2^{n-4} \le k < 3 \cdot 2^{n-3}$ open sets is then given by Table 2.

Using the same approach for three levels, we obtain two possibilities for the case $(n-2, 1, 1)$, nine for $(n-3, 2, 1)$, one for $(n-3, 1, 2)$, eight for $(n-4, 3, 1)$, one for $(n-4, 1, 3)$, two for $(n-4, 2, 2)$, two for all the cases $(n-i, i-1, 1)$, $i = 5, \ldots, n-1$, and finally one for all the cases $(n-i, i-2, 2)$, $i = 5, \ldots, n-1$. Omitting again the tedious details, the numbers $T_0^3(n, k)$ with $5 \cdot 2^{n-4} \le k < 3 \cdot 2^{n-3}$ are given by Table 3.

For four levels, we obtain only one possibility for the case $(n-3, 1, 1, 1)$. The numbers $T_0^4(n, k)$ with $5 \cdot 2^{n-4} \le k < 3 \cdot 2^{n-3}$ are given by Table 4.

TABLE 1

Possibility	Number of open sets	Number of such topologies
$U_{y_1} = \{x_1, x_2, y_1\}$, $U_{y_2} = \{x_1, x_2, x_q, \ldots, x_{q+l+1}, y_2\}$	$k = 5 \cdot 2^{n-4} + 2^{n-5-l}$	$\frac{(n)_{l+6}}{2(l+2)!}$, $l = 0, \ldots, n-6$
$U_{y_1} = \{x_1, x_2, y_1\}$, $U_{y_2} = \{x_1, x_q, \ldots, x_{q+l+1}, y_2\}$	$k = 5 \cdot 2^{n-4} + 3 \cdot 2^{n-6-l}$	$\frac{(n)_{l+6}}{(l+2)!}$, $l = 0, \ldots, n-6$
$U_{y_1} = \{x_1, x_2, y_1\}$, $U_{y_2} = \{x_q, \ldots, x_{q+l+2}, y_2\}$	$k = 5 \cdot 2^{n-4} + 5 \cdot 2^{n-7-l}$	$\frac{(n)_{l+7}}{2(l+3)!}$, $l = 0, \ldots, n-7$
$U_{y_1} = \{x_1, x_2, x_3, y_1\}$, $U_{y_2} = \{x_1, x_2, x_3, y_2\}$	$k = 5 \cdot 2^{n-4} + 2^{n-5}$	$\frac{(n)_5}{12}$
$U_{y_1} = \{x_1, x_2, x_3, y_1\}$, $U_{y_2} = \{x_1, x_2, x_4, y_2\}$	$k = 5 \cdot 2^{n-4} + 2^{n-6}$	$\frac{(n)_6}{4}$
$U_{y_1} = \{x_1, x_2, x_3, y_1\}$, $U_{y_2} = \{x_1, x_4, x_5, y_2\}$	$k = 5 \cdot 2^{n-4} + 2^{n-7}$	$\frac{(n)_7}{8}$
$U_{y_1} = \{x_1, x_2, x_3, y_1\}$, $U_{y_2} = \{x_4, x_5, x_6, y_2\}$	$k = 5 \cdot 2^{n-4} + 2^{n-8}$	$\frac{(n)_8}{72}$
$U_{y_1} = \{x_1, x_2, x_3, y_1\}$, $U_{y_2} = \{x_1, x_2, x_3, x_4, y_2\}$	$k = 5 \cdot 2^{n-4}$	$\frac{(n)_6}{6}$

TABLE 2

k	$T_0^2(n,k)$
$5 \cdot 2^{n-4}$	$\frac{n^2 + 41n - 234}{36}(n)_6$
$5 \cdot 2^{n-4} + 2^{n-8}$	$\frac{n^2 - 11n + 214}{720}(n)_8$
$5 \cdot 2^{n-4} + 2^{n-7}$	$\frac{n^2 - 10n + 66}{120}(n)_7$
$5 \cdot 2^{n-4} + 2^{n-6}$	$\frac{n^2 + 15n - 84}{24}(n)_6$
$5 \cdot 2^{n-4} + 2^{n-5}$	$\frac{n^2 - 2n - 14}{6}(n)_5$
$5 \cdot 2^{n-4} + 2^{n-i}$, $i = 9, \ldots, n$	$\frac{n-3}{(i-2)!}(n)_{i+1}$
$5 \cdot 2^{n-4} + 3 \cdot 2^{n-i}$, $i = 6, \ldots, n$	$\frac{2}{(i-4)!}(n)_i$
$5 \cdot 2^{n-4} + 5 \cdot 2^{n-7}$	$\frac{5}{6}(n)_7$
$5 \cdot 2^{n-4} + 5 \cdot 2^{n-i}$, $i = 8, \ldots, n$	$\frac{2}{(i-4)!}(n)_i$

Adding all these computations, we arrive at the following.

Theorem 11. *Let $n \geq 4$. The number $T_0(n,k)$ of all labeled T_0-topologies on n points and having $5 \cdot 2^{n-4} \leq k < 3 \cdot 2^{n-3}$ open sets is given in Table 5.*

All the other T_0-topologies have either $k < 5 \cdot 2^{n-4}$ or $k \geq 3 \cdot 2^{n-3}$ open sets.

Corollary 12. *According to Erné's formula $T(n,k) = \sum_{l=1}^{n} S_{n,l} T_0(l,k)$, the total number of non-T_0-topologies on arbitrary n points and having k open sets may be known until $k \geq 5 \cdot 2^{n-5}$.*

TABLE 3

k	$T_0^3(n,k)$
$5 \cdot 2^{n-4}$	$\frac{n^2 + 3n - 22}{4}(n)_5$
$5 \cdot 2^{n-4} + 2^{n-6}$	$\frac{n+3}{3}(n)_6$
$5 \cdot 2^{n-4} + 2^{n-5}$	$(n-1)(n)_5$
$5 \cdot 2^{n-4} + 2^{n-i}$, $i = 7, \ldots, n$	$\frac{2n-6}{(i-3)!}(n)_i$

TABLE 4

k	$T_0^4(n,k)$
$5 \cdot 2^{n-4}$	$(n)_4$

4. Computing Unlabeled Topologies

As mentioned in the Introduction, the numbers $t_0(n,k)$ and $t_{n0}(n,k)$ are computed by Stanley [13], respectively, for $k \geq 7 \cdot 2^{n-4}$ and $k \geq 3 \cdot 2^{n-3}$. In this section, we extend the range until $k \geq 5 \cdot 2^{n-4}$ for the T_0-topologies and until $k \geq 2^{n-2}$ for the non-T_0-topologies. Then we rediscover Stanley's results. The following observation is the key of our result.

Lemma 13. *Let τ and τ' be two T_0-topologies on n points. Then τ and τ' are homeomorphic if and only if their corresponding levels are, respectively, homeomorphic.*

Proof. It is trivial to see that if the corresponding levels are, respectively, homeomorphic, then the two topologies are homeomorphic. Let us prove the nontrivial sense. Suppose that f is an homeomorphism from (E,τ) into (E,τ'), and let

$$\begin{gathered}\{A_1^1, \ldots, A_{n_1}^1, A_1^2, \ldots, A_{n_2}^2, \ldots, A_1^m, \ldots, A_{n_m}^m\}, \\ \{B_1^1, \ldots, B_{n_1'}^1, B_1^2, \ldots, B_{n_2'}^2, \ldots, B_1^{m'}, \ldots, B_{n_{m'}'}^{m'}\}\end{gathered} \tag{21}$$

TABLE 5

k	$T_0(n,k)$
$5\cdot 2^{n-4}$	$\frac{n^2+41n-234}{36}(n)_6+\frac{n^2+3n-22}{4}(n)_5+(n)_4$
$5\cdot 2^{n-4}+2^{n-8}$	$\frac{n^2+n+178}{720}(n)_8$
$5\cdot 2^{n-4}+2^{n-7}$	$\frac{n^2+36}{120}(n)_7$
$5\cdot 2^{n-4}+2^{n-6}$	$\frac{n^2+23n-60}{24}(n)_6$
$5\cdot 2^{n-4}+2^{n-5}$	$\frac{n^2+4n-20}{6}(n)_5$
$5\cdot 2^{n-4}+2^{n-i}$, $i=9,\dots,n$	$\frac{n-3}{(i-2)!}(n)_{i+1}+\frac{2n-6}{(i-3)!}(n)_i$
$5\cdot 2^{n-4}+3\cdot 2^{n-i}$, $i=6,\dots,n$	$\frac{2}{(i-4)!}(n)_i$
$5\cdot 2^{n-4}+5\cdot 2^{n-7}$	$\frac{5}{6}(n)_7$
$5\cdot 2^{n-4}+5\cdot 2^{n-i}$, $i=8,\dots,n$	$\frac{2}{(i-4)!}(n)_i$

TABLE 6

k	$t_0(n,k)$
$5\cdot 2^{n-4}$	22
$5\cdot 2^{n-4}+1$	4
$5\cdot 2^{n-4}+2$	8
$5\cdot 2^{n-4}+2^{n-i}, 9\le i\le n-2$	10
$5\cdot 2^{n-4}+2^{n-8}$	16
$5\cdot 2^{n-4}+2^{n-7}$	12
$5\cdot 2^{n-4}+2^{n-6}$	18
$5\cdot 2^{n-4}+2^{n-5}$	16
$5\cdot 2^{n-4}+3\cdot 2^{n-i}, 6\le i\le n$	2
$5\cdot 2^{n-4}+5\cdot 2^{n-7}$	6
$5\cdot 2^{n-4}+5\cdot 2^{n-i}, 8\le i\le n$	4
$3\cdot 2^{n-3}$	9
$3\cdot 2^{n-3}+1$	2
$3\cdot 2^{n-3}+2^{n-j}, 7\le j\le n-1$	4
$3\cdot 2^{n-3}+3\cdot 2^{n-j}, 7\le j\le n-1$	2
$25\cdot 2^{n-6}$	7
$13\cdot 2^{n-5}$	6
$27\cdot 2^{n-6}$	3
$7\cdot 2^{n-4}$	5
$15\cdot 2^{n-5}$	2
2^{n-1}	2
$2^{n-1}+2^{n-j}, 5\le j\le n$	2
$9\cdot 2^{n-4}$	3
$5\cdot 2^{n-3}$	2
$3\cdot 2^{n-2}$	1
2^n	1

be, respectively, their corresponding partitions. Since the first blocks $A_1^1,\dots,A_{n_1}^1$ are open and singletons, then $f(\bigcup_{i=1}^{n_1}A_i^1)=\bigcup_{i=1}^{n_1'}B_i^1$ and so $n_1=n_1'$. But f is also an homeomorphism from (E^2,τ^2) into (E'^2,τ'^2) and for the same reason, we obtain $n_2=n_2'$. We repeat this process to obtain that $m=m'$, and $n_j=n_j'$ for all j. On the other hand, let $A^j=\bigcup_{i=1}^{n_j}A_i^j$, $U^j=\bigcup_{i=1}^{j}A^j$ and $B^j=\bigcup_{i=1}^{n_j}B_i^j$, $Y^j=\bigcup_{i=1}^{j}B^j$. As the above, we have $f(A^j)=B^j$, and then $f(U^j)=Y^j$ for all j. Since U^j and Y^j are open with respect to τ and τ', respectively, f is also an homeomorphism from the subspace $(U^j,\tau|_{U^j})$ into the subspace $(Y^j,\tau'|_{Y^j})$. But $\tau|_{U^j}=\bigcup_{i=0}^{j}L^i$ and $\tau'|_{Y^j}=\bigcup_{i=0}^{j}L'^i$; so for $j=1$, we obtain that L^1 is homeomorphic to L'^1. For $j=2$, we have $L^1\cup L^2$ which is homeomorphic to $L'^1\cup L'^2$ and then L^2 which is homeomorphic to L'^2. We repeat this process to obtain that L^j is homeomorphic to L'^j for all j, and the proof is done. □

Corollary 14. *Lemma 13 holds if τ and τ' are non-T_0.*

It follows that the number of unlabelled T_0-topologies is exactly the number of the possibilities given in the labelled case (see the above section). So, omitting the easy computations for $n\le 9$, we obtain the following.

Theorem 15. *Let $n\ge 10$. The number $t_0(n,k)$ of all unlabeled T_0-topologies, with $k\ge 5\cdot 2^{n-4}$ open sets, is detailed in Table 6, totaling $24n-43$ unlabeled T_0-topologies. All the other unlabelled T_0-topologies have $k<5\cdot 2^{n-4}$ open sets.*

For the non-T_0-topologies, we obtain the following.

Table 7

k	$t_{n0}(n,k)$
2^{n-2}	9
$2^{n-2}+1$	4
$2^{n-2}+2^{n-j}, 6 \le j \le n-1$	6
$2^{n-2}+2^{n-5}$	8
$2^{n-2}+2^{n-4}$	6
$2^{n-2}+2^{n-3}$	3
2^{n-1}	1

Theorem 16. *Let $n \ge 6$. The number $t_{n0}(n,k)$ of all unlabeled and non-T_0-topologies, with $k \ge 2^{n-2}$ open sets, is given in Table 7, totaling $6n-5$ unlabeled non-T_0-topologies. All the other unlabelled and non-T_0-topologies have $k < 2^{n-2}$ open sets.*

5. Conclusion

In this paper, we have computed the numbers $T_0(n,k)$, $t_0(n,k)$, and $t_{n0}(n,k)$ via the notion of level and the numbers $T_0^m(n,k)$. Unfortunately, there is no explicit formula giving the $T_0^m(n,k)$, while some cases are already known. For example, there is one T_0-topology with one level (the discrete topology) and $\sum_{n_1=1}^{n-1}\binom{n}{n_1}(2^{n_1}-1)^{n-n_1}$ T_0-topologies with two levels. In the opposite sense, there are $((n-1)^2/2)n!$ T_0-topologies with $n-1$ levels and $n!$ T_0-topologies with n levels. But for $3 \le m \le n-2$, the calculation of the $T_0^m(n,k)$ seems to be a very hard task. However, for small values of n, these calculations can be programmed and then computed by computer. This is the project of a forthcoming paper.

Conflict of Interests

The author declares that there is no conflict of interests regarding the publication of this paper.

Acknowledgments

The author is very grateful to the anonymous referee for her/his corrections, valuable suggestions, and comments. This work is supported by the second project of the King Khalid University, under the Project no. 208 (1433).

References

[1] P. S. Alexandroff, "Diskrete Räume," *Matematicheskii Sbornik*, vol. 2, pp. 501–518, 1937.

[2] G. Birkhoff, "Rings of sets," *Duke Mathematical Journal*, vol. 3, no. 3, pp. 443–454, 1937.

[3] L. Comtet, "Recouvrements, bases de filtre et topologies d'un ensemble fini," *Comptes Rendus de l'Académie des Sciences*, vol. 262, pp. A1091–A1094, 1966.

[4] J. W. Evans, F. Harary, and M. S. Lynn, "On the computer enumeration of finite topologies," *Communications of the ACM*, vol. 10, no. 5, pp. 295–297, 1967.

[5] P. Renteln, "On the enumeration of finite topologies," *Journal of Combinatorics, Information & System Sciences*, vol. 19, no. 3-4, pp. 201–206, 1994.

[6] M. Erné, "Struktur- und Anzahlformeln für Topologien auf endlichen Mengen," *Manuscripta Mathematica*, vol. 11, pp. 221–259, 1974.

[7] S. K. Das, "A machine representation of finite T_0 topologies," *Journal of the Association for Computing Machinery*, vol. 24, no. 4, pp. 676–692, 1977.

[8] M. Erné and K. Stege, "Counting finite posets and topologies," *Order*, vol. 8, no. 3, pp. 247–265, 1991.

[9] G. Brinkmann and B. D. McKay, "Posets on up to 16 points," *Order*, vol. 19, no. 2, pp. 147–179, 2002.

[10] G. Brinkmann and B. D. McKay, "Counting unlabelled topologies and transitive relations," *Journal of Integer Sequences*, vol. 8, no. 2, article 05.2.1, 2005.

[11] H. Sharp, Jr., "Cardinality of finite topologies," *Journal of Combinatorial Theory*, vol. 5, pp. 82–86, 1968.

[12] D. Stephen, "Topology on finite sets," *The American Mathematical Monthly*, vol. 75, pp. 739–741, 1968.

[13] R. P. Stanley, "On the number of open sets of finite topologies," *Journal of Combinatorial Theory A*, vol. 10, pp. 74–79, 1971.

[14] M. Kolli, "Direct and elementary approach to enumerate topologies on a finite set," *Journal of Integer Sequences*, vol. 10, no. 3, article 07.3.1, 2007.

[15] M. Erné and K. Stege, "Combinatorial applications of ordinal sum decompositions," *Ars Combinatoria*, vol. 40, pp. 65–88, 1995.

[16] M. Benoumhani, "The number of topologies on a finite set," *Journal of Integer Sequences*, vol. 9, no. 2, article 06.2.6, 2006.

[17] M. Benoumhani and M. Kolli, "Finite topologies and partitions," *Journal of Integer Sequences*, vol. 13, no. 3, article 10.3.5, 2010.

13

On Normal q-Ary Codes in Rosenbloom-Tsfasman Metric

R. S. Selvaraj and Venkatrajam Marka

Department of Mathematics, National Institute of Technology Warangal, Andhra Pradesh 506004, India

Correspondence should be addressed to R. S. Selvaraj; rsselva@nitw.ac.in

Academic Editors: A. Cossidente, E. Manstavicius, and S. Richter

The notion of normality of codes in Hamming metric is extended to the codes in Rosenbloom-Tsfasman metric (RT-metric, in short). Using concepts of partition number and l-cell of codes in RT-metric, we establish results on covering radius and normality of q-ary codes in this metric. We also examine the acceptability of various coordinate positions of q-ary codes in this metric. And thus, by exploring the feasibility of applying amalgamated direct sum method for construction of codes, we analyze the significance of normality in RT-metric.

1. Introduction

Covering properties of codes have unique significance in coding theory, and covering radius, one of the four fundamental parameters, of a code is important in several respects [1]. Considering the fact that it is a geometric property of codes that characterizes maximal error correcting capability in the case of minimum distance decoding, covering radius had been extensively studied by many researchers (see, e.g., [2, 3] and the literature therein) especially with respect to the conventional Hamming metric. In fact, it has evolved into a subject in its own right mainly because of its practical applicability in areas such as data compression, testing, and write-once memories and also because of the mathematical beauty that it possesses. More on covering radius can be found in the monograph compiled by Cohen et al. [4]. In order to improve upon the bounds on covering radius, various construction techniques that use two or more known codes to construct a new code were proposed over the last few decades. One such method is direct sum construction which is a basic yet useful construction method. To improve upon the bounds related to covering radius of codes obtained using this method, the notion of normality which facilitates a construction technique known as amalgamated direct sum (ADS) was introduced for binary linear codes by Graham and Sloane in [5]. The same concepts were extended to binary nonlinear codes by Cohen et al. in [6]. Later, Lobstein and van Wee generalized these results to q-ary codes [7].

In the present paper, we extend the notion of normality to codes in Rosenbloom-Tsfasman metric (RT-metric, in short). The present work is a generalization of our work on binary codes [8] to q-ary case (though the discussed results hold good for codes over any alphabet of size q, for simplicity, we have taken it to be the finite field $\mathbb{F}_q$). RT-metric was introduced by Rosenbloom and Tsfasman in [9] and independently by Skriganov in [10] and is more adequate than Hamming metric in dealing with channels in which errors have a tendency to occur with a periodic spikewise perturbation of period $s \in \mathbb{N}$. As a generalization of the classical Hamming metric with rich mathematical beauty and being advantageous over Hamming metric in dealing with certain channels [9, 10], this metric attracted the attention of coding theorists over the last two decades [11–13]. The covering problem in RT-metric was first dealt with by Yildiz et al. for RT-spaces over Galois Rings [14].

The organization of the present paper is as follows. In Section 2, some basic definitions and notations that are used in this paper are presented. In Section 3, we have studied the covering radius of q-ary codes in this metric by introducing two new tools, namely, *partition number* and *l-cell* which greatly reduce the difficulty in determining the covering radius. In Section 4, we investigate the normality of q-ary codes in RT-metric. In this section, we also discuss the possibilities of extending the direct sum and amalgamated direct sum (ADS) constructions from Hamming metric to RT-metric. And, finally, Section 5 provides the conclusions of this paper.

2. Definitions and Notations

For $x = (x_1, x_2, \ldots, x_s) \in \mathbb{F}_q^s$, where $\mathbb{F}_q = \mathrm{GF}(q)$ is a finite field of q elements, we define the ρ-weight (RT-weight) of x to be

$$wt_\rho(x) = \max\{i \mid x_i \neq 0, 1 \leq i \leq s\}. \quad (1)$$

The RT-distance or ρ-distance between $x = (x_1, x_2, \ldots, x_s)$ and $y = (y_1, y_2, \ldots, y_s) \in \mathbb{F}_q^s$ can be defined by $d_\rho(x, y) = wt_\rho(x-y)$. The subsets of the space equipped with this metric are called RT-metric codes over $\mathbb{F}_q$ (or q-ary RT-metric codes) and the subspaces are called linear RT-metric codes over $\mathbb{F}_q$. The value $d_{\min} = \min\{d_\rho(x, y) : x, y \in C\}$ is called the *minimum ρ-distance* of the code C. The maximum distance of any word in the ambient space from an RT-metric code is called the *covering radius* of the code. A code in which each nonzero codeword is of the same weight w_ρ is said to be a *constant weight code*. A q-ary code of length s, cardinality K, and minimum ρ-distance d_ρ is said to be a maximum distance separable code (an MDS code, in short) if $K = q^{(s-d_\rho+1)}$.

Throughout this paper, unless otherwise specified, by a code we mean a q-ary RT-metric code and by distance we mean RT-distance. Moreover, $[s, k, d_\rho]_q R$ denotes a q-ary linear code of length s, dimension k, minimum distance d_ρ, and covering radius R, $(s, K, d_\rho)_q R$ denotes a q-ary code with cardinality K, and $t_q[s, k]$ denotes the minimum covering radius that a q-ary linear code of length s and dimension k can possess.

Definition 1 (direct sum of two codes). Let C_1 and C_2 be codes with parameters $(s_1, K_1)_q$ and $(s_2, K_2)_q$, respectively. Then, the direct sum of C_1 and C_2, denoted by $C_1 \bigoplus C_2$, is defined as

$$C_1 \bigoplus C_2 = \{(c_1, c_2) \mid c_1 \in C_1, c_2 \in C_2\}, \quad (2)$$

and is a code of length $s_1 + s_2$ and cardinality $K_1 K_2$.

Definition 2 (q-ary normal codes). Let C be a q-ary RT-metric code of length s, cardinality K, and covering radius R, and, also, for each $a \in \mathbb{F}_q$, let $C_a^{(i)}$ denote the set of codewords in which the ith coordinate is a. Then, the *norm* of C with respect to the ith coordinate is defined as

$$N^{(i)} = \max_{x \in \mathbb{F}_q^s} \left\{ \sum_{a \in \mathbb{F}_q} d_\rho\left(x, C_a^{(i)}\right) \right\}, \quad (3)$$

where

$$d_\rho\left(x, C_a^{(i)}\right) = \begin{cases} \min\{d_\rho(x, y) \mid y \in C_a^{(i)}\} & \text{if } C_a^{(i)} \neq \emptyset \\ s & \text{if } C_a^{(i)} = \emptyset. \end{cases} \quad (4)$$

Now, $N = \min_i N^{(i)}$ is called the norm of C and the coordinates i for which $N^{(i)} = N$ are said to be *acceptable*. Finally, a code is said to be *normal* if its norm satisfies $N \leq qR + q - 1$. If the code C is not clear from the context, we use the notations $N^{(i)}(C)$ and $N(C)$.

3. Covering Radius of RT-Metric Codes over $\mathbb{F}_q$

Definition 3 (partition number of a q-ary RT-metric code). Let C be an $(s, K, d_\rho)_q$ code in RT-metric. The largest nonnegative integer l, for which each q-ary l-tuple can be assigned to at least one codeword whose last l coordinates are actually that l-tuple, is called the *partition number* of the code C. The code with partition number l can be partitioned into q^l parts, each of which has the property that all its members have the same q-ary l-tuple as their last l coordinates. A part obtained in the above fashion is called an *l-cell* of the code, if at least one field element is not present in the $(s-l)$th coordinate of its member codewords. As the definition of partition number suggests, a code with partition number l contains at least q^l codewords.

If C is an $[s, k, d_\rho]_q$ linear code, then each of the q^l parts does have q^{k-l} codewords in it and so does each l-cell by its definition. If C_1 is an l-cell of the linear code C, then $x + C_1$ for $x \in C \backslash C_1$ will also be an l-cell different from C_1. Thus, for linear codes, if there is one l-cell, then there will be q^l such l-cells.

Now, we will observe that covering radius of a code can be determined using the concepts of partition number and l-cell. The definition of partition number serves as a tool in determining the covering radius of an RT-metric code, as shown by the following theorem.

Theorem 4. *Let C be an $(s, K, d_\rho)_q R$ code in RT-metric. Then, the partition number of C is l if and only if its covering radius is $s - l$.*

Proof. First, let the partition number of the code be l. Partition the code such that codewords in each part contain a unique q-ary $(l+1)$-tuple as their last $l+1$ coordinates. Associate each part with the respective $(l+1)$-tuple. Since the partition number of the code is l, the number of such parts will be less than q^{l+1}. Choose an $x \in \mathbb{F}_q^s$, whose last $l+1$ coordinates constitute the $(l+1)$-tuple that does not have a part associated with it. Such an x will be at distance $s - l$ from the code, which is the maximum distance that a word can actually be from the code C. Hence, the covering radius is $s - l$.

Conversely, let the covering radius be $R = s - l$. By the definition of covering radius, any word must be at distance at most $s - l$ from the code, which means that, for each word, there is at least one codeword which agrees with the word in the last $s - (s - l) = l$ coordinate positions. This implies that the partition number of the code is greater than or equal to l. Let us assume that the partition number of the code is $m > l$. Then, to each q-ary m-tuple, we can associate a codeword whose last m coordinates coincide with that m-tuple. Thus, each word in $\mathbb{F}_q^s$ will be at distance at most $s-m$, contradicting the fact that covering radius of C is $s - l$. Hence, the partition number of the code C is l. □

Now, the above theorem can be restated in terms of l-cell in the following manner, as the definition of the l-cell suggests.

Corollary 5. *Let C be an RT-metric code of length s over $\mathbb{F}_q$. Then, the covering radius of C is $s-l$ if and only if $\exists$ an l-cell of the code C.*

Proposition 6. *Let C_1 and C_2 be any two RT-metric codes over $\mathbb{F}_q$ with parameters $(s_1, K_1, d_1)_q R_1$ and $(s_2, K_2, d_2)_q R_2$, respectively. Then, their direct sum $C = C_1 \bigoplus C_2$ is an $(s_1 + s_2, K_1K_2, d)_q R$ code with minimum distance $d = d_1$ and covering radius $R = s_1 + R_2$, provided $C_2 \neq \mathbb{F}_q^{s_2}$.*

Proof. The direct sum of C_1 and C_2 is given by

$$C_1 \bigoplus C_2 = \{(c_1, c_2) \mid c_1 \in C_1, c_2 \in C_2\}. \tag{5}$$

Clearly, this is a code of length $s_1 + s_2$ and cardinality K_1K_2. Now, since the minimum distance of C_1 is d_1, there exist two codewords c_1 and c_1' in C_1 such that $d_\rho(c_1, c_1') = d_1$. Then, the codewords $c = (c_1, c_2)$ and $c' = (c_1', c_2)$ for some $c_2 \in C_2$ will also be at distance d_1 which is minimum among all the codewords of C. From Theorem 4 and Definition 3, it is clear that the covering radius of a code in RT-metric depends on the partition number of the code and that the process of partitioning starts with the right most coordinate. Therefore, by Theorem 4, unless C_2 is the space $\mathbb{F}_q^{s_2}$, the partition number of C must be equal to that of C_2 which is $s_2 - R_2$ and, hence, the covering radius of C is $s_1 + s_2 - (s_2 - R_2) = s_1 + R_2$. □

Remark 7. If $C_2 = \mathbb{F}_q^{s_2}$, then the covering radius of the code $C = C_1 \bigoplus C_2$ is $R = R_1$.

4. Normality of Codes in RT-Metric

Proposition 8. *If C is an $(s, K, d_\rho)_q R$ RT-metric code over $\mathbb{F}_q$, then $N(C) \geq qR$.*

Proof. The proof follows directly from the definition of covering radius and that of norm of a code. □

Theorem 9. *Any q-ary RT-metric code of length s and covering radius s is normal and all the coordinates are acceptable.*

Proof. As $R = s$, the partition number is 0 which means that there exists an $\alpha \in \mathbb{F}_q$ which is not present as the last coordinate of any codeword in C. If $x = (x_1, x_2, \ldots, x_s) \in \mathbb{F}_q^s$ is such that $x_s = \alpha$, then $d_\rho(x, C_a^{(i)}) = s$ for each $a \in \mathbb{F}_q$ and for any $i \in \{1, 2, \ldots, s\}$. Thus, $N^{(i)} = qs < qs + q - 1 = qR + (q-1)$. Hence, the code is normal and all the coordinates are acceptable. □

Lemma 10. *Let C be any RT-metric code with length s and covering radius $R < s$; then*

$$N^{(i)} = (q-1)\, i + R, \tag{6}$$

for each $i \in \{R+1, R+2, \ldots, s\}$.

Proof. For a coordinate position $i \in \{R+1, R+2, \ldots, s\}$, the norm is given by

$$N^{(i)} = \max_{x \in \mathbb{F}_q^s} \left\{ \sum_{a \in \mathbb{F}_q} d_\rho\left(x, C_a^{(i)}\right) \right\}. \tag{7}$$

As the covering radius of C is R, it has partition number $s - R$. So, the definition of partition number implies that each codeword can be associated with a unique $(s-R)$-tuple which actually consists of the last $(s-R)$ coordinates of that codeword and, also, that there exists at least one $(s-R+1)$-tuple which is not the same as the last $s-R+1$ coordinates of any of the codewords. Now, when we partition the code into q parts $C_a^{(i)}$, for $a \in \mathbb{F}_q$, we can also partition the set of all $(s-R)$-tuples into q parts corresponding to the associated codewords in $C_a^{(i)}$. If we choose a word $x \in \mathbb{F}_q^s$ whose last $s-R+1$ coordinates do not match with the last $s-R+1$ coordinates of any of the codewords, then, for this word, $\sum_{a \in \mathbb{F}_q} d_\rho(x, C_a^{(i)}) = (q-1)i + R$ which is the maximum value that a word can give. Hence, the proof holds. □

Theorem 11. *Let C be any RT-metric code of length s with covering radius $R < s$. Then, the coordinates $i \in \{R+2, R+3, \ldots, s\}$ are not acceptable.*

Proof. The minimum norm is

$$N \leq N^{(i)}, \quad \forall i \in \{1, 2, \ldots, s\}. \tag{8}$$

By Lemma 10, $N^{(i)} = (q-1)i + R$, for all $i \geq R+1$, which implies

$$N^{(R+1)} < N^{(R+2)} < N^{(R+3)} < \cdots < N^{(s)}. \tag{9}$$

From (8) and (9), one can conclude that

$$N < N^{(i)}, \quad \forall i \in \{R+2, R+3, \ldots, s\}. \tag{10}$$

This completes the proof. □

The above theorem is not sufficient to arrive at a decision on the acceptability of the sth coordinate, when the code has covering radius $R = s - 1$. This can be settled by the following theorem which says that no q-ary linear RT-metric code with dimension more than 1 and covering radius $s - 1$ has acceptable last coordinate.

Theorem 12. *Let C be an $[s, k, d_\rho]_q R$ linear RT-metric code over $\mathbb{F}_q$ with $k > 1$ and $R = s - 1$. Then, the last coordinate is not acceptable.*

Proof. By Lemma 10,

$$\begin{aligned} N^{(s)} &= (q-1)\, s + R \\ &= (q-1)\, s + s - 1 = qs - 1 \\ &= q(s-1) + q - 1 = qR + (q-1) \\ &\qquad \text{(since } R = s - 1\text{)}. \end{aligned} \tag{11}$$

In order to prove this theorem, we must show that there exists at least one coordinate j for which $N^{(j)} < N^{(s)}$. As the dimension of C is k, there will be k coordinate positions $i_1, i_2, \ldots, i_k$ such that $C_a^{(i_t)} \neq \emptyset$ for all $a \in \mathbb{F}_q$ and for each $t = 1, 2, \ldots, k$. One such coordinate position is s (but not $s-1$, as the partition number is 1). As $k > 1$, there exists at least one more such coordinate position $j \in \{i_1, i_2, \ldots, i_k\}$ and $j \neq s$ wherein $C_a^{(j)} \neq \emptyset$ for all $a \in \mathbb{F}_q$, and so $N^{(j)} \leq q(s-1) < N^{(s)}$. This completes the proof. □

But, when the code is either of dimension 1 or of constant weight, all the coordinates become acceptable as the following results suggest.

Proposition 13. *Let C be an $[s, 1]_q$ linear RT-metric code. Then, C is normal and all the coordinates are acceptable.*

Proof. It can easily be seen that the covering radius of C is either s or $s-1$. If it is s, the result follows by Theorem 9. If it is $s-1$, there exists a 1-cell. Since C is of dimension 1, there exist q such 1-cells. By Lemma 10, $N^{(s)} = qs - 1$. Now for each $i \in \{1, 2, \ldots, s-1\}$, either $C_0^{(i)} = C$ or $|C_0^{(i)}| = 1$. In the former case, $C_a^{(i)} = \emptyset$ for all $a \neq 0$ and, in the latter case, $|C_a^{(i)}| = 1$ for all $a \in \mathbb{F}_q$. In any case, $N^{(i)} = qs - 1$. This proves the result stated. □

Proposition 14. *Let C be a constant weight code of length s over $\mathbb{F}_q$. Then, C is normal and all the coordinates are acceptable.*

Proof. It is easy to see that the covering radius R of the constant weight code C is either s or $s-1$, as it can have partition number of either 0 or 1. If $R = s$, then, by Lemma 10, it is done. If $R = s-1$, then $C_a^{(s)} \neq \emptyset$ for any $a \in \mathbb{F}_q$. Then, for any i, $\overline{0} \in C_0^{(i)}$ and $|C_0^{(i)}| \geq 1$. For an $x \in \mathbb{F}_q^s$ with $x_s = 0$ and $x_{s-1} \neq 0$, $N^{(i)}$ equals $qs - 1$. Hence, the result follows. □

Remark 15. If $C = \mathbb{F}_q^s$, then its partition number is s and so its covering radius R is 0. By (3), for any i, $N^{(i)} = (q-1)i + 0 = (q-1)i$. And so, $N^{(1)} < N^{(2)} < \cdots < N^{(s)}$ and $N^{(1)} = q - 1 = qR + q - 1$. Thus, the ambient space $\mathbb{F}_q^s$ is normal with the first coordinate being the only acceptable coordinate.

4.1. Normality and Direct Sum Construction

Proposition 16. *Let C_1 and C_2 be $(s_1, K_1, d_1)_q R_1$ and $(s_2, K_2, d_2)_q R_2$ RT-metric codes, respectively, such that $R_2 \neq 0$. Then,*

(i) *the direct sum $C = C_1 \bigoplus C_2$ is normal;*

(ii) *all the coordinates corresponding to C_1 are acceptable for C;*

(iii) *the coordinate i for $s_1 + 1 \leq i \leq s_2$ is acceptable for C only if the norm of C_2 with respect to the coordinate $i - s_1$ is equal to qR_2.*

Proof. Let C_1 and C_2 be any $(s_1, K_1, d_1)_q R_1$ and $(s_2, K_2, d_2)_q R_2$ q-ary RT-metric codes, respectively. Then, by Proposition 6, their direct sum $C = C_1 \bigoplus C_2$ is an $(s, K_1K_2, d_1)_q R$ code with length $s = s_1 + s_2$ and covering radius $R = s_1 + R_2$. Now, for $i \in \{1, 2, \ldots, s_1\}$, by (3), we have

$$N^{(i)}(C) = q(s_1 + R_2). \tag{12}$$

Now, in order to determine the norm $N(C)$, we have to find $N^{(j)}(C)$ for each $j \in \{s_1+1, s_1+2, \ldots, s_1+s_2\}$, the coordinate positions corresponding to C_2. But, we know by *Proposition 8* that $N(C_2) \geq qR_2$ which implies $N^{(j)}(C) \geq q(s_1 + R_2)$. Hence, $N(C) = q(s_1 + R_2) = qR < qR + q - 1$. Thus, C is normal and all the coordinate positions corresponding to C_1 are acceptable and the coordinate position $j \in \{s_1+1, s_1+2, \ldots, s_1+s_2\}$ corresponding to C_2 is acceptable if $N^{(j-s_1)}(C_2) = qR_2$. □

4.2. Normality and Amalgamated Direct Sum Construction. The amalgamated direct sum of two codes is defined as follows.

Definition 17. Let C_1 be an $[s_1, k_1]_q R_1$ normal code with the last coordinate being acceptable and let C_2 be an $[s_2, k_2]_q R_2$ normal code with the first coordinate being acceptable. Then, their amalgamated direct sum (or shortly ADS), denoted by $C_1 \dot{\bigoplus} C_2$, is an $[s_1 + s_2 - 1, k_1 + k_2 - 1]_q (s_1 + R_2 - 1)$ code and is given by

$$C_1 \dot{\bigoplus} C_2 = \{(c_1 \mid u \mid c_2) : (c_1 \mid u) \in C_1,\ (u \mid c_2) \in C_2\}. \tag{13}$$

Here, in this definition, we consider those codes C_1 and C_2 which are normal, respectively, with the last coordinate and first coordinate being acceptable such that the parts $C_{1_a}^{(s)}$ and $C_{2_a}^{(1)}$ are nonempty for all $a \in \mathbb{F}_q$.

As Theorem 12 shows the nonexistence of linear codes of dimension more than 1 whose last coordinate is acceptable, the significance of amalgamated direct sum and hence that of normality in RT-metric are less as far as the resulting covering radius is concerned. But, since the one-dimensional codes are normal and all the coordinates are acceptable, one can only use this ADS construction method to combine a 1-dimensional code with any other linear code, which may be helpful in the construction of MDS codes as seen in the following result.

Theorem 18. *Let C_1 be an $[s_1, 1]_q$ MDS code with covering radius $t_q[s_1, 1]$ (which is normal with last coordinate being acceptable) and let C_2 be an $[s_2, k_2]_q$ MDS code with covering radius $t_q[s_2, k_2]$. Then, their amalgamated direct sum $C_1 \dot{\bigoplus} C_2$ is an $[s_1 + s_2 - 1, k_2]_q$ MDS code with covering radius $t_q[s_1 + s_2 - 1, k_2]$.*

Proof. From the definition of partition number and from Theorem 4, it is easy to see that $t_q[s, k] = s - k$. So, $t_q[s_1, 1] = s_1 - 1$ and $t_q[s_2, k_2] = s_2 - k_2$. Now, as the dimension of $C_1 \dot{\bigoplus} C_2$ is k_2 (which is less than or equal to s_2), the covering radius of $C_1 \dot{\bigoplus} C_2$ depends only on the coordinates pertaining to the code C_2. But C_2 has covering radius $s_2 - k_2$ and hence has partition number k_2. So, $C_1 \bigoplus C_2$ also will have partition number k_2 and hence will have covering radius $s_1 + s_2 - 1 - k_2$ which is equal to $t_q[s_1 + s_2 - 1, k_2]$, which implies that $C_1 \bigoplus C_2$ is MDS. This completes the proof. □

5. Conclusion

We discussed the normality of codes over $\mathbb{F}_q$ in RT-metric and determined its norm with respect to various coordinate positions. We also established that the last coordinate is not acceptable for any nontrivial code in this metric which makes the ADS construction less significant as far as the higher dimensional RT-metric codes are concerned.

Conflict of Interests

The authors declare that there is no conflict of interests regarding the publication of this paper.

References

[1] P. Delsarte, "Four fundamental parameters of a code and their combinatorial significance," *Information and Control*, vol. 23, no. 5, pp. 407–438, 1973.

[2] G. D. Cohen, M. G. Karpovsky, H. F. Mattson, and J. R. Schatz, "Covering radius—survey and recent results," *IEEE Transactions on Information Theory*, vol. 31, no. 3, pp. 328–343, 1985.

[3] G. D. Cohen, S. N. Litsyn, A. C. Lobstein, and H. F. Mattson Jr., "Covering radius 1985–1994," *Applicable Algebra in Engineering, Communications and Computing*, vol. 8, no. 3, pp. 173–239, 1997.

[4] G. D. Cohen, I. Honkala, S. Litsyn, and A. Lobstein, *Covering Codes*, Elsevier, North-Holland, The Netherlands, 1997.

[5] R. L. Graham and N. J. A. Sloane, "On the covering radius of codes," *IEEE Transactions on Information Theory*, vol. 31, no. 3, pp. 385–401, 1985.

[6] G. D. Cohen, A. C. Lobstein, and N. J. A. Sloane, "Further results on the covering radius of codes," *IEEE Transactions on Information Theory*, vol. 32, no. 6, pp. 680–694, 1986.

[7] A. C. Lobstein and G. J. M. van Wee, "On normal and subnormal *q*-ary codes," *IEEE Transactions on Information Theory*, vol. 35, no. 6, pp. 1291–1295, 1989.

[8] R. S. Selvaraj and V. Marka, "Normality of binary codes in Rosenbloom-Tsfasman metric," in *Proceedings of the International Conference on Advances in Computing, Communications and Informatics (ICACCI '13)*, pp. 1370–1373, August 2013.

[9] M. Yu. Rosenbloom and M. A. Tsfasman, "Codes for the *m*-metric," *Problems of Information Transmission*, vol. 33, pp. 45–52, 1997.

[10] M. M. Skriganov, "Coding theory and uniform distributions," *St. Petersburg Mathematical Journal*, vol. 143, no. 2, pp. 301–337, 2002.

[11] I. Siap and M. Ozen, "The complete weight enumerator for codes over $M_{n \times s}(R)$ with respect to the Rosenbloom-Tsfasman metric," *Applied Mathematics Letters*, vol. 17, no. 1, pp. 65–69, 2004.

[12] S. T. Dougherty and K. Shiromoto, "Maximum distance codes in $Mat_{n \times s}(\mathbb{Z}_k)$ with a non-hamming metric and uniform distributions," *Designs, Codes, and Cryptography*, vol. 33, no. 1, pp. 45–61, 2004.

[13] S. Jain, "CT bursts—from classical to array coding," *Discrete Mathematics*, vol. 308, no. 9, pp. 1489–1499, 2008.

[14] B. Yildiz, I. Siap, T. Bilgin, and G. Yesilot, "The covering problem for finite rings with respect to the RT-metric," *Applied Mathematics Letters*, vol. 23, no. 9, pp. 988–992, 2010.

14

Crystal Bases as Tuples of Integer Sequences

Deniz Kus

Mathematisches Institut, Universität zu Köln, 50931 Köln, Germany

Correspondence should be addressed to Deniz Kus; dkus@math.uni-koeln.de

Academic Editors: M. Caramia and R. Dondi

We describe a set $\mathscr{R}^\infty$ consisting of tuples of integer sequences and provide certain explicit maps on it. We show that this defines a semiregular crystal for $\mathfrak{sl}_{n+1}$ and $\mathfrak{sp}_{2n}$, respectively. Furthermore, we define for any dominant integral weight λ a connected subcrystal $\mathscr{R}(\lambda)$ in $\mathscr{R}^\infty$, such that this crystal is isomorphic to the crystal graph $B(\lambda)$. Finally, we provide an explicit description of these connected crystals $\mathscr{R}(\lambda)$.

1. Introduction

Let $\mathfrak{g}$ be a symmetrizable Kac-Moody algebra and let $\mathfrak{U}_q(\mathfrak{g})$ be the corresponding quantum algebra. For these quantum algebras, Kashiwara developed the crystal bases theory for integrable modules in [1] and thus provided a remarkable combinatorial tool for studying these modules. In particular crystal bases can be viewed as bases at $q = 0$ and they contain structures of edge-colored oriented graphs satisfying a set of axioms, called the crystal graphs. These crystal graphs have certain nice properties; for instance, characters of $\mathfrak{U}_q(\mathfrak{g})$-modules can be computed and the decomposition of tensor products of modules into irreducible ones can also be determined from the crystal graphs, to name but a few. It is thus an important problem to have explicit realizations of crystal graphs.

There are many such realizations, combinatorial and geometrical, worked out during the last decades; for instance, we refer to [2–5]. In [2], the authors give a tableaux realization of crystal graphs for irreducible modules over the quantum algebra for all classical Lie algebras, which is a purely combinatorial model. Another significant combinatorial model for any symmetrizable Kac-Moody algebra is provided in [3], called Littelmann's path model. The underlying set here is a set of piecewise linear maps, and the crystal graph of an irreducible module of any dominant integral highest weight λ can be generated by an algorithm using the straight path connecting 0 and λ.

A geometrical realization of crystals is also known and is provided by Nakajima [5] by showing that there exists a crystal structure on the set of irreducible components of a lagrangian subvariety of the quiver variety $\mathscr{M}$. This realization can be translated into a purely combinatorial model, the set of Nakajima monomials, where the action of the Kashiwara operators can be understood as a multiplication with monomials. Moreover, it is shown in [6] that the connected component of any highest weight monomial of highest weight λ is isomorphic to the crystal graph $B(\lambda)$ obtained from Kashiwara's crystal bases theory. For special highest weight monomials these connected components are explicitly characterized for $\mathfrak{sl}_{n+1}$ in [7] and for the other classical Lie algebras in [8]. A combinatorial isomorphism from connected components corresponding to arbitrary highest weight monomials of highest weight λ and those in [7, 8] is provided in [9] for the types A and C and in [10] for the types B and D.

In this paper we introduce a set $\mathscr{R}^\infty$ consisting of tuples of integer sequences; that is, a typical element in $\mathscr{R}^\infty$ is given by

$$\mathbf{x} = (x_1, x_2, \dots) \in \mathscr{R}^\infty, \tag{1}$$

where each component x_j consists of certain ordered pairs of integers, $x_j = (i_1, i_1') \cdots (i_s, i_s')$ (see Definition 3). Furthermore, the number of nonzero components is finite. We provide certain maps on $\mathscr{R}^\infty$, the Kashiwara operators $\widetilde{e}_l$, $\widetilde{f}_l$, and maps ϵ_l, φ_l for all $l = 1, \dots, n$ and prove that $\mathscr{R}^\infty$ is a semiregular crystal if $\mathfrak{g}$ is $\mathfrak{sl}_{n+1}$ or $\mathfrak{sp}_{2n}$ (see Definition 3 and Proposition 9).

Moreover, we introduce for any dominant integral weight λ a subcrystal $\mathscr{R}(\lambda)$ as the connected component of $\mathscr{R}^{\infty}$ containing a highest weight element r_λ and prove the following theorem.

Theorem 1. *Let λ be a dominant integral weight, then there exists a crystal isomorphism*

$$\mathscr{R}(\lambda) \longrightarrow B(\lambda), \tag{2}$$

mapping r_λ to the highest weight element $b_\lambda \in B(\lambda)$.

Therefore, similar to the setting of Nakajima monomials, a natural question arises; namely, can one characterize for each dominant integral weight λ explicitly the sequences appearing in $\mathscr{R}(\lambda)$? We answer this question by describing explicitly these connected components (for the special linear Lie algebra in Theorem 18 and the symplectic Lie algebra in Theorem 19).

Our paper is organized as follows: in Section 2 we fix some notations and review briefly the crystal theory. In Section 3 we present the main definitions, especially the definition of $\mathscr{R}^{\infty}$ and we equip our main object with a crystal structure. In Section 4 Nakajima monomials are recalled. In Section 5 we introduce for any dominant integral weight λ the subcrystals $\mathscr{R}(\lambda)$ and describe them explicitly. Finally, in Section 6 we prove that they are isomorphic to $B(\lambda)$.

2. Notations and a Review of Crystal Theory

Let $\mathfrak{g}$ be a complex simple Lie algebra of rank n with index set $I = \{1, \dots, n\}$ and fix a Cartan subalgebra $\mathfrak{h}$ in $\mathfrak{g}$ and a Borel subalgebra $\mathfrak{b} \supseteq \mathfrak{h}$. We denote by $\Phi \subseteq \mathfrak{h}^*$ the root system of the Lie algebra, and corresponding to the choice of $\mathfrak{b}$ let Φ^+ be the subset of positive roots. Further, we denote by $\Pi = \{\alpha_1, \dots, \alpha_n\}$ the corresponding basis of Φ and the basis of the dual root system $\Phi^\vee \subseteq \mathfrak{h}$ is denoted by $\Pi^\vee = \{\alpha_1^\vee, \dots, \alpha_n^\vee\}$. Let $\mathfrak{g} = \mathfrak{n}^+ \oplus \mathfrak{h} \oplus \mathfrak{n}^-$ be a Cartan decomposition and for a given root $\alpha \in \Phi$ let $\mathfrak{g}_\alpha$ be the corresponding root space. For a dominant integral weight λ we denote by $V(\lambda)$ the irreducible $\mathfrak{g}$-module with a highest weight λ. Fix a highest weight vector $v_\lambda \in V(\lambda)$, then $V(\lambda) = \mathfrak{U}(\mathfrak{n}^-)v_\lambda$, where $\mathfrak{U}(\mathfrak{n}^-)$ denotes the universal enveloping algebra of $\mathfrak{n}^-$. For an indetermined element q we denote by $\mathfrak{U}_q(\mathfrak{g})$ the corresponding quantum algebra. The theory of studying modules of quantum algebras is quite parallel to that of Kac-Moody algebras and the irreducible modules are classified again in terms of highest weights (see [11]). Using the crystal bases theory, introduced by Kashiwara in [1], we can compute the character of an integrable module M in the category $\mathscr{O}^q$ as follows:

$$\operatorname{ch} M = \sum_{\mu} \sharp \left(B_\mu\right) e^{\mu}, \tag{3}$$

whereby (L, B) is the crystal bases of M (see [11]). The crystal graph associated with the irreducible module of highest weight λ is denoted by $B(\lambda)$. So finding expressions for the characters can be achieved by finding explicit combinatorial description of crystal bases. For some examples we refer to [2–4].

From now on we assume that $\mathfrak{g}$ is a classical Lie algebra of type A_n or C_n. Note that the positive roots are all of the following form:

$$\begin{aligned} &\text{Type } A_n : \alpha_{i,j} = \alpha_i + \alpha_{i+1} + \cdots + \alpha_j, \\ &\qquad \text{for } 1 \le i \le j \le n, \\ &\text{Type } C_n : \alpha_{i,j} = \alpha_i + \alpha_{i+1} + \cdots + \alpha_j, \\ &\qquad \text{for } 1 \le i \le j \le n, \\ &\alpha_{i,\bar{j}} = \alpha_i + \alpha_{i+1} + \cdots + \alpha_n + \alpha_{n-1} + \cdots + \alpha_j, \\ &\qquad \text{for } 1 \le i \le j \le n. \end{aligned} \tag{4}$$

Furthermore, let $P = \bigoplus_{i \in I} \mathbb{Z}\omega_i$ be the set of classical integral weights and $P^+ = \bigoplus_{i \in I} \mathbb{Z}_+\omega_i$ be the set of classical dominant integral weights. Before we discuss the crystal bases theory in detail we review first the notion of abstract crystals.

2.1. Abstract Crystals. Crystal bases of integrable $\mathfrak{U}_q(\mathfrak{g})$-modules in the category $\mathscr{O}^q$ are characterized by certain maps satisfying some properties. One can define the abstract notion of crystals associated with a Cartan datum as follows.

Definition 2. Let I be a finite index set and let $A = (a_{i,j})_{i,j \in I}$ be a generalized Cartan matrix with the Cartan datum $(A, \Pi, \Pi^\vee, P, P^\vee)$. A crystal associated with the Cartan datum $(A, \Pi, \Pi^\vee, P, P^\vee)$ is a set B together with maps $wt : B \to P$, $\tilde{e}_l, \tilde{f}_l : B \to B \cup \{0\}$, and $\epsilon_l, \varphi_l : B \to \mathbb{Z} \cup \{-\infty\}$ satisfying the following properties for all $l \in I$:

(1) $\varphi_l(b) = \epsilon_l(b) + \langle \alpha_l^\vee, wt(b) \rangle$,

(2) $wt(\tilde{e}_l b) = wt(b) + \alpha_l$ if $\tilde{e}_l b \in B$,

(3) $wt(\tilde{f}_l b) = wt(b) - \alpha_l$ if $\tilde{f}_l b \in B$,

(4) $\epsilon_l(\tilde{e}_l b) = \epsilon_l(b) - 1$, $\varphi_l(\tilde{e}_l b) = \varphi_l(b) + 1$ if $\tilde{e}_l b \in B$,

(5) $\epsilon_l(\tilde{f}_l b) = \epsilon_l(b) + 1$, $\varphi_l(\tilde{f}_l b) = \varphi_l(b) - 1$ if $\tilde{f}_l b \in B$,

(6) $\tilde{f}_l b = b'$ if and only if $\tilde{e}_l b' = b$ for $b, b' \in B$,

(7) if $\varphi_l(b) = -\infty$ for $b \in B$, then $\tilde{f}_l b = \tilde{e}_l b = 0$.

Furthermore, a crystal B is said to be semiregular if the equalities

$$\begin{aligned} \epsilon_l(b) &= \max\left\{k \ge 0 \mid \tilde{e}_l^k b \ne 0\right\}, \\ \varphi_l(b) &= \max\left\{k \ge 0 \mid \tilde{f}_l^k b \ne 0\right\} \end{aligned} \tag{5}$$

hold.

The maps $\tilde{e}_l$ and $\tilde{f}_l$ are called Kashiwara's crystal operators and the map wt is called the weight function. So, on the one hand, one can associate with any integrable $\mathfrak{U}_q(\mathfrak{g})$-module a set B satisfying the properties from Definition 3, and, on the other hand, one can study the notion of abstract crystals. A natural question which arises at this point is therefore the following: can one determine whether an abstract crystal is the crystal of a module? Stembridge [12] gave a set of local axioms to characterize the set of crystals of module in the class of all crystals when $\mathfrak{g}$ is simply laced and a list of local axioms for B_2-crystals is provided in [13]. In the following sections we define our underlying set and realize the crystal obtained from Kashiwara's crystal bases theory for the types A_n and C_n. We start by equipping our underlying set with an abstract crystal structure and later we prove that this crystal is the crystal of a module.

3. Tuples of Integer Sequences as Crystals

In this section we introduce a set $\mathscr{R}^\infty$ consisting of tuples of integer sequences (see Definition 16) and a crystal structure on it in the sense of Definition 3. Our purpose is to identify for any dominant integral weight λ certain subcrystals $\mathscr{R}(\lambda)$; that is, $\bigcup_{\lambda\in P^+}\mathscr{R}(\lambda) \subseteq \mathscr{R}^\infty$, such that $\mathscr{R}(\lambda)$ has a strong connection to the crystal graph $B(\lambda)$ (see Corollary 21).

3.1. Set of Tuples of Integer Sequences. In order to define $\mathscr{R}^\infty$ we consider a total order on $\mathbf{I} = \{1,\dots,n\}$ if $\mathfrak{g}$ is of type A_n and a total order on $\mathbf{I} = \{1,\dots,n,\overline{n-1},\dots,\overline{1}\}$ if $\mathfrak{g}$ is of type C_n, namely,

$$\begin{gathered}1<2<\cdots<n,\\ 1<2<\cdots<n<\overline{n-1}<\cdots<\overline{1},\end{gathered} \tag{6}$$

respectively. Furthermore, especially in Section 5, we need for type C_n the following bijective map:

$$\begin{gathered}\bar{\ }:\mathbf{I}\longrightarrow\mathbf{I},\\ n\longmapsto n,\quad \bar{i}\longmapsto i,\quad i\longmapsto\bar{i},\quad \text{for } i\in\{1,\dots,n-1\}.\end{gathered} \tag{7}$$

For $1\le i\le n$, $s\in\mathbb{Z}_{\ge0}$ we set $\mathscr{R}^i_s$ to be the set of all sequences $(i_1,i'_1)\cdots(i_s,i'_s)$ with $i_j,i'_j\in\mathbf{I}$, such that

$$\begin{gathered}1\le i_s<i_{s-1}<\cdots<i_1\le i\le i'_1<i'_2<\cdots<i'_s\le\max\mathbf{I},\\ i'_j\le\overline{i_j},\quad j=1,\dots,s,\end{gathered} \tag{8}$$

where $\max\mathbf{I}$ denotes the maximal element in $\mathbf{I}$ with respect to $<$. We denote by $\emptyset_i$ the unique element in $\mathscr{R}^i_0$.

Definition 3. We define $\mathscr{R}^\infty$ to be the set of all infinite sequences $\mathbf{x} = (x_1,x_2,x_3,\dots)$ where each component x_j is contained in $\mathscr{R} = \bigcup_{s,i}\mathscr{R}^i_s \cup \{0\}$ and only finitely many components are nonzero. We identify $\mathscr{R}^k$ with the sequences of the form $(x_1,\dots,x_k,0,0,\dots)$.

Before we mention the crystal structure on $\mathscr{R}^\infty$ we will initially introduce a list of properties. We need these to define the Kashiwara operators. Let $x=(i_1,i'_1)\cdots(i_s,i'_s)\in\mathscr{R}^i_s$ be an arbitrary element and fix $l\in I$:

(a)

$$\begin{gathered}l\notin\{i_1,\dots,i_s,i'_1,\dots,i'_s\},\\ l+1\in\{i_1,\dots,i_s\},\quad \text{if } l<i,\\ l-1\in\{i'_1,\dots,i'_s\},\quad \text{if } l>i,\\ /,\quad \text{if } l=i,\\ \left(\overline{l+1}\in\{i'_1,\dots,i'_s\}\vee\bar{l}\notin\{i'_1,\dots,i'_s\}\right),\end{gathered} \tag{9}$$

(a′) replace in (a) $\vee$ by $\wedge$,

(b)

$$\begin{gathered}\bar{l}\notin\{i'_1,\dots,i'_s\},\qquad \overline{l+1}\in\{i'_1,\dots,i'_s\},\\ l+1\notin\{i_1,\dots,i_s\}\vee l\in\{i_1,\dots,i_s\},\quad \text{if } l<i,\\ l-1\notin\{i'_1,\dots,i'_s\}\vee l\in\{i'_1,\dots,i'_s\},\quad \text{if } l>i,\\ l=i_1\vee l=i'_1,\quad \text{if } l=i,\end{gathered} \tag{10}$$

(c)

$$\begin{gathered}l\in\{i_1,\dots,i_s,i'_1,\dots,i'_s\},\\ l+1\notin\{i_1,\dots,i_s\},\quad \text{if } l<i,\\ l-1\notin\{i'_1,\dots,i'_s\},\quad \text{if } l>i,\\ (i_1,i'_1)=(l,l),\quad \text{if } l=i,\\ \left(\overline{l+1}\in\{i'_1,\dots,i'_s\}\vee\bar{l}\notin\{i'_1,\dots,i'_s\}\right),\end{gathered} \tag{11}$$

(d)

$$\begin{gathered}\bar{l}\in\{i'_1,\dots,i'_s\},\qquad \overline{l+1}\notin\{i'_1,\dots,i'_s\},\\ l+1\notin\{i_1,\dots,i_s\}\vee l\in\{i_1,\dots,i_s\},\quad \text{if } l<i,\\ l-1\notin\{i'_1,\dots,i'_s\}\vee l\in\{i'_1,\dots,i'_s\},\quad \text{if } l>i,\\ l=i_1\vee l=i'_1,\quad \text{if } l=i,\end{gathered} \tag{12}$$

(d′) replace everywhere in (d) $\vee$ by $\wedge$.

Let us consider an example.

Example 4. (1) Let $\mathfrak{g}=A_5$, $l=2$, and

$$x_1=(3,4)(1,5)\in\mathscr{R}^3_2,\qquad x_2=(2,3)\in\mathscr{R}^2_1, \tag{13}$$

then x_1 satisfies (a) while x_2 violates (a).

(2) Let $\mathfrak{g}=C_3$, $l=1$, and

$$x=\left(1,\bar{2}\right)\in\mathscr{R}^2_1, \tag{14}$$

then x satisfies (c).

Remark 5. If x satisfies (a′) and (d′), respectively, then it satisfies also (a) and (d), respectively. If $\mathfrak{g}$ is further of type A_n, these properties can be simplified. In particular, the properties (a′), (b), (d), and (d′) are superfluous.

Henceforth we define a crystal structure on $\mathscr{R}^\infty$, such that the semiregularity holds. For this let $\mathbf{x} = (x_1, x_2, x_3, \ldots)$ be such a sequence with finitely many components different from zero; recall that each component is a sequence as in (8). The weight function is given by

$$wt(\mathbf{x}) = \sum_{i=1}^{n} c_i \omega_i - \sum_j wt(x_j), \quad (15)$$

where

$$wt(x_j) = \begin{cases} \sum_{j=1}^{s} \alpha_{i_j, i'_j} & \text{if } x_j = (i_1, i'_1) \cdots (i_s, i'_s), \\ 0 & \text{if } x_j = 0 \text{ or } x_j \in \{\emptyset_1, \ldots, \emptyset_n\} \end{cases} \quad (16)$$

and $c_i = \sharp\{x_j \neq 0 \mid x_j \in \bigcup_s \mathscr{R}_s^i\}$. Suppose that the nonzero components in $\mathbf{x}$ are given by $x_{q_1} \in \mathscr{R}_{s_1}^{j_1}, \ldots, x_{q_k} \in \mathscr{R}_{s_k}^{j_k}$. For fixed $l \in I$ we define the following maps.

(i) For $2 \le j \le k+1$, let $\sigma_l^j : \mathscr{R}^\infty \to \mathbb{Z}_{\ge 0}$ be the map given by

$$\sigma_l^j(\mathbf{x}) = a_l^j(\mathbf{x}) + b_l^j(\mathbf{x}), \quad (17)$$

where

$$a_l^j(\mathbf{x}) = \sharp\{x_{q_p} \mid 1 \le p \le j-1, x_{q_p} \text{ satisfies (a) or (b)}\},$$
$$b_l^j(\mathbf{x}) = \sharp\{x_{q_p} \mid 1 \le p \le j-1, x_{q_p} \text{ satisfies } (\mathrm{a}')\}. \quad (18)$$

Furthermore, we define $\theta_l(x_{q_p})$ to be the sequence which arises out of x_{q_p} by

$$\begin{aligned} &\text{replacing } l+1 \text{ by } l, \quad \text{if } l < j_p, \\ &\text{replacing } l-1 \text{ by } l, \quad \text{if } l > j_p, \\ &\text{adding } (l,l), \quad \text{if } l = j_p, \end{aligned} \quad (19)$$

if x_{q_p} satisfies (a). If x_{q_p} satisfies (b), let $\theta_l(x_{q_p})$ be the sequence which arises out of x_{q_p} by replacing $\overline{l+1}$ by $\bar{l}$. If neither (a) nor (b) is fulfilled, we set $\theta_l(x_{q_p}) = 0$.

(ii) For $2 \le j \le k$, let $\tau_l^j : \mathscr{R}^\infty \to \mathbb{Z}_{\ge 0}$ be the map given by

$$\tau_l^j(\mathbf{x}) = c_l^j(\mathbf{x}) + d_l^j(\mathbf{x}), \quad (20)$$

where

$$c_l^j(\mathbf{x}) = \sharp\{x_{q_p} \mid 2 \le p \le j, x_{q_p} \text{ satisfies (c) or (d)}\},$$
$$d_l^j(\mathbf{x}) = \sharp\{x_{q_p} \mid 2 \le p \le j, x_{q_p} \text{ satisfies } (\mathrm{d}')\}. \quad (21)$$

We define $\rho_l(x_{q_p})$ to be the sequence which arises out of x_{q_p} by

$$\begin{aligned} &\text{replacing } l \text{ by } l+1, \quad \text{if } l < j_p, \\ &\text{replacing } l \text{ by } l-1, \quad \text{if } l > j_p, \\ &\text{erasing } (l,l), \quad \text{if } l = j_p, \end{aligned} \quad (22)$$

if x_{q_p} satisfies (c). If x_{q_p} satisfies (d), let $\rho_l(x_{q_p})$ be the sequence which arises out of x_{q_p} by replacing $\bar{l}$ by $\overline{l+1}$. If neither (c) nor (d) is fulfilled, we set $\rho_l(x_{q_p}) = 0$.

Remark 6. (1) For $j = 1$, we set $\sigma_l^j = \tau_l^j = 0$.

(2) Note that the image of $x \in \bigcup_s \mathscr{R}_s^i$ under the maps θ_l and ρ_l, respectively, is contained in $\bigcup_s \mathscr{R}_s^i \cup \{0\}$, that is,

$$\begin{aligned} \theta_l : \bigcup_s \mathscr{R}_s^i &\longrightarrow \bigcup_s \mathscr{R}_s^i \cup \{0\}, \\ \rho_l : \bigcup_s \mathscr{R}_s^i &\longrightarrow \bigcup_s \mathscr{R}_s^i \cup \{0\}. \end{aligned} \quad (23)$$

One important fact about these maps is described in the next lemma.

Lemma 7. *Let x, $\tilde{x}$ be nonzero sequences as in (8), then one has*

$$\theta_l(x) = \tilde{x} \quad \text{iff } \rho_l(\tilde{x}) = x. \quad (24)$$

Proof. One can easily show that x satisfies (a) if and only if $\tilde{x}$ satisfies (c). Hence, we can suppose that x does not fulfill all properties enumerated in (a). By observing the action we see that $\tilde{x}$ arises from x by replacing $\overline{l+1}$ by $\bar{l}$, which means that (c) is violated. In particular $\overline{l+1}$ does not appear in $\tilde{x}$ and $\bar{l}$ appears in $\tilde{x}$, which means that the properties in (d) hold. Hence, $\rho(\tilde{x}) = x$. The arguments for the reverse direction are the same. □

Let

$$f_l(\mathbf{x}) := \max\{1 \le p \le k \mid \sigma_l^p(\mathbf{x}) - \tau_l^p(\mathbf{x}) = \min\{\sigma_l^j(\mathbf{x}) - \tau_l^j(\mathbf{x}) \mid 1 \le j \le k\}\}, \quad (25)$$

$$e_l(\mathbf{x}) := \min\{1 \le p \le k \mid \sigma_l^p(\mathbf{x}) - \tau_l^p(\mathbf{x}) = \min\{\sigma_l^j(\mathbf{x}) - \tau_l^j(\mathbf{x}) \mid 1 \le j \le k\}\}. \quad (26)$$

Now we are able to define the Kashiwara operators,

$$\tilde{f}_l\mathbf{x} = \begin{cases} 0, & \text{if } \theta_l\left(x_{q_{f_l(\mathbf{x})}}\right) = 0, \\ \left(\ldots, \theta_l\left(x_{q_{f_l(\mathbf{x})}}\right), \ldots\right), & \text{else,} \end{cases}$$
$$\tilde{e}_l\mathbf{x} = \begin{cases} 0, & \text{if } \rho_l\left(x_{q_{e_l(\mathbf{x})}}\right) = 0, \\ \left(\ldots, \rho_l\left(x_{q_{e_l(\mathbf{x})}}\right), \ldots\right), & \text{else.} \end{cases} \quad (27)$$

Let us consider an example.

Example 8. (1) Let $\mathfrak{g} = A_4$ and $\mathbf{x} = (x_1, x_2, x_3, 0, 0, \ldots)$ with

$$x_1 = (2,2)(1,3) \in \mathcal{R}_2^2,$$
$$x_2 = (2,4) \in \mathcal{R}_1^4, \quad (28)$$
$$x_3 = (3,3)(2,4) \in \mathcal{R}_2^3.$$

For $l = 2$, we get $\sigma_2^1(\mathbf{x}) = \sigma_2^2(\mathbf{x}) = \sigma_2^3(\mathbf{x}) = \tau_2^1(\mathbf{x}) = 0$, $\tau_2^2(\mathbf{x}) = 1$, $\tau_2^3(\mathbf{x}) = 2$ and hence

$$\tilde{f}_2\mathbf{x} = 0. \quad (29)$$

(2) Let $\mathfrak{g} = C_3$ and $\mathbf{x} = (x_1, x_2, 0, 0, \ldots)$ with

$$x_1 = \left(1, \overline{2}\right) \in \mathcal{R}_1^2, \qquad x_2 = (3,3)\left(1,\overline{2}\right) \in \mathcal{R}_2^3, \quad (30)$$

then $\sigma_3^2(\mathbf{x}) = 0$, $\tau_3^2(\mathbf{x}) = 1$, and

$$\tilde{e}_3\mathbf{x} = (x_1, x_1, 0, 0\cdots). \quad (31)$$

It remains to define the maps φ_l and ϵ_l. These maps are given by the next formula:

$$\epsilon_l(\mathbf{x}) = \tau_l^2\left(\left(\emptyset_1, x_{q_1}, 0, 0\cdots\right)\right) - \min\left\{\sigma_l^j(\mathbf{x}) - \tau_l^j(\mathbf{x}) \mid 1 \le j \le k\right\},$$
$$\varphi_l(\mathbf{x}) = \sigma_l^{k+1}(\mathbf{x}) - \tau_l^k(\mathbf{x}) - \min\left\{\sigma_l^j(\mathbf{x}) - \tau_l^j(\mathbf{x}) \mid 1 \le j \le k\right\}. \quad (32)$$

Proposition 9. *The set $\mathcal{R}^\infty$ becomes a semiregular crystal.*

Proof. In Lemma 10 we proof the semiregularity of $\mathcal{R}^\infty$, which ensures that (4) and (5) from Definition 3 hold. So, to verify the proposition, it is sufficient to prove (1), (2), (3), and (6), where (2) and (3) are easily checked with the help of Remark 6. Let us start by proving (1); so let $\mathbf{x} \in \mathcal{R}^\infty$ be arbitrary with finitely many nonzero components, say, $x_{q_1} \in \mathcal{R}_{s_1}^{j_1}, \ldots, x_{q_k} \in \mathcal{R}_{s_k}^{j_k}$. Then we order these components in a way such that the first components are contained in $\bigcup_s \bigcup_{m<l} \mathcal{R}_s^m$ followed by components in $\bigcup_s \bigcup_{m>l} \mathcal{R}_s^m$ and the last ones are contained in $\bigcup_s \mathcal{R}_s^l$. So we can write the set of nonzero components of $\mathbf{x}$ as a disjoint union of three subsets $A_{<l} \cup A_{>l} \cup A_{=l}$. Further let $\mathbf{x}_1$ be the element in $\mathcal{R}^\infty$ obtained from $\mathbf{x}$ by replacing all components not belonging to $A_{<l}$ by 0. And $\mathbf{x}_2$ and $\mathbf{x}_3$, respectively, are similarly defined using $A_{>l}$ and $A_{=l}$, respectively. Then we get

$$\begin{aligned} &\varphi_l(\mathbf{x}) - \epsilon_l(\mathbf{x}) \\ &= \sigma_l^{k+1}(\mathbf{x}) - \tau_l^k(\mathbf{x}) - \tau_l^2\left(\left(\emptyset_1, x_{q_1}, 0, 0\cdots\right)\right) \\ &= \sigma_l^{|A_{<l}|+1}(\mathbf{x}_1) - \tau_l^{|A_{<l}|}(\mathbf{x}_1) + \sigma_l^{|A_{>l}|+1}(\mathbf{x}_2) - \tau_l^{|A_{>l}|}(\mathbf{x}_2) \\ &\quad + \sigma_l^{|A_{=l}|+1}(\mathbf{x}_3) - \tau_l^{|A_{=l}|}(\mathbf{x}_3) - \sum_{i=1}^{3}\tau_l^2\left(\left(\emptyset_1, y_i, 0, 0\cdots\right)\right), \end{aligned} \quad (33)$$

where the sequence y_i is the first nonzero element in $\mathbf{x}_i$. A short calculation shows

$$\begin{aligned} &\sigma_l^{|A_{<l}|+1}(\mathbf{x}_1) - \tau_l^{|A_{<l}|}(\mathbf{x}_1) - \tau_l^2\left(\left(\emptyset_1, y_1, 0, 0\cdots\right)\right) = -\sum_{1\le i\le k: x_{q_i}\in A_{<l}} \left\langle \alpha_l^\vee, wt\left(x_{q_i}\right)\right\rangle, \\ &\sigma_l^{|A_{>l}|+1}(\mathbf{x}_2) - \tau_l^{|A_{>l}|}(\mathbf{x}_2) - \tau_l^2\left(\left(\emptyset_1, y_2, 0, 0\cdots\right)\right) = -\sum_{1\le i\le k: x_{q_i}\in A_{>l}} \left\langle \alpha_l^\vee, wt\left(x_{q_i}\right)\right\rangle, \\ &\sigma_l^{|A_{=l}|+1}(\mathbf{x}_3) - \tau_l^{|A_{=l}|}(\mathbf{x}_3) - \tau_l^2\left(\left(\emptyset_1, y_3, 0, 0\cdots\right)\right) = c_l - \sum_{1\le i\le k: x_{q_i}\in A_{=l}} \left\langle \alpha_l^\vee, wt\left(x_{q_i}\right)\right\rangle. \end{aligned} \quad (34)$$

Now we proceed to prove (6). By the definition of the Kashiwara operators and Lemma 7, it is enough to show that $f_l(\tilde{e}_l\mathbf{x}) = e_l(\mathbf{x})$ and $e_l(\tilde{f}_l\mathbf{x}) = f_l(\mathbf{x})$. Since the proofs are similar, we prove only the latter equation. Assume that $\tilde{f}_l\mathbf{x} \ne 0$ and $f_l(\mathbf{x}) \ne 1$, then

$$\sigma_l^j\left(\tilde{f}_l\mathbf{x}\right) = \begin{cases} \sigma_l^j(\mathbf{x}), & \text{if } j \le f_l(\mathbf{x}), \\ \sigma_l^j(\mathbf{x}) - 1, & \text{if } j > f_l(\mathbf{x}), \end{cases}$$
$$\tau_l^j\left(\tilde{f}_l\mathbf{x}\right) = \begin{cases} \tau_l^j(\mathbf{x}), & \text{if } j \le f_l(\mathbf{x}) - 1, \\ \tau_l^j(\mathbf{x}) + 1, & \text{if } j \ge f_l(\mathbf{x}). \end{cases} \quad (35)$$

Subsequently we have

$$\begin{aligned} &e_l\left(\tilde{f}_l\mathbf{x}\right) \\ &= \min\left\{1 \le p \le k \mid \sigma_l^p\left(\tilde{f}_l\mathbf{x}\right) - \tau_l^p\left(\tilde{f}_l\mathbf{x}\right) = \min\left\{\sigma_l^j\left(\tilde{f}_l\mathbf{x}\right) - \tau_l^j\left(\tilde{f}_l\mathbf{x}\right) \mid 1 \le j \le k\right\}\right\} \\ &= \min\left\{1 \le p \le k \mid \sigma_l^p\left(\tilde{f}_l\mathbf{x}\right) - \tau_l^p\left(\tilde{f}_l\mathbf{x}\right) = \sigma_l^{f_l(\mathbf{x})}(\mathbf{x}) - \tau_l^{f_l(\mathbf{x})}(\mathbf{x}) - 1\right\} = f_l(\mathbf{x}). \end{aligned} \quad (36)$$

In the case where $f_l(\mathbf{x}) = 1$, we obtain

$$e_l\left(\tilde{f}_l\mathbf{x}\right) = \min\left\{1 \le p \le k \mid \sigma_l^p\left(\tilde{f}_l\mathbf{x}\right) - \tau_l^p\left(\tilde{f}_l\mathbf{x}\right) = 0\right\} = 1. \quad (37)$$

□

Hence we have shown that the set $\mathscr{R}^{\infty}$ is an abstract crystal provided that the semiregularity is shown. Thus, our aim now is to verify that the maps φ_l and ϵ_l, respectively, determine how often one can act with $\widetilde{f}_l$ and $\widetilde{e}_l$, respectively. The semiregularity is a necessary condition of a crystal B, if one wants to identify it with the crystal graph $B(\lambda)$.

Lemma 10. *Let ϵ_l and φ_l be as in (32). For a given element $\mathbf{x} \in \mathscr{R}^{\infty}$ one obtains*

$$\begin{aligned}\epsilon_l(\mathbf{x}) &= \max\left\{k \geq 0 \mid \widetilde{e}_l^k \mathbf{x} \neq 0\right\},\\ \varphi_l(\mathbf{x}) &= \max\left\{k \geq 0 \mid \widetilde{f}_l^k \mathbf{x} \neq 0\right\}.\end{aligned} \tag{38}$$

Proof. We proof the statement by induction on $z := \max\{k \geq 0 \mid \widetilde{e}_l^k \mathbf{x} \neq 0\}$; so let $z = 0$ and suppose first that $e_l(\mathbf{x}) \neq 1$. By the definition of $e_l(\mathbf{x})$ and ρ_l we have

$$\begin{aligned}&\sigma_l^{e_l(\mathbf{x})}(\mathbf{x}) - \tau_l^{e_l(\mathbf{x})}(\mathbf{x}) < \sigma_l^{e_l(\mathbf{x})-1}(\mathbf{x}) - \tau_l^{e_l(\mathbf{x})-1}(\mathbf{x})\\ &\Longleftrightarrow \sigma_l^{e_l(\mathbf{x})}(\mathbf{x}) < \sigma_l^{e_l(\mathbf{x})-1}(\mathbf{x}),\end{aligned} \tag{39}$$

which is a contradiction to $e_l(\mathbf{x}) \neq 1$. Hence, we have $e_l(\mathbf{x}) = 1$ and as a consequence we obtain $\epsilon_l(\mathbf{x}) = \tau_l^2((\emptyset_1, x_{q_1}, 0, 0, \cdots)) = 0$, which proves the initial step. Now assume that $z > 0$ and consider the element $\widetilde{e}_l\mathbf{x}$, where we again presume initially $e_l(\mathbf{x}) \neq 1$. By applying the induction hypothesis and using

$$\begin{aligned}\sigma_l^j(\widetilde{e}_l\mathbf{x}) &= \begin{cases}\sigma_l^j(\mathbf{x}), & \text{if } j \leq e_l(\mathbf{x}),\\ \sigma_l^j(\mathbf{x}) + 1, & \text{if } j > e_l(\mathbf{x}),\end{cases}\\ \tau_l^j(\widetilde{e}_l\mathbf{x}) &= \begin{cases}\tau_l^j(\mathbf{x}), & \text{if } j \leq e_l(\mathbf{x}) - 1,\\ \tau_l^j(\mathbf{x}) - 1, & \text{if } j \geq e_l(\mathbf{x}),\end{cases}\end{aligned} \tag{40}$$

we arrive at

$$\begin{aligned}&\max\left\{k \geq 0 \mid \widetilde{e}_l^{k+1}\mathbf{x} \neq 0\right\}\\ &= \epsilon_l(\widetilde{e}_l\mathbf{x})\\ &= \tau_l^2\left(\left(\emptyset_1, x_{q_1}, 0, 0\cdots\right)\right) - \left(\sigma_l^{e(\mathbf{x})}(\mathbf{x}) - \tau_l^{e(\mathbf{x})}(\mathbf{x}) + 1\right)\\ &= \epsilon_l(\mathbf{x}) - 1.\end{aligned} \tag{41}$$

If $e_l(\mathbf{x}) = 1$,

$$\begin{aligned}&\epsilon_l(\widetilde{e}_l(\mathbf{x}))\\ &= \tau_l^2\left(\left(\emptyset_1, \rho_l\left(x_{q_1}\right), 0, 0\cdots\right)\right) + \sigma_l^{e(\mathbf{x})}(\mathbf{x}) - \tau_l^{e(\mathbf{x})}(\mathbf{x})\\ &= \epsilon_l(\mathbf{x}) - 1.\end{aligned} \tag{42}$$

The proof of the remaining equality $\varphi_l(\mathbf{x}) = \max\{k \geq 0 \mid \widetilde{f}_l^k\mathbf{x} \neq 0\}$ is quite parallel. □

Before we introduce the subcrystals $\mathscr{R}(\lambda)$ we need some facts about the theory of tensor products of crystals. The tensor product rule is a very nice combinatorial feature and important to realize the crystal bases of a tensor product of two $\mathfrak{U}_q(\mathfrak{g})$-modules.

4. Tensor Products and Nakajima Monomials

In this section, we want to recall tensor products of crystals and investigate the action of Kashiwara operators on tensor products. With the aim of having a different realization of $B(\lambda)$ from our approach, we want to introduce the set of all Nakajima monomials, such that we can think of $B(\lambda)$ in terms of certain monomials. This theory is discovered by Nakajima [14] and generalized by Kashiwara [6].

4.1. Tensor Product of Crystals. Suppose that we have two abstract crystals B_1, B_2 in the sense of Definition 3, then we can construct a new crystal which is as a set $B_1 \times B_2$. This crystal is denoted by $B_1 \otimes B_2$ and the Kashiwara operators are given as follows:

$$\begin{aligned}\widetilde{f}_l(b_1 \otimes b_2) &= \begin{cases}\left(\widetilde{f}_l b_1\right) \otimes b_2, & \text{if } \varphi_l(b_1) > \epsilon_l(b_2),\\ b_1 \otimes \left(\widetilde{f}_l b_2\right), & \text{if } \varphi_l(b_1) \leq \epsilon_l(b_2),\end{cases}\\ \widetilde{e}_l(b_1 \otimes b_2) &= \begin{cases}\left(\widetilde{e}_l b_1\right) \otimes b_2, & \text{if } \varphi_l(b_1) \geq \epsilon_l(b_2),\\ b_1 \otimes \left(\widetilde{e}_l b_2\right), & \text{if } \varphi_l(b_1) < \epsilon_l(b_2).\end{cases}\end{aligned} \tag{43}$$

Furthermore, one can describe explicitly the maps wt, φ_l, and ϵ_l on $B_1 \otimes B_2$, namely,

$$\begin{aligned}wt(b_1 \otimes b_2) &= wt(b_1) + wt(b_2),\\ \varphi_l(b_1 \otimes b_2) &= \max\{\varphi_l(b_2), \varphi_l(b_1) + \varphi_l(b_2) - \epsilon_l(b_2)\},\\ \epsilon_l(b_1 \otimes b_2) &= \max\{\epsilon_l(b_1), \epsilon_l(b_1) + \epsilon_l(b_2) - \varphi_l(b_1)\}.\end{aligned} \tag{44}$$

One of the most important interpretations of the tensor product rule is the following theorem (for more details see [11]).

Theorem 11. *Let M_j be an integrable module in the category $\mathcal{O}^q$ and let (L_j, B_j) be a crystal bases of M_j, $(j = 1, 2)$. Set $L = L_1 \otimes L_2$ and $B = B_1 \otimes B_2$. Then (L, B) is a crystal bases of $M_1 \otimes M_2$.*

4.2. Nakajima Monomials. For $i \in I$ and $n \in \mathbb{Z}$, we consider monomials in the variables $Y_i(n)$; that is, we obtain the set of Nakajima monomials $\mathscr{M}$ as follows:

$$\begin{aligned}&\mathscr{M}\\ &:= \Bigg\{\prod_{i\in I, n\in\mathbb{Z}} Y_i(n)^{y_i(n)} \mid y_i(n) \in \mathbb{Z}\\ &\qquad \text{vanish except for finitely many } (i, n)\Bigg\}.\end{aligned} \tag{45}$$

With the goal of defining a crystal structure on $\mathscr{M}$, we take some integers $c = (c_{i,j})_{i \neq j}$ such that $c_{i,j} + c_{j,i} = 1$. Let now

$M = \prod_{i\in I, n\in\mathbb{Z}} Y_i(n)^{y_i(n)}$ be an arbitrary monomial in $\mathcal{M}$ and $l \in I$; then we set.

$$\begin{aligned} wt(M) &= \sum_i \left(\sum_n y_i(n)\right)\omega_i, \\ \varphi_l(M) &= \max\left\{\sum_{k\le n} y_l(k) \mid n \in \mathbb{Z}\right\}, \\ \epsilon_l(M) &= \max\left\{-\sum_{k>n} y_l(k) \mid n \in \mathbb{Z}\right\}, \\ n_f^l &= \min\left\{n \mid \varphi_l(M) = \sum_{k\le n} y_l(k)\right\}, \\ n_e^l &= \max\left\{n \mid \epsilon_l(M) = -\sum_{k>n} y_l(k)\right\}. \end{aligned} \tag{46}$$

The Kashiwara operators are defined as follows:

$$\begin{aligned} \tilde{f}_l M &= \begin{cases} A_l\left(n_f^l\right)^{-1} M, & \text{if } \varphi_l(M) > 0, \\ 0, & \text{if } \varphi_l(M) = 0, \end{cases} \\ \tilde{e}_l M &= \begin{cases} A_l\left(n_e^l\right) M, & \text{if } \epsilon_l(M) > 0, \\ 0, & \text{if } \epsilon_l(M) = 0, \end{cases} \end{aligned} \tag{47}$$

whereby

$$A_l(n) := Y_l(n) Y_l(n+1) \prod_{i\ne l} Y_i(n + c_{i,l})^{\langle \alpha_i^\vee, \alpha_l\rangle}. \tag{48}$$

The following two results are shown by Kashiwara [6].

Proposition 12. *With the maps wt, φ_l, ϵ_l, $\tilde{f}_l$, $\tilde{e}_l$, $l \in I$, the set $\mathcal{M}$ becomes a semiregular crystal.*

Remark 13. A *priori* the crystal structure depends on c; hence, we will denote this crystal by $\mathcal{M}_c$. But it is easy to see that the isomorphism class of $\mathcal{M}_c$ does not depend on this choice. In the literature c is often chosen as

$$c_{i,j} = \begin{cases} 0, & \text{if } i > j, \\ 1, & \text{else} \end{cases} \quad \text{or} \quad c_{i,j} = \begin{cases} 0, & \text{if } i < j, \\ 1, & \text{else.} \end{cases} \tag{49}$$

Proposition 14. *Let M be a monomial in $\mathcal{M}$, such that $\tilde{e}_l M = 0$ for all $l \in I$. Then the connected component of $\mathcal{M}$ containing M is isomorphic to $B(wt(M))$.*

According to the latter proposition, it is of great interest to describe these connected components explicitly. This is worked out for special highest weight monomials for all classical Lie algebras in [7, 8] and for the affine Lie algebra $A_n^{(1)}$ in [15]. We recall the results here only for type C_n stated originally in [8].

Proposition 15. *Let $\lambda = \sum_{i=1}^n m_i\omega_i$ be a dominant integral weight and consider the highest weight monomial $M = Y_1(1)^{m_1} \cdots Y_n(1)^{m_n}$. Then the connected component of $\mathcal{M}$ containing M is characterized as the set of monomials of the form*

$$\begin{aligned} X_{t_{1,1}}(1) \cdots X_{t_{1,\alpha_1}}(1) \cdots X_{t_{n,1}}(n) \cdots X_{t_{n,\alpha_n}}(n), \\ t_{r,s} \in \left\{1 < \cdots < n < \overline{n} < \cdots < \overline{1}\right\}, \end{aligned} \tag{50}$$

satisfying

(1) $\alpha_j = m_j + \cdots + m_n$ *for all* $j = 1, \ldots, n$,

(2) $t_{j,1} \ge \cdots \ge t_{j,\alpha_j}$ *for all* $j = 1, \ldots, n$,

(3) $t_{j-1,k} > t_{j,k}$ *for all* $j = 2, \ldots, n$ *and* $k = 1, \ldots, \alpha_j$, *where*

$$\begin{aligned} X_i(m) &= Y_{i-1}(m+1)^{-1} Y_i(m), \\ X_{\bar{i}}(m) &= Y_{i-1}(m + (n-i+1)) Y_i(m + (n-i+1))^{-1}. \end{aligned} \tag{51}$$

Summarized, we have a semiregular crystal $\mathcal{M}$and for each dominant integral weight λ certain connected subcrystals contained in $\mathcal{M}$. These are isomorphic to $B(\lambda)$ and an explicit description of these components is worked out for the classical simple Lie algebras and $A_n^{(1)}$. In the remaining parts of this paper we prove a similar result as Propositions 14 and 15, whereby our "big" semiregular crystal is $\mathcal{R}^\infty$.

5. Explicit Description of the Connected Components

In this section we define for the dominant integral weights $\lambda = \sum_{i=1}^n m_i\omega_i$ certain connected subcrystals $\mathcal{R}(\lambda) \subseteq \mathcal{R}^\infty$. Furthermore, we provide an explicit description of these crystals in Theorems 18 and 19, respectively; that is, we give a set of conditions describing $\mathcal{R}(\lambda)$.

Definition 16. For a dominant integral weight $\lambda = \sum_{i=1}^n m_i\omega_i$, let $\mathcal{R}(\lambda)$ be the connected component of $\mathcal{R}^\infty$ containing

$$r_\lambda = \Big(\underbrace{\emptyset_1, \ldots, \emptyset_1}_{m_1}, \ldots, \underbrace{\emptyset_n, \ldots, \emptyset_n}_{m_n}, 0, 0 \cdots\Big). \tag{52}$$

Note that the weight of r_λ is precisely λ. Furthermore, by definition, $\mathcal{R}(\lambda)$ is connected and for $\lambda = \omega_i$ we can immediately provide a description of $\mathcal{R}(\omega_i)$. To be more accurate we prove as a first step the following proposition:

Proposition 17. *Let $\lambda = \omega_i$; then one obtains*

$$\mathcal{R}(\lambda) = \left(\bigcup_s R_s^i, 0, 0, \ldots\right) = \bigcup_s R_s^i. \tag{53}$$

Proof. Since $(\bigcup_s R_s^i, 0, 0, \ldots)$ is stable under the Kashiwara operators $\tilde{f}_l$ and $\tilde{e}_l$ (Remark 6), it is enough to prove that $\emptyset_i$ is the unique highest weight element in $\bigcup_s R_s^i$, that is,

$$\begin{aligned} \tilde{e}_l(x, 0, 0, \ldots) = 0 \quad \forall l \in I \\ \Longrightarrow x = (\emptyset_i, 0, 0 \cdots). \end{aligned} \tag{54}$$

Assume $x = (i_1, i_1') \cdots (i_s, i_s')$ and $\tilde{e}_l(x, 0, 0, \ldots) = 0$ for all $l \in I$. If $i_1 < i$, we have $\tilde{e}_{i_1}(x, 0, 0, \ldots) \neq 0$ and if $i < i_1' \leq n$, we have $\tilde{e}_{i_1'}(x, 0, 0, \ldots) \neq 0$. So let $\overline{n-1} \leq i_1'$ (this case cannot appear if $\mathfrak{g} = A_n$); then since $i_1' \in \{i_1', \ldots, i_s'\}$, $\overline{\overline{i_1'}} + 1 \notin \{i_1', \ldots, i_s'\}$ and $i_1' \leq \overline{i_1} = \overline{i}$, we obtain that $\overline{i_1'} \in \{i, \ldots, n-1\}$. Thus, x satisfies (d) and hence $\tilde{e}_{\overline{i_1'}}(x, 0, 0, \ldots) \neq 0$. According to these calculations, the pair (i, i) must appear in x. This implies $\tilde{e}_i(x, 0, 0, \ldots) \neq 0$.

For general λ's, we can describe the connected components and refer to the following two subsections. □

5.1. Explicit Description of $\mathscr{R}(\lambda)$ in Type A_n. In this subsection we give an explicit characterization of $\mathscr{R}(\lambda)$ if $\mathfrak{g}$ is the special linear Lie algebra; that is, we give conditions whether a sequence $\mathbf{x}$ is contained in $\mathscr{R}(\lambda)$ or not (Theorem 18). Initially we note that $\mathscr{R}(\lambda)$ lives in $\mathscr{R}^{k=\sum_i m_i}$ and the first k components are nonzero. For simplicity we set $e_l(\mathbf{x}) = e(\mathbf{x})$, $f_l(\mathbf{x}) = f(\mathbf{x})$, $\theta_l = \theta$, and $\rho_l = \rho$.

Theorem 18. *The crystal $\mathscr{R}(\lambda)$ consists of all sequences*

$$\mathbf{x} = (x_1, \ldots, x_k, 0, 0, \ldots), \tag{55}$$

such that

(1) *for all pairs $(x_q \neq \emptyset_i, x_{q+1} \neq \emptyset_j)$, say, $x_q = (i_1, i_1') \cdots (i_s, i_s') \in \mathscr{R}_s^i$ and $x_{q+1} = (j_1, j_1') \cdots (j_t, j_t') \in \mathscr{R}_t^j$ one has*

 (i) $\sharp\{i_k \mid 1 \leq i_k \leq j_{t-p}\} \geq p+1$, *for all* $0 \leq p \leq u := \max\{0 \leq r \leq t-1 \mid j_{t-r} \leq i\}$,

 (ii) $j_{t-p} \leq i'_{j_{t-p}-p+s-i}$, *for all* $p \in \{u+1 \leq r \leq t-1 \mid j_{t-r} \leq r+i\}$,

 (iii) $j_p' \leq i'_{j-i+s-t+p}$, *for all* $1 \leq p \leq i-j+t$,

(2) *there is no pair (x_q, x_{q+1}) of the form $(\emptyset_i, x_{q+1} \neq \emptyset_j)$ with $i \geq j_t$.*

Proof. First we note that the element r_λ is contained in $\mathscr{R}(\lambda)$ and is a highest weight element. Furthermore, we claim that r_λ is the unique highest weight element. So suppose that we have another element $\mathbf{x} = (x_1, \ldots, x_k, 0, 0, \ldots)$ satisfying $\tilde{e}_l\mathbf{x} = 0$ for all $l \in I$. Let z be the lowest integer which appears in one of the sequences $x_1, \ldots, x_k$ and let p be the minimal integer such that z appears in x_p, say, $x_p = (j_1, j_1') \cdots (j_t = z, j_t') \in \mathscr{R}_t^j$. In the case where such a z does not exist, we have $\mathbf{x} = r_\lambda$. We remark that

$$\tilde{e}_l(x_p, 0, 0 \cdots) = 0 \quad \forall l \geq z \tag{56}$$

would imply the claim, whereby the reason is the following: assume that (56) holds and let $r = \max\{1 \leq r \leq t \mid j_{r-1} \neq j_r + 1\}$. If $j_1 = j$ and $r = 1$, we set $i = j_1' + 1$ and $i = j_r + 1$else. In either case we obtain $\epsilon_{i-1}(x_p, 0, 0, \ldots) \neq 0$, which is a contradiction to (56). So if suffices to show (56).

Primarily we claim that $x_1, \ldots, x_{p-1} \in \{\emptyset_1, \ldots, \emptyset_n\}$. Assume that the element $x_{p-1} = (i_1, i_1') \cdots (i_s, i_s') \in \mathscr{R}_s^i$ is not in the aforementioned set; then the property (1)(i) is violated by the choice of z and p because

$$0 - \sharp\{i_k \mid 1 \leq i_k \leq z\} \geq 1. \tag{57}$$

Hence, there exists a pair $(\emptyset_j, x_p)$ which forces $j < z$ and thus, again by the choice of z, the remaining sequences are again contained in the set $\{\emptyset_1, \ldots, \emptyset_n\}$. As a consequence, a short calculation by using Lemma 10 shows the uniqueness of the a highest weight element, namely,

$$\begin{aligned}
&\epsilon_l(\mathbf{x}) = 0 \quad \forall l \in I \\
&\Longrightarrow \epsilon_l\left(\left(x_p, \ldots, x_k, 0, 0 \cdots\right)\right) = 0 \quad \forall l \geq z \\
&\Longrightarrow \epsilon_l\left(\left(x_p, 0, 0 \cdots\right)\right) = 0 \quad \forall l \geq z.
\end{aligned} \tag{58}$$

In order to obtain a connected crystal it remains to show that $\tilde{e}_l\mathbf{x}, \tilde{f}_l\mathbf{x} \in \mathscr{R}(\lambda) \cup \{0\}$. Assume that $\tilde{f}_l\mathbf{x} \neq 0$, say,

$$\tilde{f}_l\mathbf{x} = \left(\ldots, x_{f(\mathbf{x})-1}, \theta\left(x_{f(\mathbf{x})}\right), x_{f(\mathbf{x})+1}, \ldots\right), \tag{59}$$

where we set for simplicity $f(\mathbf{x}) = q$. Our goal here is to show that the properties (1) and (2) hold for $\tilde{f}_l\mathbf{x}$, where we start by proving (2).

It is easy to see that (2) can only be violated if one of the following two cases occurs:

(i) $(x_{q-1}, x_q) = (\emptyset_i, \emptyset_i)$ and $l = i$,

(ii) $x_q = (i_1, i_1') \cdots (i_s, i_s') \in \mathscr{R}_s^i$, $x_{q-1} = \emptyset_{i_s-1}$ and $l = i_s - 1 < i$.

In either case we obtain $\sigma_l^{q-1}(\mathbf{x}) - \tau_l^{q-1}(\mathbf{x}) < \sigma_l^q(\mathbf{x}) - \tau_l^q(\mathbf{x})$, which is a contradiction to the choice of q. The proof of the fact that (1) holds will proceed in several cases. In the remaining parts we denote the entries of $\theta(x_q)$ by $\theta(i_k), \theta(i_k')$, that is,

$$\theta\left(x_q\right) = \left(\theta\left(i_1\right), \theta\left(i_1'\right)\right) \cdots \left(\theta\left(i_{s+\delta_{l,i}}\right), \theta\left(i'_{s+\delta_{l,i}}\right)\right). \tag{60}$$

Case 1.1 ((1) is violated for the pair $(\theta(x_q), x_{q+1})$). Let $x_q = (i_1, i_1') \cdots (i_s, i_s') \in \mathscr{R}_s^i$ and $x_{q+1} = (j_1, j_1') \cdots (j_t, j_t') \in \mathscr{R}_t^j$. We first consider the case $l < i$, which means that we replace $l+1$ by l. Since the entries i_p' stay unchanged, only property (1)(i) can be violated. However, property (1)(i) is still fulfilled because

$$\sharp\left\{i_k \mid 1 \leq i_k \leq j_{t-p}\right\} \leq \sharp\left\{\theta\left(i_k\right) \mid 1 \leq \theta\left(i_k\right) \leq j_{t-p}\right\}, \quad \forall 0 \leq p \leq u. \tag{61}$$

If $l > i$ we replace $l-1$ by l and hence (1) is obviously fulfilled. So suppose that $l = i$, which means that we add the entry (l, l). The equality

$$\sharp\left\{i_k \mid 1 \leq i_k \leq j_{t-p}\right\} = \sharp\left\{\theta\left(i_k\right) \mid 1 \leq \theta\left(i_k\right) \leq j_{t-p}\right\}, \quad \forall 0 \leq p \leq u-1 \tag{62}$$

and inequality

$$\sharp\{i_k \mid 1 \le i_k \le j_{t-u}\} \le \sharp\{\theta(i_k) \mid 1 \le \theta(i_k) \le j_{t-u}\} \quad (63)$$

imply that property (1)(i) is still fulfilled. Furthermore, since $\theta(i'_k) = i'_{k-1}$ for $k > 1$, $\theta(i'_1) = i$, and $\theta(x_q) \in \mathscr{R}^i_{s+1}$, we obtain

$$\begin{aligned} &j_{t-p} \le i'_{j_{t-p}-p+s-i} = \theta\left(i'_{j_{t-p}-p+s+1-i}\right), \\ &\forall p \in \{u+1 \le r \le t-1 \mid j_{t-r} \le r+i\}, \\ &j'_p \le i'_{j-i+s-t+p} = \theta\left(i'_{j-i+s+1-t+p}\right), \\ &\forall 1 \le p \le i-j+t, \end{aligned} \quad (64)$$

which particularly means that Case 1.1 can never appear.

Case 1.2 ((1)(i) is violated for the pair $(x_{q-1}, \theta(x_q))$). For simplicity, we set $x_{q-1} = (i_1, i'_1)\cdots(i_s, i'_s) \in \mathscr{R}^i_s$ and $x_q = (j_1, j'_1)\cdots(j_t, j'_t) \in \mathscr{R}^j_t$. In that case there exists at least one $0 \le p \le \theta(u) = \max\{r \mid \theta(j_{t-r}) \le i\}$, such that

$$\sharp\{i_k \mid 1 \le i_k \le \theta(j_{t-p})\} < p+1. \quad (65)$$

Necessarily we must have $l \le i$ and in the case where $l < j$ (65) implies

$$\begin{aligned} \sharp\{i_k \mid 1 \le i_k \le l+1\} &> \sharp\{i_k \mid 1 \le i_k \le l\}, \\ \sharp\{i_k \mid 1 \le i_k \le l-1\} &= \sharp\{i_k \mid 1 \le i_k \le l\}. \end{aligned} \quad (66)$$

As a consequence we get that $l+1$ appears in x_{q-1} while l does not appear and thus $\sigma^{q-1}_l(\mathbf{x}) - \tau^{q-1}_l(\mathbf{x}) < \sigma^q_l(\mathbf{x}) - \tau^q_l(\mathbf{x})$, which is a contradiction to the choice of q.

Eventually if $l = j = i$, we obtain in a similar way

$$\begin{aligned} t &\ge \sharp\{i_k \mid 1 \le i_k \le \theta(j_1) = j\} \\ &= s \ge \sharp\{i_k \mid 1 \le i_k \le j_1\} \ge t, \end{aligned} \quad (67)$$

which forces on the one hand $s = t$ and on the other hand that j does not appear in x_{q-1}. To be more precise, we can conclude the latter statement with the help of (1)(iii) and (1)(i), namely,

$$\begin{aligned} &j < j'_1 \le i'_{s-t+1} = i'_1, \\ &s \ge \sharp\{i_k \mid 1 \le i_k < i\} \ge \sharp\{i_k \mid 1 \le i_k \le j_1\} \ge t = s. \end{aligned} \quad (68)$$

Thus, we get again $\sigma^{q-1}_l(\mathbf{x}) - \tau^{q-1}_l(\mathbf{x}) < \sigma^q_l(\mathbf{x}) - \tau^q_l(\mathbf{x})$, which is once more a contradiction to the choice of q.

Case 1.3 ((1)(ii) is violated for the pair $(x_{q-1}, \theta(x_q))$). Here we have $l \le j$ and if $l = j$, we must have $j > i$ because otherwise (1)(ii) would not be violated. We first consider the case where $l = j$ and notice that the only possible violation is given by the following inequality:

$$j > i'_{j-t+s-i}. \quad (69)$$

We can conclude that j does not occur in x_{q-1} because either $j = t+i$ and hence

$$i'_s < j \quad (70)$$

or $j+1 \le t+i$ and (1)(iii) is applicable, which yields $j < j'_1 \le i'_{j-t+s-i+1}$. To obtain a contradiction we have to show that $j-1$ appears in x_{q-1}. If $j_1 > i$, this follows by the subsequent calculation:

$$\begin{aligned} i'_{j-t+s-i} &\le j-1 \le j_1 + (j-1-j_1) \\ &\le i'_{j_1-t+1+s-i} + (j-1-j_1) \le i'_{j-t+s-i}. \end{aligned} \quad (71)$$

If $j_1 \le i$, we obtain with property (1)(i) that $s \ge \{i_k \mid 1 \le i_k \le j_1\} \ge t$. In particular we actually have $s = t$ because otherwise we would get

$$j \le j-t+s-1 \le i'_{j-t+s-i}. \quad (72)$$

Eventually we can conclude again that $j-1$ must appear in x_{q-1}

$$i'_{j-i} \le j-1 \le i'_{j-i}. \quad (73)$$

Now we suppose $l < j$. Then an easy consideration shows that (1)(ii) can only be violated if $l > i$ and thus we obtain similarly as before that the only violation which can occur is the following:

$$l > i'_{l-r+s-i}, \quad (74)$$

where we expect $j_{t-r} = l+1$. We would like to show as before that l does not appear in x_{q-1} while $l-1$ appears. We either have $l = r+i$ and thus $l > i'_s$ or $l+1 \le r+i$. In the latter case we apply property (1)(ii) and obtain

$$l < i'_{l-r+s-i+1}. \quad (75)$$

In either case we notice that l does not occur in x_{q-1}.

In order to prove the remaining part we consider the element $j_{t-(r-1)}$ which is considerable, since

$$l-r+s-1 \le i'_{l-r+s-i} \le l-1 \Longrightarrow 1 \le s \le r. \quad (76)$$

In the case where $j_{t-(r-1)}$ is greater than i, we obtain

$$\begin{aligned} i'_{l-r+s-i} &\le l-1 \le j_{t-(r-1)} + (l-1-j_{t-(r-1)}) \\ &\le i'_{j_{t-(r-1)}-r+1+s-i} + (l-1-j_{t-(r-1)}) \le i'_{l-r+s-i} \end{aligned} \quad (77)$$

and otherwise by using property (1)(i) we get $r = s$. Thus,

$$l-1 \ge i'_{l-i} \ge l-1. \quad (78)$$

Case 1.4 ((1)(iii) is violated for the pair $(x_{q-1}, \theta(x_q))$). We suppose that (1)(iii) is violated, which forces $l \ge j$. In the case where $l > j$, we have

$$\exists p : j'_p = l-1 = i'_{j-i+s-t+p}. \quad (79)$$

It follows that $l-1$ appears and l does not appear in x_{q-1} because $l = i'_{j-i+s-t+p+1}$ would imply

$$l = i'_{j-i+s-t+p+1} \geq j'_{p+1} > j'_p = l-1 \Longrightarrow j'_{p+1} = l. \tag{80}$$

Consequently, $\sigma_l^{q-1}(\mathbf{x}) - \tau_l^{q-1}(\mathbf{x}) < \sigma_l^q(\mathbf{x}) - \tau_l^q(\mathbf{x})$ and we obtain as usual a contradiction to the choice of q.

If the remaining case $l = j$ occurs, then one of the inequalities

$$\theta\left(j'_p\right) \leq i'_{j-i+s-t-1+p}, \quad 1 \leq p \leq i-j+t+1 \tag{81}$$

must be violated. Clearly, the only possibility is that $j \leq i'_{j-i+s-t}$ does not hold. If $t = j-i$, we get

$$j > i'_s \tag{82}$$

and else we can assume $j-i+1 \leq t$ so that (1)(iii) is applicable, which yields

$$i \leq i'_{j-i+s-t} < j < j'_1 \leq i'_{j-i+s-t+1}. \tag{83}$$

Therefore, j does not appear in x_{q-1}. In what follows, we finish our proof by showing that $j-1$ appears in x_{q-1}. If $j_1 > i$, we can apply (1)(ii) and get

$$\begin{aligned} i'_{j-i+s-t} &\leq j_1 + (j-1-j_1) \\ &\leq i'_{j_1-t+1+s-i} + (j-1-j_1) \leq i'_{j-i+s-t}. \end{aligned} \tag{84}$$

If $j_1 \leq i$, we can verify with (1)(i) that $s \geq t$, but the assumption $s-1 \geq t$ yields in a contradiction, namely,

$$i'_{j-i+s-t} \geq j+s-t-1 \geq j. \tag{85}$$

So $s = t$ and we obtain the required equality $j-1 \geq i'_{j-i} \geq j-1$.

The proof of $\tilde{e}_l\mathbf{x} \in \mathcal{R}(\lambda) \cup \{0\}$ is similar, which completes the proof. □

5.2. Explicit Description of $\mathcal{R}(\lambda)$ in Type C_n. In this subsection we would like to give an explicit characterization of $\mathcal{R}(\lambda)$ if $\mathfrak{g}$ is a symplectic Lie algebra. In order to state the main theorem we fix some notation.

For an arbitrary subset $A \subseteq \mathbf{I}$, $n \in \{0,1\}$, and $y \in \mathbf{I}$, we set

$$\delta_{y,A} = \begin{cases} 1, & \text{if } y \in A, \\ 0, & \text{if } y \notin A, \end{cases} \qquad nA = \begin{cases} A, & \text{if } n = 1, \\ \emptyset, & \text{if } n = 0. \end{cases} \tag{86}$$

The analogue result to Theorem 18 for type C_n is the following.

Theorem 19. *The crystal $\mathcal{R}(\lambda)$ consists of all sequences*

$$\mathbf{x} = (x_1, \ldots, x_k, 0, 0, \ldots), \tag{87}$$

such that

(1) *for all pairs $(x_q \neq \emptyset_i, x_{q+1} \neq \emptyset_j)$, say, $x_q = (i_1, i'_1) \cdots (i_s, i'_s) \in \mathcal{R}^i_s$ and $x_{q+1} = (j_1, j'_1) \cdots (j_t, j'_t) \in \mathcal{R}^j_t$ one has*

   (i) $\sharp\{i_k \mid 1 \leq i_k \leq j_{t-p}\} \geq p+1$, *for all* $0 \leq p \leq u := \max\{0 \leq r \leq t-1 \mid j_{t-r} \leq i\}$,

   (ii) $j_{t-p} \leq i'_{j_{t-p}-p+s-i}$, *for all* $p \in \{u+1 \leq r \leq t-1 \mid j_{t-r} \leq r+i\}$,

   (iii) $j'_p \leq i'_{j-i+s-t+p}$, *for all* $1 \leq p \leq i-j+t$,

(2) *there is no pair (x_q, x_{q+1}) of the form $(\emptyset_i, x_{q+1} \neq \emptyset_j)$ with $i \geq j_t$,*

(3) *there is no pair $(x_q \neq \emptyset_i, x_{q+1})$ with the following property: there exists $i'_p \geq i'_r \geq n$ with*

   (a) $\overline{i'_r} - \delta_{\overline{i'_r},\{i+1,\ldots,n\}} \notin (1-\delta_{\overline{i'_r},\{i+1,\ldots,n\}})\{i_1,\ldots,i_s\} \cup \delta_{\overline{i'_r},\{i+1,\ldots,n\}}(\mathbf{I} - \{i'_1,\ldots,i'_s\})$,

   (b) $\overline{i'_p} - \delta_{\overline{i'_p},\{j+1,\ldots,n\}} \notin (1-\delta_{\overline{i'_p},\{j+1,\ldots,n\}})\{j_1,\ldots,j_t\} \cup \delta_{\overline{i'_p},\{j+1,\ldots,n\}}(\mathbf{I} - \{j'_1,\ldots,j'_t\})$,

   (c) $(p-r) + \sharp\{j_k \mid j_k < \overline{i'_p}\} - \sharp\{i_k \mid i_k < \overline{i'_r}\} + \sharp\{i'_k \mid i'_k < \overline{i'_r}\} - \sharp\{j'_k \mid j'_k < \overline{i'_p}\} \geq \max\{0, \overline{i'_r} - i\} - \max\{0, \overline{i'_p} - j\}$,

(4) *there is no pair $(x_q \neq \emptyset_i, x_{q+1} \neq \emptyset_j)$ with the following property: there exists $i'_p \geq j'_r \geq n$ with*

   (a) $\overline{i'_p} - \delta_{\overline{i'_p},\{j+1,\ldots,n\}} \notin (1-\delta_{\overline{i'_p},\{j+1,\ldots,n\}})\{j_1,\ldots,j_t\} \cup \delta_{\overline{i'_p},\{j+1,\ldots,n\}}(\mathbf{I} - \{j'_1,\ldots,j'_t\})$,

   (b) $\overline{j'_r} - \delta_{\overline{j'_r},\{j+1,\ldots,n\}} \notin (1-\delta_{\overline{j'_r},\{j+1,\ldots,n\}})\{j_1,\ldots,j_t\} \cup \delta_{\overline{j'_r},\{j+1,\ldots,n\}}(\mathbf{I} - \{j'_1,\ldots,j'_t\})$,

   (c) $(i-j) + (t-s) + (p-r) + \sharp\{j'_k \mid \overline{i'_p} \leq j'_k < \overline{j'_r}\} - \sharp\{j_k \mid \overline{i'_p} \leq j_k < \overline{j'_r}\} \geq \max\{0, \overline{j'_r} - j\} - \max\{0, \overline{i'_p} - j\}$.

Proof. First of all we note as in the A_n case that the element r_λ is contained in $\mathcal{R}(\lambda)$ and is a highest weight element. In order to prove that r_λ is the unique highest weight element in $\mathcal{R}(\lambda)$ we assume $\mathbf{x} = (x_1, \ldots, x_k, 0, 0, \ldots)$ to be another one. Let z be the lowest integer which appears in one of the sequences $x_1, \ldots, x_k$ and let p be the minimal integer such that z appears in x_p, say, $x_p = (i_1, i'_1) \cdots (i_s = z, i'_s) \in \mathcal{R}^i_s$. We can prove similar to Theorem 18 that the elements $x_1, \ldots, x_p$ are contained in $\{\emptyset_1, \ldots, \emptyset_n\}$. Accordingly we get once more with Lemma 10 (similar to the A_n case)

$$\begin{aligned} \tilde{e}_l\mathbf{x} = 0 \quad \forall l \in I &\Longrightarrow \tilde{e}_l\left(\left(x_p, \ldots, x_k, 0, 0 \cdots\right)\right) = 0 \\ \forall l \geq z &\Longrightarrow \tilde{e}_l\left(\left(x_p, 0, 0 \cdots\right)\right) = 0 \quad \forall l \geq z. \end{aligned} \tag{88}$$

Our aim is again to prove the impossibility of $\tilde{e}_l((x_p, 0, 0 \cdots)) = 0$ for all $l \geq z$. Let $r = \max\{1 \leq r \leq s \mid i_{r-1} \neq i_r + 1\}$. If $i_1 = i$, $r = 1$, and $i'_1 \notin \{\overline{n-1}, \ldots, \overline{1}\}$, we set $j = i'_1 + 1$ and if $i_1 = i, r = 1$ is not satisfied, we set $j = i_r + 1$ and obtain similar to Theorem 18 that $\epsilon_{j-1}(x_p, 0, 0, \ldots) \neq 0$. Thus, the only remaining case which can appear is when $i_1 = i$, $r = 1$, and $i'_1 \in \{\overline{n-1}, \ldots, \overline{1}\}$. In this particular case we set

$j = \overline{i'_1}$ and claim $\epsilon_j(x_p, 0, 0, \ldots) \neq 0$. The latter claim is true because on the one hand we have

$$i'_1 \in \{i'_1, \ldots, i'_s\}, \overline{\overline{i'}_1 + 1} \notin \{i'_1, \ldots, i'_s\} \quad (89)$$

and on the other hand we got $\overline{j} = i'_1 \leq \overline{i_1} = \overline{i} \Rightarrow j \geq i$, which verifies the properties listed in (d). To be more precise, if $j > i$, we have $j - 1 \notin \{i'_1, \ldots, i'_s\}$ and if $j = i$, we have $j = i = i_1$.

In order to finish the theorem it remains to show that $\mathscr{R}(\lambda)$ is stable under the Kashiwara operators, that is, $\widetilde{e}_l\mathbf{x}, \widetilde{f}_l\mathbf{x} \in \mathscr{R}(\lambda) \cup \{0\}$. Assume that $\widetilde{f}_l\mathbf{x} \neq 0$, say,

$$\widetilde{f}_l\mathbf{x} = (\ldots, x_{q-1}, \theta(x_q), x_{q+1}, \ldots). \quad (90)$$

Our aim here is to prove that the properties (1)–(4) hold for $\widetilde{f}_l\mathbf{x}$, whereby the verification of the first and the second properties proceeds almost similar to Theorem 18. Nevertheless we will demonstrate some parts of it in Case 1. In the remaining parts of our proof, we set $x_q = (i_1, i'_1) \cdots (i_s, i'_s) \in \mathscr{R}^i_s$ and $x_{q+1} = (j_1, j'_1) \cdots (j_t, j'_t) \in \mathscr{R}^j_t$, whenever they are contained in $\mathscr{R} - \{\emptyset_1, \ldots, \emptyset_n\}$. We will divide our proof into several cases.

Case 1 ($l \pm 1 \mapsto l$ or $+(l, l)$). Here we assume that the action of the Kashiwara operator on $\mathbf{x}$ is given in a way such that x_q satisfies property (a). It means that we either replace $l \pm 1$ by l or add the pair (l, l) as described in Section 3. Here we consider again several cases, where each case assumes that a condition described in Theorem 19 is violated.

Subcase 1.1 ((1)(i) is violated for the pair $(x_{q-1}, \theta(x_q))$. Since (1)(i) is violated, there must exist an element $\theta(j_{t+\delta_{l,j}-r}) = l$ such that

$$\sharp\{i_k \mid 1 \leq i_k \leq l\} \leq r. \quad (91)$$

Further, as in Subcase 1.2 of Theorem 18, one can verify that l does not appear in x_{q-1} and if in addition $l \leq i < j$ holds, then $l+1$ occurs in x_{q-1}. Consequently $\overline{l}$ must be contained in $\{i'_1, \ldots, i'_s\}$ because otherwise we would obtain a contradiction to the choice of q. Hence, if we take $i'_p = i'_r = \overline{l}$, it follows immediately that the properties (a), (b), and (c) in (3) hold for the pair (x_{q-1}, x_q), which is impossible. For instance (c) is fulfilled with (91), since

$$\sharp\{i_k \mid 1 \leq i_k \leq l\} = \sharp\{i_k \mid i_k < l\} \leq r \leq \sharp\{j_k \mid j_k < l\}. \quad (92)$$

The other violations of the properties in (1) or (2) can be proven similarly, so that we consider as a next step the following case.

Subcase 1.2 ((3) is violated for the pair $(\theta(x_q), x_{q+1})$). A simple case-by-case observation shows that this case can only occur if there exists $\theta(i'_p) \geq \theta(i'_r) = \overline{l+1}$, such that (a), (b) and (c) are satisfied. Suppose that there exists an element $\overline{m}$ in the set $\{\theta(i'_{r+1}), \ldots, \theta(i'_p)\}$ ($\overline{m} = \theta(i'_{r+u})$), such that

$$\begin{aligned} m - \delta_{m,\{i+1,\ldots,n\}} &\notin (1 - \delta_{m,\{i+1,\ldots,n\}})\{i_1, \ldots, i_s\} \\ &\cup \delta_{m,\{i+1,\ldots,n\}}(\mathbf{I} - \{i'_1, \ldots, i'_s\}) \end{aligned} \quad (93)$$

and $\overline{m}$ is minimal with this property. In the case where such an element does not exist we set $\overline{m} = \theta(i'_p)$. We claim the following.

Claim. Let $\overline{m}$ be as described before; then

$$\begin{aligned} &u + \min\{l, i\} - \min\{m, i\} - \sharp\{i_k \mid m \leq i_k < l\} \\ &\quad + \sharp\{i'_k \mid m \leq i'_k < l\} \leq l + 1 - m. \end{aligned} \quad (94)$$

Proof of the Claim. We consider again various cases, starting with

(i) $l \leq i$: the minimality of $\overline{m}$ implies $\overline{\theta(i'_{r+1})}, \ldots, \overline{\theta(i'_{r+u-1})} \in \{i_1, \ldots, i_s\}$, which yields

$$\begin{aligned} &\underbrace{\sharp\{i'_k \mid \theta(i'_r) < i'_k < \overline{m}\}}_{=u-1} \leq \sharp\{i_k \mid m \leq i_k < l\} \\ &\Longleftrightarrow \sharp\{i'_k \mid m \leq i'_k < l\} - \sharp\{i_k \mid m \leq i_k < l\} \\ &\leq -u + 1, \end{aligned} \quad (95)$$

(ii) $l > i$ and $m \leq i$: using the minimality of $\overline{m}$ we obtain similarly $\overline{\theta(i'_{r+u-1})}, \ldots, \overline{\theta(i'_{r+x})} \in \{i_1, \ldots, i_s\}$ and $\overline{\theta(i'_{r+1})} - 1, \ldots, \overline{\theta(i'_{r+x-1})} - 1 \notin \{i'_1, \ldots, i'_s\}$, for some integer x. Therefore, the following calculation implies (94):

$$\begin{aligned} &i - \sharp\{i_k \mid m \leq i_k\} + \sharp\{i'_k \mid i'_k < l\} \\ &\leq l - (x - 1) - \sharp\{i_k \mid m \leq i_k\} \\ &\leq l - \sharp\{i'_k \mid \theta(i'_r) < i'_k < \overline{m}\}, \end{aligned} \quad (96)$$

(iii) $m > i$: the minimality of $\overline{m}$ provides $\overline{\theta(i'_{r+1})} - 1, \ldots, \overline{\theta(i'_{r+u-1})} - 1 \notin \{i'_1, \ldots, i'_s\}$. Hence,

$$\sharp\{i'_k \mid m \leq i'_k < l\} \leq l - m - \sharp\{i'_k \mid \theta(i'_r) < i'_k < \overline{m}\}, \quad (97)$$

which finishes the proof of the claim.

As a consequence of (94) we obtain

$$\begin{aligned}
&\sharp\left\{j_k \mid j_k < \overline{\theta\left(i'_p\right)}\right\} - \sharp\left\{\theta\left(i_k\right) \mid \theta\left(i_k\right) < \overline{\theta\left(i'_r\right)}\right\} \\
&\quad + \sharp\left\{\theta\left(i'_k\right) \mid \theta\left(i'_k\right) < \overline{\theta\left(i'_r\right)}\right\} - \sharp\left\{j'_k \mid j'_k < \overline{\theta\left(i'_p\right)}\right\} \\
&\le \sharp\left\{j_k \mid j_k < \overline{i'_{p-\delta_{l,i}}}\right\} - \sharp\left\{i_k \mid m \le i_k < l\right\} \\
&\quad - \sharp\left\{i_k \mid i_k < m\right\} \\
&\quad + \sharp\left\{i'_k \mid m \le i'_k < l\right\} + \sharp\left\{i'_k \mid i'_k < m\right\} \\
&\quad - \sharp\left\{j'_k \mid j'_k < \overline{i'_{p-\delta_{l,i}}}\right\} \\
&< l + 1 - m - \min\{l, i\} + \min\{m, i\} + \max\{0, m - i\} \\
&\quad - \max\left\{0, \overline{i'_{p-\delta_{l,i}}} - j\right\} - (p - r) \\
&= l + 1 - \min\{l, i\} - \max\left\{0, \overline{i'_{p-\delta_{l,i}}} - j\right\} - (p - r) \\
&\le \max\{0, l + 1 - i\} - \max\left\{0, \overline{i'_{p-\delta_{l,i}}} - j\right\} - (p - r) \\
&\quad + \delta_{l,\{1,\dots,i-1\}},
\end{aligned} \tag{98}$$

where the first estimation is strict if $l < i$. Hence, we have a contradiction to the assumption that (c) holds.

Subcase 1.3 ((3) is violated for the pair $(x_{q-1}, \theta(x_q))$). In that case there exists $i'_p = \overline{l+1} \ge i'_r$, such that (a), (b), and (c) is satisfied. Therefore, we must have $\overline{l} = i'_{p+1}$ because otherwise we obtain a contradiction to the choice of q. It follows that

$$\begin{aligned}
&(p - r) + \sharp\left\{\theta\left(j_k\right) \mid \theta\left(j_k\right) < \overline{i'_p}\right\} - \sharp\left\{i_k \mid i_k < \overline{i'_r}\right\} \\
&\quad + \sharp\left\{i'_k \mid i'_k < \overline{i'_r}\right\} - \sharp\left\{\theta\left(j'_k\right) \mid \theta\left(j'_k\right) < \overline{i'_p}\right\} \\
&\le (p + 1 - r) + \sharp\left\{j_k \mid j_k < \overline{i'_{p+1}}\right\} - \sharp\left\{i_k \mid i_k < \overline{i'_r}\right\} \\
&\quad + \sharp\left\{i'_k \mid i'_k < \overline{i'_r}\right\} - \sharp\left\{j'_k \mid j'_k < \overline{i'_{p+1}}\right\} \\
&< \max\left\{0, \overline{i'_r} - i\right\} - \max\{0, l - j\} \le \max\left\{0, \overline{i'_r} - i\right\} \\
&\quad - \max\{0, l + 1 - j\} + \delta_{l,\{j,\dots,n\}},
\end{aligned} \tag{99}$$

where the first estimation is strict whenever $l \ge j$ and thus provides a contradiction to (c).

Subcase 1.4 ((4) is violated for the pair $(\theta(x_q), x_{q+1})$). It is easy to see that this case can never appear.

Subcase 1.5 ((4) is violated for the pair $(x_{q-1}, \theta(x_q))$). In that case we have two possibilities, where we start by supposing that there is $i'_p = \overline{l+1} \ge \theta(j'_r)$, such that the properties (a), (b), and (c) are fulfilled. Similar to Subcase 1.3, we must have an element $\overline{l} = i'_{p+1}$ because otherwise we would obtain a contradiction to the choice of q. Subsequently we get

$$\begin{aligned}
&(i - j) + \left(t + \delta_{l,j} - s\right) + (p - r) \\
&\quad + \sharp\left\{\theta\left(j'_k\right) \mid \overline{i'_p} \le \theta\left(j'_k\right) < \overline{\theta\left(j'_r\right)}\right\} \\
&\quad - \sharp\left\{\theta\left(j_k\right) \mid \overline{i'_p} \le \theta\left(j_k\right) < \overline{\theta\left(j'_r\right)}\right\} \\
&\le (i - j) + \left(t + \delta_{l,j} - s\right) + (p + 1 - r) \\
&\quad + \sharp\left\{j'_k \mid l < j'_k \le \overline{\theta\left(j'_r\right)}\right\} - \sharp\left\{j_k \mid l \le j_k < \overline{\theta\left(j'_r\right)}\right\} \\
&< \max\left\{0, \overline{\theta\left(j'_r\right)} - j\right\} - \max\{0, l - j\} \\
&\le \max\left\{0, \overline{\theta\left(j'_r\right)} - j\right\} - \max\{0, l + 1 - j\} + \delta_{l,\{j,\dots,n\}},
\end{aligned} \tag{100}$$

where the first estimation is strict provided that $l \ge j$ meaning that this calculation contradicts once more property (c).

The last and second possibility which can occur is that there exists $i'_p \ge \theta(j'_r) = \overline{l+1}$, such that the properties (a), (b), and (c) are fulfilled.

Then we make a similar construction as in Subcase 1.2; namely, we suppose that

$$\overline{m} \in \left\{\theta\left(j'_{r+1}\right), \dots, \theta\left(j'_{t+\delta_{l,j}}\right)\right\}, \quad \text{say, } \overline{m} = \theta\left(j'_u\right) = j'_{u-\delta_{l,j}}, \tag{101}$$

such that

$$\begin{aligned}
m - \delta_{m,\{j+1,\dots,n\}} \notin &\left(1 - \delta_{m,\{j+1,\dots,n\}}\right)\{j_1, \dots, j_t\} \\
&\cup \delta_{m,\{j+1,\dots,n\}}\left(\mathbf{I} - \left\{j'_1, \dots, j'_t\right\}\right)
\end{aligned} \tag{102}$$

and $m \ge \overline{i'_p}$. If such an element exists we choose m maximal with this property and otherwise we set $\overline{m} = i'_p$. Using the maximality, we can verify similar to (94) the correctness of

$$\begin{aligned}
&(u - r) - \sharp\left\{j_k \mid m \le j_k < l\right\} + \sharp\left\{j'_k \mid m \le j'_k < l\right\} \\
&\quad + \min\{l, j\} - \min\{m, j\} \le l + 1 - m.
\end{aligned} \tag{103}$$

As a corollary we obtain as usual a contradiction to property (c) (recall that $\theta(x_q) \in \mathscr{R}^{j}_{t+\delta_{l,j}}$)

$$\begin{aligned}
&(i-j)+\left(t+\delta_{l,j}-s\right)+(p-r)\\
&\quad+\sharp\left\{\theta\left(j'_k\right)\mid \overline{i'_p}\leq\theta\left(j'_k\right)<\overline{\theta\left(j'_r\right)}\right\}\\
&\quad-\sharp\left\{\theta\left(j_k\right)\mid \overline{i'_p}\leq\theta\left(j_k\right)<\overline{\theta\left(j'_r\right)}\right\}\\
&\leq(i-j)+\left(t+\delta_{l,j}-s\right)+(p-u)+(u-r)\\
&\quad+\sharp\left\{j'_k\mid \overline{i'_p}\leq j'_k<m\right\}\\
&\quad+\sharp\left\{j'_k\mid m\leq j'_k<l\right\}-\sharp\left\{j_k\mid \overline{i'_p}\leq j_k<m\right\}\\
&\quad-\sharp\left\{j_k\mid m\leq j_k<l\right\}\\
&<\max\left\{0,m-j\right\}-\max\left\{0,\overline{i'_p}-j\right\}\\
&\quad+l+1-m-\min\left\{l,j\right\}+\min\left\{m,j\right\}\\
&\leq l+1-\min\left\{l,j\right\}-\max\left\{0,\overline{i'_p}-j\right\}\\
&\leq\max\left\{0,l+1-j\right\}-\max\left\{0,\overline{i'_p}-j\right\}+\delta_{l,\{1,\dots,j-1\}},
\end{aligned}\tag{104}$$

where the first estimation is strict whenever $l<j$ is satisfied.

Case 2 ($\overline{l+1}\mapsto\overline{l}$). Now we assume that the action of the Kashiwara operator on $\mathbf{x}$ is given in a way such that x_q satisfies property (b) while (a) is violated, which in particular means that we replace the entry $\overline{l+1}$ by $\overline{l}$. Since the proofs are similar to Case 1, we do not give them in full details. We only consider the case where we presume that property (3) is violated, that is, one has the following.

Subcase 2.1 ((3) is violated for the pair $(x_{q-1},\theta(x_q))$). It is easy to see that this case can never appear.

Subcase 2.2 ((3) is violated for the pair $(\theta(x_q),x_{q+1})$). The first possibility which can occur is that there exists $\theta(i'_p)=i'_p\geq\theta(i'_r)=\overline{l}$, such that the properties (a), (b), and (c) are fulfilled. Because of (a) and $i'_r=\overline{l+1}$, we must have

$$\begin{aligned}
l+1-\delta_{l+1,\{i+1,\dots,n\}}\notin{}&\left(1-\delta_{l+1,\{i+1,\dots,n\}}\right)\left\{i_1,\dots,i_s\right\}\\
&\cup\,\delta_{l+1,\{i+1,\dots,n\}}\left(\mathbf{I}-\left\{i'_1,\dots,i'_s\right\}\right),
\end{aligned}\tag{105}$$

since otherwise we would obtain that x_q satisfies (a) (from Section 3) and thus the Kashiwara operator would act as in Case 1. Accordingly we can apply our assumptions to $i'_p\geq i'_r=\overline{l+1}$ and obtain a contradiction to (c), namely,

$$\begin{aligned}
&\sharp\left\{j_k\mid j_k<\overline{i'_p}\right\}-\sharp\left\{\theta\left(i_k\right)\mid\theta\left(i_k\right)<l\right\}\\
&\quad+\sharp\left\{\theta\left(i'_k\right)\mid\theta\left(i'_k\right)<l\right\}-\sharp\left\{j'_k\mid j'_k<\overline{i'_p}\right\}\\
&=\sharp\left\{j_k\mid j_k<\overline{i'_p}\right\}+\min\left\{l,i\right\}-\min\left\{l,i\right\}\\
&\quad-\sharp\left\{i_k\mid i_k<l\right\}+\sharp\left\{i'_k\mid i'_k<l\right\}-\sharp\left\{j'_k\mid j'_k<\overline{i'_p}\right\}\\
&<\sharp\left\{j_k\mid j_k<\overline{i'_p}\right\}+\min\left\{l+1,i\right\}\\
&\quad-\min\left\{l,i\right\}-\sharp\left\{i_k\mid i_k<l+1\right\}\\
&\quad+\sharp\left\{i'_k\mid i'_k<l+1\right\}-\sharp\left\{j'_k\mid j'_k<\overline{i'_p}\right\}\\
&<\min\left\{l+1,i\right\}-\min\left\{l,i\right\}+\max\left\{0,l+1-i\right\}\\
&\quad-\max\left\{0,\overline{i'_p}-j\right\}-(p-r)\\
&=1+\max\left\{0,l-i\right\}-\max\left\{0,\overline{i'_p}-j\right\}-(p-r).
\end{aligned}\tag{106}$$

The second and last possibility which can occur in that case isthat there exists $\theta(i'_p)=\overline{l}\geq\theta(i'_r)=i'_r$, such that the properties (a), (b), and (c) are fulfilled. As before we can assume $i'_p=\overline{l+1}$. Suppose that there exists $\overline{m}\in\{i'_r,\dots,i'_p\}$ ($\overline{m}=i'_{r+u}$), such that

$$\begin{aligned}
m-\delta_{m,\{j+1,\dots,n\}}\notin{}&\left(1-\delta_{m,\{j+1,\dots,n\}}\right)\left\{j_1,\dots,j_t\right\}\\
&\cup\,\delta_{m,\{j+1,\dots,n\}}\left(\mathbf{I}-\left\{j'_1,\dots,j'_t\right\}\right)
\end{aligned}\tag{107}$$

and m is minimal with this property. If such an element does not exist, we set $\overline{m}=i'_r$. Similar to (94) one gets

$$\begin{aligned}
&p-(r+u)-\min\left\{l,j\right\}+\min\left\{m,j\right\}\\
&\quad-\sharp\left\{j_k\mid l\leq j_k<m\right\}+\sharp\left\{j'_k\mid l\leq j'_k<m\right\}\leq m-l.
\end{aligned}\tag{108}$$

Using this inequality we arrive once more at a contradiction, namely,

$$\begin{aligned}
&\sharp\left\{j_k\mid j_k<l\right\}-\sharp\left\{\theta\left(i_k\right)\mid\theta\left(i_k\right)<\overline{i'_r}\right\}\\
&\quad+\sharp\left\{\theta\left(i'_k\right)\mid\theta\left(i'_k\right)<\overline{i'_r}\right\}-\sharp\left\{j'_k\mid j'_k<l\right\}\\
&=\sharp\left\{j_k\mid j_k<m\right\}-\sharp\left\{j_k\mid l\leq j_k<m\right\}\\
&\quad-\sharp\left\{i_k\mid i_k<\overline{i'_r}\right\}+\sharp\left\{i'_k\mid i'_k<\overline{i'_r}\right\}\\
&\quad-\sharp\left\{j'_k\mid j'_k<m\right\}+\sharp\left\{j'_k\mid l\leq j'_k<m\right\}\\
&<\max\left\{0,\overline{i'_r}-i\right\}-\max\left\{0,m-j\right\}-(p-r)\\
&\quad-\min\left\{m,j\right\}+\min\left\{l,j\right\}+m-l\\
&=\max\left\{0,\overline{i'_r}-i\right\}-\max\left\{0,l-j\right\}-(p-r).
\end{aligned}\tag{109}$$

The proof of $\widetilde{e}_l \mathbf{x} \in \mathscr{R}(\lambda) \cup \{0\}$ is similar, which completes the proof. □

6. Crystal Bases as Tuples of Integer Sequences

In this section we will verify with Theorem 20 that the crystal $\mathscr{R}(\lambda)$ can be identified with the crystal graph $B(\lambda)$ obtained from Kashiwara's crystal bases theory. Our strategy here is to show that there exists an isomorphism of $\mathscr{R}(\lambda)$ onto the connected component of $\bigotimes_{i=1}^{n} B(\omega_i)^{\otimes m_i}$ containing $\bigotimes_{i=1}^{n} b_i^{\otimes m_i}$, where b_i denotes the highest weight element in $B(\omega_i)$. For the proofs in type A_n we will need a result stated in [16], where the affine type A Kirillov-Reshetikhin crystals are realized via polytopes. We will need the realization of level 1 KR-crystals especially, since they are as classical crystals isomorphic to $B(\omega_i)$. In type C_n we will use a short induction argument to prove our results.

Theorem 20. *Let λ be an arbitrary dominant integral weight and set $k = \sum_i m_i$ as before. Thenone has the following.*

(1) *If $k = 1$, one has an isomorphism of crystals $\mathscr{R}(\lambda) \xrightarrow{\sim} B(\lambda)$.*

(2) *If $k > 1$ and j is the maximal integer such that $m_j \neq 0$, one obtains a strict crystal morphism*

$$\phi : \mathscr{R}(\lambda) \longrightarrow \mathscr{R}(\lambda - \omega_j) \otimes \mathscr{R}(\omega_j) \tag{110}$$

mapping $\mathbf{x} = (x_1, \dots, x_k, 0, 0 \cdots)$ to the tensor product $\mathbf{x}_{k-1} \otimes x_k$, where $\mathbf{x}_{k-1} = (x_1, \dots, x_{k-1}, 0, 0 \cdots)$.

Proof. With the help of Proposition 17, the crystal $\mathscr{R}(\omega_i)$ is characterized as $\bigcup_s \mathscr{R}_s^i$. In the case where $\mathfrak{g}$ is of type A_n, we will consider the map

$$x = (i_1, i_1') \cdots (i_s, i_s') \longmapsto X, \tag{111}$$

whereby X is a pattern as in [16, Definition 2], with filling

$$a_{p,q} = \begin{cases} 1, & \text{if } (p,q) = (i_r, i_r') \text{ for some } r \\ 0, & \text{else.} \end{cases} \tag{112}$$

By an inspection of the crystal structure on the KR-crystal $B^{1,i} \cong B(\omega_i)$ ([16, Section 3.2]), it is easy to see that this map becomes an isomorphism of crystals.

If $\mathfrak{g}$ is of type C_n, the proof of part (1) will proceed by upward induction on i. An observation of the crystal graph of $R(\omega_1)$

$$\begin{gathered} \emptyset_1 \xrightarrow{1} (1,1) \xrightarrow{2} \cdots \xrightarrow{n-1} (1, n-1) \xrightarrow{n} (1,n) \\ \xrightarrow{n-1} \left(1, \overline{n-1}\right) \xrightarrow{n-2} \cdots \xrightarrow{1} \left(1, \overline{1}\right) \end{gathered} \tag{113}$$

proves the initial step. Now we assume the correctness of the claim for all integers less than i, especially we have $\mathscr{R}(\omega_{i-1}) \cong B(\omega_{i-1})$. For the purpose of completing the induction we consider the injective map

$$\eta : \mathscr{R}(\omega_i) \longrightarrow \mathscr{R}(\omega_{i-1}) \otimes \mathscr{R}(\omega_1) \tag{114}$$

given by

$$\eta(x) = \begin{cases} \emptyset_{i-1} \otimes (1, i-1), & \text{if } s = 0 \\ \emptyset_{i-1} \otimes (1, i_1'), & \text{if } s = 1,\ i_1 = i \\ (i_1, i-1)^{(1-\delta_{i_1,i})} (i_2, i_1') \cdots (i_s, i_{s-1}') \otimes (1, i_s'), & \text{else.} \end{cases} \tag{115}$$

If η would be a strict crystal morphism, we would get $\mathscr{R}(\omega_i) \cong \mathscr{I}m(\eta) \cong B(\omega_i)$, which finishes the induction. Therefore, we prove that η is a strict crystal morphism, where we consider the cases $s = 0$ and $s = 1, i_1 = i$ separately. In the separated cases we draw a part of the crystal graph in order to see that the properties of a crystal morphism hold. If $s = 0$, we obtain

$$\begin{gathered} \emptyset_i \xrightarrow{i} (i,i), \\ \emptyset_{i-1} \otimes (1, i-1) = \eta(\emptyset_i) \xrightarrow{i} \emptyset_{i-1} \otimes (1, i) = \eta((i,i)) \end{gathered} \tag{116}$$

and if $s = 1, i_1 = i$, we get

(1) if $i_1' \leq n-1$ (see Figure 1),

(2) if $i_1' \geq n$ and $i_1' \neq \overline{i}$ (see Figure 2),

(3) if $i_1' \geq n$ and $i_1' = \overline{i}$ (see Figure 3).

Thus, from now on we can presume that $s > 1$ or $s = 1$, $i_1 \neq i$. Let $l \in I$ be an arbitrary integer. We set for convenience $\eta(x) = x_1 \otimes x_2$; then

$$\begin{aligned} \varphi_l(\eta(x)) &= \max\{\varphi_l(x_2), \varphi_l(x_1) + \varphi_l(x_2) - \epsilon_l(x_2)\} \\ &= \begin{cases} \varphi_l(x_1) + \varphi_l(x_2), & \text{if } l \neq i_s', \overline{i_s'} \\ \max\{0, \varphi_l(x_1) - 1\}, & \text{else.} \end{cases} \end{aligned} \tag{117}$$

The proof of $\varphi_l(\eta(x)) = \varphi_l(x)$ is an intensive investigation of the properties (a)–(d) listed in Section 3. To avoid confusion with indices we consider only the following case:

(i) $l = i_s'$ or $l = \overline{i_s'}$: the case $l = i_s'$ is very simple because x satisfies neither (a) nor (b), which yields $\varphi_l(x) = 0$. Furthermore, since $\overline{l+1} = \overline{i_s'+1} \notin \{i_1', \dots, i_s'\}$, we get $\varphi_l(x_1) \leq 1$. Consequently $\varphi_l(\eta(x)) = 0$. So it remains to consider the case $l = \overline{i_s'}$. Here we claim the following: let $(\widetilde{a})$ be the property which arises from (a) by erasing the condition $\overline{l} \notin \{i_1', \dots, i_s'\}$, then x satisfies $(\widetilde{a})$ if and only if x_1 satisfies $(\widetilde{a})$.

First we observe that the property $\overline{l+1} \notin \{i_1', \dots, i_s'\}$ is equivalent to $\overline{l+1} \notin \{i_1', \dots, i_{s-1}'\}$, so that we can ignore it. We start the proof by assuming that x_1 satisfies $(\widetilde{a})$ and x does not satisfy $(\widetilde{a})$, for example, $l \in \{i_1, \dots, i_s, i_1', \dots, i_s'\}$. This case is only possible if $l = i_1 = i$ and thus $l - 1 = i - 1 \in \{i_1', \dots, i_{s-1}'\}$, which is a contradiction. Since $l - 1 \in \{i_1', \dots, i_s'\}$ is always fulfilled if $l > i > i - 1$, we can assume that $l < i$ and additionally $l + 1 \notin \{i_1, \dots, i_s\}$. Without loss of generality we

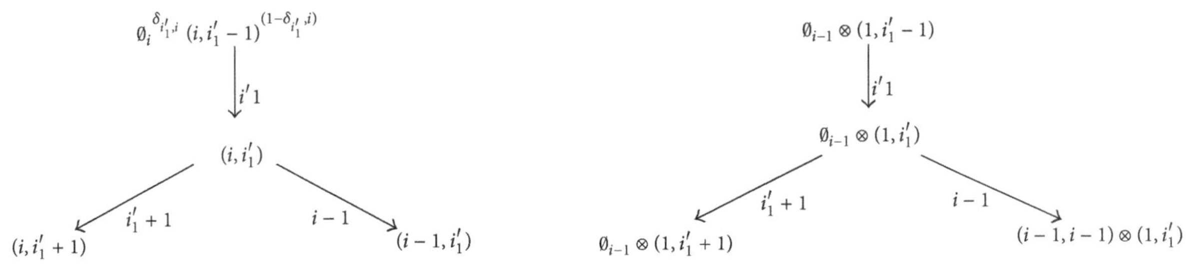

Figure 1

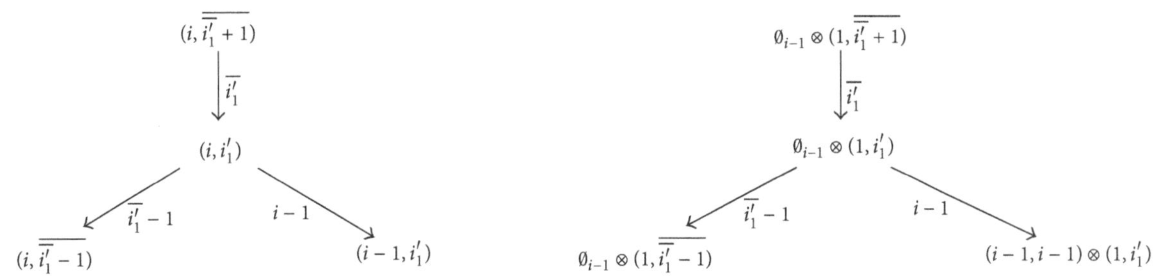

Figure 2

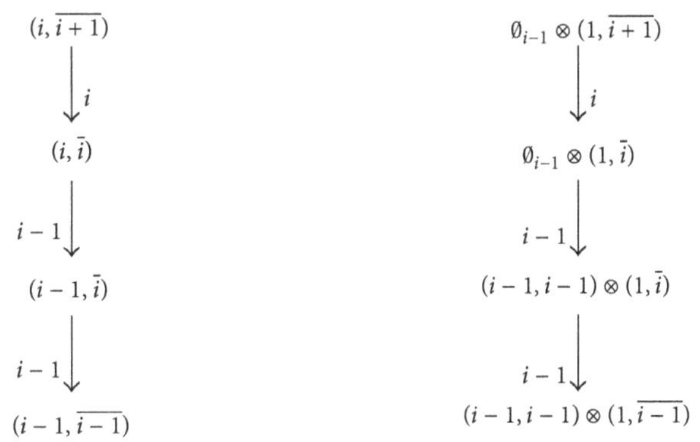

Figure 3

set $l = i - 1$ because else we have $l + 1 \in \{i_1^{(1-\delta_{i_1,i})}, i_2, \ldots, i_s\}$. As a consequence we get

$$l + 1 = i \notin \{i_1, \ldots, i_s\} \Longrightarrow i_1 \leq i - 1 \Longrightarrow l \neq i - 1, \quad (118)$$

which is again a contradiction.

According to these calculations, x must satisfy $(\tilde{a})$. To show the other direction, let x_1 violate one of the properties in $(\tilde{a})$, for example, $l \in \{(i_1, i - 1)^{(1-\delta_{i_1,i})}, i_2, \ldots, i_s, i_1', \ldots, i_{s-1}'\}$. Then we must have $l = i - 1$ and $i_1 < i$ which violates $l + 1 \in \{i_1, \ldots, i_s\}$. The only additional possibility which can occur, such that x_1 does not satisfy $(\tilde{a})$, is $l < i - 1$ and $l + 1 \notin \{i_1^{(1-\delta_{i_1,i})}, i_2, \ldots, i_s\}$. Then we get $l + 1 = i_1 = i$, which is a contradiction to $l < i - 1$ and so the reverse direction is also completed.

According to this we have $\varphi_l(x) = 1 \Rightarrow \varphi_l(x_1) = 2$ and $\varphi_l(x) = 0 \Rightarrow \varphi_l(x_1) \leq 1$.

To show the existence of a morphism of crystals we have to show (among others) the weight invariance of η, which is proven by the following calculation:

$$\begin{aligned} &wt\,(\eta\,(x)) \\ &= \omega_{i-1} + \omega_1 - \sum_{j=1}^{s-1} \alpha_{i_{j+1},i_j'} - \alpha_{1,i_s'} - \left(1 - \delta_{i_1,i}\right) \alpha_{i_1,i-1} \\ &= \omega_{i-1} + \omega_1 - \alpha_{1,i-1} - \sum_{j=1}^{s} \alpha_{i_j,i_j'} \\ &= \omega_i - \sum_{j=1}^{s} \alpha_{i_j,i_j'}. \end{aligned} \quad (119)$$

The verification of $\epsilon_l(\eta(x)) = \epsilon_l(x)$ is therefore proven with Definition 3 (1) and (119). Suppose now that $\tilde{f}_l \eta(x) \neq 0$ and $\tilde{f}_l$ acts on the second tensor. A short investigation of the crystal graph (113) yields

$$l = \begin{cases} i_s' + 1, & \text{if } i_s' \leq n - 1 \\ \overline{i_s'} - 1, & \text{else.} \end{cases} \quad (120)$$

Since $\varphi_l(\eta(x)) = \varphi_l(x) = 1 \neq 0$, we obtain that

$$\tilde{f}_l x = \begin{cases} \left(i_1, i_1'\right) \cdots \left(i_s, i_s' + 1\right), & \text{if } l = i_s' + 1 \\ \left(i_1, i_1'\right) \cdots \left(i_s, \overline{\overline{i_s'} - 1}\right), & l = \overline{i_s'} - 1 \end{cases} \quad (121)$$

because if $l = \overline{i_s'} - 1$ and x would satisfy (a), it would automatically follow that x satisfies (a$'$) which is impossible since $\varphi_l(x) = 1 \neq 2$. Thus, $\tilde{f}_l \eta(x) = \eta(\tilde{f}_l x)$.

If $\tilde{f}_l\eta(x) \neq 0$ and $\tilde{f}_l$ acts on the first tensor, we get with the tensor product rule $\varphi_l(x_1) > \epsilon_l(x_2)$. Note that the operation with the Kashiwara operator $\tilde{f}_l$ would change the entry i'_s in x if and only if either $l = i'_s + 1$ or $l = \overline{i'_s} - 1$ and x is not subject to (a). Our goal is to show here that $\tilde{f}_l$ has no effect on i'_s. If $l = i'_s + 1$, we would get $\varphi_l(x_1) = 0$ and thus a contradiction to $\varphi_l(x_1) > \epsilon_l(x_2)$. In the case where $l = \overline{i'_s} - 1$, we have that (b) is not fulfilled for x_1. Therefore, x_1 must be subject to (a). Consequently we obtain that x must fulfill (a) as well because otherwise we would end in a contradiction; namely, the only property in (a) which can be violated is $l \notin \{i_1, \ldots, i_s, i'_1, \ldots, i'_s\}$. So if l is contained in the aforementioned set, we get by the definition of η that $l = i_1 = i$. Hence, $l - 1 = i - 1$ must be contained in the set $\{i \leq i'_1, \ldots, i'_{s-1}\}$, which is impossible.

Thus, the entry i'_s stays unchanged in $\tilde{f}_l x$ which provides $\tilde{f}_l\eta(x) = \eta(\tilde{f}_l x)$. The proof of $\tilde{e}_l\eta(x) = \eta(\tilde{e}_l x)$ is similar, which completes the proof of (1).

In order to prove (2) we will check as in (1) step for step the properties of a morphism of crystals. We get

$$\begin{aligned}
&\varphi_l(\mathbf{x}_{k-1} \otimes x_k) \\
&= \max\{\varphi_l(x_k), \varphi_l(\mathbf{x}_{k-1}) + \varphi_l(x_k) - \epsilon_l(x_k)\} \\
&= \max\Big\{\varphi_l(x_k), \sigma_l^k(\mathbf{x}_{k-1}) - \tau_l^{k-1}(\mathbf{x}_{k-1}) \\
&\qquad - \min\{\sigma_l^j(\mathbf{x}_{k-1}) - \tau_l^j(\mathbf{x}_{k-1}) \mid 1 \leq j \leq k-1\} \\
&\qquad + \varphi_l(x_k) - \epsilon_l(x_k)\Big\} \\
&= \max\Big\{\varphi_l(x_k), \varphi_l(x_k) + \sigma_l^k(\mathbf{x}) - \tau_l^k(\mathbf{x}) \\
&\qquad - \min\{\sigma_l^j(\mathbf{x}) - \tau_l^j(\mathbf{x}) \mid 1 \leq j \leq k-1\}\Big\} \\
&= \sigma_l^{k+1}(\mathbf{x}) - \tau_l^k(\mathbf{x}) - \min\{\sigma_l^j(\mathbf{x}) - \tau_l^j(\mathbf{x}) \mid 1 \leq j \leq k\} \\
&= \varphi_l(\mathbf{x}).
\end{aligned} \tag{122}$$

The same is trivially fulfilled for ϵ_l because of Definition 3 (1) and the weight invariance of ϕ. Now suppose that $f_l(\mathbf{x}) = p$ is as in (25). If we apply the Kashiwara operator $\tilde{f}_l$ to the tensor product $\mathbf{x}_{k-1} \otimes x_k$, we obtain with the above calculations

$$\begin{aligned}
&\tilde{f}_l(\mathbf{x}_{k-1} \otimes x_k) \\
&= \begin{cases} \tilde{f}_l\mathbf{x}_{k-1} \otimes x_k, & \text{if } \varphi_l(\mathbf{x}_{k-1}) > \epsilon_l(x_k) \\ \mathbf{x}_{k-1} \otimes \tilde{f}_l x_k, & \text{if } \varphi_l(\mathbf{x}_{k-1}) \leq \epsilon_l(x_k) \end{cases} \\
&= \begin{cases} \tilde{f}_l\mathbf{x}_{k-1} \otimes x_k, & \text{if } \sigma_l^k(\mathbf{x}) - \tau_l^k(\mathbf{x}) > \sigma_l^p(\mathbf{x}) - \tau_l^p(\mathbf{x}) \\ \mathbf{x}_{k-1} \otimes \tilde{f}_l x_k, & \text{if } \sigma_l^k(\mathbf{x}) - \tau_l^k(\mathbf{x}) \leq \sigma_l^p(\mathbf{x}) - \tau_l^p(\mathbf{x}). \end{cases}
\end{aligned} \tag{123}$$

According to this we get that $p \in \{1, \ldots, k-1\}$ if $\tilde{f}_l$ acts on the first tensor and $p = k$ if $\tilde{f}_l$ acts on the second tensor. The proof for the Kashiwara operator $\tilde{e}_l$ works similarly. □

Corollary 21. *One has an isomorphism of crystals*

$$\mathcal{R}(\lambda) \cong B(\lambda). \tag{124}$$

Proof. The proof will proceed by induction on k, where the initial step is exactly part (1) of Theorem 20. If $k > 1$ and j is the maximal integer where m_j is non zero, we can assume with the induction hypothesis that $\mathcal{R}(\lambda - \omega_j) \cong B(\lambda - \omega_j)$ and $\mathcal{R}(\omega_j) \cong B(\omega_j)$. The rest of the proof is done with part (2) of Theorem 20, since the map ϕ is injective and the image is a connected crystal containing the highest weight element $r_{\lambda-\omega_j} \otimes r_{\omega_j}$. □

Acknowledgment

The author was sponsored by the SFB/TR 12-Symmetries and Universality in Mesoscopic Systems.

References

[1] M. Kashiwara, "On crystal bases of the Q-analogue of universal enveloping algebras," *Duke Mathematical Journal*, vol. 63, no. 2, pp. 465–516, 1991.

[2] M. Kashiwara and T. Nakashima, "Crystal graphs for representations of the q-analogue of classical Lie algebras," *Journal of Algebra*, vol. 165, no. 2, pp. 295–345, 1994.

[3] P. Littelmann, "A Littlewood-Richardson rule for symmetrizable Kac-Moody algebras," *Inventiones Mathematicae*, vol. 116, no. 1–3, pp. 329–346, 1994.

[4] M. Kashiwara and Y. Saito, "Geometric construction of crystal bases," *Duke Mathematical Journal*, vol. 89, no. 1, pp. 9–36, 1997.

[5] H. Nakajima, "Quiver varieties and tensor products," *Inventiones Mathematicae*, vol. 146, no. 2, pp. 399–449, 2001.

[6] M. Kashiwara, "Realizations of crystals," in *Combinatorial and Geometric Representation Theory (Seoul, 2001)*, vol. 325 of *Contemporary Mathematics*, pp. 133–139, American Mathematical Society, Providence, RI, USA, 2003.

[7] S.-J. Kang, J.-A. Kim, and D.-U. Shin, "Monomial realization of crystal bases for special linear Lie algebras," *Journal of Algebra*, vol. 274, no. 2, pp. 629–642, 2004.

[8] S.-J. Kang, J.-A. Kim, and D.-U. Shin, "Crystal bases for quantum classical algebras and Nakajima's monomials," *Kyoto University. Research Institute for Mathematical Sciences. Publications*, vol. 40, no. 3, pp. 757–791, 2004.

[9] M. Meng, "Compression of Nakajima monomials in type A and C," *Journal of Algebraic Combinatorics*, vol. 35, no. 4, pp. 649–690, 2012.

[10] G. Fourier and D. Kus, "Compression of Nakajima monomials in type B and D," In preparation.

[11] J. Hong and S.-J. Kang, *Introduction to Quantum Groups and Crystal Bases*, vol. 42 of *Graduate Studies in Mathematics*, American Mathematical Society, Providence, RI, USA, 2002.

[12] J. R. Stembridge, "A local characterization of simply-laced crystals," *Transactions of the American Mathematical Society*, vol. 355, no. 12, pp. 4807–4823, 2003.

[13] V. I. Danilov, A. V. Karzanov, and G. A. Koshevoy, "B_2-crystals: axioms, structure, models," *Journal of Combinatorial Theory A*, vol. 116, no. 2, pp. 265–289, 2009.

[14] H. Nakajima, "t-analogs of q-characters of quantum affine algebras of type A_n, D_n," in *Combinatorial and Geometric*

Representation Theory (Seoul, 2001), vol. 325 of *Contemporary Mathematics*, pp. 141–160, American Mathematical Society, Providence, RI, USA, 2003.

[15] J.-A. Kim, "Monomial realization of crystal graphs for $U_q(A_n^{(1)})$," *Mathematische Annalen*, vol. 332, no. 1, pp. 17–35, 2005.

[16] D. Kus, "Realization of affine type A Kirillov-Reshetikhin crystals via polytopes," http://arxiv.org/abs/1209.6019.

15

Generalized Pattern-Matching Conditions for $C_k \wr S_n$

Sergey Kitaev,[1] Andrew Niedermaier,[2] Jeffrey Remmel,[2] and Manda Riehl[3]

[1] *Department of Computer and Information Sciences, University of Strathclyde, Livingstone Tower, 26 Richmond Street, Glasgow G1 1XH, UK*
[2] *Department of Mathematics, University of California, San Diego, La Jolla, CA 92093-0112, USA*
[3] *Department of Mathematics, University of Wisconsin, Eau Claire, WI 54702-4004, USA*

Correspondence should be addressed to Sergey Kitaev; sergey.kitaev@gmail.com

Academic Editors: S. D. Georgiou and J. Siemons

We derive several multivariable generating functions for a generalized pattern-matching condition on the wreath product $C_k \wr S_n$ of the cyclic group C_k and the symmetric group S_n. In particular, we derive the generating functions for the number of matches that occur in elements of $C_k \wr S_n$ for any pattern of length 2 by applying appropriate homomorphisms from the ring of symmetric functions over an infinite number of variables to simple symmetric function identities. This allows us to derive several natural analogues of the distribution of rises relative to the product order on elements of $C_k \wr S_n$. Our research leads to connections to many known objects/structures yet to be explained combinatorially.

1. Introduction

The goal of this paper is to study pattern-matching conditions on the wreath product $C_k \wr S_n$ of the cyclic group C_k and the symmetric group S_n. $C_k \wr S_n$ is the group of $k^n n!$ signed permutations where we allow k signs of the form $1 = \omega^0, \omega, \omega^2, \ldots, \omega^{k-1}$ for some primitive kth root of unity ω. We can think of the elements $C_k \wr S_n$ as pairs $\gamma = (\sigma, \epsilon)$ where $\sigma = \sigma_1 \cdots \sigma_n \in S_n$ and $\epsilon = \epsilon_1 \cdots \epsilon_n \in \{1, \omega, \ldots, \omega^{k-1}\}^n$. For ease of notation, if $\epsilon = (\omega^{w_1}, \omega^{w_2}, \ldots, \omega^{w_n})$ where $w_i \in \{0, 1, \ldots, k-1\}$ for $i = 1, 2, \ldots, n$, then we simply write $\gamma = (\sigma, w)$ where $w = w_1 w_2 \cdots w_n \in \{0, 1, \ldots, k-1\}^n$.

Given a sequence $\sigma = \sigma_1 \cdots \sigma_n$ of distinct integers, let $\text{red}(\sigma)$ be the permutation found by replacing the ith largest integer that appears in σ by i. For example, if $\sigma = 2\ 7\ 5\ 4$, then $\text{red}(\sigma) = 1\ 4\ 3\ 2$. Given a permutation τ in the symmetric group S_j, we say a permutation $\sigma = \sigma_1 \cdots \sigma_n \in S_n$ has a *τ-match starting at position i* provided $\text{red}(\sigma_i \cdots \sigma_{i+j-1}) = \tau$. Let $\tau\text{-mch}(\sigma)$ be the number of τ-matches in the permutation σ. Similarly, we say that τ *occurs* in σ if there exist $1 \le i_1 < \cdots < i_j \le n$ such that $\text{red}(\sigma_{i_1} \cdots \sigma_{i_j}) = \tau$. We say that σ *avoids* τ if there are no occurrences of τ in σ.

We can define similar notions for words over a finite alphabet $[k] = \{0, 1, \ldots, k-1\}$. Given a word $w = w_1 \cdots w_n \in [k]^n$, let $\text{red}(w)$ be the word found by replacing the ith largest integer that appears in w by $i-1$. For example, if $w = 2\ 7\ 2\ 4\ 7$, then $\text{red}(w) = 0\ 2\ 0\ 1\ 2$. Given a word $u \in [k]^j$ such that $\text{red}(u) = u$, we say a word $w \in [k]^n$ has a *u-match starting at position i* provided $\text{red}(w_i \cdots w_{i+j-1}) = u$. Let $u\text{-mch}(w)$ be the number of u-matches in the word w. Similarly, we say that u *occurs* in a word w if there exist $1 \le i_1 < \cdots < i_j \le n$ such that $\text{red}(w_{i_1} \cdots w_{i_j}) = u$. We say that w *avoids* u if there are no occurrences of u in w.

There are a number of papers on pattern matching and pattern avoidance in $C_k \wr S_n$ [1–4]. For example, the following pattern matching condition was studied in [2–4].

Definition 1. Let $(\tau, u) \in C_k \wr S_j$, Υ be a subset of $C_k \wr S_j$ and $(\sigma, w) \in C_k \wr S_n$.

(1) One says that (σ, w) has an *exact occurrence* of (τ, u) (resp., Υ) if there are $1 \le i_1 < i_2 < \cdots < i_j \le n$ such that $(\text{red}(\sigma_{i_1} \cdots \sigma_{i_j}), w_{i_1} \cdots w_{i_j}) = (\tau, u)$ (resp., $(\text{red}(\sigma_{i_1} \cdots \sigma_{i_j}), w_{i_1} \cdots w_{i_j}) \in \Upsilon$).

(2) One says that $(\sigma, w) \in C_k \wr S_n$ *avoids an exact occurrence of* (τ, u) (resp., Υ) if there are no exact occurrences of (τ, u) (resp., Υ) in (σ, w).

(3) One says that there is an *exact* (τ, u)*-match in* (σ, w) *starting at position* i (resp., *exact* Υ*-match in* (σ, w) *starting at position* i) if

$$\left(\text{red}\left(\sigma_i\sigma_{i+1}\cdots\sigma_{i+j-1}\right), w_iw_{i+1}\cdots w_{i+j-1}\right) = (\tau, u)$$
$$\left(\text{resp.}, \left(\text{red}\left(\sigma_i\sigma_{i+1}\cdots\sigma_{i+j-1}\right), w_iw_{i+1}\cdots w_{i+j-1}\right) \in \Upsilon\right). \quad (1)$$

That is, an exact occurrence or an exact match of $(\tau, u) \in C_k \wr S_j$ in an element $(\sigma, w) \in C_k \wr S_n$ is just an ordinary occurrence or match of τ in σ where the corresponding signs agree *exactly*. For example, Mansour [3] proved via recursion that for any $(\tau, u) \in C_k \wr S_2$, the number of elements in $C_k \wr S_n$ which avoid exact occurrences of (τ, u) is $\sum_{j=0}^{n} j!(k-1)^j\binom{n}{j}^2$. This generalized a result of Simion [5] who proved the same result for the hyperoctahedral group $C_2 \wr S_n$. Similarly, Mansour and West [4] determined the number of permutations in $C_2 \wr S_n$ that avoid all possible exact occurrences of 2 or 3 element sets of patterns of elements of $C_2 \wr S_2$. For example, let K_n^1 be the number of $(\sigma, \epsilon) \in C_2 \wr S_n$ that avoid all exact occurrences of the patterns in the set $\{(1\ 2, 0\ 0), (1\ 2, 0\ 1), (2\ 1, 1\ 0)\}$, let K_n^2 be the number of $(\sigma, \epsilon) \in C_2 \wr S_n$ that avoid all exact occurrences of the patterns in the set $\{(1\ 2, 0\ 1), (1\ 2, 1\ 0), (2\ 1, 0\ 1)\}$, and let K_n^3 be the number of $(\sigma, \epsilon) \in C_2 \wr S_n$ that avoid all exact occurrences of the patterns in the set $\{(1\ 2, 0\ 0), (1\ 2, 0\ 1), (2\ 1, 0\ 0)\}$. Then Mansour and West [4] proved that

$$K_n^1 = F_{2n+1},$$
$$K_n^2 = n!\sum_{j=0}^{n}\binom{n}{j}^{-1}, \quad (2)$$
$$K_n^3 = n! + n!\sum_{j=1}^{n}\frac{1}{j},$$

where F_n is the nth Fibonacci number.

An alternative matching condition arises when we drop the requirement of the exact matching of signs and replace it by the condition that the two sequences of signs match in the sense of words described above. That is, we will consider the following matching conditions.

Definition 2. Let $(\tau, u) \in C_k \wr S_j$ where $\text{red}(u) = u$, Υ be a subset of $C_k \wr S_j$ where for all $(\tau, u) \in \Upsilon$, $\text{red}(u) = u$, and $(\sigma, w) \in C_k \wr S_n$.

(1) One says that (σ, w) has an *occurrence* of (τ, u) (resp., Υ) if there are $1 \le i_1 < i_2 < \cdots < i_j \le n$ such that $(\text{red}(\sigma_{i_1}\cdots\sigma_{i_j}), \text{red}(w_{i_1}\cdots w_{i_j})) = (\tau, u)$ (resp., $(\text{red}(\sigma_{i_1}\cdots\sigma_{i_j}), \text{red}(w_{i_1}\cdots w_{i_j})) \in \Upsilon$).

(2) We say that $(\sigma, w) \in C_k \wr S_n$ *avoids* (τ, u) (resp., Υ) if there are no occurrences of (τ, u) (resp., Υ) in (σ, w).

(3) One says that there is a (τ, u)*-match in* (σ, w) *starting at position* i (resp., Υ*-matchin* (σ, w)*starting at position* i) if $(\text{red}(\sigma_i\sigma_{i+1}\cdots\sigma_{i+j-1}), w_iw_{i+1}\cdots w_{i+j-1}) = (\tau, u)$ (resp., $(\text{red}(\sigma_i\sigma_{i+1}\cdots\sigma_{i+j-1}), \text{red}(w_iw_{i+1}\cdots w_{i+j-1})) \in \Upsilon$).

For example, suppose that $(\tau, u) = (1\ 2, 0\ 0)$ and $(\sigma, w) = (1\ 3\ 2\ 4, 1\ 2\ 2\ 2)$. Then there are no exact occurrences or exact matches of (τ, u) in (σ, w). However, there are two occurrences of (τ, u), one in positions 2 and 4 and one in positions 3 and 4. Thus, there are two occurrences of (τ, u) in (σ, w), and there is a (τ, u)-match in (σ, w) starting at position 3.

Finally, we will consider a more general matching condition which generalizes both occurrences and matches and exact occurrences and exact matches in $C_k \wr S_n$.

Definition 3. Let $(\tau, u) \in C_k \wr S_j$, let Υ be a subset of $C_k \wr S_j$, and let $\vec{A} = (A_1, \ldots, A_j)$ be a sequence of subsets of $[k]$, and $(\sigma, w) \in C_k \wr S_n$.

(1) One says that (σ, w) has an *occurrence* of $(\tau, u, \vec{A})$ (resp., $(\Upsilon, \vec{A})$) if there are $1 \le i_1 < i_2 < \cdots < i_j \le n$ such that $(\text{red}(\sigma_{i_1}\cdots\sigma_{i_j}), \text{red}(w_{i_1}\cdots w_{i_j})) = (\tau, \text{red}(u))$ (resp., $(\text{red}(\sigma_{i_1}\cdots\sigma_{i_j}), \text{red}(w_{i_1}\cdots w_{i_j}))$ is equal to $(\tau, \text{red}(u))$ for some (τ, u) in Υ) and $w_{i_s} \in A_s$ for $s = 1, 2, \ldots, j$.

(2) One says that $(\sigma, w) \in C_k \wr S_n$ *avoids* (τ, u) (resp., Υ) if there are no occurrences of (τ, u) (resp., Υ) in (σ, w).

(3) We say that there is a (τ, u)*-match in* (σ, w) *starting at position* i (resp., Υ*-match in* (σ, w) *starting at position* i) if $(\text{red}(\sigma_i\sigma_{i+1}\cdots\sigma_{i+j-1}), \text{red}(w_iw_{i+1}\cdots w_{i+j-1})) = (\tau, \text{red}(u))$ (resp., $(\text{red}(\sigma_i\sigma_{i+1}\cdots\sigma_{i+j-1}), \text{red}(w_iw_{i+1}\cdots w_{i+j-1}))$ is equal to $(\tau, \text{red}(u))$ for some (τ, u) in Υ) and $w_{i+s} \in A_{s+1}$ for $s = 0, 1, \ldots, j-1$.

Thus, a $(\tau, u, \vec{A})$-occurrence or $(\tau, u, \vec{A})$-match where $u = u_1\cdots u_j$ and $\vec{A} = (A_1, \ldots, A_j)$ is such that $A_i = \{u_i\}$ for $i = 1, \ldots, j$ is just an exact occurrence or exact match of (τ, u). Similarly, a $(\tau, u, \vec{A})$-occurrence or $(\tau, u, \vec{A})$-match where $u = u_1\cdots u_j$, $\text{red}(u) = u$, and $\vec{A} = (A_1, \ldots, A_j)$ is such that $A_i = [k]$ for $i = 1, \ldots, j$ is just an occurrence or match of (τ, u).

Suppose we are given $(\tau, u) \in C_k \wr S_j$, $\Upsilon \subseteq C_k \wr S_j$, $\vec{A} = (A_1, \ldots, A_j)$ where $A_i \subseteq [k]$ for $i = 1, \ldots, j$, and $(\sigma, w) \in C_k \wr S_n$. We let (τ, u)-mch$((\sigma, w))$ (resp., (τ, u)-Emch$((\sigma, w))$) denote the number of (τ, u)-matches (resp., exact (τ, u)-matches) in (σ, w). We let Υ-mch$((\sigma, w))$ (resp., Υ-Emch$((\sigma, w))$) denote the number of Υ-matches (resp., exact Υ-matches) in (σ, w). We let $(\tau, u, \vec{A})$-mch$((\sigma, w))$ be the number of $(\tau, u, \vec{A})$-matches in (σ, w) and let $(\Upsilon, \vec{A})$-mch$((\sigma, w))$ be the number of $(\Upsilon, \vec{A})$-matches in (σ, w).

The main result of this paper is to derive a generating function for the distribution of $(\tau, u, \vec{A})$-matches where (τ, u) is any element of $C_k \wr S_2$. To state our main result, we first need some notation. We define the p, q-analogues of n, $n!$, $\binom{n}{k}$, and $\binom{n}{a_1, \ldots, a_m}$ by

$$[n]_{p,q} = \frac{p^n - q^n}{p - q} = p^{n-1} + p^{n-2}q + \cdots + pq^{n-2} + q^{n-1},$$
$$[n]_{p,q}! = [n]_{p,q}[n-1]_{p,q}\cdots[2]_{p,q}[1]_{p,q},$$

$$\begin{bmatrix} n \\ k \end{bmatrix}_{p,q} = \frac{[n]_{p,q}!}{[k]_{p,q}![n-k]_{p,q}!},$$
$$\begin{bmatrix} n \\ a_1,\dots,a_m \end{bmatrix}_{p,q} = \frac{[n]_{p,q}!}{[a_1]_{p,q}!\cdots[a_m]_{p,q}!}, \tag{3}$$

respectively. We define the q-analogues of n, $n!$, $\binom{n}{k}$, and $\binom{n}{a_1,\dots,a_m}$ by $[n]_{1,q}$, $[n]_{1,q}!$, $\left[\begin{smallmatrix} n \\ k \end{smallmatrix}\right]_{1,q}$, and $\left[\begin{smallmatrix} n \\ a_1,\dots,a_m \end{smallmatrix}\right]_{1,q}$, respectively.

Next suppose that $\Upsilon \subseteq C_k \wr S_2$ and $\vec{A} = (A_1, A_2)$ where $A_1, A_2 \subseteq [k]$. Then we will say that $(\sigma, w) \in C_k \wr S_n$ is a *maximum packing* for $(\Upsilon, \vec{A})$ if (σ, w) has $(\Upsilon, \vec{A})$-matches starting at positions $1, 2, \dots, n-1$. We let $\mathcal{MP}_{(\Upsilon,\vec{A}),n}$ denote the set of $(\sigma, w) \in C_k \wr S_n$ which are maximum packings for $(\Upsilon, \vec{A})$. Given any word $w = w_1 \cdots w_n \in [k]^n$, we let $z(w) = \prod_{i=1}^{n} z_{w_i}$. For any $\sigma = \sigma_1 \cdots \sigma_n \in S_n$, we let $\mathrm{inv}(\sigma)$ (resp., $\mathrm{coinv}(\sigma)$) equal the number of pairs (i, j) such that $1 \le i < j \le n$ and $\sigma_i > \sigma_j$ (resp., $\sigma_i < \sigma_j$). We then define

$$\mathrm{mp}_{(\Upsilon,\vec{A}),n}(p,q;z_0,\dots,z_{k-1}) = \sum_{(\sigma,w)\in\mathcal{MP}_{(\Upsilon,\vec{A}),n}} p^{\mathrm{coinv}(\sigma)} q^{\mathrm{inv}(\sigma)} z(w). \tag{4}$$

We shall also be interested in the specializations

$$\mathrm{mp}_{(\Upsilon,\vec{A}),n}(p,q;1,\dots,1) = \sum_{(\sigma,w)\in\mathcal{MP}_{(\Upsilon,\vec{A}),n}} p^{\mathrm{coinv}(\sigma)} q^{\mathrm{inv}(\sigma)},$$
$$\mathrm{mp}_{(\Upsilon,\vec{A}),n}\left(p,q;1,r,\dots,r^{k-1}\right) = \sum_{(\sigma,w)\in\mathcal{MP}_{(\Upsilon,\vec{A}),n}} p^{\mathrm{coinv}(\sigma)} q^{\mathrm{inv}(\sigma)} r^{\|w\|}, \tag{5}$$

where for any word $w = w_1 \cdots w_n \in [k]^n$, $\|w\| = w_1 + \cdots + w_n$. In the special case, where $\Upsilon = \{(\tau, u)\}$ where $(\tau, u) \in C_k \wr S_2$, we will denote $\mathrm{mp}_{(\Upsilon,\vec{A}),n}(p,q;z_0,\dots,z_{k-1})$ as $\mathrm{mp}_{(\tau,u,\vec{A}),n}(p,q;z_0,\dots,z_{k-1})$. For example, if $\tau = 12$, $u = 01$, and $\vec{A} = ([k],[k])$, then $(\sigma, w) \in C_k \wr S_n$ is a maximum packing for $(\tau, u, \vec{A})$ if and only if $\sigma = 12\cdots n$ is the identity permutation and $w = w_1 \cdots w_n$ is such that $0 \le w_1 < \cdots < w_n \le k-1$. Thus,

$$\mathrm{mp}_{(\tau,u,\vec{A}),n}(p,q;z_0,\dots,z_{k-1}) = p^{\binom{n}{2}} \left(\prod_{i=0}^{k-1}(1+z_i t)\right)\Bigg|_{t^n}, \tag{6}$$

where for any power series $f(t) = \sum_{n\ge0} f_n t^n$, we write $f(t)|_{t^n} = f_n$ for the coefficient of t^n in $f(t)$. Then it is easy to prove that

$$\mathrm{mp}_{(\tau,u,\vec{A}),n}(p,q;1,\dots,1) = p^{\binom{n}{2}} \binom{k}{n},$$
$$\mathrm{mp}_{(\tau,u,\vec{A}),n}\left(p,q;1,r,\dots,r^{k-1}\right) = p^{\binom{n}{2}} r^{\binom{n}{2}} \begin{bmatrix} k \\ n \end{bmatrix}_r. \tag{7}$$

If $\tau = 12$, $u = 00$, and $\vec{A} = ([k],[k])$, then $(\sigma, w) \in C_k \wr S_n$ is a maximum packing for $(\tau, u, \vec{A})$ if and only if $\sigma = 12\cdots n$ is the identity permutation and $w = i^n$ for some $0 \le i \le k-1$. Thus,

$$\mathrm{mp}_{(\tau,u,\vec{A}),n}(p,q;z_0,\dots,z_{k-1}) = p^{\binom{n}{2}} \sum_{i=0}^{k-1} z_i^n. \tag{8}$$

It then follows that

$$\mathrm{mp}_{(\tau,u,\vec{A}),n}(p,q;1,\dots,1) = k p^{\binom{n}{2}},$$
$$\mathrm{mp}_{(\tau,u,\vec{A}),n}\left(p,q;1,r,\dots,r^{k-1}\right) = [k]_{r^n} p^{\binom{n}{2}}. \tag{9}$$

Similarly if $\Upsilon = \{(12,00),(12,01)\}$ and $\vec{A} = ([k],[k])$, then $(\sigma, w) \in C_k \wr S_n$ is a maximum packing for $(\Upsilon, \vec{A})$ if and only if $\sigma = 12\cdots n$ is the identity permutation and $w = w_1 \cdots w_n$ is such that $0 \le w_1 \le \cdots \le w_n \le k-1$. Thus,

$$\mathrm{mp}_{(\Upsilon,\vec{A}),n}(p,q;z_0,\dots,z_{k-1}) = p^{\binom{n}{2}} \left(\prod_{i=0}^{k-1} \frac{1}{1-z_i t}\right)\Bigg|_{t^n}. \tag{10}$$

We shall then prove that

$$\mathrm{mp}_{(\Upsilon,\vec{A}),n}(p,q;1,\dots,1) = p^{\binom{n}{2}} \binom{n+k-1}{n},$$
$$\mathrm{mp}_{(\Upsilon,\vec{A}),n}\left(p,q;1,r,\dots,r^{k-1}\right) = p^{\binom{n}{2}} \begin{bmatrix} n+k-1 \\ n \end{bmatrix}_r. \tag{11}$$

The main result of this paper is to prove the following.

Theorem 4. *For any $\Upsilon \subseteq C_k \wr S_2$ and $\vec{A} = (A_1, A_2)$ where $A_1, A_2 \subseteq [k]$,*

$$\begin{aligned}
&\sum_{n\ge0} \frac{t^n}{[n]_{p,q}!} \sum_{(\sigma,w)\in C_k\wr S_n} p^{\mathrm{coinv}(\sigma)} q^{\mathrm{inv}(\sigma)} x^{(\Upsilon,\vec{A})\text{-}\mathrm{mch}((\sigma,w))} z(w) \\
&\quad = 1 \times \Big(1 - \Big((z_0 + \cdots + z_{k-1})t \\
&\qquad + \sum_{n\ge2} \frac{t^n}{[n]_{p,q}!} (x-1)^{n-1} \mathrm{mp}_{(\Upsilon,\vec{A}),n} \\
&\qquad \times (p,q;z_0,\dots,z_{k-1})\Big)\Big)^{-1}.
\end{aligned} \tag{12}$$

We note that (12) can be specialized to give natural analogues for generating functions of rises and descents in $C_k \wr S_n$ where we compare pairsusing the product order. That is, instead of thinking of an element of $C_k \wr S_n$ as a pair $(\sigma_1 \cdots \sigma_n, w_1 \cdots w_n)$, we can think of it as a sequence of pairs $(\sigma_1, w_1)(\sigma_2, w_2)\cdots(\sigma_n, w_n)$. We then define a partial order on such pairs by the usual product order. That is, $(i_1, j_1) \le (i_2, j_2)$

if and only if $i_1 \leq i_2$ and $j_1 \leq j_2$. Then we define the following sets and statistics for elements $(\sigma, w) \in C_k \wr S_n$:

$$\begin{aligned}
&\mathrm{Des}((\sigma, w)) = \{i : \sigma_i > \sigma_{i+1} \,\&\, w_i \geq w_{i+1}\},\\
&\mathrm{des}((\sigma, w)) = |\mathrm{Des}((\sigma, w))|,\\
&\mathrm{Ris}((\sigma, w)) = \{i : \sigma_i < \sigma_{i+1} \,\&\, w_i \leq w_{i+1}\},\\
&\mathrm{ris}((\sigma, w)) = |\mathrm{Ris}((\sigma, w))|,\\
&W\mathrm{Des}((\sigma, w)) = \{i : \sigma_i > \sigma_{i+1} \,\&\, w_i = w_{i+1}\},\\
&w\,\mathrm{des}((\sigma, w)) = |W\mathrm{Des}((\sigma, w))|,\\
&W\mathrm{Ris}((\sigma, w)) = \{i : \sigma_i < \sigma_{i+1} \,\&\, w_i = w_{i+1}\},\\
&w\mathrm{ris}((\sigma, w)) = |W\mathrm{Ris}((\sigma, w))|,\\
&S\mathrm{Des}((\sigma, w)) = \{i : \sigma_i > \sigma_{i+1} \,\&\, w_i > w_{i+1}\},\\
&s\,\mathrm{des}((\sigma, w)) = |S\mathrm{Des}((\sigma, w))|,\\
&S\mathrm{Ris}((\sigma, w)) = \{i : \sigma_i < \sigma_{i+1} \,\&\, w_i < w_{i+1}\},\\
&s\mathrm{ris}((\sigma, w)) = |S\mathrm{Ris}((\sigma, w))|.
\end{aligned} \tag{13}$$

We shall refer to $\mathrm{Des}((\sigma, w))$ as the *descent set* of (σ, w), $W\mathrm{Des}((\sigma, w))$ as the *weak descent set* of (σ, w), and $S\mathrm{Des}((\sigma, w))$ as the *strict descent set* of (σ, w). Similarly, we will refer to $\mathrm{Ris}((\sigma, w))$ as the *rise set* of (σ, w), $W\mathrm{Ris}((\sigma, w))$ as the *weak rise set* of (σ, w), and $S\mathrm{Ris}((\sigma, w))$ as the *strict rise set* of (σ, w). It is easy to see that

$i \in W\mathrm{Des}((\sigma, w))$ if and only if there is a $(2\ 1, 0\ 0)$-match starting at position i,

$i \in S\mathrm{Des}((\sigma, w))$ if and only if there is a $(2\ 1, 1\ 0)$-match starting at position i, and

$i \in \mathrm{Des}((\sigma, w))$ if and only if there is a Υ-match starting at position i where $\Upsilon = \{(2\ 1, 0\ 0), (2\ 1, 1\ 0)\}$.

Similarly,

$i \in W\mathrm{Ris}((\sigma, w))$ if and only if there is a $(1\ 2, 0\ 0)$-match starting at position i,

$i \in S\mathrm{Ris}((\sigma, w))$ if and only if there is a $(1\ 2, 0\ 1)$-match starting at position i, and

$i \in \mathrm{Ris}((\sigma, w))$ if and only if there is a Υ-match starting at position i where $\Upsilon = \{(1\ 2, 0\ 0), (1\ 2, 0\ 1)\}$.

If $\sigma = \sigma_1 \cdots \sigma_n \in S_n$, then we define the reverse of σ, σ^r by $\sigma^r = \sigma_n \cdots \sigma_1$. Similarly, if $w = w_1 \cdots w_n \in [k]^n$, then we define $w^r = w_n \cdots w_1$. It is easy to see that

$$\begin{aligned}
&\mathrm{ris}((\sigma, w)) = \mathrm{des}((\sigma^r, w^r)),\\
&w\mathrm{ris}((\sigma, w)) = w\,\mathrm{des}((\sigma^r, w^r)),\\
&s\mathrm{ris}((\sigma, w)) = s\,\mathrm{des}((\sigma^r, w^r)).
\end{aligned} \tag{14}$$

Thus, we need to find the distributions for only one of the corresponding pairs. Substituting $z_0 = z_1 = \cdots = z_{k-1} = 1$ into (12) in these special cases will yield the following generating functions for rises, strict rises, and weak rises in $C_k \wr S_n$.

Theorem 5. *Consider the following:*

$$\begin{aligned}
&\sum_{n\geq 0} \frac{t^n}{n!} \sum_{(\sigma,w)\in C_k \wr S_n} x^{\mathrm{ris}((\sigma,w))}\\
&\quad = \frac{1-x}{1-x+\sum_{n\geq 1}\left(((x-1)t)^n/n!\right)\binom{n+k-1}{n}},
\end{aligned} \tag{15}$$

$$\begin{aligned}
&\sum_{n\geq 0} \frac{t^n}{n!} \sum_{(\sigma,w)\in C_k \wr S_n} x^{w\mathrm{ris}((\sigma,w))}\\
&\quad = \frac{1-x}{1-x+k\left(e^{(x-1)t}-1\right)},\\
&\sum_{n\geq 0} \frac{t^n}{n!} \sum_{(\sigma,w)\in C_k \wr S_n} x^{s\mathrm{ris}((\sigma,w))}\\
&\quad = \frac{1-x}{1-x+\sum_{n\geq 1}\left(((x-1)t)^n/n!\right)\binom{k}{n}}.
\end{aligned} \tag{16}$$

Other distribution results for (τ, u)-matches follow from these results. For example, if $\sigma = \sigma_1 \cdots \sigma_n \in S_n$, then we define the complement of σ, σ^c by

$$\sigma^c = (n+1-\sigma_1)\cdots(n+1-\sigma_n). \tag{17}$$

If $w = w_1 \cdots w_n \in [k]^n$, then we define the complement of w, w^c by

$$w^c = (k-1-w_1)\cdots(k-1-w_n). \tag{18}$$

We can then consider maps $\phi_{a,b} : C_k \wr S_n \rightarrow C_k \wr S_n$ where $\phi_{a,b}((\sigma, w)) = (\sigma^a, w^b)$ for $a, b \in \{r, c\}$. Such maps will easily allow us to establish that the distribution of (τ, u)-matches is the same for various classes of (τ, u)'s. For example, one can use such maps to show that the distributions of $(1\ 2, 0\ 1)$-matches, $(2\ 1, 0\ 1)$-matches, $(1\ 2, 1\ 0)$-matches, and $(2\ 1, 1\ 0)$-matches are all the same.

Another interesting case is when we let $\Upsilon = \{(1\ 2, 0\ 1), (1\ 2, 1\ 0)\}$. In this case, we have a Υ-match in (σ, w) starting at i if and only if $\sigma_i < \sigma_{i+1}$ and $w_i \neq w_{i+1}$. In that case, (12) can be specialized to prove the following.

Theorem 6. *Consider the following:*

$$\begin{aligned}
&\sum_{n\geq 0} \frac{t^n}{n!} \sum_{(\sigma,w)\in C_k \wr S_n} x^{\Upsilon\text{-mch}((\sigma,w))}\\
&\quad = \frac{(k-1)(1-x)}{(k-1)(1-x)+k\left(e^{(k-1)(x-1)t}-1\right)}.
\end{aligned} \tag{19}$$

Substituting $z_i = r^i$ for $i = 0, \ldots, k-1$ into (12) will yield the following generating function:

$$\begin{aligned}
&D_k^{\Upsilon}(x, p, q, r, t)\\
&\quad = \sum_{n\geq 0} \frac{t^n}{[n]_{p,q}!} \sum_{(\sigma,w)\in C_k \wr S_n} q^{\mathrm{inv}(\sigma)} p^{\mathrm{coinv}(\sigma)} r^{\|w\|} x^{\Upsilon\text{-mch}((\sigma,w))}.
\end{aligned} \tag{20}$$

Let

$$\begin{aligned}\Upsilon_{\mathbf{r}} &= \{(1\ 2, 0\ 0), (1\ 2, 0\ 1)\},\\ \Upsilon_{\mathbf{w}} &= \{(1\ 2, 0\ 0)\},\\ \Upsilon_{\mathbf{s}} &= \{(1\ 2, 0\ 1)\},\\ \Upsilon_{\mathbf{d}} &= \{(1\ 2, 0\ 1), (1\ 2, 1\ 0)\}.\end{aligned} \tag{21}$$

Thus, $\Upsilon_{\mathbf{r}}$-matches correspond to rises, $\Upsilon_{\mathbf{w}}$-matches correspond to weak rises, and $\Upsilon_{\mathbf{s}}$-matches correspond to strict rises. We shall find $D_k^{\Upsilon_{\mathbf{a}}}(x, p, q, r, t)$ for $\mathbf{a} \in \{\mathbf{r}, \mathbf{w}, \mathbf{s}\}$ and find $D_k^{\Upsilon_{\mathbf{d}}}(x, p, q, 1, t)$. For example, we will show that the following generating function for the distribution of inversions, coinversions, and rises over $C_k \wr S_n$ is an immediate consequence of (12).

Lemma 7. *Consider the following:*

$$\begin{aligned}&D_k^{\Upsilon_{\mathbf{r}}}(x, p, q, r, t)\\ &= \sum_{n\geq 0} \frac{t^n}{[n]_{p,q}!} \sum_{(\sigma,w)\in C_k\wr S_n} q^{\mathrm{inv}(\sigma)} p^{\mathrm{coinv}(\sigma)} r^{\|w\|} x^{\mathrm{ris}((\sigma,w))}\\ &= \frac{1-x}{1-x+\sum_{n\geq 1}\left(p^{\binom{n}{2}}((x-1)t)^n/[n]_{p,q}!\right)\left[{n+k-1 \atop n}\right]_r},\end{aligned} \tag{22}$$

which reduces to (15) *when one sets* $p = q = r = 1$.

We shall prove (12) by applying a ring homomorphism, defined on the ring Λ of symmetric functions over infinitely many variables $x_1, x_2, \ldots$, to a simple symmetric function identity. There has been a long line of research, [6–15], which shows that a large number of generating functions for permutation statistics can be obtained by applying homomorphisms defined on the ring of symmetric functions Λ over infinitely many variables $x_1, x_2, \ldots$ to simple symmetric function identities. For example, the nth elementary symmetric function, e_n, and the nth homogeneous symmetric function, h_n, are defined by the generating functions

$$\begin{aligned}E(t) &= \sum_{n\geq 0} e_n t^n = \prod_i (1 + x_i t),\\ H(t) &= \sum_{n\geq 0} h_n t^n = \prod_i \frac{1}{1 - x_i t}.\end{aligned} \tag{23}$$

We let $P(t) = \sum_{n\geq 0} p_n t^n$ where $p_n = \sum_i x_i^n$ is the nth power symmetric function. A partition of n is a sequence $\mu = (\mu_1, \ldots, \mu_k)$ such that $0 < \mu_1 \leq \cdots \leq \mu_k$ and $\mu_1 + \cdots + \mu_k = n$. We write $\mu \vdash n$ if μ is a partition of n, and we let $\ell(\mu)$ denote the number of parts of μ. If $\mu \vdash n$, we set $h_\mu = \prod_{i=1}^{\ell(\mu)} h_{\mu_i}$, $e_\mu = \prod_{i=1}^{\ell(\mu)} e_{\mu_i}$, and $p_\mu = \prod_{i=1}^{\ell(\mu)} p_{\mu_i}$. Let Λ_n denote the space of homogeneous symmetric functions of degree n over infinitely many variables $x_1, x_2, \ldots$ so that $\Lambda = \bigoplus_{n\geq 0}\Lambda_n$. It is well known that $\{e_\lambda : \lambda \vdash n\}$, $\{h_\lambda : \lambda \vdash n\}$, and $\{p_\lambda : \lambda \vdash n\}$ are all bases of Λ_n. It follows that $\{e_0, e_1, \ldots\}$ is an algebraically independent set of generators for Λ and, hence, we can define a ring homomorphism $\xi : \Lambda \to R$ where R is a ring by simply specifying $\xi(e_n)$ for all $n \geq 0$.

Now it is well known that

$$H(t) = \frac{1}{E(-t)}, \tag{24}$$

$$P(t) = \frac{\sum_{n\geq 1}(-1)^{n-1} n e_n t^n}{E(-t)}. \tag{25}$$

A surprisingly large number of results on generating functions for various permutation statistics in the literature and large number of new generating functions can be derived by applying homomorphisms on Λ to simple identities such as (24) and (25). We shall show that (12) can be proved by applying appropriate ring homomorphisms to identity (24).

The outline of this paper is as follows. In Section 2, we will provide the necessary background in symmetric functions that we will need to derive our generating functions. In Section 3, we will prove (12) and give a number of cases where we can compute $\mathrm{mp}_{(\Upsilon,\vec{A})}(p, q; z_0, \ldots, z_{k-1})$ or its specializations which will prove all of the formulas in the special cases described above. We shall also show that if $NM_{(\Upsilon,\vec{A}),n}$ denotes the number of $(\sigma, w) \in C_k \wr S_n$ such that (σ, w) has no $(\Upsilon, \vec{A})$-matches, then the sequence $(NM_{(\Upsilon,\vec{A}),n})_{n\geq 0}$ appears in the OIES [16] in several special cases. In Section 4, we will study $NM_{(\Upsilon,\vec{A}),n}$ as a function of k, the size of the underlying cyclic group C_k. We will show that in several cases, $NM_{(\Upsilon,\vec{A}),n}$ is a polynomial in k whose coefficients have interesting combinatorial properties. Finally, in Section 5, we will discuss some related results and directions for future research.

2. Symmetric Functions

In this section, we give the necessary background on symmetric functions needed for our proofs of the generating functions.

Let Λ denote the ring of symmetric functions over infinitely many variables $x_1, x_2, \ldots$ with coefficients in the field of complex numbers $\mathbb{C}$. The nth elementary symmetric function e_n in the variables $x_1, x_2, \ldots$ is defined by

$$E(t) = \sum_{n\geq 0} e_n t^n = \prod_i (1 + x_i t), \tag{26}$$

and the nth homogeneous symmetric function h_n in the variables $x_1, x_2, \ldots$ is defined by

$$H(t) = \sum_{n\geq 0} h_n t^n = \prod_i \frac{1}{1 - x_i t}. \tag{27}$$

Thus,

$$H(t) = \frac{1}{E(-t)}. \tag{28}$$

Let $\lambda = (\lambda_1, \ldots, \lambda_\ell)$ be an integer partition, that is, λ is a finite sequence of weakly increasing positive integers. Let $\ell(\lambda) = l$ denote the number of parts of λ. If the sum of these integers is n, we say that λ is a partition of n and write $\lambda \vdash n$. For

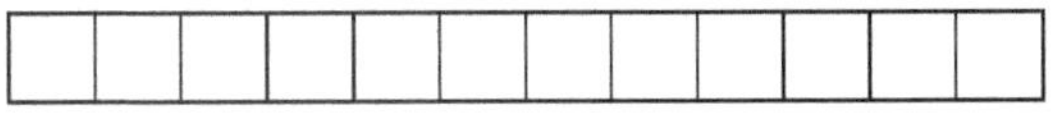

FIGURE 1: A brick tabloid of shape (12) and type (1, 1, 2, 3, 5).

4	6	12	1	5	7	8	10	11	2	3	9

FIGURE 2: An element of IF(3, 6, 3).

any partition $\lambda = (\lambda_1, \ldots, \lambda_\ell)$, let $e_\lambda = e_{\lambda_1} \cdots e_{\lambda_\ell}$. The well known fundamental theorem of symmetric functions says that $\{e_\lambda : \lambda \text{ is a partition}\}$ is a basis for Λ or that $\{e_0, e_1, \ldots\}$ is an algebraically independent set of generators for Λ. Similarly, if we define $h_\lambda = h_{\lambda_1} \cdots h_{\lambda_\ell}$, then $\{h_\lambda : \lambda \text{ is a partition}\}$ is also a basis for Λ. Since $\{e_0, e_1, \ldots\}$ is an algebraically independent set of generators for Λ, we can specify a ring homomorphism θ on Λ by simply defining $\theta(e_n)$ for all $n \geq 0$.

Since the elementary symmetric functions e_λ and the homogeneous symmetric functions h_λ are both bases for Λ, it makes sense to talk about the coefficient of the homogeneous symmetric functions when written in terms of the elementary symmetric function basis. These coefficients have been shown to equal the sizes of certain sets of combinatorial objects up to a sign. A *brick tabloid* of shape (n), and type $\lambda = (\lambda_1, \ldots, \lambda_k)$ is a filling of a row of n squares of cells with bricks of lengths $\lambda_1, \ldots, \lambda_k$ such that bricks do not overlap. One brick tabloid of shape (12) and type (1, 1, 2, 3, 5) is displayed in Figure 1.

Let $\mathscr{B}_{\lambda,n}$ denote the set of all λ-brick tabloids of shape (n) and let $B_{\lambda,n} = |\mathscr{B}_{\lambda,n}|$. Through simple recursions stemming from (28), Eğecioğlu and Remmel proved in [17] that

$$h_n = \sum_{\lambda \vdash n} (-1)^{n-\ell(\lambda)} B_{\lambda,n} e_\lambda. \tag{29}$$

We end this section with two lemmas that will be needed in later sections. Both of the lemmas follow from simple codings of a basic result of Carlitz [18] that

$$\begin{bmatrix} n \\ k \end{bmatrix}_q = \sum_{r \in \mathscr{R}(1^k 0^{n-k})} q^{\mathrm{inv}(r)}, \tag{30}$$

where $\mathscr{R}(1^k 0^{n-k})$ is the number of rearrangements of k 1's and $n-k$ 0's. We start with a lemma from [19]. Fix a brick tabloid $T = (b_1, \ldots, b_{\ell(\mu)}) \in \mathscr{B}_{\mu,n}$. Let IF$(T)$ denote the set of all fillings of the cells of $T = (b_1, \ldots, b_{\ell(\mu)})$ with the numbers $1, \ldots, n$ so that the numbers increase within each brick reading from left to right. We then think of each such filling as a permutation of S_n by reading the numbers from left to right in each row. For example, Figure 2 pictures an element of IF(3, 6, 3) whose corresponding permutation is 4 6 12 1 5 7 8 10 11 2 3 9.

Then the following lemma which is proved in [19] gives a combinatorial interpretation to $p^{\sum_{i=1}^{\ell(\mu)} \binom{b_i}{2}} \begin{bmatrix} n \\ b_1, \ldots, b_{\ell(\mu)} \end{bmatrix}_{p,q}$.

Lemma 8. *If $T = (b_1, \ldots, b_{\ell(\mu)})$ is a brick tabloid in $\mathscr{B}_{\mu,n}$, then*

$$p^{\sum_{i=1}^{\ell(\mu)} \binom{b_i}{2}} \begin{bmatrix} n \\ b_1, \ldots, b_{\ell(\mu)} \end{bmatrix}_{p,q} = \sum_{\sigma \in IF(T)} q^{\mathrm{inv}(\sigma)} p^{\mathrm{coinv}(\sigma)}. \tag{31}$$

Another well known combinatorial interpretation for $\begin{bmatrix} n+k-1 \\ k-1 \end{bmatrix}_q$ ([13]) is that it is equal to the sum of the sizes of the partitions that are contained in an $n \times (k-1)$ rectangle. Thus one has the following lemma.

Lemma 9. *Consider the following:*

$$\sum_{0 \leq a_1 \leq \cdots \leq a_n \leq k-1} q^{a_1 + \cdots + a_n} = \begin{bmatrix} n+k-1 \\ n \end{bmatrix}_q. \tag{32}$$

3. Generating Functions

The main goal of this section is to prove (12). That is, we will prove the following theorem.

Theorem 10. *Let $\Upsilon \subseteq C_k \wr S_2$ and $\vec{A} = (A_1, A_2)$ where $A_1, A_2 \subseteq [k]$. For all $k \geq 2$,*

$$\begin{aligned} &\sum_{n \geq 0} \frac{t^n}{[n]_{p,q}!} \sum_{(\sigma,w) \in C_k \wr S_n} p^{\mathrm{coinv}(\sigma)} q^{\mathrm{inv}(\sigma)} x^{(\Upsilon,\vec{A})\text{-mch}((\sigma,w))} z(w) \\ &\quad = 1 \times \Big(1 - \Big((z_0 + \cdots + z_{k-1}) t \\ &\qquad + \sum_{n \geq 2} \frac{t^n}{[n]_{p,q}!} (x-1)^{n-1} \mathrm{mp}_{(\Upsilon,\vec{A}),n} \\ &\qquad \times (p, q; z_0, \ldots, z_{k-1})\Big)\Big)^{-1}. \end{aligned} \tag{33}$$

Proof. Define a ring homomorphism $\Gamma : \Lambda \to \mathbb{Q}(p, q, z_0, \ldots, z_{k-1}, x)$ by setting

$$\begin{aligned} \Gamma(e_0) &= 1, \\ \Gamma(e_1) &= z_0 + \cdots + z_{k-1}, \\ \Gamma(e_n) &= (-1)^{n-1} (x-1)^{n-1} \frac{1}{[n]_{p,q}!} \mathrm{mp}_{(\Upsilon,\vec{A}),n} \\ &\quad \times (p, q; z_0, \ldots z_{k-1}) \quad \text{for } n \geq 1. \end{aligned} \tag{34}$$

Then we claim that

$$\begin{aligned} &[n]_{p,q}! \Gamma(h_n) \\ &\quad = \sum_{(\sigma,w) \in C_k \wr S_n} q^{\mathrm{inv}(\sigma)} p^{\mathrm{coinv}(\sigma)} z(w) x^{(\Upsilon,\vec{A})\text{-mch}((\sigma,w))}, \end{aligned} \tag{35}$$

for all $n \geq 1$. That is,

$$\begin{aligned} &[n]_{p,q}! \Gamma(h_n) \\ &\quad = [n]_{p,q}! \sum_{\mu \vdash n} (-1)^{n-\ell(\mu)} B_{\mu,(n)} \Gamma(e_\mu) \\ &\quad = [n]_{p,q}! \sum_{\mu \vdash \text{ÂŘ} n} (-1)^{n-\ell(\mu)} \end{aligned}$$

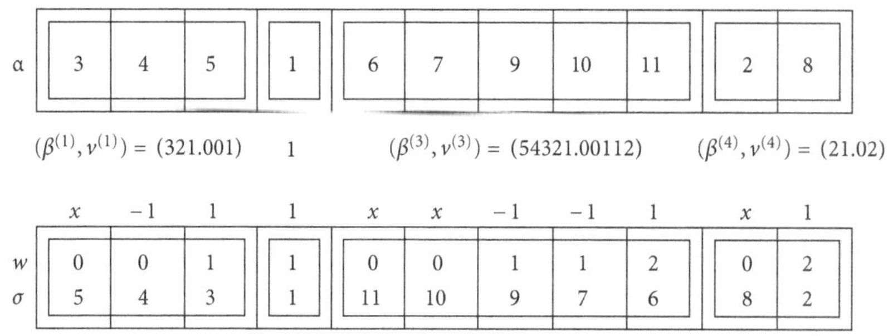

FIGURE 3: A composite object $C \in \mathscr{F}_{12}$.

$$\times \sum_{(b_1,\ldots,b_{\ell(\mu)})\in\mathscr{B}_{\mu,n}} \prod_{j=1}^{\ell(\mu)} (-1)^{b_j-1}(x-1)^{b_j-1}$$
$$\times \Big(\left(z_0+\cdots+z_{k-1}\right)\chi\left(b_j=1\right)$$
$$+ \frac{1}{\left[b_j\right]_{p,q}!}\mathrm{mp}_{(\Upsilon,\vec{A}),b_j}\left(p,q;z_0,z_1,\ldots,z_{k-1}\right)$$
$$\times\chi\left(b_j>1\right)\Big)$$
$$= \sum_{\mu\vdash \text{ÂŘ} n} \sum_{(b_1,\ldots,b_{\ell(\mu)})\in\mathscr{B}_{\mu,n}} p^{\sum_{j=1}^{\ell(\mu)}\binom{b_j}{2}} \begin{bmatrix} n \\ b_1,\ldots,b_{\ell(\mu)} \end{bmatrix}_{p,q}$$
$$\times \prod_{j=1}^{\ell(\mu)} \Big(\left(z_0+\cdots+z_{k-1}\right)\chi\left(b_j=1\right)$$
$$+ \frac{p^{-\binom{b_j}{2}}}{\left[b_j\right]_{p,q}!}\mathrm{mp}_{(\Upsilon,\vec{A}),b_j}\left(p,q;z_0,z_1,\ldots,z_{k-1}\right)$$
$$\times\chi\left(b_j>1\right)\Big). \tag{36}$$

Next we want to give a combinatorial interpretation to (36). By Lemma 8, for each brick tabloid $T = (b_1,\ldots,b_{\ell(\mu)})$, we can interpret $p^{\sum_{j=1}^{\ell(\mu)}\binom{b_j}{2}}\left[\begin{smallmatrix} n \\ b_1,\ldots,b_{\ell(\mu)}\end{smallmatrix}\right]_{p,q}$ as the sum of the weights of all fillings of T with a permutation $\alpha \in S_n$ such that α is increasing in each brick, and we weight α with $q^{\mathrm{inv}(\alpha)}p^{\mathrm{coinv}(\alpha)}$. If $b_j = 1$ and covers cell r, then we will interpret the factor $z_0+\cdots+z_{k-1}$ as picking an element $w_r \in [k]$ to put on top of α_r. Next suppose that $b_j > 1$ and covers cells $r, r+1,\ldots,r+s$. Let $\alpha_r < \alpha_{r+1} < \cdots < \alpha_{r+s}$ be the elements of α in cells $r, r+1,\ldots,r+s$, respectively. Then we interpret $p^{-\binom{n}{2}}\mathrm{mp}_{(\Upsilon,\vec{A}),b_j}(p,q;z_0,\ldots,z_{k-1})$ as the numbers of ways of picking a maximum packing $(\beta^{(j)},\upsilon^{(j)}) \in C_k \wr S_{b_j}$ for $(\Upsilon,\vec{A})$ where we weight $(\beta^{(j)},\upsilon^{(j)})$ by $p^{-\binom{b_j}{2}}p^{\mathrm{coinv}(\beta^{(j)})}q^{\mathrm{inv}(\beta^{(j)})}z(\upsilon^{(j)})$. We then reorder by the proper contribution after reordering which is $\mathrm{coinv}(\beta^{(j)})$ to the coinversion count of the resulting permutation and $\mathrm{inv}(\beta^{(j)})$ to the inversion count of the resulting permutation. Finally, we interpret $\prod_{j=1}^{\ell(\mu)}(x-1)^{b_j-1}$ as all ways of picking a label of the cells of each brick except the final cell with either an x or a -1. For completeness, we label the final cell of each brick with 1. For example, suppose that $\Upsilon = \{(21,00),(21,01)\}$, $k = 3$, and $\vec{A} = ([3],[3])$. Thus (σ,w) is a maximum packing for $(\Upsilon,\vec{A})$ only if σ is strictly decreasing and w is a weakly increasing word over the alphabet $\{0,1,2\}$. Then at the top of Figure 3, we have pictured the brick tabloid $T = (3,1,5,2)$ along with a permutation α which is increasing within bricks. Below that, we have picked our choices of $(\beta^{(j)},\upsilon^{(j)})$ for the bricks b_j for $j = 1,3,4$ and choice of 1 for brick b_2 which is of length 1. These choices result in the filled brick tabloid pictured at the bottom of Figure 3. We have also specified a labeling of the cells with either x, -1, or 1 so that the last cell of each brick is labeled with 1 and the remaining cells are labeled with either x of -1. We shall call all such objects created in this way filled labeled brick tabloids, and let $\mathscr{F}_n$ denote the set of all filled labeled brick tabloids that arise in this way. Thus, a $C \in \mathscr{F}_n$ consists of a brick tabloid $T = (b_1,\ldots,b_t)$, an element $(\sigma,w) \in C_k \wr S_n$, and a labeling L of the cells of T with elements from $\{x,1,-1\}$ such that

(1) if $b_j = 1$ and covers cell r, then w_r is an arbitrary element of $[k]$,

(2) if $b_j > 1$ and covers cells $r, r+1,\ldots,r+s$, then $(\mathrm{red}(\sigma_r\sigma_{r+1}\cdots\sigma_{r+s}), w_rw_{r+1}\cdots w_{r+s})$ is an element of $\mathscr{MP}_{(\Upsilon,\vec{A}),b_j}$,

(3) the final cell of each brick is labeled with 1,

(4) each cell which is not a final cell of a brick is labeled with x or -1.

We then define the weight $w(C)$ of C to be $q^{\mathrm{inv}(\sigma)}p^{\mathrm{coinv}(\sigma)}z(w)$ times the product of all the x labels in L and the sign $\mathrm{sgn}(C)$ of C to be the product of all the -1 labels in L. For example, if $C = (T,(\sigma,w),L)$ is the filled labeled brick tabloid pictured at the bottom of Figure 3, then $w(C) = q^{28}p^{27}z_0^5z_1^4z_2^2x^4$ and $\mathrm{sgn}(C) = -1$. It follows that

$$[n]_{p,q}!\Gamma\left(h_n\right) = \sum_{C\in\mathscr{F}_n} \mathrm{sgn}\left(C\right)w\left(C\right). \tag{37}$$

Next we define a weight-preserving sign-reversing involution $I : \mathscr{F}_n \to \mathscr{F}_n$. To define $I(C)$, we scan the cells of $C = (T,(\sigma,w),L)$ from left to right looking for the leftmost cell t

such that either (i) t is labeled with -1 or (ii) t is at the end of a brick b_j, and the brick b_{j+1} immediately following b_j has the property that if b_j together with b_{j+1} cover cells $r, r+1, \ldots, r+s$, then $(\mathrm{red}(\sigma_r\sigma_{r+1}\cdots\sigma_{r+s}), w_r w_{r+1}\cdots, w_{r+s})$ is a maximum packing for $(\Upsilon, \vec{A})$. In case (i), $I(C) = (T', (\sigma', w'), L')$ where T' is the result of replacing the brick b in T containing t by two bricks b^* and b^{**} where b^* contains the cell t plus all the cells in b to the left of t and b^{**} contains all the cells of b to the right of t, $\sigma = \sigma'$, $w = w'$, and L' is the labeling that results from L by changing the label of cell t from -1 to 1. In case (ii), $I(C) = (T', (\sigma', w'), L')$ where T' is the result of replacing the bricks b_j and b_{j+1} in T by a single brick b, $\sigma = \sigma'$, $w = w'$, and L' is the labeling that results from L by changing the label of cell t from 1 to -1. If neither case (i) or case (ii) applies, then we let $I(C) = C$. For example, if C is the element of $\mathscr{F}_{12}$ pictured in Figure 3, then $I(C)$ is pictured in Figure 4.

It is easy to see that I is a weight-preserving sign-reversing involution and hence I shows that

$$[n]_{p,q}!\Gamma(h_n) = \sum_{C\in\mathscr{F}_n, I(C)=C} \operatorname{sgn}(C)\, w(C). \qquad (38)$$

Thus, we must examine the fixed points $C = (T, \sigma, w, L)$ of I. First there can be no -1 labels in L so that $\operatorname{sgn}(C) = 1$. Moreover, if b_j and b_{j+1} are two consecutive bricks in T and t is the last cell of b_j, then it cannot be the case that there is $(\Upsilon, \vec{A})$-match starting at position t in (σ, w) since otherwise we could combine b_j and b_{j+1}. For any such fixed point, we associate an element $(\sigma, w) \in C_k \wr S_n$. For example, a fixed point of I is pictured in Figure 5 where $\sigma = 5\ 4\ 3\ 1\ 11\ 10\ 9\ 7\ 6\ 8\ 2$ and $w = 0\ 0\ 1\ 1\ 0\ 0\ 1\ 1\ 2\ 0\ 2$.

It follows that if cell t is at the end of a brick, then t is not the start of a $(\Upsilon, \vec{A})$-match. However, if v is a cell which is not at the end of a brick, then our definitions force v to be the start of a $(\Upsilon, \vec{A})$-match. Since each such cell v must be labeled with an x, it follows that $\operatorname{sgn}(C)w(C) = q^{\mathrm{inv}(\sigma)} p^{\mathrm{coinv}(\sigma)} z(w) x^{(\Upsilon,\vec{A})\text{-mch}((\sigma,w))}$. Vice versa, if $(\sigma, w) \in C_k \wr S_n$, then we can create a fixed point $C = (T, (\sigma, w), L)$ by having the bricks in T end at cells of the form t where t is not the start of a $(\Upsilon, \vec{A})$-match, labeling each cell t which is the start of a $(\Upsilon, \vec{A})$-match with x, and labeling the remaining cells with 1. Thus, we have shown that

$$[n]_{p,q}!\Gamma(h_n) = \sum_{(\sigma,w)\in C_k\wr S_n} q^{\mathrm{inv}(\sigma)} p^{\mathrm{coinv}(\sigma)} z(w)\, x^{(\Upsilon,\vec{A})\text{-mch}((\sigma,w))} \qquad (39)$$

as desired.

	x	1	1	1	x	x	-1	-1	1	x	1
w	0	0	1	1	0	0	1	1	2	0	2
σ	5	4	3	1	11	10	9	7	6	8	2

FIGURE 4: $I(C)$ for C in Figure 3.

	x	x	x	1	x	x	x	x	1	x	1
w	0	0	1	1	0	0	1	1	2	0	2
σ	5	4	3	1	11	10	9	7	6	8	2

FIGURE 5: A fixed point of I.

Applying Γ to the identity $H(t) = (E(-t))^{-1}$, we get

$$\begin{aligned}
\sum_{n\geq 0}\Gamma(h_n)t^n &= \sum_{n\geq 0}\frac{t^n}{[n]_{p,q}!} \times \sum_{(\sigma,w)\in C_k\wr S_n} q^{\mathrm{inv}(\sigma)} p^{\mathrm{coinv}(\sigma)} z(w)\, x^{(\Upsilon,\vec{A})\text{-mch}((\sigma,w))} \\
&= \frac{1}{1+\sum_{n\geq 1}(-t)^n\Gamma(e_n)} \\
&= 1\times\Bigg(1+(z_0+\cdots+z_{k-1})(-t) + \sum_{n\geq 1}(-1)^n t^n \times \frac{(-1)^{n-1}(x-1)^{n-1}}{[n]_{p,q}!} \times \mathrm{mp}_{(\Upsilon,\vec{A}),n}(p,q;z_0,\ldots,z_{k-1})\Bigg)^{-1} \qquad (40)\\
&= 1\times\Big(1-\big((z_0+\cdots+z_{k-1})t + \sum_{n\geq 2}\big((x-1)^{n-1}t^n/[n]_{p,q}!\big) \times \mathrm{mp}_{(\Upsilon,\vec{A}),n}(p,q;z_0,\ldots,z_{k-1})\big)\Big)^{-1},
\end{aligned}$$

which proves (33). □

To be able to use Theorem 10, we need to be able to compute $\mathrm{mp}_{(\Upsilon,\vec{A}),n}(p,q;z_0,\ldots,z_{k-1})$. In fact, this is easy to do in most cases. Thus, we end this section by giving several such examples. We will examine the cases when

(i) $\Upsilon = \Upsilon_{\mathbf{s}} = \{(12, 01)\}$ and $\vec{A} = ([k], [k])$,

(ii) $\Upsilon = \Upsilon_{\mathbf{r}} = \{(12, 00), (12, 01)\}$ and $\vec{A} = ([k], [k])$,

(iii) $\Upsilon = \Upsilon_{\mathbf{w}} = \{(12, 00)\}$ and $\vec{A} = ([k], [k])$,

(iv) $\Upsilon = \Upsilon_{\mathbf{d}} = \{(12, 01), (12, 10)\}$ and $\vec{A} = ([k], [k])$,

(v) $\Upsilon = \{(12, ab)\}$ and $\vec{A} = (\{a\}, \{b\})$,

(vi) $\Upsilon = \{(12, 00), (12, 01)\}$ and $\vec{A} = ([k], \{k-1\})$,

(vii) $\Upsilon = \{(12,00),(12,01)\}$ and $\vec{A} = ([k],\{k-2,k-1\})$,

(viii) $\Upsilon = \{(12,01)\}$ and $\vec{A} = ([k],\{k-2,k-1\})$.

Example 11. Let $\Upsilon = \Upsilon_{\mathbf{s}} = \{(12,01)\}$ and $\vec{A} = ([k],[k])$.

In this case for any $(\sigma,w) \in C_k \wr S_n$, $(\Upsilon,\vec{A})\text{-mch}((\sigma,w)) = \text{sris}((\sigma,w))$. It is easy to see that for $n \geq 2$, $(\sigma,w) \in \mathcal{MP}_{(\Upsilon,\vec{A}),n}$ if and only if $\sigma = 12\cdots n$ is the identity permutation and $w = w_1\cdots w_n$ where $0 \leq w_1 < \cdots < w_n \leq k-1$. As we pointed out in the introduction, it follows that

$$\text{mp}_{(\Upsilon,\vec{A}),n}(p,q;z_0,\ldots,z_{k-1}) = p^{\binom{n}{2}}\left(\prod_{i=0}^{k-1}(1+z_it)\right)\Bigg|_{t^n},$$
$$\text{mp}_{(\Upsilon_{\mathbf{s}},\vec{A}),n}(p,q;1,\ldots,1) = p^{\binom{n}{2}}\binom{k}{n}. \tag{41}$$

Note that if $0 \leq w_1 < \cdots < w_n \leq k-1$ and $a_i = w_i-(i-1)$ for $i = 1,\ldots,n$, then $0 \leq a_1 \leq a_2 \leq \cdots \leq a_n \leq k-n$ and $\sum_{i=1}^n w_i = \binom{n}{2} + \sum_{i=1}^n a_i$. Hence,

$$\begin{aligned}\text{mp}_{(\Upsilon_{\mathbf{s}},\vec{A}),n}(p,q;1,r\ldots,r^{k-1}) &= p^{\binom{n}{2}}\sum_{0\leq w_1<\cdots<w_n\leq k-1} r^{w_1+\cdots+w_n}\\ &= p^{\binom{n}{2}}r^{\binom{n}{2}}\sum_{0\leq a_1\leq\cdots\leq a_n\leq k-n} r^{\sum_{i=1}^n a_i}\\ &= p^{\binom{n}{2}}r^{\binom{n}{2}}\begin{bmatrix}k\\n\end{bmatrix}_r.\end{aligned} \tag{42}$$

Here the last equality follows from Lemma 9. Thus, it follows that for all $k \geq 2$,

$$\begin{aligned}&D_k^{\Upsilon_{\mathbf{s}}}(x,p,q,r,t)\\ &= \sum_{n\geq 0}\frac{t^n}{[n]_{p,q}!}\sum_{(\sigma,w)\in C_k\wr S_n} q^{\text{inv}(\sigma)}p^{\text{coinv}(\sigma)}r^{\|w\|}x^{\text{sris}((\sigma,w))}\\ &= 1\times\Bigg(1-\Bigg(\left(1+r+\cdots+r^{k-1}\right)t\\ &\qquad+\sum_{n\geq 2}\frac{(x-1)^{n-1}t^n}{[n]_{p,q}!}\\ &\qquad\times \text{mp}_{(\Upsilon_{\mathbf{s}},A),n}\left(p,q;1,r,\ldots,r^{k-1}\right)\Bigg)\Bigg)^{-1}\\ &= \frac{1-x}{1-x+\sum_{n\geq 1}\left(p^{\binom{n}{2}}((x-1)t)^n/[n]_{p,q}!\right)r^{\binom{n}{2}}\left[\begin{smallmatrix}k\\n\end{smallmatrix}\right]_r}.\end{aligned} \tag{43}$$

Example 12. Let $\Upsilon = \Upsilon_{\mathbf{r}} = \{(12,00),(12,01)\}$ and $\vec{A} = ([k],[k])$.

In this case for any $(\sigma,w) \in C_k \wr S_n$, $(\Upsilon,\vec{A})\text{-mch}((\sigma,w)) = \text{ris}((\sigma,w))$. It is easy to see that for $n \geq 2$, $(\sigma,w) \in \mathcal{MP}_{(\Upsilon,\vec{A}),n}$ if and only if $\sigma = 12\cdots n$ is the identity permutation and $w = w_1\cdots w_n$ where $0 \leq w_1 \leq \cdots \leq w_n \leq k-1$. As we pointed out in the introduction, it follows that

$$\begin{aligned}\text{mp}_{(\Upsilon,\vec{A}),n}(p,q;z_0,\ldots,z_{k-1}) &= p^{\binom{n}{2}}\left(\prod_{i=0}^{k-1}\frac{1}{(1-z_it)}\right)\Bigg|_{t^n},\\ \text{mp}_{(\Upsilon_{\mathbf{r}},\vec{A}),n}(p,q;1,r\ldots,r^{k-1}) &= p^{\binom{n}{2}}\sum_{0\leq w_1\leq\cdots\leq w_n\leq k-1} r^{w_1+\cdots+w_n}\\ &= p^{\binom{n}{2}}\begin{bmatrix}n+k-1\\n\end{bmatrix}_r,\\ \text{mp}_{(\Upsilon_{\mathbf{r}},\vec{A}),n}(p,q;1,\ldots,1) &= p^{\binom{n}{2}}\binom{n+k-1}{n}.\end{aligned} \tag{44}$$

Hence, for all $k > 2$,

$$\begin{aligned}&D_k^{\Upsilon_{\mathbf{r}}}(x,p,q,r,t)\\ &= \sum_{n\geq 0}\frac{t^n}{[n]_{p,q}!}\sum_{(\sigma,w)\in C_k\wr S_n} q^{\text{inv}(\sigma)}p^{\text{coinv}(\sigma)}r^{\|w\|}x^{\text{ris}((\sigma,w))}\\ &= 1\times\Bigg(1-\Bigg(\left(1+r+\cdots+r^{k-1}\right)t\\ &\qquad+\sum_{n\geq 2}\frac{(x-1)^{n-1}t^n}{[n]_{p,q}!}\text{mp}_{(\Upsilon_{\mathbf{r}},A),n}\\ &\qquad\times \left(p,q;1,r,\ldots,r^{k-1}\right)\Bigg)\Bigg)^{-1}\\ &= \frac{1-x}{1-x+\sum_{n\geq 1}\left(p^{\binom{n}{2}}((x-1)t)^n/[n]_{p,q}!\right)\left[\begin{smallmatrix}n+k-1\\n\end{smallmatrix}\right]_r}.\end{aligned} \tag{45}$$

Example 13. Let $\Upsilon = \Upsilon_{\mathbf{w}} = \{(12,00)\}$ and $\vec{A} = ([k],[k])$.

In this case for any $(\sigma,w) \in C_k \wr S_n$, $(\Upsilon,\vec{A})\text{-mch}((\sigma,w)) = \text{wris}((\sigma,w))$. It is easy to see that for $n \geq 2$, $(\sigma,w) \in \mathcal{MP}_{(\Upsilon,\vec{A}),n}$ if and only if $\sigma = 12\cdots n$ is the identity permutation and $w = w_1\cdots w_n$ where $0 \leq w_1 = \cdots = w_n \leq k-1$. As we pointed out in the introduction, it follows that

$$\text{mp}_{(\Upsilon,\vec{A}),n}(p,q;z_0,\ldots,z_{k-1}) = p^{\binom{n}{2}}\sum_{i=0}^{k-1}z_i^n. \tag{46}$$

Hence,

$$\begin{aligned}\text{mp}_{(\Upsilon_{\mathbf{w}},\vec{A}),n}(p,q;1,\ldots,1) &= kp^{\binom{n}{2}},\\ \text{mp}_{(\Upsilon_{\mathbf{w}},\vec{A}),n}(p,q;1,r,\ldots,r^{k-1}) &= p^{\binom{n}{2}} = p^{\binom{n}{2}}[k]_{r^n}.\end{aligned} \tag{47}$$

Thus, for all $k \geq 2$,

$$\begin{aligned}
&D_k^{\Upsilon_w}(x,p,q,r,t) \\
&= \sum_{n\geq 0} \frac{t^n}{[n]_{p,q}!} \sum_{(\sigma,w)\in C_k\wr S_n} q^{\mathrm{inv}(\sigma)} p^{\mathrm{coinv}(\sigma)} r^{\|w\|} x^{\mathrm{wris}((\sigma,w))} \\
&= 1\times\Big(1-\Big(\big(1+r+\cdots+r^{k-1}\big)t \\
&\quad + \sum_{n\geq 2}\frac{(x-1)^{n-1}t^n}{[n]_{p,q}!}\mathrm{mp}_{(\Upsilon_r,A),n} \\
&\quad \times \big(p,q;1,r,\ldots,r^{k-1}\big)\Big)\Big)^{-1} \\
&= \frac{1-x}{1-x+\sum_{n\geq 1}\left(p^{\binom{n}{2}}((x-1)t)^n/[n]_{p,q}!\right)[k]_{r^n}}.
\end{aligned} \tag{48}$$

Example 14. Let $\Upsilon = \Upsilon_{\mathbf{d}} = \{(12,01),(12,10)\}$ and $\vec{A} = ([k],[k])$.

In this case, it is easy to see that for $n \geq 2$, $(\sigma,w) \in \mathcal{MP}_{(\Upsilon,\vec{A}),n}$ if and only if $\sigma = 12\cdots n$ and $w = w_1\cdots w_n$ where $w_i \neq w_{i+1}$ for $i = 1,\ldots,n$. In this case, one cannot find simple compact expressions for $\mathrm{mp}_{(\Upsilon_r,A),n}(p,q;z_0,\ldots,z_{k-1})$ or $\mathrm{mp}_{(\Upsilon_r,A),n}(p,q;1,r,\ldots,r^{k-1})$. However, it is clear that $\mathrm{mp}_{(\Upsilon_r,A),n}(p,q;1,\ldots,1) = k(k-1)^{n-1}$. Thus, it follows that, for all $k \geq 2$,

$$\begin{aligned}
&D_k^{\Upsilon_d}(x,p,q,t) \\
&= \sum_{n\geq 0} \frac{t^n}{[n]_{p,q}!} \sum_{(\sigma,w)\in C_k\wr S_n} q^{\mathrm{inv}(\sigma)} p^{\mathrm{coinv}(\sigma)} x^{\mathrm{wris}((\sigma,w))} \\
&= 1\times\left(1-\left(kt+\sum_{n\geq 2}\frac{(x-1)^{n-1}t^n}{[n]_{p,q}!}\right.\right. \\
&\quad \left.\left.\times\, \mathrm{mp}_{(\Upsilon_d,A),n}(p,q;1,1,\ldots,1)\right)\right)^{-1} \\
&= \frac{1-x}{1-x+\sum_{n\geq 1}\left(p^{\binom{n}{2}}((x-1)t)^n/[n]_{p,q}!\right)k(k-1)^{n-1}}.
\end{aligned} \tag{49}$$

Example 15. Let $\Upsilon = \{(12,ab)\}$ and $\vec{A} = (\{a\},\{b\})$.

In this case, $(\Upsilon,\vec{A})$-mch$((\sigma,w))$ $=(\tau,u)$-Emch$((\sigma,w))$ where $(\tau,u) = (12,ab)$. There are two cases that we have to consider depending on whether a equals b or not. If $a = b$, then for $n \geq 2$, $(\sigma,w) \in C_k \wr S_n$ is an element of $\mathcal{MP}_{(\Upsilon,\vec{A}),n}$ if and only if $\sigma = 12\cdots n$ and $w = a^n$. Thus, $\mathrm{mp}_{(\Upsilon,\vec{A}),n}(p,q;z_0,\ldots,z_{k-1}) = p^{\binom{n}{2}}z_a^n$. It follows that if $(\tau,u) = (12,aa)$, then

$$\begin{aligned}
&\sum_{n\geq 0}\frac{t^n}{[n]_{p,q}!}\sum_{(\sigma,w)\in C_k\wr S_n} q^{\mathrm{inv}(\sigma)}p^{\mathrm{coinv}(\sigma)}z(w)\,x^{(\tau,u)\text{-Emch}((\sigma,w))} \\
&= 1\times\left(1-\left((z_0+\cdots+z_{k-1})t\right.\right. \\
&\quad \left.\left.+\sum_{n\geq 2}\frac{(x-1)^{n-1}t^n}{[n]_{p,q}!}p^{\binom{n}{2}}z_a^n\right)\right)^{-1}.
\end{aligned} \tag{50}$$

In particular,

$$\begin{aligned}
&\sum_{n\geq 0}\frac{t^n}{[n]_{p,q}!}\sum_{(\sigma,w)\in C_k\wr S_n} q^{\mathrm{inv}(\sigma)}p^{\mathrm{coinv}(\sigma)}x^{(\tau,u)\text{-Emch}((\sigma,w))} \\
&= \frac{1-x}{1-x+k(x-1)t+\sum_{n\geq 2}\left(p^{\binom{n}{2}}((x-1)t)^n/[n]_{p,q}!\right)}.
\end{aligned} \tag{51}$$

If $a \neq b$, then the only possible maximum packings for $(12,ab)$ are when $n = 2$ in which case $(\sigma,w) \in C_k \wr S_2$ is an element of $\mathcal{MP}_{(\Upsilon,\vec{A}),n}$ if and only if $(\sigma,w) = (12,ab)$. Thus, $\mathrm{mp}_{(\Upsilon,\vec{A}),2}(p,q;z_0,\ldots,z_{k-1}) = pz_az_b$. It follows that if $(\tau,u) = (12,ab)$ with $a \neq b$, then

$$\begin{aligned}
&\sum_{n\geq 0}\frac{t^n}{[n]_{p,q}!}\sum_{(\sigma,w)\in C_k\wr S_n} q^{\mathrm{inv}(\sigma)}p^{\mathrm{coinv}(\sigma)}z(w)\,x^{(\tau,u)\text{-Emch}((\sigma,w))} \\
&= \frac{1}{1-\left((z_0+\cdots+z_{k-1})t+\left((x-1)t^2/[2]_{p,q}!\right)pz_az_b\right)}.
\end{aligned} \tag{52}$$

In particular,

$$\begin{aligned}
&\sum_{n\geq 0}\frac{t^n}{[n]_{p,q}!}\sum_{(\sigma,w)\in C_k\wr S_n} q^{\mathrm{inv}(\sigma)}p^{\mathrm{coinv}(\sigma)}x^{(\tau,u)\text{-Emch}((\sigma,w))} \\
&= \frac{1-x}{1-x+k(x-1)t+\left(p(x-1)^2t^2/[2]_{p,q}!\right)}.
\end{aligned} \tag{53}$$

Example 16. Let $\Upsilon = \{(12,00),(12,01)\}$ and $\vec{A} = ([k],\{k-1\})$.

In this case, it is easy to see that for all $n \geq 2$, $(\sigma,w) \in \mathcal{MP}_{(\Upsilon,\vec{A}),n}$ if and only if $\sigma = 12\cdots n$ and $w = a(k-1)^{n-1}$ where $a \in [k]$. It follows that for all $n \geq 2$,

$$\mathrm{mp}_{(\Upsilon,\vec{A}),n}(p,q;z_0,\ldots,z_{k-1}) = p^{\binom{n}{2}}(z_0+\cdots+z_{k-1})z_{k-1}^{n-1}. \tag{54}$$

Hence,

$$\sum_{n\geq 0}\frac{t^n}{[n]_{p,q}!}\sum_{(\sigma,w)\in C_k\wr S_n} q^{\mathrm{inv}(\sigma)}p^{\mathrm{coinv}(\sigma)}z(w)\,x^{(\Upsilon,\vec{A})\text{-mch}((\sigma,w))}$$
$$= 1\times\Big(1-\Big((z_0+\cdots+z_{k-1})t + \sum_{n\geq 2}\frac{(x-1)^{n-1}t^n}{[n]_{p,q}!} \times p^{\binom{n}{2}}(z_0+\cdots+z_{k-1})z_{k-1}^{n-1}\Big)\Big)^{-1}. \quad (55)$$

Setting $p=q=z_0=\cdots=z_{k-1}=1$ in (55), we obtain that

$$\sum_{n\geq 0}\frac{t^n}{[n]_{p,q}!}\sum_{(\sigma,w)\in C_k\wr S_n} x^{(\Upsilon,\vec{A})\text{-mch}((\sigma,w))}$$
$$=\frac{1}{1-k\sum_{n\geq 1}\left((x-1)^{n-1}t^n/n!\right)}$$
$$=\frac{1-x}{1-x+k\sum_{n\geq 1}\left(((x-1)t)^n/n!\right)} \quad (56)$$
$$=\frac{1-x}{1-x+k\left(e^{(x-1)t}-1\right)}$$
$$=\frac{1-x}{1-k-x+ke^{(x-1)t}}.$$

Let $NM_{(\{(12,00),(12,01)\},([k],\{k-1\})),n}$ denote the number of $\sigma\in C_k\wr S_n$ that have no $(\{(12,00),(12,01)\},([k],\{k-1\}))$-matches. Then setting $x=0$ in (56), we see that

$$\sum_{n\geq 0}\frac{t^n}{n!}NM_{(\{(12,00),(12,01)\},([k],\{k-1\})),n} = \frac{1}{1-k+ke^{-t}}=\frac{e^t}{k-(k-1)e^t}. \quad (57)$$

One can easily calculate that the sequence $(NM_{(\{(12,00),(12,01)\},(\{0,1\},\{1\})),n})_{n\geq 0}$ starts out with

$$1, 2, 6, 26, 150, 1082, 9366, 94586, 1091670, \ldots \quad (58)$$

which is sequence A000629 in the OEIS and which also counts the number of necklaces of partitions of $n+1$ labeled beads. Similarly, the sequence $(NM_{(\{(12,00),(12,01)\},(\{0,1,2\},\{2\})),n})_{n\geq 0}$ starts out with

$$1, 3, 15, 111, 1095, 13503, 199815, 3449631, 68062695, \ldots, \quad (59)$$

and the sequence $(NM_{(\{(12,00),(12,01)\},(\{0,1,2,3\},\{3\})),n})_{n\geq 0}$ starts out with

$$1, 4, 28, 292, 4060, 70564, 1471708, 35810212, 995827420, \ldots. \quad (60)$$

Neither of these two sequences appear in the OEIS.

Example 17. Let $\Upsilon=\{(12,00),(12,01)\}$ and $\vec{A}=([k],\{k-2,k-1\})$.

In this case, it is easy to see that for all $k\geq 2$ and $n\geq 2$, $(\sigma,w)\in\mathscr{MP}_{(\Upsilon,\vec{A}),n}$ if and only if $\sigma=12\cdots n$ and w is either of the form $w=a(k-2)^i(k-1)^j$ where $0\leq a\leq k-3$ and $i+j=n-1$ or $w=(k-2)^r(k-1)^s$ where $r+s=n$. Hence, for all $n\geq 2$,

$$\mathrm{mp}_{(\Upsilon,\vec{A}),n}(p,q;1,\ldots,1)=((k-1)n+1)\,p^{\binom{n}{2}}. \quad (61)$$

It follows that

$$\sum_{n\geq 0}\frac{t^n}{[n]_{p,q}!}\sum_{(\sigma,w)\in C_k\wr S_n} q^{\mathrm{inv}(\sigma)}p^{\mathrm{coinv}(\sigma)}x^{(\Upsilon,\vec{A})\text{-mch}((\sigma,w))}$$
$$=1\times\Big(1-\Big((z_0+\cdots+z_{k-1})t+\sum_{n\geq 2}\frac{(x-1)^{n-1}t^n}{[n]_{p,q}!}\times((k-1)n+1)\,p^{\binom{n}{2}}\Big)\Big)^{-1}. \quad (62)$$

Setting $p=q=1$ in (62), we obtain that

$$\sum_{n\geq 0}\frac{t^n}{[n]_{p,q}!}\sum_{(\sigma,w)\in C_k\wr S_n} x^{(\Upsilon,\vec{A})\text{-mch}((\sigma,w))}$$
$$=\frac{1}{1-\sum_{n\geq 1}((k-1)n+1)\left((x-1)^{n-1}t^n/n!\right)}$$
$$=\frac{1-x}{1-x+\sum_{n\geq 1}((k-1)n+1)\left(((x-1)t)^n/n!\right)}$$
$$=1-x\left(1-x+(k-1)(x-1)t\times\left(\sum_{n\geq 1}\frac{((x-1)t)^{n-1}}{(n-1)!}\right)+\sum_{n\geq 1}\frac{((x-1)t)^n}{n!}\right)^{-1} \quad (63)$$
$$=\frac{1-x}{1-x+(k-1)(x-1)te^{(x-1)t}+e^{(x-1)t}-1}$$
$$=\frac{1-x}{-x+(1+(k-1)(x-1)t)\,e^{(x-1)t}}.$$

Let $NM_{(\{(12,00),(12,01)\},([k],\{k-2,k-1\})),n}$ denote the number of $\sigma\in C_k\wr S_n$ that have no $(\{(12,00),(12,01)\},([k],\{k-2,k-1\}))$-matches. Then setting $x=0$ in (63), we see that

$$\sum_{n\geq 0}\frac{t^n}{n!}NM_{(\{(12,00),(12,01)\},([k],\{k-2,k-1\})),n} = \frac{1}{(1-(k-1)t)e^{-t}}=\frac{e^t}{1-(k-1)t}. \quad (64)$$

Note that it follows that for $n \geq 1$,

$$\begin{aligned}
&NM_{(\{(12,00),(12,01)\},([k],\{k-2,k-1\})),n} \\
&= n!\sum_{j=0}^{n}\frac{(k-1)^j}{(n-j)!} \\
&= 1 + n!\sum_{j=1}^{n}\frac{(k-1)^j}{(n-j)!} \qquad (65)\\
&= 1 + (k-1)\,n\left((n-1)!\sum_{i=0}^{n-1}\frac{(k-1)^i}{(n-1-i)!}\right) \\
&= 1 + (k-1)\,nNM_{(\{(12,00),(12,01)\},([k],\{k-2,k-1\})),n-1}.
\end{aligned}$$

It would be interesting to find a direct combinatorial proof of this recursion, which we leave as an open problem.

One can easily calculate that the sequence $(NM_{(\{(12,00),(12,01)\},(\{0,1\},\{0,1\})),n})_{n\geq 0}$ starts out with

$$1, 2, 5, 16, 65, 326, 1957, 13700, 109601, 986410\ldots, \quad (66)$$

which is sequence A000522 in the OEIS having multiple combinatorial interpretations including the number of permutations $\sigma = \sigma_1 \cdots \sigma_{n+1}$ with no $1 \leq i < j \leq n$ such that $\sigma_i < \sigma_j < \sigma_{n+1}$, that is, in the *Babson-Steingrimsson notation*, the number of σ in S_{n+1} which avoid the pattern 1-2-3] (the same pattern is denoted 123] in the recently introduced notation [20]).

The sequence $(NM_{(\{(12,00),(12,01)\},(\{0,1,2\},\{1,2\})),n})_{n\geq 0}$ starts out with

$$1, 3, 13, 79, 633, 6331, 75973, 1063623, 18017969, 306323443\ldots, \quad (67)$$

which is sequence A010844 of the OEIS. This sequence does not have a combinatorial interpretation listed in the OEIS so that we have obtained a simple combinatorial interpretation of this sequence. However, if $a(n)$ is the nth term of this sequence where $a(0) = 1$, then $a(n)$ is equal to 2^n times the permanent of the $n \times n$ matrix with 3/2 on the main diagonal and 1s everywhere else.

The sequence $(NM_{(\{(12,00),(12,01)\},(\{0,1,2,3\},\{2,3\})),n})_{n\geq 0}$ starts out with

$$1, 4, 25, 226, 2713, 40696, 732529, 15383110, 369194641, 9968255308,\ldots \quad (68)$$

is sequence A010844 of the OEIS. This sequence does not have a combinatorial interpretation listed in the OEIS. Thus, we have obtained a new combinatorial interpretation of this sequence. In this case, if $a(n)$ is the nth term of the sequence where $a(0) = 1$, then $a(n)$ is equal to 3^n times the permanent of the $n \times n$ matrix with 4/3 on the main diagonal and 1's everywhere else.

Similarly, the sequences $(NM_{(\{(12,00),(12,01)\},([k],\{k-2,k-1\})),n})_{n\geq 0}$ for $k = 5, 6,$ and 7 appear in the OEIS as sequences as A056545, A056546, and A056547, respectively.

Example 18. Let $\Upsilon = \{(12, 01)\}$ and $\vec{A} = ([k], \{k-2, k-1\})$.

In this case, it is easy to see that $\mathcal{MP}_{(\Upsilon,\vec{A}),n} = \emptyset$ for $k > 3$. That is, for $(\sigma, w) \in \mathcal{MP}_{(\Upsilon,\vec{A}),n}$, we must have $\sigma = 12\cdots n$ and $w = w_1 \cdots w_n$ where $0 \leq w_1 < \cdots < w_n \leq k-1$ and $w_i \in \{k-2, k-1\}$ for $i > 1$. Now if $n = 2$, our choices for w are either $a(k-2)$ or $a(k-1)$ where $a \in \{0,\ldots,k-3\}$ or $w = (k-2)(k-1)$ so that $\mathrm{mp}_{(\Upsilon,\vec{A}),2}(p, q; 1,\ldots,1) = p(2(k-2)+1) = p(2k-3)$. Now if $n = 3$, our choices for w are $a(k-2)(k-1)$ where $a \in \{0,\ldots,k-3\}$, so that $\mathrm{mp}_{(\Upsilon,\vec{A}),3}(p, q; 1,\ldots,1) = p^3(k-2)$.

It follows that

$$\begin{aligned}
&\sum_{n\geq 0}\frac{t^n}{[n]_{p,q}!}\sum_{(\sigma,w)\in C_k\wr S_n} q^{\mathrm{inv}(\sigma)}p^{\mathrm{coinv}(\sigma)}x^{(\Upsilon,\vec{A})\text{-}\mathrm{mch}((\sigma,w))} \\
&= 1 \times \left(1 - \left(kt + p(2k-3)(x-1)\frac{t^2}{[2]_{p,q}!}\right.\right. \qquad (69)\\
&\qquad \left.\left. + p^3(x-1)^2(k-2)\frac{t^3}{[3]_{p,q}!}\right)\right)^{-1}.
\end{aligned}$$

Setting $p = q = 1$ in (69), we obtain that for $k \geq 2$

$$\begin{aligned}
&\sum_{n\geq 0}\frac{t^n}{[n]_{p,q}!}\sum_{(\sigma,w)\in C_k\wr S_n} x^{(\Upsilon,\vec{A})\text{-}\mathrm{mch}((\sigma,w))} \\
&= 1 \times \left(1 - \left(kt + (2k-3)(x-1)\left(\frac{t^2}{2}\right)\right.\right. \qquad (70)\\
&\qquad \left.\left. + (x-1)^2(k-2)\left(\frac{t^3}{6}\right)\right)\right)^{-1}.
\end{aligned}$$

Let $NM_{(12,01),([k],\{k-2,k-1\}),n}$ denote the number of $\sigma \in C_k \wr S_n$ that have no $((12, 01), ([k], \{k-2, k-1\}))$-matches. Then setting $x = 0$ in (70), we see that

$$\begin{aligned}
&\sum_{n\geq 0}\frac{t^n}{n!}NM_{(12,01),([k],\{k-2,k-1\}),n} \\
&= \frac{1}{1 - kt + (2k-3)(t^2/2) - (k-2)(t^3/6)}. \qquad (71)
\end{aligned}$$

One can easily calculate the initial terms of the sequences $(NM_{(12,01),([k],\{k-2,k-1\}),n})_{n\geq 0}$. For example, the sequence $(NM_{(12,01),([2],\{0,1\}),n})_{n\geq 0}$ starts out

$$1, 2, 7, 36, 246, 2100, 21150, 257040, 3510360, 53933040,\ldots. \quad (72)$$

The sequence $(NM_{(12,01),([3],\{1,2\}),n})_{n\geq 0}$ starts out

$$1, 3, 15, 109, 1050, 12630, 182270, 3068730, 59046120, 1278133080,\ldots. \quad (73)$$

The sequence $(NM_{(12,01),([4],\{2,3\}),n})_{n\geq 0}$ starts out

$$1, 4, 27, 266, 3478, 56800, 1112990, 25443180, 664724760, 19537301289,\ldots. \quad (74)$$

However, none of these sequences appear in the OEIS.

TABLE 1: $\mathrm{NM}^{\Upsilon_r}_{n,k}$ for $k, n \le 5$.

	$n=0$	$n=1$	$n=2$	$n=3$	$n=4$	$n=5$
$k=2$	1	2	5	16	65	326
$k=3$	1	3	12	64	441	3771
$k=4$	1	4	22	164	1589	19136
$k=5$	1	5	35	335	4180	64876

4. Expressions for $NM_{(\Upsilon,\vec{A}),n}$ as Polynomials in k and Connections to Other Objects

Recall that for any $\Upsilon \subseteq C_k \wr S_2$ and $\vec{A} = (A_1, A_2)$ where $A_i \subseteq [k]$, $NM_{(\Upsilon,\vec{A}),n}$ is the number of $(\sigma, w) \in C_k \wr S_n$ such that (σ, w) has no $(\Upsilon, \vec{A})$-matches. We start out by considering the special case where $\Upsilon \subseteq C_2 \wr S_2$ and $\vec{A} = ([k],[k])$. In that case, we will simply write $A^{\Upsilon}_{n,k}$ for $NM_{(\Upsilon,\vec{A}),n}$. Thus, $A^{\Upsilon}_{n,k}$ is the number of permutations $(\sigma, w) \in C_k \wr S_n$ such that (σ, w) has no Υ-matches. By setting $p = q = z_0 = \cdots = z_{k-1} = 1$ in (33), we see that

$$\sum_{n\ge 0} A^{\Upsilon}_{n,k} \frac{t^n}{n!} = 1 \times \left(1 - \left(kt + \sum_{n\ge 2} \mathrm{mp}_{(\Upsilon,([k],[k])),n} \times (1,1;1,\ldots,1)\left((-1)^{n-1} t^n/n!\right)\right)\right)^{-1}. \quad (75)$$

We can easily compute the initial sequences of values for these generating functions using any computer algebra system such as Mathematica or Maple. In this section, we will demonstrate these values in several cases.

4.1. $NM^{\Upsilon_r}_{n,k}$. For example, if $\Upsilon = \Upsilon_r = \{(1\ 2, 0\ 0), (1\ 2, 0\ 1)\}$, $NM^{\Upsilon_r}_{n,k}$ equals the number of $(\sigma, \epsilon) \in C_k \wr S_n$ such that $\mathrm{ris}((\sigma, \epsilon)) = 0$. Table 1 gives initial values of $NM^{\Upsilon_r}_{n,k}$.

Several of these sequences appear in the OEIS [16]. In fact, we can easily calculate $NM^{\Upsilon_r}_{n,k}$ as a polynomial in k. For example, we have

$$NM^{\Upsilon_r}_{0,k} = 1,$$
$$NM^{\Upsilon_r}_{1,k} = k,$$
$$NM^{\Upsilon_r}_{2,k} = \frac{1}{2}k(3k-1),$$
$$NM^{\Upsilon_r}_{3,k} = \frac{1}{6}k\left(19k^2 - 15k + 2\right),$$
$$NM^{\Upsilon_r}_{4,k} = \frac{1}{24}k\left(211k^3 - 270k^2 + 89k - 6\right),$$
$$NM^{\Upsilon_r}_{5,k} = \frac{1}{120}k\left(3651k^4 - 6490k^3 + 3585k^2 - 650k + 24\right). \quad (76)$$

We point out that $NM^{\Upsilon_r}_{2,k}$ forms the familiar sequence of pentagonal numbers (A000326 in [16]). Other previously documented sequences appearing in Table 1 include the structured octagonal antiprism numbers (A100184 in [16]) for $NM^{\Upsilon_r}_{3,k}$, as well as $NM^{\Upsilon_r}_{n,2}$ (A000522 in [16]), for which there are many known combinatorial interpretations, including the total number of arrangements of all subsets of $[n]$.

We conjecture that for $n \ge 1$ and $k \ge 2$, $NM^{\Upsilon_r}_{n,k}$ is always of the form $(1/n!)kP_n(k)$ where $P_n(k)$ is a polynomial of degree $n-1$ whose leading coefficient is positive and such that signs of the remaining coefficients alternate. Now we can prove that $NM^{\Upsilon_r}_{n,k}$ is always of the form $(1/n!)kP_n(k)$ where $P_n(k)$ is a polynomial of degree $n-1$ and the term of degree 1 in k is $(-1)^{n-1}(n-1)!$. That is, for any $k \ge 2$, if we set $p = q = z_0 = \cdots = z_{k-1} = 1$ and $x = 0$ in the proof of Theorem 10 and use the fact that $\mathrm{mp}_{(\Upsilon_r,([k],[k])),n}(1,1;1,\ldots,1) = \binom{n+k-1}{n}$, we see that

$$n!\overline{\Gamma}(h_n) = NM^{\Upsilon_r}_{n,k}, \quad (77)$$

where

$$\overline{\Gamma}(e_n) = \frac{\binom{n+k-1}{n}}{n!} = \frac{(k)\uparrow_n}{(n!)^2}. \quad (78)$$

Here we let $(q)\uparrow_0 = 1$ and $(q)\uparrow_n = q(q+1)\cdots(q+n-1)$ for $n \ge 1$. But then

$$\begin{aligned} n!\overline{\Gamma}(h_n) &= n!\sum_{\mu\vdash n}(-1)^{n-\ell(\mu)}B_{\mu,n}\overline{\Gamma}(e_\mu) \\ &= \frac{1}{n!}\sum_{\mu \text{ÂŘ}\vdash n}(-1)^{n-\ell(\mu)}B_{\mu,n}\binom{n}{\mu_1,\ldots,\mu_{\ell(\mu)}}\prod_{i=1}^{2\ell(\mu)}(k)\uparrow_{\mu_i} \\ &= \frac{1}{n!}\sum_{\mu \text{ÂŘ}\vdash n}(-1)^{n-\ell(\mu)}\sum_{(b_1,\ldots,b_{\ell(\mu)})\in\mathscr{B}_{\mu,n}}\binom{n}{b_1,\ldots,b_{\ell(\mu)}}\prod_{i=1}^{2\ell(\mu)}(k)\uparrow_{b_i}. \end{aligned} \quad (79)$$

It is easy to see that the right-hand side of (79) is a polynomial of degree n, and the lowest degree term comes from the term $(-1)^{n-1}k(k+1)\cdots(k+n-1)$ corresponding to $\mu = (n)$ which is of the form $(-1)^{n-1}(n-1)!k + O(k^2)$. On the other hand, the highest-degree term arises when we pick the highest power of k from the products $\prod_{i=1}^{\ell(\mu)}(k)\uparrow_{b_i}$ which is clearly just k^n. It follows that the highest power of k in $A^{\Upsilon_r}_{n,k}$ is just

$$\frac{1}{n!}\sum_{\mu\vdash \text{ÂŘ}n}(-1)^{n-\ell(\mu)}\sum_{(b_1,\ldots,b_{\ell(\mu)})\in\mathscr{B}_{\mu,n}}\binom{n}{b_1,\ldots,b_{\ell(\mu)}}^2. \quad (80)$$

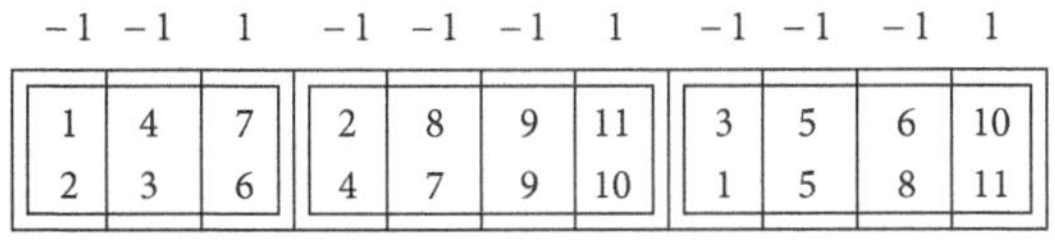

FIGURE 6: A brick tabloid of shape (12) and type (1, 1, 2, 3, 5).

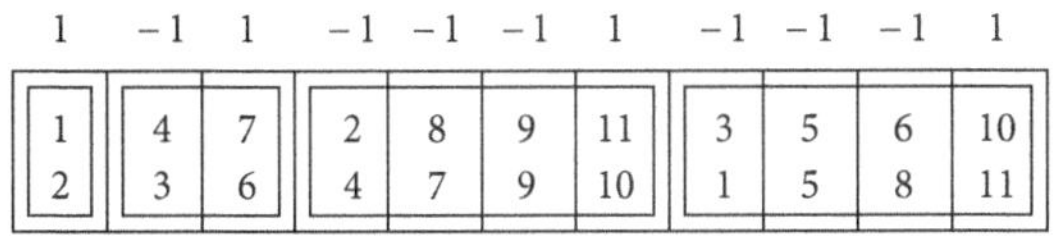

FIGURE 7: $I_r(O)$ for O in Figure 6.

We can give a combinatorial interpretation to

$$F_n = \sum_{\mu \vdash \text{ÂŘ}n} (-1)^{n-\ell(\mu)} \sum_{(b_1,\ldots,b_{\ell(\mu)}) \in \mathscr{B}_{\mu,n}} \binom{n}{b_1,\ldots,b_{\ell(\mu)}}^2. \quad (81)$$

That is, we will show that F_n counts the number of pairs of permutations $(\alpha,\beta) \in S_n \times S_n$ such that if $\alpha = \alpha_1 \cdots \alpha_n$ and $\beta = \beta_1,\ldots,\beta_n$, then there is no $1 \le i \le n-1$ such that $\alpha_i < \alpha_{i+1}$ and $\beta_i < \beta_{i+1}$. That is, F_n is the number of pairs $(\alpha,\beta) \in S_n \times S_n$ that have no common rises. To prove our claim, we can give a combinatorial interpretation to the right-hand side of (81). That is, if we start with a brick tabloid $T = (b_1,\ldots,b_{\ell(\mu)})$, then by Lemma 8, we can interpret $\binom{n}{b_1,\ldots,b_{\ell(\mu)}}^2$ as all ways to fill the cells of T with pairs of permutations (α,β) such that both α and β are increasing in each brick. Then we interpret $(-1)^{n-\ell(\mu)}$ as labeling each cell of T which is not at the end of brick with -1 and labeling each cell which is at the end of the brick with 1. We let $\mathscr{OR}_n$ denote the set of all filled brick tabloids constructed in this way. For example, Figure 6 pictures a typical element of $\mathscr{OR}_{11}$. If $O \in \mathscr{OR}_n$, we define the sign of O, sgn(O), as the product of labels at the top of the cells of O.

We now can define a sign-reversing involution I_r on $\mathscr{OR}_n$.

To define I_r, suppose that $O \in \mathscr{OR}_n$ and α and β are the two permutations in O with α on the bottom and β on the top. Then we scan the cells of O from left to right looking for the leftmost cell t such that either (i) t is labeled with -1 or (ii) t is at the end of a brick b_j, and the brick b_{j+1} immediately following b_j has the property that both α and β are strictly increasing in all the cells corresponding to b_j and b_{j+1}. In case (i), $I_r(O)$ is the result of replacing the brick b in O containing t by two bricks b^* and b^{**} where b^* contains the cell t plus all the cells in b to the left of t and changing the label of cell t from -1 to 1. In case (ii), $I_r(O)$ is the result of replacing the bricks b_j and b_{j+1} in T by a single brick b and changing the label of cell t from 1 to -1. If neither case (i) or case (ii) applies, then we let $I(O) = O$. For example, if O is the element of $\mathscr{F}_{12}$ pictured in Figure 6, then $I_r(O)$ is pictured in Figure 7.

It is easy to see that I_r is a weight-preserving sign-reversing involution and, hence, I_r shows that

$$F_n = \sum_{O \in \mathscr{OR}_n, I_r(O)=O} \operatorname{sgn}(O). \quad (82)$$

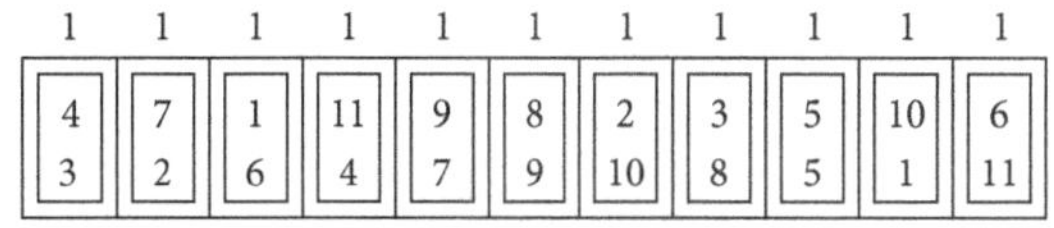

FIGURE 8: A fixed point of I_r.

TABLE 2: $NM_{n,k}^{\Upsilon_s}$ for $k, n \le 5$.

	$n=0$	$n=1$	$n=2$	$n=3$	$n=4$	$n=5$
$k=2$	1	2	7	36	246	2100
$k=3$	1	3	15	109	1050	12630
$k=4$	1	4	26	244	3031	47000
$k=5$	1	5	40	460	6995	132751

Thus, we must examine the fixed points O of I_r. First there can be no -1 labels in O so that all bricks in O must be of size 1 and $\operatorname{sgn}(O) = 1$. Moreover, if b_j and b_{j+1} are two consecutive bricks in T and t is the last cell of b_j, then it cannot be the case that $\alpha_t < \alpha_{t+1}$ and $\beta_t < \beta_{t+1}$ where α and β are the two permutations in O since otherwise we could combine b_j and b_{j+1}. Thus, (α,β) is a pair of permutations in S_n with no common rises. Vice versa, if (α,β) is a pair of permutations in S_n with no common rises, we can create a fixed point O of I_r by having α be the bottom permutation of O, having β be the top permutation of O, and having all bricks be of size 1. For example, Figure 8 pictures a fixed point of I_r.

It follows that F_n is the number of pairs $(\alpha,\beta) \in S_n^2$ with no common rises so that the leading coefficient of $A_{k,r}^{\Upsilon_r}$ is $F_n/n!$.

4.2. $NM_{n,k}^{\Upsilon_s}$. For $\Upsilon_s = \{(1\ \ 2, 0\ \ 1)\}$, $NM_{n,k}^{\Upsilon_s}$ equals the number of $(\sigma,\epsilon) \in C_k \wr S_n$ such that $sris((\sigma,\epsilon)) = 0$. Table 2 gives initial values of $NM_{n,k}^{\Upsilon_s}$.

In fact, we can easily calculate $NM_{n,k}^{\Upsilon_s}$ as a polynomial in k. For example, we have

$$NM_{0,k}^{\Upsilon_s} = 1,$$

$$NM_{1,k}^{\Upsilon_s} = k,$$

$$NM_{2,k}^{\Upsilon_s} = \frac{1}{2}k(3k+1),$$

$$NM_{3,k}^{\Upsilon_s} = \frac{1}{6}k\left(19k^2 + 15k + 2\right),$$

$$NM_{4,k}^{\Upsilon_s} = \frac{1}{24}k\left(211k^3 + 270k^2 + 89k + 6\right),$$

$$NM_{5,k}^{\Upsilon_s} = \frac{1}{120}k\left(3651k^4 + 6490k^3 + 3585k^2 + 650k + 24\right). \quad (83)$$

We point out that $NM_{2,k}^{\Upsilon_s}$ forms the familiar sequence of the second pentagonal numbers (A005449 in [16]). None of the other rows or columns in Table 2 matched any previously known sequences in [16].

TABLE 3: $\mathrm{NM}_{n,k}^{\Upsilon_w}$ for $k, n \le 5$.

	$n = 0$	$n = 1$	$n = 2$	$n = 3$	$n = 4$	$n = 5$
$k = 2$	1	2	6	26	150	1082
$k = 3$	1	3	15	111	1095	13503
$k = 4$	1	4	28	292	4060	70564
$k = 5$	1	5	45	605	10845	243005

We conjecture that for $n \ge 1$ and $k \ge 2$, $NM_{n,k}^{\Upsilon_s}$ is always of the form $(1/n!)kR_n(k)$ where $R_n(k)$ is a polynomial of degree $n-1$ with positive coefficients. In fact, we see that the coefficients of $P_{n,k}$ and $R_{n,k}$ are the same up to a sign for all n. we can prove this. That is, for any $k \ge 2$, if we set $p = q = z_0 = \cdots = z_{k-1} = 1$ and $x = 0$ in the proof of Theorem 10 and use the fact that $\mathrm{mp}_{(\Upsilon_s,([k],[k])),n} = \binom{k}{r}$, we see that

$$n!\overline{\Gamma}_s(h_n) = NM_{n,k}^{\Upsilon_s}, \tag{84}$$

where

$$\overline{\Gamma}_s(e_n) = \frac{\binom{k}{n}}{n!} = \frac{(k)\downarrow_n}{(n!)^2}. \tag{85}$$

Here we let $(q)\downarrow_0 = 1$ and $(q)\downarrow_n = q(q-1)\cdots(q-n+1)$ for $n \ge 1$. But then

$$\begin{aligned} n!\overline{\Gamma}_s(h_n) &= n!\sum_{\mu\vdash n}(-1)^{n-\ell(\mu)}B_{\mu,n}\overline{\Gamma}(e_\mu) \\ &= \frac{1}{n!}\sum_{\mu\text{ÂŘ}\vdash n}(-1)^{n-\ell(\mu)}B_{\mu,n}\binom{n}{\mu_1,\ldots,\mu_{\ell(\mu)}}\prod_{i=1}^{2\ell(\mu)}(k)\downarrow_{\mu_i} \\ &= \frac{1}{n!}\sum_{\mu\vdash n}(-1)^{n-\ell(\mu)}\sum_{(b_1,\ldots,b_{\ell(\mu)})\in\mathscr{B}_{\mu,n}}\binom{n}{b_1,\ldots,b_{\ell(\mu)}}\prod_{i=1}^{2\ell(\mu)}(k)\downarrow_{b_i}. \end{aligned} \tag{86}$$

Since for any $n \ge 1$, $(k)\downarrow_n = (-1)^n(k)\uparrow_n$, it is easy to see that the right-hand side of (86) is obtained from the right-hand side of (79) by replacing k by $-k$ and multiplying by $(-1)^n$. Thus, conjecture that $R_n(k)$ has positive coefficients is equivalent to the conjecture that the signs of the coefficients of $P_n(k)$ alternate.

4.3. $NM_{n,k}^{\Upsilon_w}$. For $\Upsilon_w = \{(1\ 2, 0\ 0)\}$, $NM_{n,k}^{\Upsilon_w}$ equals the number of $(\sigma,\epsilon) \in C_k \wr S_n$ such that $wris((\sigma,\epsilon)) = 0$. Table 3 gives initial values of $NM_{n,k}^{\Upsilon_w}$.

In fact, we can easily calculate $NM_{n,k}^{\Upsilon_w}$ as a polynomial in k. For example, we have

$$\begin{aligned} NM_{0,k}^{\Upsilon_w} &= 1, \\ NM_{1,k}^{\Upsilon_w} &= k, \\ NM_{2,k}^{\Upsilon_w} &= k(2k-1), \\ NM_{3,k}^{\Upsilon_w} &= k(6k^2-6k+1), \\ NM_{4,k}^{\Upsilon_w} &= k(24k^3-36k^2+14k-1), \\ NM_{5,k}^{\Upsilon_w} &= k(120k^4-240k^3+150k^2-30k+1). \end{aligned} \tag{87}$$

We point out that $NM_{2,k}^{\Upsilon_w}$ forms the familiar sequence of hexagonal numbers (A000384 in [16]). Additionally, $NM_{n,2}^{\Upsilon_w}$ matches the sequence counting the number of necklaces on the set of labeled beads (A000629 in [16]). In fact, in this case, we can give a completely combinatorial interpretation of the coefficients of $NM_{n,k}^{\Upsilon_w}$ as a polynomial in k. Let $\mathrm{Ostpn}(n)$ denote the set of ordered set partitions of $\{1,2,\ldots,n\}$. For any set partition $\pi \in \mathrm{Ostpn}(n)$, let $\ell(\pi)$ denote the number of parts of π. Then we claim that

$$NM_{n,k}^{\Upsilon_w} = \sum_{\pi\in\mathrm{Ostpn}(n)}(-1)^{n-\ell(\pi)}k^{\ell(\pi)}, \tag{88}$$

so that the coefficient of k^j in $NM_{n,k}^{\Upsilon_w}$ is equal to $(-1)^{n-j}j!S_{n,j}$ where $S_{n,j}$ is the Stirling number of the second kind which is the number of set partitions of $\{1,\ldots,n\}$ into j parts. That is, for any $k \ge 2$, if we set $p = q = z_0 = \cdots = z_{k-1} = 1$ and $x = 0$ in Theorem 10 and use the fact that $\mathrm{mp}_{\Upsilon_w,n} = k$ for all $n \ge 2$, then we see that

$$n!\overline{\Gamma}_w(h_n) = NM_{n,k}^{\Upsilon_w}, \tag{89}$$

where

$$\overline{\Gamma}_w(e_n) = \frac{k}{n!}. \tag{90}$$

But then

$$\begin{aligned} n!\overline{\Gamma}_s(h_n) &= n!\sum_{\mu\text{ÂŘ}\vdash n}(-1)^{n-\ell(\mu)}\overline{\Gamma}_s(e_\mu) \\ &= n!\sum_{\mu\vdash\text{ÂŘ}n}(-1)^{n-\ell(\mu)}\sum_{(b_1,\ldots,b_{\ell(\mu)})\in\mathscr{B}_{\mu,n}}\prod_{i=1}^{\ell(\mu)}\frac{k}{b_i!} \\ &= \sum_{\mu\vdash\text{ÂŘ}n}(-1)^{n-\ell(\mu)}\sum_{(b_1,\ldots,b_{\ell(\mu)})\in\mathscr{B}_{\mu,n}}\binom{n}{b_1,\ldots,b_{\ell(\mu)}}k^{\ell(\mu)}. \end{aligned} \tag{91}$$

Since $\binom{n}{b_1,\ldots,b_{\ell(\mu)}}$ counts the number of ordered set partitions $\pi = (\pi_1,\ldots,\pi_{\ell(\mu)})$ such that $|\pi_j| = b_j$, it is easy to see that the right-hand side of (91) equals the right-hand side of (88).

TABLE 4: $\mathrm{NM}_{n,k}^{\Upsilon_d}$ for $k, n \le 5$.

	$n=0$	$n=1$	$n=2$	$n=3$	$n=4$	$n=5$
$k=2$	1	2	6	26	150	1082
$k=3$	1	3	12	66	480	4368
$k=4$	1	4	20	132	1140	12324
$k=5$	1	5	30	230	2280	28280

L	1	$-k$	1	k	1	$-k$	k	1	1	k	$-k$	k
σ	2	3	4	11	6	9	10	1	8	12	5	7

FIGURE 9: A composite object $C \in \mathcal{D}_{12}$.

4.4. $NM_{n,k}^{\Upsilon_d}$. Recall that $\Upsilon_{\mathbf{d}} = \{(1\ 2, 0\ 1), (1\ 2, 1\ 0)\}$. Table 4 gives initial values of $NM_{n,k}^{\Upsilon_d}$.

In fact, we can easily calculate $NM_{n,k}^{\Upsilon_d}$ as a polynomial in k. For example, we have

$$\begin{aligned} NM_{0,k}^{\Upsilon_d} &= 1, \\ NM_{1,k}^{\Upsilon_d} &= k, \\ NM_{2,k}^{\Upsilon_d} &= k^2 + k, \\ NM_{3,k}^{\Upsilon_d} &= k^3 + 4k^2 + k, \\ NM_{4,k}^{\Upsilon_d} &= k^4 + 11k^3 + 11k^2 + k, \\ NM_{5,k}^{\Upsilon_d} &= k^5 + 26k^4 + 66k^3 + 26k^2 + k. \end{aligned} \tag{92}$$

In this case, we will show that $NM_{n,k}^{\Upsilon_d}$ is just the Eulerian polynomial

$$NM_{n,k}^{\Upsilon_d} = \sum_{\sigma \in S_n} x^{\mathrm{des}(\sigma)+1}. \tag{93}$$

That is, for any $k \ge 2$, if we set $p = q = z_0 = \cdots = z_{k-1} = 1$ and $x = 0$ in the proof of Theorem 10 and the fact that $\mathrm{mp}_{(\Upsilon_d,([k],[k])),n} = k(k-1)^{n-1}$, we see that

$$n!\overline{\Gamma}_U(h_n) = NM_{n,k}^{\Upsilon_d}, \tag{94}$$

where

$$\overline{\Gamma}_U(e_n) = \frac{k(k-1)^{n-1}}{n!}. \tag{95}$$

But then

$$\begin{aligned} n!\overline{\Gamma}_U(h_n) &= n! \sum_{\mu \vdash \text{ÂŘ}n} (-1)^{n-\ell(\mu)} B_{\mu,n} \overline{\Gamma}_U(e_\mu) \\ &= n! \sum_{\mu \vdash \text{ÂŘ}n} (-1)^{n-\ell(\mu)} \sum_{(b_1,\dots,b_{\ell(\mu)}) \in \mathcal{B}_{\mu,n}} \prod_{i=1}^{\ell(\mu)} \frac{k(k-1)^{b_i-1}}{b_i!} \\ &= \sum_{\mu \text{ÂŘ} \vdash n} \sum_{(b_1,\dots,b_{\ell(\mu)}) \in \mathcal{B}_{\mu,n}} \binom{n}{b_1,\dots,b_{\ell(\mu)}} \prod_{i=1}^{\ell(\mu)} k(1-k)^{b_i-1}. \end{aligned} \tag{96}$$

L	1	k	1	k	1	$-k$	k	1	1	k	$-k$	k
σ	2	3	4	11	6	9	10	1	8	12	5	7

FIGURE 10: $J(C)$ for C in Figure 9.

Next we want to give a combinatorial interpretation to (96). For any brick tabloid $T = (b_1, \dots, b_{\ell(\mu)}) \in \mathcal{B}_{\mu,n}$, we can interpret $\binom{n}{b_1,\dots,b_{\ell(\mu)}}$ as the set of all fillings of T with a permutation $\sigma \in S_n$ such that σ is increasing in each brick. We then interpret $\prod_{j=1}^{\ell(\mu)} k(1-k)^{b_j-1}$ as all ways of picking a label of the cells of each brick except the final cell with either an 1 or a $-k$ and letting the label of the last cell of each brick be k. We let $\mathcal{D}_n$ denote the set of all filled labeled brick tabloids that arise in this way. Thus, a $C \in \mathcal{D}_n$ consists of a brick tabloid T, a permutation $\sigma \in S_n$, and a labeling L of the cells of T with elements from $\{k, -k, 1\}$ such that

(1) σ is strictly increasing in each brick,

(2) the final cell of each brick is labeled with k,

(3) each cell which is not a final cell of a brick is labeled with 1 or $-k$.

We then define the weight $w(C)$ of C to be the product of all the k labels in L and the sign $\mathrm{sgn}(C)$ of C to be the product of all the -1 labels in L. For example, if $n = 12$, $k = 4$, and $T = (4,3,3,2)$, then Figure 9 pictures such a composite object $C \in \mathcal{D}_{12}$ where $w(C) = k^7$ and $\mathrm{sgn}(C) = -1$.

Thus,

$$n!\overline{\Gamma}_U(h_n) = \sum_{C \in \mathcal{D}_n} \mathrm{sgn}(C)\, w(C). \tag{97}$$

Next we define a weight-preserving sign-reversing involution $J : \mathcal{D}_n \to \mathcal{D}_n$. To define $J(C)$, we scan the cells of $C = (T, \sigma, L)$ from left to right looking for the leftmost cell t such that either (i) t is labeled with $-k$ or (ii) t is at the end of a brick b_j, and the brick b_{j+1} immediately following b_j has the property that σ is strictly increasing in all the cells corresponding to b_j and b_{j+1}. In case (i), $J(C) = (T', \sigma', L')$ where T' is the result of replacing the brick b in T containing t by two bricks b^* and b^{**} where b^* contains the cell t plus all the cells in b to the left of t and b^{**} contains all the cells of b to the right of t, $\sigma = \sigma'$, and L' is the labeling that results from L by changing the label of cell t from $-k$ to k. In case (ii), $J(C) = (T', \sigma', L')$ where T' is the result of replacing the bricks b_j and b_{j+1} in T by a single brick b, $\sigma = \sigma'$, and L' is the labeling that results from L by changing the label of cell t from k to $-k$. If neither case (i) or case (ii) applies, then we let $J(C) = C$. For example, if C is the element of $\mathcal{D}_{12}$ pictured in Figure 9, then $J(C)$ is pictured in Figure 10.

It is easy to see that J is a weight-preserving sign-reversing involution, and hence J shows that

$$n!\overline{\Gamma}_U(h_n) = \sum_{C \in \mathcal{D}_n, J(C)=C} \mathrm{sgn}(C)\, w(C). \tag{98}$$

Thus, we must examine the fixed points $C = (T, \sigma, L)$ of J. First there can be no $-k$ labels in L so that $\mathrm{sgn}(C) = 1$.

L	1	1	1	k	1	1	k	1	1	k	1	k
σ	2	3	4	11	6	9	10	1	8	12	5	7

FIGURE 11: A fixed point of J.

Moreover, if b_j and b_{j+1} are two consecutive bricks in T and t is the last cell of b_j, then it cannot be the case that $\sigma_t < \sigma_{t+1}$ since otherwise we could combine b_j and b_{j+1}. For any such fixed point, we associate an element $(\sigma, w) \in C_k \wr S_n$. For example, a fixed point of I is pictured in Figure 11 where

$$\sigma = 2\ 3\ 4\ 11\ 6\ 9\ 10\ 1\ 8\ 12\ 5\ 7. \quad (99)$$

It follows that if cell t is at the end of a brick which is not the last brick, then $\sigma_t > \sigma_{t1}$. However, if v is a cell which is not at the end of a brick, then our definitions force $\sigma_v < \sigma_{v+1}$. Since each such cell v must be labeled with an 1, it follows that $\mathrm{sgn}(C)w(C) = k^{\mathrm{des}(\sigma)+1}$ where the +1 comes from the fact that the last cell of the last brick is also labeled with k. Vice versa, if $\sigma \in S_n$, then we can create a fixed point $C = (T, \sigma, L)$ by having the bricks in T end at cells of the form t where $\sigma_t > \sigma_{t+1}$ and labeling each such cell with k, labeling the last cell with k, and labeling the remaining cells with 1. Thus, we have shown that

$$n!\Gamma_U(h_n) = \sum_{\sigma \in S_n} k^{\mathrm{des}(\sigma)+1} \quad (100)$$

as desired.

4.5. $NM_{(\Upsilon,\vec{A}),n}$. Next we consider another case where we can show that as a polynomial in k, $NM_{(\Upsilon,\vec{A}),n}$ has coefficients whose signs alternate. That is, it follows from (64) that

$$\begin{aligned} NM_{(\{(12,00),(12,01)\},([k],\{k-2,k-1\})),n} &= n! \sum_{r=0}^{n} \frac{(k-1)^r}{(n-r)!} \\ &= \sum_{r=0}^{n} (n)\downarrow_r (k-1)^r. \end{aligned} \quad (101)$$

We can easily calculate $NM_{(\{(12,00),(12,01)\},([k],\{k-2mk-1\})),n}$ as a polynomial in k. For example, we have

$$NM_{(\{(12,00),(12,01)\},([k],\{k-2,k-1\})),0} = 1,$$

$$NM_{(\{(12,00),(12,01)\},([k],\{k-2,k-1\})),1} = k,$$

$$NM_{(\{(12,00),(12,01)\},([k],\{k-2,k-1\})),2} = 2k^2 - 2k + 1,$$

$$NM_{(\{(12,00),(12,01)\},([k],\{k-2,k-1\})),3} = 6k^3 - 12k^2 + 9k - 2,$$

$$NM_{(\{(12,00),(12,01)\},([k],\{k-2,k-1\})),4} = 24k^4 - 72k^3 + 84k^2 - 44k + 9,$$

$$NM_{(\{(12,00),(12,01)\},([k],\{k-2,k-1\})),5} = 120k^5 - 480k^4 + 780k^3 + -640k^2 + 265k - 44. \quad (102)$$

Thus, we see that the signs of the polynomial $NM_{(\{(12,00),(12,01)\},([k],\{k-2,k-1\})),n}$ seem to alternate and that the leading term seems to be $n!k^n$. We can prove both of these facts. Taking the coefficient of k^s on both sides of (101), we see that

$$NM_{(\{(12,00),(12,01)\},([k],\{k-2,k-1\})),n}\big|_{k^s} = \sum_{r=s}^{n} (n)\downarrow_r \binom{r}{s} (-1)^{r-s}. \quad (103)$$

Now we can give a combinatorial interpretation to the right-hand side of (103) as follows. First for each r, we interpret $(n)\downarrow_r$ as the number of ways to pick a sequence $a_1 \cdots a_r$ of pairwise distinct elements from $[n]$. Then we interpret $\binom{r}{s}$ as picking s of the elements which we indicate by putting an x above each of the s elements picked. Finally, we interpret $(-1)^{r-s}$ as placing -1 above the elements in the sequence. We let $M_{n,s,r}$ denote the set of all labeled sequences constructed in this way. For example, if $n = 10$, $s = 3$, and $r = 6$, then the following labeled sequence would be in $M_{10,4,6}$:

$$(a, L) = \begin{matrix} -1 & x & -1 & -1 & x & x \\ 6 & 3 & 10 & 4 & 5 & 8 \end{matrix}. \quad (104)$$

Given a labeled sequence $(a, L) \in M_{n,s,r}$, we let the sign of (a, L), $\mathrm{sgn}(a, L)$, be the product of the -1 labels in L. We let $M_{n,s} = \bigcup_{r=s}^{n} M_{n,s,r}$. Then it is easy to see that

$$NM_{(\{(12,00),(12,01)\},([k],\{k-2,k-1\})),n}\big|_{k^s} = \sum_{r=s}^{n} (n)\downarrow_r \binom{r}{s} (-1)^{r-s} = \sum_{(a,L)\in M_{n,s}} \mathrm{sgn}(a, L). \quad (105)$$

Given a sequence $(a, L) = (a_1 \cdots a_r, L_1 \cdots L_r)$, we say that the minimal sequence of (a, L) is the longest string $a_t a_{t+1} \cdots a_r$ such that for all $j = t, \ldots, r$, $a_j = \min([n] - \{a_1, \ldots, a_{j-1}\})$ and $L_i = -1$ for $j = t, \ldots, r$. The minimal sequence for (a, L) might be empty. For example, it could be the last label $L_r = x$ or it could be that $a_r \neq \min([n] - \{a_1, \ldots, a_{r-1}\})$. We let $\mathrm{ms}(a, L)$ denote the length of the minimal sequence of (a, L). For example, the minimal sequence of

$$(a, L) = \begin{matrix} -1 & x & x & x & -1 & -1 \\ 6 & 3 & 10 & 4 & 5 & 8 \end{matrix} \quad (106)$$

is empty so that $\mathrm{ms}(a, L) = 0$. The minimal sequence of

$$(a', L') = \begin{matrix} -1 & x & x & x & -1 & -1 \\ 1 & 3 & 10 & 4 & 2 & 5 \end{matrix} \quad (107)$$

is 25 so that $\mathrm{ms}(a, L) = 2$.

Now suppose that $(a, L) \in M_{n,s,r}$ is such that $a = a_1 \cdots a_r$, $L = L_1 \cdots L_r$, and its minimal sequence is empty.

Let $b_1 < \cdots < b_{n-r}$ be the elements of $[n] - \{a_1, \ldots, a_r\}$. Then we define the minimal sequence chain of (a, L) to be $(a^{(r)}, L^{(r)}), (a^{(r+1)}, L^{(r+1)}), \ldots, (a^{(n)}, L^{(n)})$ where $(a^{(r)}, L^{(r)}) = (a, L)$ and $(a^{(r+p)}, L^{(r+p)}) = (a_1 \cdots a_r b_1 \cdots b_p, L_1 \cdots L_r (-1)^p)$ for $1 \le p \le n - r$. Note that the signs of the elements of a minimal sequence chain of (a, L) alternate. Thus, if the minimal sequence chain of (a, L) is of even length, then $\text{ms}(a^{(n)}, L^{(n)})$ is odd, and the sum of the signs of elements in the minimal sequence chain of (a, L) is 0. If the minimal sequence chain of (a, L) is of odd length, then $\text{ms}(a^{(n)}, L^{(n)})$ is even, and the sum of the signs of elements in the minimal sequence chain of (a, L) is $\text{sgn}(a^{(n)}, L^{(n)})$. Clearly every element of $M_{n,r}$ is an element of a unique minimal sequence chain for some $(a, L) \in M_{n,s}$ whose minimal sequence is empty so that

$$\sum_{(a,L)\in M_{n,s}} \text{sgn}(a, L) = \sum_{(a,L)\in M_{n,s,n}, \text{ms}(a,L) \text{ is even}} \text{sgn}(a, L) = (-1)^{n-s} \left|\{(a, L) \in M_{n,s,n} : \text{ms}(a, L) \text{ is even}\}\right|. \tag{108}$$

Thus, we have shown that

$$NM_{(\{(12,00),(12,01)\},([k],\{k-2,k-1\})),n}|_{k^s} = (-1)^{n-s} \left|\{(a, L) \in M_{n,s,n} : \text{ms}(a, L) \text{ is even}\}\right|. \tag{109}$$

This shows that as a polynomial in k, $NM_{(\{(12,00),(12,01)\},([k],\{k-2,k-1\})),n}$ is sign alternating starting with the leading term $n!k^n$. That is, by (109),

$$NM_{(\{(12,00),(12,01)\},([k],\{k-2,k-1\})),n}|_{k^n} = \left|\{(a, L) \in M_{n,n,n} : \text{ms}(a, L) \text{ is even}\}\right|. \tag{110}$$

But the elements of $M_{n,n,n}$ have no -1 labels since every element of $M_{n,n,n}$ has an empty minimal sequences and, hence, $|\{(a, L) \in M_{n,n,n} : \text{ms}(a, L) \text{ is even}\}| = n!$.

We can also compute several other coefficients of $NM_{(\{(12,00),(12,01)\},([k],\{k-2,k-1\})),n}$ as a polynomial in k. For example, we can explicitly compute the coefficient of k^{n-1} in $NM_{(\{(12,00),(12,01)\},([k],\{k-2,k-1\})),n}$. Note that

$$NM_{(\{(12,00),(12,01)\},([k],\{k-2,k-1\})),n}|_{k^{n-1}} = -\left|\{(a, L) \in M_{n,n-1,n} : \text{ms}(a, L) \text{ is even}\}\right|. \tag{111}$$

Now an element $(a, L) \in M_{n,n-1,n}$ has exactly one label which is -1. If $L_n = -1$, then the last element of a is automatically its minimal sequence so that it is not counted in $|\{(a, L) \in M_{n,n-1,n} : \text{ms}(a, L) \text{ is even}\}|$. Thus, $|\{(a, L) \in M_{n,n-1,n} : \text{ms}(a, L) \text{ is even}\}|$ counts all $(a, L) \in M_{n,n-1,n}$ such that $L_n = x$ which is clearly $(n - 1)n!$.

We claim that

$$NM_{(\{(12,00),(12,01)\},([k],\{k-2,k-1\})),n}|_{k^0} = (-1)^n D_n, \tag{112}$$

where D_n is the number of derangements of S_n, that is, the number of permutations with no fixed points. It is well known that $D_n = n! \sum_{k=0}^{n} (-1)^k / k!$ for $n \ge 1$. But by (105),

$$NM_{(\{(12,00),(12,01)\},([k],\{k-2,k-1\})),n}|_{k^0} = \sum_{r=0}^{n} (n) \downarrow_r (-1)^r = (-1)^n n! \sum_{r=0}^{n} \frac{(-1)^{n-r}}{(n-r)!} = D_n. \tag{113}$$

One can also use our combinatorial interpretation of $NM_{(\{(12,00),(12,01)\},([k],\{k-2,k-1\})),n}|_{k^0}$ to give a combinatorial proof of this fact. That is, it is well known that $D_1 = 0, D_2 = 1$ for $n \ge 2 D_{n+1} = nD_n + nD_{n-1}$. In our case, we know that

$$NM_{(\{(12,00),(12,01)\},([k],\{k-2,k-1\})),n}|_{k^0} = (-1)^n \left|\{(a, L) \in M_{n,0,n} : \text{ms}(a, L) \text{ is even}\}\right|. \tag{114}$$

Thus, we must show that $D_n = |\{(a, L) \in M_{n,0,n} : \text{ms}(a, L) \text{ is even}\}|$. Now if $(a, L) \in M_{n,0,n}$, then all the labels in L are -1 so that we can identify each element $(a, L) \in M_{n,0,n}$ with just a permutation $a = \sigma = \sigma_1 \cdots \sigma_n \in S_n$. Now consider a $\sigma \in S_{n+1}$ whose minimal sequence has even length. If $\sigma_1 \ne 1$, then clearly the minimal sequence of $\text{red}(\sigma_2 \cdots \sigma_{n+1}) \in S_n$ has the same length as the minimal sequence of σ. Vice versa, given $\tau_1 \cdots \tau_n \in S_n$, let $\tau^{(i)} = i\tau_1^* \cdots \tau_n^*$ be the permutation of S_{n+1} such that $\text{red}(\tau_1^* \cdots \tau_n^*) = \tau$. Clearly for $i = 2, \ldots, n+1$, $\tau^{(i)}$ is a permutation in S_{n+1} whose minimal sequences have the same length as the minimal sequence of τ. It follows that

$$\left|\{(a, L) \in M_{n+1,0,n+1} : \text{ms}(\sigma, L) \text{ is even } \& \ \sigma_1 \ne 1\}\right| = n\left|\{(a, L) \in M_{n,0,n} : \text{ms}(a, L) \text{ is even}\}\right|. \tag{115}$$

Next suppose that $\sigma \in S_{n+1}$ is such that $\sigma_1 = 1$, and its minimal sequence has even length. Then we claim that $\text{red}(\sigma_3 \cdots \sigma_{n+1})$ has a minimal sequence whose length has the same parity as the length of the minimal sequence of σ. That is, if 1 is an element of the minimal sequence of σ, then it must be the case that $\sigma = 12 \cdots (n+1)$ is the identity so that $\text{red}(\sigma_3 \cdots \sigma_{n+1}) = 1 \cdots n-1$ has minimal sequence of length $n - 1$. If 1 is not in the minimal sequence of σ, then it cannot be the case that σ_2 is in the minimal sequence of σ so that the lengths of the minimal sequences of σ and $\text{red}(\sigma_3 \cdots \sigma_{n+1})$ are the same. Vice versa, given $\tau_1 \cdots \tau_{n-1} \in S_{n-1}$, let $\tau^{(1,i)} = 1i\tau_1^* \cdots \tau_{n-1}^*$ be the permutation of S_{n+1} such that $\text{red}(\tau_1^* \cdots \tau_{n-1}^*) = \tau$. Clearly for $i = 2, \ldots, n+1$, $\tau^{(1,i)}$ is a permutation in S_{n+1} whose lengths of the minimal sequences have the same parity as the length of the minimal sequence of τ. It follows that

$$\left|\{(a, L) \in M_{n+1,0,n+1} : \text{ms}(\sigma, L) \text{ is even and } \sigma_1 = 1\}\right| = n\left|\{(a, L) \in M_{n-1,0,n-1} : \text{ms}(a, L), \text{ is even}\}\right| \tag{116}$$

Hence, $|\{(a, L) \in M_{n+1,0,n+1} : \text{ms}(\sigma, L) \text{ is even}|$ satisfies the same recursion as the recursion for D_n.

Finally, it is not difficult to see that

$$NM_{(\{(12,00),(12,01)\},([k],\{k-2,k-1\})),n}|_{k^1} = NM_{(\{(12,00),(12,01)\},([k],\{k-2,k-1\})),n+1}|_{k^0}. \tag{117}$$

To prove this, we will construct a bijection $\theta : M_{n,1,n} \to M_{n+1,0,n+1}$ which has the property that for all $(a, L) \in M_{n,1,n}$, $\mathrm{ms}(a, L)$ and $\mathrm{ms}(\theta(a, L))$ have the same parity. Now suppose we are given an $(a, L) \in M_{n,1,n}$ where $a = a_1 \cdots a_n$ and $L = L_1 \cdots L_n$. Then is must be the case that there is a unique j such that $L_j = x$. Let $\theta(a, L) = (a', L') = (a_1 + 1, \ldots, a_{j-1} + 1, 1, a_j + 1, \ldots, a_n + 1, (-1)^{n+1})$. That is, (a', L') arises from (a, L) by inserting 1 just before the element of a that is marked with x and adding 1 to all the elements of a and removing the x and replacing it with -1 and labeling 1 with -1. It is easy to see that θ is a bijection. Thus, all we need to do is to show that for all $(a, L) \in M_{n,1,n}$, $\mathrm{ms}(a, L)$ and $\mathrm{ms}(\theta(a, L))$ have the same parity. The minimal sequence of (a, L) does not contain a_j. In (a', L'), the minimal sequence does not contain $a'_j = a_{j-1} + 1 > 1$ since $1 = \min([n + 1] - \{a'_1, \ldots, a'_{j-2}\})$. Now if the minimal sequence of (a, L) does not contain a_{j+1}, then it is easy to see that the minimal sequence of (a', L') does not contain a'_{j+2} which means that the lengths of the minimal sequences of (a, L) and (a', L') are the same. So suppose that the minimal sequence of (a, L) is $a_{j+1} \cdots a_n$. Then there are two cases. Namely, either (i) $a_j = \min([n] - \{a_1, \ldots, a_{j-1}\})$ in which case the minimal sequence of (a', L') is $1, a_j + 1, a_{j+1} + 1, \ldots, a_n + 1$ or (ii) $a_j \neq \min([n] - \{a_1, \ldots, a_{j-1}\})$ in which case the minimal sequence of (a', L') is $a_{j+1} + 1, \ldots, a_n + 1$. Hence in all cases, the lengths of the minimal sequences of (a, L) and (a', L') have the same parity. Thus, θ shows that

$$\begin{aligned} &\left|\{(a, L) \in M_{n,1,n} : \mathrm{ms}(\sigma, L) \text{ is even}\}\right| \\ &\quad = \left|\{(a, L) \in M_{n+1,0,n+1} : \mathrm{ms}(\sigma, L) \text{ is even}\}\right|, \end{aligned} \tag{118}$$

which by (109) establishes (117).

5. Further Research

An obvious question is to ask whether the results of this paper can be extended to patterns of length ≥ 3. There are some partial results in this direction due to Duane and Remmel [21]. Given a word $u \in \{0, 1, \ldots, k-1\}^j$ such that $\mathrm{red}(u) = u$ and $\tau \in S_j$, Duane and Remmel say that (τ, u) has the $C_k \wr S_n$-*minimal overlapping property* if the smallest i such that there exists a $(\sigma, w) \in C_k \wr S_i$ with (τ, u)-$\mathrm{mch}(\sigma, w) = 2$ is $2j-1$. This means that in a k-colored permutation, (σ, w), two (τ, u)-matches in (σ, w) can share at most one pair of letters, and this pair of letters must occur at the end of the first (τ, u)-match and at the start of the second (τ, u)-match. Similarly, we say that (τ, u) has the $C_k \wr S_n$-*exact match minimal overlapping property* if the smallest i such that there exists a $(\sigma, w) \in C_k \wr S_i$ with (τ, u)-$\mathrm{Emch}(\sigma, w) = 2$ is $2j - 1$. Now if (τ, u) has the $C_k \wr S_n$-minimal overlapping property, then the shortest k-colored permutations, (σ, w), such that (τ, u)-$\mathrm{mch}(\sigma, w) = n$, have length $n(j-1) + 1$. We let $\mathscr{MP}^k_{(\tau,u),n(j-1)+1}$ equal the set of k-colored permutations, $(\sigma, w) \in C_k \wr S_{n(j-1)+1}$, such that (τ, u)-$\mathrm{mch}(\sigma, w) = n$. We refer to the k-colored permutations in $\mathscr{MP}^k_{(\tau,u),n(j-1)+1}$ as maximum packings for (τ, u). We let

$$\begin{aligned} \mathrm{mp}^k_{(\tau,u),n(j-1)+1} &= \left|\mathscr{MP}^k_{(\tau,u),n(j-1)+1}\right|, \\ \mathrm{mp}^k_{(\tau,u),n(j-1)+1}(p, q, r) &= \sum_{(\sigma,w)\in\mathscr{MP}^k_{(\tau,u),n(j-1)+1}} p^{\mathrm{coinv}(\sigma)} q^{\mathrm{inv}(\sigma)} r^{\Sigma w} \\ \mathrm{mp}^k_{(\tau,u),n(j-1)+1}(p, q, z_0, \ldots, z_{k-1}) &= \sum_{(\sigma,w)\in\mathscr{MP}^k_{(\tau,u),n(j-1)+1}} p^{\mathrm{coinv}(\sigma)} q^{\mathrm{inv}(\sigma)} z(w). \end{aligned} \tag{119}$$

Similarly, we let $\mathscr{EMP}^k_{(\tau,u),n(j-1)+1}$ equal the set of k-colored permutations, $(\sigma, w) \in C_k \wr S_{n(j-1)+1}$, such that (τ, u)-$\mathrm{Emch}(\sigma, w) = n$, and refer to the k-colored permutations in $\mathscr{EMP}^k_{(\tau,u),n(j-1)+1}$ as exact match maximum packings for (τ, u). We let

$$\begin{aligned} \mathrm{emp}^k_{(\tau,u),n(j-1)+1} &= \left|\mathscr{EMP}^k_{(\tau,u),n(j-1)+1}\right|, \\ \mathrm{emp}^k_{(\tau,u),n(j-1)+1}(p, q, r) &= \sum_{(\sigma,w)\in\mathscr{EMP}^k_{(\tau,u),n(j-1)+1}} p^{\mathrm{coinv}(\sigma)} q^{\mathrm{inv}(\sigma)} r^{\Sigma w}, \\ \mathrm{emp}^k_{(\tau,u),n(j-1)+1}(p, q, z_0, \ldots, z_{k-1}) &= \sum_{(\sigma,w)\in\mathscr{EMP}^k_{(\tau,u),n(j-1)+1}} p^{\mathrm{coinv}(\sigma)} q^{\mathrm{inv}(\sigma)} z(w). \end{aligned} \tag{120}$$

Then Duane and Remmel [21] proved that (I) if $\mathrm{red}(u) = u$ and (τ, u) has the $C_k \wr S_n$-minimal overlapping property, then

$$\begin{aligned} &\sum_{n\geq 0} \frac{t^n}{n!} \sum_{(\sigma,w)\in C_k\wr S_n} x^{(\tau,u)\text{-}\mathrm{mch}(\sigma,u)} p^{\mathrm{coinv}(\sigma)} q^{\mathrm{inv}(\sigma)} z(w) \\ &= 1 \times \Bigg(1 - \Bigg((z_0 + \cdots + z_{k-1})t \\ &\qquad + \sum_{n\geq 1} \frac{t^{n(j-1)+1}}{[n(j-1)+1]_{p,q}!} (x-1)^n \\ &\qquad \times \mathrm{mp}^k_{(\tau,u),n(j-1)+1}(p, q, z_0, \ldots, z_{k-1})\Bigg)\Bigg)^{-1}, \end{aligned} \tag{121}$$

(II) if $(\tau, u) \in C_k \wr S_j$ has the $C_k \wr S_n$-exact match minimal overlapping property, then

$$\sum_{n\geq 0} \frac{t^n}{n!} \sum_{(\sigma,w)\in C_k \wr S_n} x^{(\tau,u)\text{-Emch}(\sigma,u)} p^{\text{coinv}(\sigma)} q^{\text{inv}(\sigma)} z(w)$$
$$= 1 \times \Bigg(1 - \bigg(\left(z_0 + \cdots + z_{k-1}\right) t + \sum_{n\geq 1} \frac{t^{n(j-1)+1}}{\left[n(j-1)+1\right]_q!} (x-1)^n \times \text{emp}^k_{(\tau,u),n(j-1)+1}\left(p, q, z_0, \ldots, z_{k-1}\right)\bigg)\Bigg)^{-1}. \quad (122)$$

Duane and Remmel's proof of (I) and (II) works equally well for $(\Upsilon, \vec{A})$-matches.

There are, however, several examples in the literature of generating functions for the number of τ-matches in permutations where τ does not have the minimal overlapping property, and we have found some analogues of such results for $C_k \wr S_n$. In addition, there are many formulas in the literature for the number of permutations that avoid certain patterns in permutations. We have found several examples of such formulas for the matching conditions for $C_k \wr S_n$ discussed in this paper. Such results will appear in subsequent papers.

Acknowledgments

J. Remmel was partially supported by NSF Grant DMS 0654060. M. Riehl was partially supported by a grant from the Office of Research and Sponsored Programs, UWEC. We define the q-analogues of n, $n!$, $\binom{n}{k}$,

References

[1] E. S. Egge, "Restricted colored permutations and Chebyshev polynomials," *Discrete Mathematics*, vol. 307, no. 14, pp. 1792–1800, 2007.

[2] T. Mansour, "Pattern avoidance in coloured permutations," *Séminaire Lotharingien de Combinatoire*, vol. 46, article B46g, 12 pages, 2001.

[3] T. Mansour, "Coloured permutations containing and avoiding certain patterns," *Annals of Combinatorics*, vol. 7, no. 3, pp. 349–355, 2003.

[4] T. Mansour and J. West, "Avoiding 2-letter signed patterns," *Séminaire Lotharingien de Combinatoire*, vol. 49, article B49a, 11 pages, 2002.

[5] R. Simion, "Combinatorial statistics on type-B analogues of noncrossing partitions and restricted permutations," *Electronic Journal of Combinatorics*, vol. 7, no. 1, article R9, pp. 1–27, 2000.

[6] F. Brenti, "Permutation enumeration, symmetric functions, and unimodality," *Pacific Journal of Mathematics*, vol. 157, pp. 1–28, 1993.

[7] F. Brenti, "A class of q-symmetric functions arising from plethysm," *Journal of Combinatorial Theory Series A*, vol. 91, no. 1-2, pp. 137–170, 2000.

[8] D. A. Beck, J. B. Remmel, and T. Whitehead, "The combinatorics of transition matrices between the bases of the symmetric functions and the B_n analogues," *Discrete Mathematics*, vol. 153, no. 1–3, pp. 3–27, 1996.

[9] T. M. Langley, "Alternative transition matrices for Brenti's q-symmetric functions and a class of q, t-symmetric functions on the hyperoctahedral group," in *Proceedings of the Formal Power Series and Algebraic Combinatorics Conference*, Melbourne, Australia, 2002.

[10] T. M. Langley and J. B. Remmel, "Enumeration of m-tuples of permutations and a new class of power bases for the space of symmetric functions," *Advances in Applied Mathematics*, vol. 36, no. 1, pp. 30–66, 2006.

[11] A. Mendes and J. Remmel, "Permutations and words counted by consecutive patterns," *Advances in Applied Mathematics*, vol. 37, no. 4, pp. 443–480, 2006.

[12] A. Mendes and J. B. Remmel, "Generating functions from symmetric functions," preprint.

[13] A. Ram, J. B. Remmel, and T. Whitehead, "Combinatorics of the q-basis of symmetric functions," *Journal of Combinatorial Theory Series A*, vol. 76, no. 2, pp. 231–271, 1996.

[14] J. D. Wagner, "The permutation enumeration of wreath products $C_k \wr S_n$ of cyclic and symmetric groups," *Advances in Applied Mathematics*, vol. 30, no. 1-2, pp. 343–368, 2003.

[15] A. Mendes, *Building generating functions brick by brick [Ph.D. thesis]*, University of California, San Diego, Calif, USA, 2004.

[16] N. J. A. Sloane, Ed., *Online Encyclopedia of Integer Sequences*, 2009, http://oeis.org/.

[17] Ö. Eğecioğlu and J. B. Remmel, "Brick tabloids and the connection matrices between bases of symmetric functions," *Discrete Applied Mathematics*, vol. 34, no. 1–3, pp. 107–120, 1991.

[18] L. Carlitz, "Sequences and inversions," *Duke Mathematical Journal*, vol. 37, pp. 193–198, 1970.

[19] A. Mendes, J. B. Remmel, and A. Riehl, "Permutations with k-regular descent patterns," *Annals of Combinatorics*. In press.

[20] S. Kitaev, *Patterns in Permutations and Words*, EATCS Monographs in Theoretical Computer Science Book Series, Springer, 2011.

[21] A. Duane and J. B. Remmel, "Minimal overlapping patterns in colored permutations," *The Electronic Journal of Combinatorics*, vol. 18, no. 2, 2011.

Permissions

The contributors of this book come from diverse backgrounds, making this book a truly international effort. This book will bring forth new frontiers with its revolutionizing research information and detailed analysis of the nascent developments around the world.

We would like to thank all the contributing authors for lending their expertise to make the book truly unique. They have played a crucial role in the development of this book. Without their invaluable contributions this book wouldn't have been possible. They have made vital efforts to compile up to date information on the varied aspects of this subject to make this book a valuable addition to the collection of many professionals and students.

This book was conceptualized with the vision of imparting up-to-date information and advanced data in this field. To ensure the same, a matchless editorial board was set up. Every individual on the board went through rigorous rounds of assessment to prove their worth. After which they invested a large part of their time researching and compiling the most relevant data for our readers. Conferences and sessions were held from time to time between the editorial board and the contributing authors to present the data in the most comprehensible form. The editorial team has worked tirelessly to provide valuable and valid information to help people across the globe.

Every chapter published in this book has been scrutinized by our experts. Their significance has been extensively debated. The topics covered herein carry significant findings which will fuel the growth of the discipline. They may even be implemented as practical applications or may be referred to as a beginning point for another development. Chapters in this book were first published by Hindawi Publishing Corporation; hereby published with permission under the Creative Commons Attribution License or equivalent.

The editorial board has been involved in producing this book since its inception. They have spent rigorous hours researching and exploring the diverse topics which have resulted in the successful publishing of this book. They have passed on their knowledge of decades through this book. To expedite this challenging task, the publisher supported the team at every step. A small team of assistant editors was also appointed to further simplify the editing procedure and attain best results for the readers.

Our editorial team has been hand-picked from every corner of the world. Their multi-ethnicity adds dynamic inputs to the discussions which result in innovative outcomes. These outcomes are then further discussed with the researchers and contributors who give their valuable feedback and opinion regarding the same. The feedback is then collaborated with the researches and they are edited in a comprehensive manner to aid the understanding of the subject.

Apart from the editorial board, the designing team has also invested a significant amount of their time in understanding the subject and creating the most relevant covers. They scrutinized every image to scout for the most suitable representation of the subject and create an appropriate cover for the book.

The publishing team has been involved in this book since its early stages. They were actively engaged in every process, be it collecting the data, connecting with the contributors or procuring relevant information. The team has been an ardent support to the editorial, designing and production team. Their endless efforts to recruit the best for this project, has resulted in the accomplishment of this book. They are a veteran in the field of academics and their pool of knowledge is as vast as their experience in printing. Their expertise and guidance has proved useful at every step. Their uncompromising quality standards have made this book an exceptional effort. Their encouragement from time to time has been an inspiration for everyone.

The publisher and the editorial board hope that this book will prove to be a valuable piece of knowledge for researchers, students, practitioners and scholars across the globe.

List of Contributors

Ângela Mestre
Centro de Estruturas Lineares e Combinat´orias, Universidade de Lisboa, Av. Prof. Gama Pinto 2, 1649-003 Lisboa, Portugal

Shyamashree Upadhyay
School of Mathematics, Tata Institute of Fundamental Research, Homi Bhabha Road, Colaba, Mumbai 400005, India

Sezer Sorgun
Department of Mathematics, Faculty of Sciences and Arts, Nevsehir University, 50300 Nevsehir, Turkey

Jun-Ming Xu
School of Mathematical Sciences, University of Science and Technology of China, Wentsun Wu Key Laboratory of CAS, Hefei, Anhui 230026, China

Donald K. Wagner
Mathematical, Computer and Information Sciences Division, Office of Naval Research, Arlington, VA 22203, USA

Bibin K. Jose
Department of Mathematics, S.D. College Alappuzha, University of Kerala, Kerala 688003, India

John P. McSorley and Philip Feinsilver
Department of Mathematics, Southern Illinois University, Carbondale, IL 62901-4408, USA

Mrinal Nandi
Department of Statistics, West Bengal State University, Barasat, North 24 Parganas 700126, India

Subrata Parui
IMBIC, AH 317, Salt Lake City, Calcutta 700091, India

Avishek Adhikari
Department of Pure Mathematics, University of Calcutta, 35 Ballygunge Circular Road, Calcutta 700019, India

Emeric Deutsch
Polytechnic Institute of New York University, Brooklyn, NY 11201, USA

Juan Alberto Rodríguez-Velázquez
Departament d'Enginyeria Inform`atica i Matem`atiques, Universitat Rovira i Virgili, Avenida Pa¨ısos Catalans 26, 43007 Tarragona, Spain

Thomas Ernst
Department of Mathematics, Uppsala University, P.O. Box 480, SE-751 06 Uppsala, Sweden

Messaoud Kolli
Department of Mathematics, Faculty of Science, King Khalid University, Abha, Saudi Arabia

R. S. Selvaraj and Venkatrajam Marka
Department of Mathematics, National Institute of Technology, Warangal, Andhra Pradesh 506004, India

Deniz Kus
Mathematisches Institut, Universit¨at zu K¨oln, 50931K¨oln, Germany

Sergey Kitaev
Department of Computer and Information Sciences, University of Strathclyde, Livingstone Tower, 26 Richmond Street, Glasgow G11XH,UK

Andrew Niedermaier and Jeffrey Remmel
Department of Mathematics, University of California, San Diego, La Jolla, CA 92093-0112,USA

Manda Riehl
Department of Mathematics, University of Wisconsin, Eau Claire, WI 54702-4004, USA